International Symposium on Intelligent Manufacturing and Automation (ISIMA 2024)

Journal of Physics: Conference Series Volume 2992

Nanjing, China
13-15 December 2024

ISBN: 979-8-3313-2032-4
ISSN: 1742-6588

Printed from e-media with permission by:

Curran Associates, Inc.
57 Morehouse Lane
Red Hook, NY 12571

CURRAN ASSOCIATES INC. proceedings.com

Some format issues inherent in the e-media version may also appear in this print version.

Institute of Physics
Dirac House, Temple Back
Bristol BS1 6BE UK

Phone: 44 1 17 929 7481
Fax: 44 1 17 920 0979

techtracking@iop.org

Additional copies of this publication are available from:

Curran Associates, Inc.
57 Morehouse Lane
Red Hook, NY 12571 USA
Phone: 845-758-0400
Fax: 845-758-2633
Email: curran@proceedings.com
Web: www.proceedings.com

TABLE OF CONTENTS

Preface

Peer Review Statement

Development of a new type of variable-diameter wheel 1
Yang Ming, Feiyun Xu

Design of two-stage parallel shock cam reducer 11
Yajuan Chen, Yang Chao, Xiaoping Wang

Reliability design of mechanical structures with small samples 19
Shenye Zou, He Zhao

Study of application of genetic aggregation in optimizing the dimensions of EDM pillar 26
Hui Li, Jianyong Liu, Dianxin Li, Jianrong Yu, Wenming Zhang, Dan Liu

Experimental Research on ultrasonic-assisted direct writing metal additive manufacturing technique 36
Guofang Hu, Ning Wang, Yanzhen Zhang

Simulation analysis and optimization design of forklift AGV fork rack 43
Yafeng Jiang, Shuaimei Lian, Senyao Wu

Computer simulation of the auxiliary feed mechanism and opening closing boxes in a ginning machine 50
Shuai Cui, Xuan Sun, Quancheng Dong, Haiqiang Gao, Yue Zhang, Huaikun Li

Design and application of four-wheel steering system for integration of theory and practice teaching 57
En Wei, Jinggong Wei, Dongfang Zhang

Optimization design of wind turbine blade structure 72
Yuhua Hao, Xiujuan Zhang

Research on the modeling of thin-walled casing bolted flange connections considering nonlinear effects 78
Yansheng Wang, Yuefang Wang

Optimization design of asymmetric fork structure for shifting mechanism based on topology method 89
Zhonglin Jiang, Yong Li, Mingge Wu, Jie Ren

Autofrettage design and study of ultrahigh-pressure emergency pressure relief valves 98
Yuehao Gao, Xuesheng Wang, Qinzhu Chen

Topology optimization-based lightweight design of the winch bracket for charged fire-connecting robots 105
Decheng Meng, Xuan Sun, Bo Chen

The design and analysis of an industrial five-axis laser elbow cutting machine 114
Wentao Yang, Junwei Liu, Tao Fan, Yong Wang

Two-dimensional ultrasonic vibration assisted WEDM technology and simulation analysis 122
Minggang Xu, Zuhuang Liu, Shengzhi Guan, Zhen Wang

Experimental setup adapted to the ultrasonic vibration-assisted preparation process for thermoplastic FMLs and its validation 132
Hang Yin, Yuqin Guo, Dongjie Yang, Fuzhu Li

Optimization design of stiffened plates for crossbeam of a certain type of gantry machining center 141
Huaming Mai, Bo Li, Weiwei Zhang, Ling Yin

Structural performance study on aluminum foil coating steel roller shaft based on finite element method 148
Bo Shen, Xin'an Zhou, Zhenyu Guo, Jiazhong Xu

Research and analysis on the reliability of large vertical CNC lathe based on improved FMECA.................. 155
Yiran Dong, Dechao Wang, Chengdao Piao, Shengkui Zheng

Tool wear detection of milling cutter based on heterogeneous data decision-level fusion............................... 164
Jingshu Wang, Weihan Zhou, Jinghua Ma, Qingfeng Wang

Acoustic emission monitoring of the hot temperature tensile behavior of 20CrMo steel................................ 171
Qingshan Li, Xumeng Xie, Wenjie Pan, Huiquan Cao, Hao Wu, Xinhua Xia

Simulation study on the power performance of LPG-NH3 dual-fuel engine 177
Zan Huang, Xiuhua Ma, Hua Xiao, Guoqing Zhang

The buffer protection mechanism of heterogeneous metal laminated structures during severe overloading 184
Yu'Nan Chen, Peng Zhang, Hao'Nan Guo, Yunbo Shi, Rui Zhao, Xiangyu Li, Yu Wang, Tao Guo

Performance study of 430 stainless steel/6061 aluminum alloy laser welded joints 191
Haiting Peng, Zhanlei Wang, Qiaobo Feng

Damping performance analysis of torsional Shear thickening fluid damper based on FLUENT.................... 200
Weijie Zhu, Kaihui Chen, Chuanze Liang, Yafeng Qiu

Workspace analysis of an R (2-UPR/2-UPS) hybrid machine tool.. 208
Zuwang Yin, Dong Wang, Xiaoguang Zhou, Jiahua Wu

Attitude wind stabilization control of inspection vehicle based on Stackelberg game theory.......................... 219
Chuanze Liang, Weijie Zhu, Yafeng Qiu

Research on nonlinear vibration characteristics of counter-rotating dual-rotor with multiple squeeze film dampers 228
Yan Li, Yan Ouyang, Huisong Wu, Xiping Kou, Su Jia

Study on slope parameter of fatigue SN curve for 2524-T3 aluminum alloy sheet... 237
Tong Wang, Xianchao Li, Fu Huang

Rolling bearing fault diagnosis based on fusing composite multiscale entropy and random forests................ 243
Hongyuan Chu, Hua Liu

Research on the whole life cycle attribute decision method for gearbox product design based on WLS and TOPSIS 250
Yu Liu, Juan Li, Jingying Li, Ying Wang

Influence of increasing temperature deformation on the removal uniformity of chemical mechanical polishing plate and improvement method 260
Guoqing Ye, Zhenqiang Yao

Research on stress distribution and dispersion of cylinder head 270
Xuanyi Zhao, Weiqing Huang, Peirong Ren, Pengyu Xia, Jiayu Kuang, Jiaheng Tang

Investigation of floating ball mandrel application in tube free bending 283
Zhenhua Xu, Jing Huang, Hua Wang, Rongjing Hong

Analysis of thermal shock failures in ceramic materials under varying temperature conditions based on phase field model 290
Qiang Shen, Haoyu Bai, Yu Xue

Research on the physical characteristics and rock-breaking effect of high-energy plasma drill bit 298
Yancong Han, Chao Zheng, Yonghong Liu, Yuming Liu, Wenhao Zhao

Research on the variation law and characterization model of kinetic energy block penetration overload 305
Zhe Yang, Fei Shang

Numerical analysis of bursting pressure of a brazed plate heat exchanger 312
Wei Wang, Yadong Zhao, Dao Chi

Experimental and simulation study of cutting forces on a porous workpiece based on Abaqus surfaces 319
Xinyu Liang, Zhenfeng Jiang, Fengyang Gao, Yanan Pan

Research on explosion dispersion based on near-field and far-field coupling calculation method 326
Yanyu Jiang, Zheng Mei, Zhenggang Xiao

Numerical simulation analysis of rolling groove forming of fire cooling pipe for offshore platforms based on ANSYS/LS-DYNA 336
Wei Chen, Guohe Jiang, Zhongliang Zhou

Dynamic modeling and analysis of tool magazine reversing mechanism 341
Wenbo Zheng, Manxin Wang, Kun Liu, Jianguo Zhu, Kangjun Xia

Application of constitutive model considering stress state and spatial gradient in the simulation of cast aluminum wheel testing 351
Yazhou Jiang, Wuguo Chen, Taisong Cui, Zihan Jiang, Bin Zhang, Dai Sun, Feng Gao, Tengteng Wang

Numerical study on flow and temperature rise characteristics of low-pressure cylinder of steam turbine under variable working conditions 360
Hongwu Wang, Huan Liu, Tingshan Ma, Tian Xie, Ting Wang, Rongzu Yang, Yaowen Wang, Pengcheng Zhai, Longwen Yu, Xinjing Jing

Numerical simulation and experimental study on heat treatment of automotive riveted nuts 370
Le Yu, Mingliang Hu

Research on the calculation method of contact stress between variable-axis diameter closed-type linear bearing and guideway 381
Weiming Li, Yuan Xue, Yuhong Liu, Yukuan Xie

Gap flow field and experiment in electrical discharge machining using water-in-oil dielectric 391
Renpeng Bian, Andong Sun, Yonghong Liu, Bo Qiao

Study on the drag and heat transfer characteristics of near-wall oblate spheroids 398
Ting Xiong, Yongkang Yang, Chenchen Chen, Jingdun Pang, Hongsheng Zhang, Yong Chen

Thermal dynamic simulation study of the aircraft ECS laboratory air source 404
Zheng Dai, Yi Cui

Author Index

Journal of Physics: Conference Series **2992** (2025) 011001 doi:10.1088/1742-6596/2992/1/011001

Preface

The 2024 International Symposium on Intelligent Manufacturing and Automation (ISIMA 2024) was convened from December 13th to 15th, 2024 in Nanjing, China. This symposium served as a premier interdisciplinary platform for researchers, practitioners, and educators to present and discuss the most recent innovations, trends, and concerns in the fields of intelligent manufacturing and automation.

In the era of Industry 4.0, the integration of advanced manufacturing technologies with intelligent systems has become a pivotal driver for global industrial transformation. ISIMA 2024 was held against this backdrop, aiming to foster collaboration and knowledge exchange among professionals from diverse backgrounds. The conference focused on a wide range of topics, including intelligent production systems, industrial Internet of Things (IIoT), applications of artificial intelligence in manufacturing, robotics, data analytics, and optimization. These themes reflect the current challenges and opportunities in the field, emphasizing the need for innovative solutions to enhance productivity, sustainability, and competitiveness in modern industries.

The program of ISIMA 2024 was meticulously designed to maximize engagement and learning opportunities for all participants. The conference kicked off with a series of keynote speeches delivered by distinguished experts in the field. These presentations provided valuable insights into the latest developments and future directions of intelligent manufacturing and automation. The keynote speakers shared their expertise on cutting-edge technologies and practical applications, inspiring the audience with their vision and innovative ideas.

Following the keynote addresses, the conference featured oral presentations and poster sessions, which allowed participants to delve deeper into specific research topics. The oral presentations showcased a selection of high-quality papers that underwent a rigorous peer-review process. Authors had the opportunity to present their work, receive feedback from peers, and engage in discussions that could lead to further collaboration and refinement of their research.

The poster session was equally vibrant, with researchers presenting their work in a more interactive format. This session enabled detailed discussions between presenters and attendees, facilitating the exchange of ideas and fostering a collaborative atmosphere. The combination of oral presentations and poster sessions ensured that a diverse range of research was highlighted, covering both theoretical advancements and practical applications in the field.

The proceedings of ISIMA 2024 represents a comprehensive collection of the latest research contributions in intelligent manufacturing and automation. It serves as a valuable resource for academics, researchers, and industry professionals who are interested in staying abreast of the latest developments in this rapidly evolving field. The papers included in this volume have undergone a thorough review process to ensure their quality and relevance. They reflect the current state of research and provide a glimpse into the future trends and innovations that will shape the industry.

In a word, ISIMA 2024 has provided a unique opportunity for the global community of researchers and practitioners to come together and share their knowledge and experiences. The proceedings of this conference will undoubtedly serve as a valuable reference for years to come, inspiring further research and development in the field of intelligent manufacturing and automation. We hope that the insights and connections made during this symposium will lead to innovative solutions and advancements that will benefit industries worldwide.

The Committee of ISIMA 2024

Journal of Physics: Conference Series **2992** (2025) 011001 doi:10.1088/1742-6596/2992/1/011001

Committee Member

Conference General Chair
ALAM MD MAHBUB, Harbin Institute of Technology (Shenzhen), China
Wenjie Hen, Anhui University, China

Publication Chair
Richard (Chunhui) Yang, Western Sydney University, Australia
Shubin Yan, Zhejiang University of Water Resources and Electric Power, China
Yunqing Rao, Huazhong University of Science and Technology, China
Sanjay Yadav, CSIR-NPL, New Delhi, India

Program Committee Chair
Liang Yan, Beihang University, China
Mehran Emadi Andani, University of Verona, Italy
David Bassir, Université de Technologie de Belfort-Montbéliard, France
Zeashan Khan, King Fahd University of Petroleum & Minerals, Saudi Arabia
Ata Moshayedi, Jiangxi University of Science and Technology, China

Organizing Committee Chair
Jiufei Luo, Chongqing University of Posts and Telecommunications, China

Organizing Committee
Xiantao Sun, Anhui University, China
Sheng Lu, Chongqing University of Posts and Telecommunications, China
Pingan Yang, Chongqing University of Posts and Telecommunications, China
Haiqing Li, Chongqing University of Posts and Telecommunications, China

Technical Program Committees
Xuegong Yu, Zhejiang University, China
Sheng-Rui Jian, I-Shou University, China
Wanmi Chen, Shanghai University, China
Ming-Yang Chen, Jiangsu University, China
Ke Chen, Hefei University of Technology, China
Jing Zhao, Xi'an Mingde Institute of Technology, China
Rongyun Zhang, Anhui Polytechnic University, China
Hao Peng, Yangtze Normal University, China
Xue Han, Chongqing Jiaotong University, China
Ziad Moumni, ENSTA ParisTech, France
Yue Zhang, Georgia Southern University, US
Kuan Yew Wong, Universiti Teknologi Malaysia, Malaysia
Hazizan Md Akil, Universiti Sains Malaysia, Malaysia
Seeram Ramakrishna, National University of Singapore, Singapore
Yasuto Hijikata, Saitama University, Japan
Eckart H. Meiburg, University of California, Santa Barbara
DAVID BASSIR, French University of Technology, France
Mohamed Kamal Ahmed Ali (M.K.A. Ali), Wuhan University of Technology, Egypt
Gang Li, University of Maryland Baltimore County, USA
Sajid Iqbal, University of Engineering and Technology (UET) Lahore, Pakistan
Ajay Kumar Agarwal, JB Knowledge Park, India
Volkan Kahya, Karadeniz Technical University, Turkey
David E. Esezobor, University of Lagos, Nigeria
Nour F. Attia, National Institute of Standards (NIS), Egypt
Yipin Wang, Chang'an University, China
Mohamed Ahmed Ramadan, Helwan University, Egypt
Mukuna Patrick Mubiayi, University of Johannesburg, South Africa
Quah Hock Jin, University Sains Malaysia, Malaysia

Journal of Physics: Conference Series **2992** (2025) 011001 doi:10.1088/1742-6596/2992/1/011001

Alexander Lukin, Western-Caucasus Research Center, Russia
R. RANJITH KUMAR, Viyen Biotech LLP, India
Hicham Bakkali, University of Cadiz, Spain
Rusdi Nur, Politeknik Negeri Ujung Pandang, Malaysia
Ang Zi Yang Adrian, Newcastle University (International Singapore), Singapore

Peer Review Statement

All papers published in this volume have been reviewed through processes administered by the Editors. Reviews were conducted by expert referees to the professional and scientific standards expected of a proceedings journal published by IOP Publishing.

- **Type of peer review:** Single Anonymous
- **Conference submission management system:** Morressier
- **Number of submissions received:** 108
- **Number of submissions sent for review:** 94
- **Number of submissions accepted:** 64
- **Acceptance Rate (Submissions Accepted / Submissions Received × 100):** 59.3
- **Average number of reviews per paper:** 3
- **Total number of reviewers involved:** 16
- **Contact person for queries:**
  Name: Xuexia
  Email: xx.ye@keoaeic.org
  Affiliation: AEIC Academic Exchange Information Centre

Development of a new type of variable-diameter wheel

Yang Ming[a*] and Feiyun Xu[b*]

Mechanical Engineering, Southeast University, Nanjing, China

[*]Corresponding author's e-mail: [a]847418781@qq.com, [b]fyxu@seu.edu.cn

Abstract. Wheeled robots have the advantages of simple structure, convenient control, and high motion stability, but their ability to traverse complex terrain is widely criticized. In this paper, a new type of variable-diameter wheel is designed. The diameter of the wheel can be changed by simple operations; Such being the case, the ability of the wheel to cross obstacles can be improved. In this paper, the application size of the wheel, which includes the number of spokes, the diameter change ratio of the wheel and the continuity of the wheel rim, was established to analyze the performance of obstacle crossing before and after the diameter change. The results show that the number of spokes, the diameter change ratio and the continuity of the wheel rim are all bound by the design size of the wheel. In other words, for different types of this variable-diameter wheel, the application size can be quickly obtained through calculation so as to facilitate the selection of other parts. Moreover, the stability of wheel motion is simulated. The reliability of the calculation results is verified. These analyses also provide the basis for wheel performance optimization design.

1. Introduction

Mobile robots have been widely used in planetary exploration, agriculture, mining, detection in dangerous environments and other work scenarios [1]. Wheeled mobile robots, because of their simple structure, convenient control, high stability and fast movement, have attracted people's attention [2]. However, in its application scenarios, it is inevitable that the wheeled robot is tired of dealing with the rough road surface, which may lead to the damage of the robot or the failure of the overall task [3]. Therefore, the adaptability of the wheel to the complex ground becomes an important criterion for the wheel design of the wheeled mobile robot.

In the design of the wheel, the research shows that by increasing the diameter of the wheel, the traction performance of the mobile robot on soft terrain will be significantly improved [4]. Its obstacle-crossing performance and its stability on the inclined plane will also be greatly enhanced [5]. So we designed a new type of variable-diameter wheel. This diameter of the wheel can be changed by users through the differential motion of both sides of the wheel. In this paper, the structure of the wheel and the concept of the variable-diameter wheel are introduced, the constraint relationship among the design parameters of the wheel is analyzed, and the obstacle crossing movement before and after the diameter change is simulated. According to the constraint relation, users can calculate various parts suitable for the wheel according to its design size.

2. The principle of diameter change

The wheel consists of a driving wheel disc, a fixed wheel disc, a state locking device, and a number of identical spokes. Every two spokes are hinged to each other to form a spoke rod pair (Figure 1). The two wheel discs and each spoke rod pair of the wheel can be regarded as a crank-rocker mechanism.

The crank corresponds to the driving wheel disc, and the connecting rod and the rocker respectively correspond to the back hinged two-spoke rod. When the driving wheel disc is moving relative to the fixed wheel disc, the wheel will drive the spoke rod to shrink and expand, so as to achieve the effect of changing the wheel diameter.

Therefore, the two use states of the wheel are as follows:

2.1 Expansion

The state of the wheel when the outer rim of the wheel is expanded into a complete circular arc is defined as the state of the expansion, as shown in Figure 2(a). At this time, each spokes of the wheel is combined into a complete circle. This state is related to the design of the spoke rod. In this paper, the expansion state of the wheel is designed as follows: The two spokes of each pair of spokes are hinged to each wheel disc at a point where the phase difference between the two points in the plane of the wheel disc is $\frac{360^{\circ}}{\mathrm{n}}$. n is the number of spoke

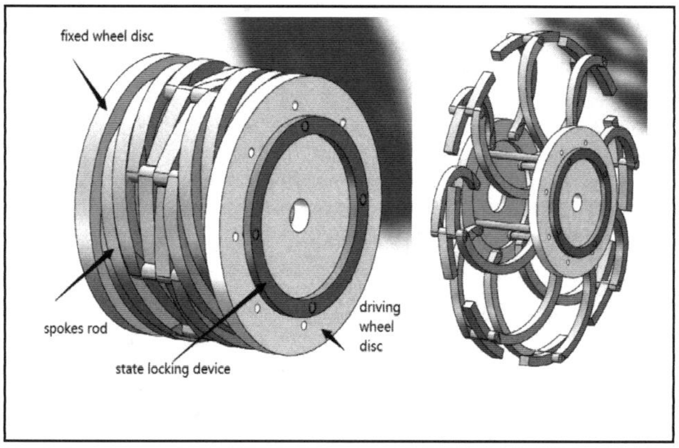

Figure 1. Three-dimensional solid view of the wheel.

rod pairs. However, it can be adjusted in the design of the spoke rod, as shown in Figure 2(b). The Angle of θ_1 determines the condition after the wheel is fully expanded. Its value can be adjusted according to the needs of the user. The selection of different angles will also have a certain extent of influence on the wheel change ratio.

2.2 Contraction

The state in which the spoke rod retracts just inside the wheel is defined as contraction. Since we choose bending rods with a certain bending radius, in the actual design, the bending rod radius Ri will have a certain influence on the contraction state. The distance between two hinge points of one spoke rod is defined as L, the diameter of the wheel discs is d_y, the thickness of the spoke rod touching the ground is b, and the distance between the hinge point on the wheel disc and the center of the circle is defined as r_j.

- $\frac{\mathrm{L}}{2} < \mathrm{R}_i < \frac{2L^2 \cdot (d_y - b)}{4L^2 + (d_y - b)^2 - 4r_j^2}$

As shown in Figure 3(a), the contraction state of the wheel at this time is marked by the inward cutting of the bent rod edge and the edge of the wheel disc.

- $\frac{2L^2 \cdot (d_y - b)}{4L^2 + (d_y - b)^2 - 4r_j^2} < \mathrm{R}_i$

The edge of the curved rod of the wheel and the edge of the wheel disc are not able to be tangent. The sign of wheel contraction is that the hinge point between the main layer of the spoke rod and the contact layer is just taken in by the wheel disc, as shown in Figure 3(b).

3. Structure design of variable-diameter wheel

The structure of the variable-diameter wheel depends on the size and quantity of the spoke rod. In the case of different sizes and quantities of the spoke rod, the parameters of the wheel will change accordingly.

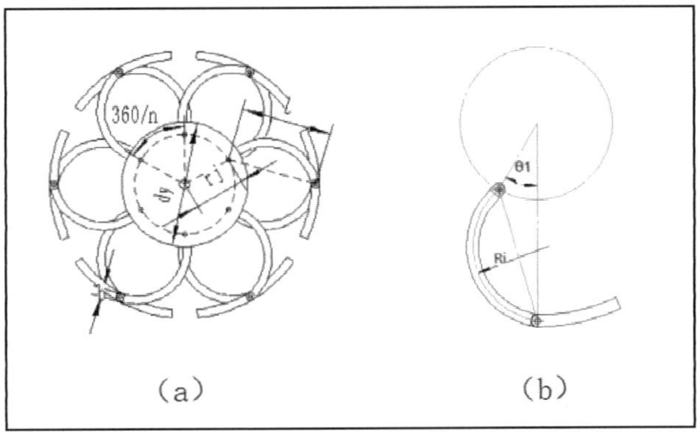

Figure 2. Expansion.

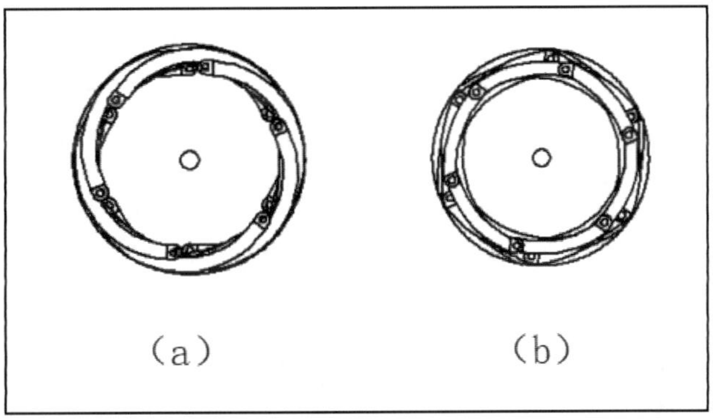

Figure 3. Contraction.

3.1 Number of spokes

The number of spokes of the variable-diameter wheel is not limited to 12 as shown in Figure 2, but can be selected according to the needs of users. The variable-diameter wheels with different number of spoke rods under the same design size given in other cases were designed and simulated, as shown in Figure 4. The wheel diameter change ratio, axial width, the continuity were calculated and the results were listed in Table 1.

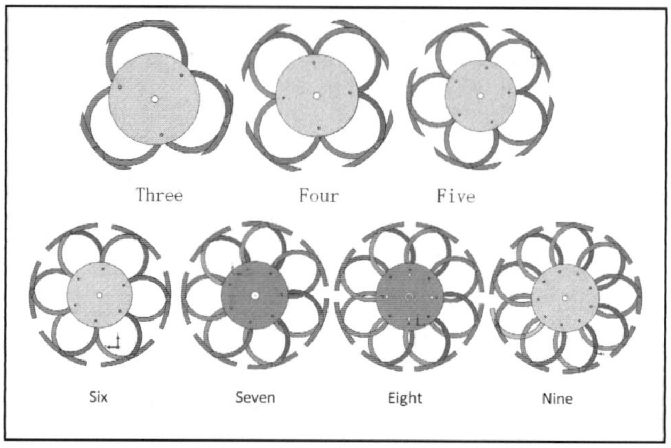

Figure 4. Variable-diameter wheel with different number of spokes.

Table 1. Application size with different numbers of spoke pairs.

Number of spokes	Application sizes	
	Diameter change ratio	***Continuity***
3	1.73	35.4%
4	1.95	48.8%
5	2.07	64.6%
6	2.14	81.5%
7	2.18	81.7%
8	2.21	79.0%
9	2.23	72.5%

According to the data above, with the increase of the spoke rod pairs, the wheel diameter change ratio gradually increased, but the range of change gradually decreased. The axial width increases uniformly, while the proportion of the whole circle occupied by the outer rim of the wheel increases at first, reaches a peak in the number of certain spoke rod pairs, and then decreases slowly. The specific mathematical relationship and reasons for its change will be explained later.

3.2 Diameter change ratio

The diameter change ratio i of a variable-diameter wheel can be expressed as:

$$i = \frac{D_y}{d_y} \tag{1}$$

where D_y is the rim diameter when the wheel is expanded, and d_y is the rim diameter when the wheel is contracted, namely, the diameter of the two wheel discs. d_y can be selected by users according to their own needs.

The expanded rim diameter D_y can be expressed by the design size of the spoke rods according to the size relationship in Figure 5:

$$D_y = 2\sqrt{r_j^2 \cdot \cos^2\theta + L^2 - r_j^2} + 2 \cdot r_j \cdot \cos\theta + b \tag{2}$$

So l is going to be in $\left(0, r_j + \frac{d_y}{2}\right)$, $\theta = \frac{180^{\circ}}{\text{n}}$. According to the above relation, it is not difficult to see that with the increase in the number of spoke rod pairs, the wheel diameter change ratio also gradually increases. Theoretically, when θ=0, the maximum reducing ratio can be obtained:

$$i_{\max} = \frac{2(r_j + \text{L}) + b}{d_y} \tag{3}$$

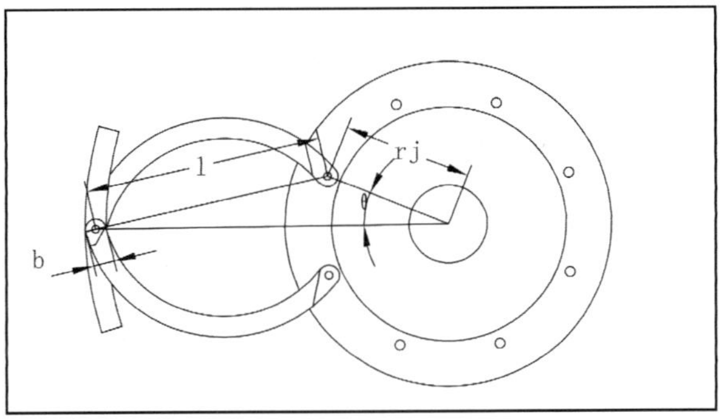

Figure 5. Diameter change ratio related design dimensions.

3.3 Continuity of wheel rim in expanded state

In order to make the variable-diameter wheel in the actual movement have the performance of a normal wheel, the rim of the wheel should be as close to a full circle as possible under the expanded state. Namely, the central angle of the circular arc of the ground touching part of each spoke rod should be $\frac{180^{\circ}}{\text{n}}$. However, when the wheel is designed according to the above method, there will be some problems such as the spoke rod exceeding the wheel disc and the interference between the spoke rods during the wheel contraction. We chose to cut off part of the arc of the spoke rod touching the ground. At the same time, we tried to ensure that the central angle of the remaining part was as close to $\frac{180^{\circ}}{\text{n}}$ as possible, so as to ensure the smoothness of wheel movement as far as possible. Here, the percentage of the arc formed by the contour of the spoke rod touching the ground in a complete arc is used to describe the degree of continuity of the wheel rim in the expansion state. To calculate this continuity, two things need to be considered as follows:

1) As shown in Figure 6, in the contraction state, the outermost end of the part where the spoke rod touches the ground goes beyond the range of the wheel disc, leading to the wheel not becoming a regular round wheel after contracting. So the part beyond the wheel disc will be cut. After cutting, $\angle BPK$ is the central angle corresponding to a single spoke rod after expansion. Then l can be obtained by the following formula:

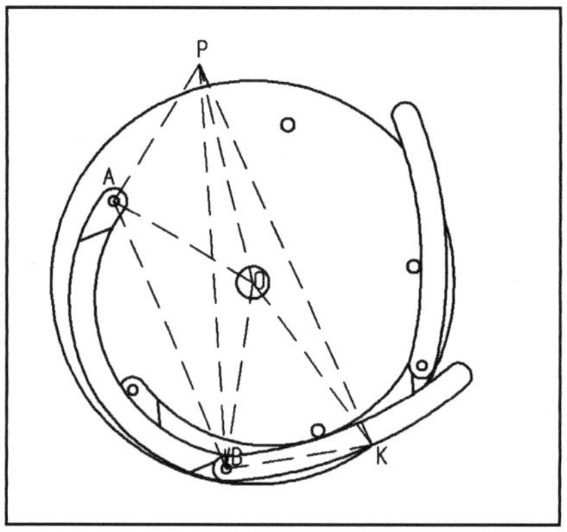

Figure 6. The spoke rod exceeds the wheel disc in the contraction state.

$$I = \frac{n \cdot \angle BPK}{180^0} \tag{4}$$

$$\angle BPK = \angle BPO + \angle OPK \tag{5}$$

$$\angle BPO = 90^0 - \frac{1}{2}\angle OAP - \arccos\frac{r_j^2 + \frac{(D_y - b)^2}{4} - L^2}{r_j \cdot (D_y - b)} \tag{6}$$

$$\angle OAP = \arcsin\left(\frac{D_y - b}{2L}\sin\frac{180}{n}\right) - \arccos\frac{r_j^2 + L^2 - OB^2}{2r_j L} \tag{7}$$

$$\angle OPK = \arccos\frac{4r_j^2 \sin^2\frac{\angle OAP}{2} + \frac{D_y^2}{4} - \frac{d_y^2}{4}}{2r_j D_y \sin\frac{\angle OAP}{2}} \tag{8}$$

According to the different bending radii of spokes and the different contraction ends, different bending radius Ri are classified and discussed:

- $\frac{L}{2} < R_i < \frac{2L^2 \cdot (d_y - b)}{4L^2 + (d_y - b)^2 - 4r_j^2}$

$$OB = \frac{d_y - b}{2} \tag{9}$$

- $\frac{2L^2 \cdot (d_y - b)}{4L^2 + (d_y - b)^2 - 4r_j^2} < R_i$

$$OB = \sqrt{r_j^2 + L^2 - 2r_j L\cos\left[\arccos\frac{r_j^2 + R_i^2 - (\frac{d_y}{2} - R_i - b)^2}{2r_j R_i} + \arccos\frac{L}{2R_i}\right]} \tag{10}$$

2) As shown in Figure 7, there is interference between spokes and spokes after wheel retraction. Now all the spokes of the wheel are shrunk to the edge of the spokes are tangent to the edge of the wheel. The critical condition of no interference was set as the point F at the most edge of the spoke rod would not contact the adjacent rod. $\angle BPF$ is the central angle occupied

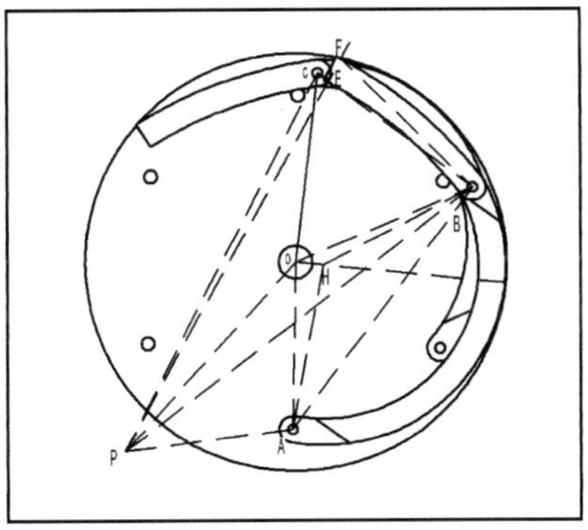

Figure 7. Interference of two spokes under contraction.

by a single spoke bar in the expansion state. Then I can be obtained by the formula below:

$$I = \frac{n \cdot \angle BPF}{180} \tag{11}$$

$$\angle BPF = \angle BPC - \angle CPF \tag{12}$$

$$\angle BPC = \arccos \frac{PC^2 + \frac{(D_y - b)^2}{4} - 2OB^2(1 - \cos\frac{360}{n})}{PC \cdot (D_y - b)} \tag{13}$$

$$\angle CPF = \arcsin \frac{b}{2PC} \tag{14}$$

$$PC = \sqrt{OB^2 + \frac{(D_y - b)^2}{4} - OB \cdot (D_y - b) \cdot \cos\angle CBP} \tag{15}$$

$$\angle CBP = \arccos \frac{\sqrt{2 - 2\cos\frac{360}{n}}}{2} + \arccos \frac{OB^2 + L^2 - r_j^2}{2OB \cdot L} - \arcsin\left(\frac{r_j}{L}\sin\frac{180}{n}\right) \tag{16}$$

According to the situation of different bending radii of spokes and different contraction ends, different bending radius R_i were classified and discussed:

- $\frac{L}{2} < R_i < \frac{2L^2 \cdot (d_y - b)}{4L^2 + (d_y - b)^2 - 4r_j^2}$

$$OB = \frac{d_y - b}{2} \tag{17}$$

- $\frac{2L^2 \cdot (d_y - b)}{4L^2 + (d_y - b)^2 - 4r_j^2} < R_i$

$$OB = \sqrt{r_j^2 + L^2 - 2r_j L \cos\left[\arccos \frac{r_j^2 + R_i^2 - (\frac{d_y}{2} - R_i - b)^2}{2r_j R_i} + \arccos \frac{L}{2R_i}\right]} \tag{18}$$

4. Performance analysis

In order to verify the practical performance of the wheel, we take a conventional type of wheel for performance analysis. The wheel parameters are as follows:

$L = 90mm, d_y = 125mm,\ n = 8,\ \theta_1 = 22.5^0$

$r_j = 50mm, b = 8mm, R_i = 50mm$.

4.1 Stationary analysis

Due to the contraction requirement, the outer rim of the wheel cannot form a complete circle after expansion. Therefore, in its actual motion, the center of mass will inevitably change in the vertical direction. According to the calculation in Chapter 3, the circular arc of the rim contour accounts for 79% of the complete circular arc of the conventional wheel in the state of expansion. At this point, the wheel was imported into ADAMS for kinematics simulation to simulate the movement of the wheel on a flat road surface as well as its centroid motion curve, as shown in Figure 8.

As can be seen from Figure 8, in the above cases, the wheel's centroid is stable most of the time, and the change of the centroid is no more than 0.8mm when the wheel moves to the gap between the spoke rod pairs, accounting for only 0.29% of the diameter of the wheel in expansion state, which is almost negligible. Therefore, its stability is good, almost the same as the ordinary round wheel.

4.2 Obstacle crossing performance analysis

In order to verify the obstacle crossing performance before and after the wheel diameter change, the obstacle crossing ability of the wheel was analyzed. As shown in Figure 9, a quasi static model is established. It is assumed that the wheel moves at a slow speed smoothly through obstacles. At this moment, the contact force between the wheels and the ground is 0. The height of the barrier is H. The radius of the wheel is R. G is the gravity of the wheel. Its driving moment is M. It receives friction f with a coefficient of friction μ, which is α degree from the ground. The corresponding mechanical relations are listed as follows:

$$f\cos\alpha + F\sin\alpha = G \tag{19}$$

$$f\sin\alpha = F\cos\alpha \tag{20}$$

$$GR\cos\alpha = M \tag{21}$$

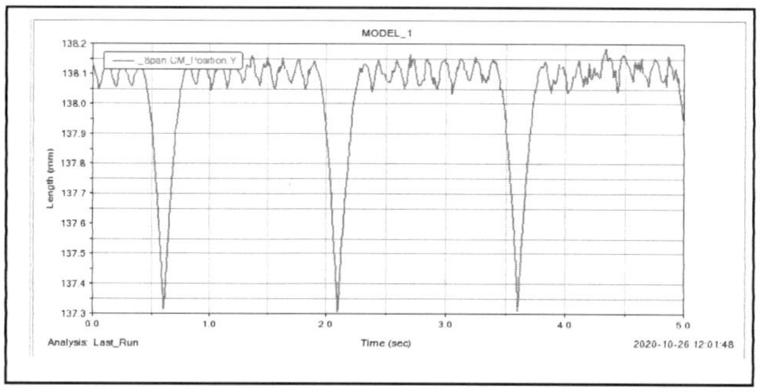

Figure 8. The trajectory of the center of mass in the vertical direction.

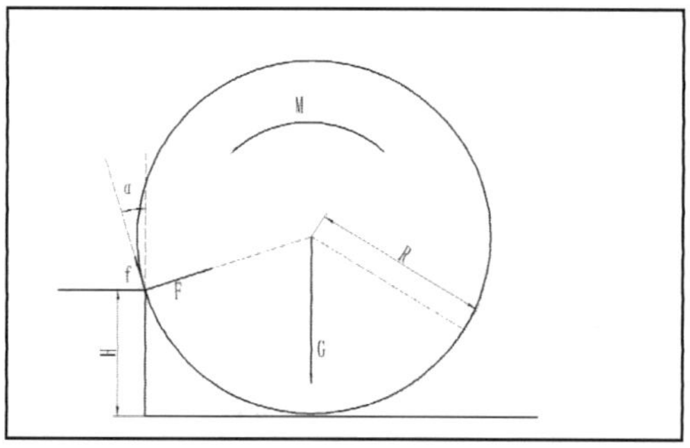

Figure 9. Simulation diagram of wheel passing through obstacles.

In the above formula:

$$\sin\alpha = \frac{R-H}{R} \tag{22}$$

$$\cos\alpha = \frac{\sqrt{2RH-H^2}}{R} \tag{23}$$

$$f = \mu F \tag{24}$$

According to the relation between the driving force and the driving torque, f can be expressed as:

$$f = \frac{M}{R} \tag{25}$$

Based on the formulas above, F , f and G are eliminated:

$$\mu = \frac{\sqrt{2RH-H^2}}{R-H} \tag{26}$$

According to Formula (26), function images of R and H are drawn in Figure 10. According to the function images, the height of the obstacle that the wheel can overcome increases with the increase of the wheel radius and friction coefficient between the wheels and the ground. Thus it can be seen that the enlargement of diameter has a great effect on the improvement of the obstacle crossing ability of the wheel.

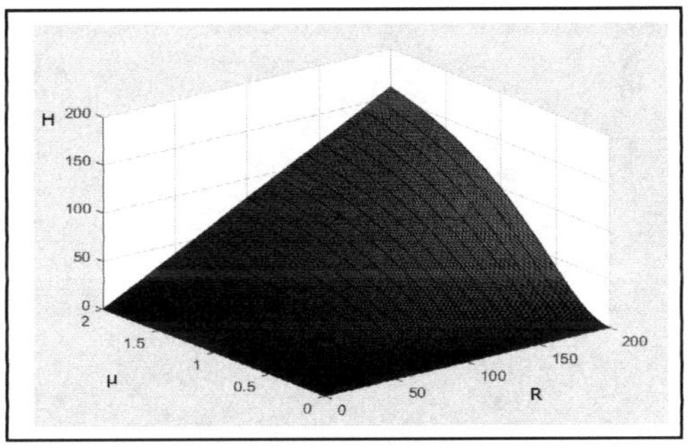

Figure 10. The maximum height the wheel can pass.

5. Conclusion

According to the above analysis, the variable-diameter wheel can change its diameter through simple operation. The operating size can be calculated according to the design and installation size of the spoke rods. In the contraction state, the rim is completely round. In the expanding state, although the rim is not completely round, its motion performance is similar to that of the circular wheel. A larger wheel diameter can help it pass more obstacles, and a smaller wheel diameter can allow the wheel to pass through narrow terrain. Therefore, the adaptability of this wheel to terrain is better than that of ordinary wheels, and it has stronger applicability. The simulation results show that the prediction results of this wheel perform well.

Reference

[1] Zheng J, Gao H, Yuan B, Liu Z, Yu H, Ding L and Deng Z Oct. 2018 Design and terramechanics analysis of a Mars rover utilising active suspension Mech. Mach. Theory 128 125-149.

[2] YAN Y,WANG S, XIE Y, et al. Obstacle-circumventing adaptive control of a four-wheeled mobile robot subjected to motion uncertainties [J]. Frontiers of Mechanical Engineering,2023,18(03):125-145.
[3] Inotsume H, Sutoh M, Nagaoka K, Nagatani K and Yoshida K Sep. 2013 Modeling, analysis, and control of an actively reconfigurable planetary rover for traversing slopes covered with loose soil J. Field. Robot 30 875-896.
[4] Aoki T, Murayama Y and Hirose S Feb. 2014 Development of a transformable three-wheeled lunar rover: Tri-Star IV J. Field. Robot 31 206-223.
[5] Wen Z H, Jiang G Y, Xu F, et al. "Research on a New Analysis Method of Traction Performance for Variable-Diameter Wheels." Proceedings of the ASME 2022 International Mechanical Engineering Congress and Exposition. Volume 5: Dynamics, Vibration, and Control. Columbus, Ohio, USA. October 30–November 3, 2022

Journal of Physics: Conference Series **2992** (2025) 012002 doi:10.1088/1742-6596/2992/1/012002

Design of two-stage parallel shock cam reducer

Yajuan Chen [1], Yang Chao [3] and Xiaoping Wang [2]*

[1, 3] Chongqing Vocational College of Architecture and Technology, Chongqing 400034, China
[2] Chongqing Yuxinpingrui Electronics Co., Ltd. Chongqing, China

*Corresponding author's e-mail: 365379562@qq.com

Abstract: In this paper, the double-click parallel shock cam reducer is taken as the research object. Based on the structural design, the virtual simulation model is established by Solidworks, and the reducer simulation and performance evaluation are carried out based on MSC. adams. Through analysis, the angular backlash and idle angle of the reducer are obtained, which can meet the structural and performance requirements.

1. Introduction

The original structure of the shock cam reducer was put forward by Germany in the 1930s. In the 1940s, Germany first applied the movable tooth transmission technology to automobiles. In the 1950s, Soviet scientists put forward a kind of "plunger drive" type and analyzed the kinematics and force. Then American scholars put forward a push rod movable tooth device and a reducer with less tooth difference. In the 1970s, American and Soviet scholars actively developed a new type of movable tooth transmission and developed "sine ball transmission" and "toothless gear transmission technology", which attracted great interest from scientists from various countries. After the 1980s, international research on movable teeth transmission technology gradually increased. Many countries have published patents and theories on movable teeth, which shows that the research on movable teeth technology was a very active field at that time.

Domestic scholars have studied the movable teeth transmission technology relatively late. Since the 1970s, China has gradually started to study movable teeth transmission technology. There is still a big gap between China and foreign countries in this field, but through continuous in-depth research by researchers, many movable teeth transmission structures have been innovated, which has made great contributions to China's movable teeth transmission technology. In the 1980s, Zhang Guorui of Shanghai Jiaotong University completed the principle and mechanical analysis of roller movable teeth planetary transmission. Chen Yangxian of Gansu University designed a multiphase cam-type movable tooth reducer with simple processing technology. From 1996 to 2000, Professor An Zijun used error theory to analyze the related structural parameters of oscillating tooth transmission and the influence of its error on tooth profile. For the first time in China, the cycloidal steel ball planetary transmission was systematically studied, and the types of cycloidal steel ball planetary transmission were comprehensively analyzed. A prototype of the transmission mechanism was manufactured.

In recent years, many experts and scholars at home and abroad have done a lot of research on different types of movable teeth reducers, such as modifying and designing the tooth profile of the central gear by using the envelope principle and the relative motion method, and calculating the meshing stiffness between the movable teeth and the central gear based on the elastic small deformation theory of the meshing pair, and related meshing characteristics.

Based on the above research, the structure of the reducer is determined to be a two-stage parallel shock cam reducer. The virtual model is established by Solidworks. The simulation and performance evaluation of the reducer are carried out based on MSC.adams.

2. Working principle and structure of two-stage parallel shock wave reducer

Cam shock wave oscillating movable teeth transmission device is mainly composed of a shock wave cam, oscillating movable teeth, central internal gear, and movable tooth frame， shown as figure1[1-3]. Among them, the cam profile of the shock wave is an elliptic curve. The oscillating tooth line is a standard circular curve. The center wheel is an internal gear ring with an envelope curve tooth shape and is fixedly connected to the frame: Each movable tooth is eccentrically sleeved on the movable tooth frame through movable tooth pins. When working, the shock wave device periodically pushes the movable teeth, and the meshing points of these movable teeth and the tooth profile of the fixed gear form a serpentine tangential wave, thus forming a continuous driving relationship with the fixed gear. The formation condition of this tangential wave is that the movable teeth and the fixed gear have different teeth numbers and their tooth pitches are unequal. It is precisely because of the different tooth pitches that “staggered tooth movement” occurs during meshing, which makes the transmission between the movable gear and the outer ring possible.

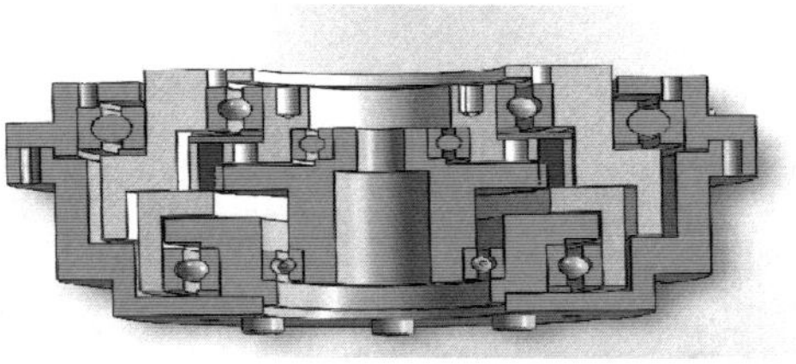

Figure 1 Gear structure.

3. Parameter selection and simulation modeling of two-stage parallel shock wave reducer

3.1 Analysis of clutch operating system

Figure 2 is a schematic diagram of the gear size. According to the theoretical curvature radius Equation (4) of the internal gear, the curvature radius Equation (5) with the working tooth profile of the internal gear is obtained[4-8].

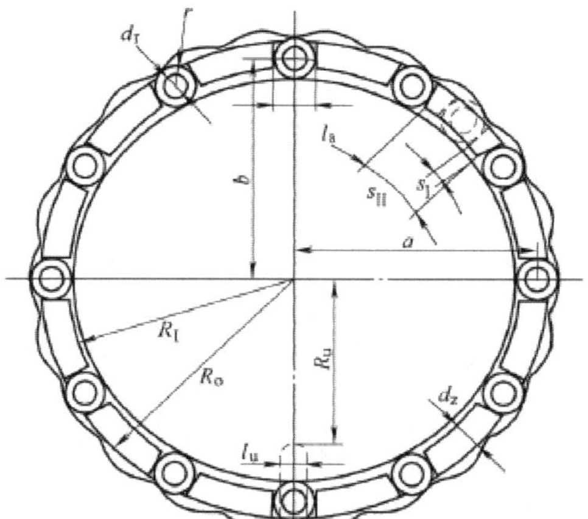

Figure 2 Schematic diagram of gear size.

 doi:10.1088/1742-6596/2992/1/012002

$$\rho = \frac{1}{K} = \frac{ab(4Q^2 + A^2)^{3/2}}{2Q^{3/2}[4Q(M' \cos\theta - N' \sin\theta) - 5A^2}$$

In the formula:

$$A = (i-1)(a^2 - b^2)\sin 2(i-1)\theta$$
$$Q = a^2 \sin^2 (i-1)\theta + b^2 \cos^2 (i-1)\theta$$
$$M' = F \sin 2(i-1)\theta \sin\theta + I \cos\theta + b^2 \cos\theta$$
$$N' = F \sin 2(i-1)\theta \cos\theta - I \sin\theta - b^2 \sin\theta$$
$$F = 0.5(a^2 - b^2)(i-1)$$
$$I = (a^2 - b^2)[(i-1)^2 \cos 2(i-1)\theta + sin^2 (i-1)\theta \quad (1)$$
$$\rho_w = \rho \pm r \qquad (2)$$

In order to ensure that there is no continuous profile of interference cusp in the working tooth profile of the internal gear, its curvature radius should meet the following conditions:

$$\rho_w \geq 0 \qquad (3)$$

According to the above curvature requirements and the requirements of transmission ratio, axle shaft length difference, and movable tooth radius on long and short axle shafts, the movable tooth radius r is initially taken as 2mm. In order to ensure that the internal gear does not have tooth interference, combined with the external dimension requirements of the reducer, the results of the two-stage transmission are shown in Table 1.

Table 1 Values of Two-stage Transmission.

	Single-stage transmission ratio	a	b	Radius of curvature	Difference between the long and short-axis
First stage transmission	11	25.25	24.25	2.70	1
Second stage transmission	9.5	19.75	18.75	2.27	1

The theoretical curvatures of the first stage and the second stage are shown in Figures 3 and 4, respectively.

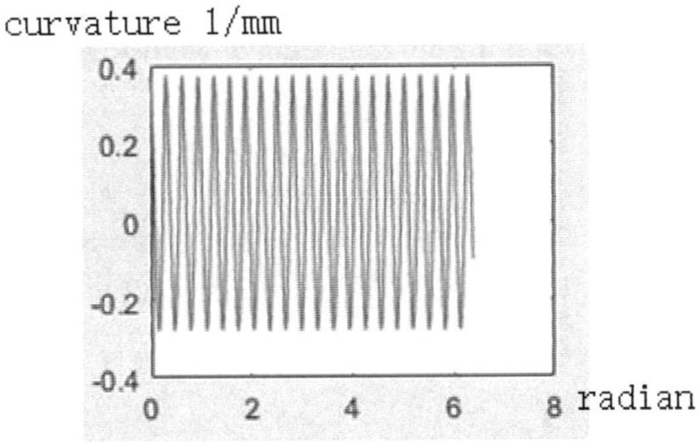

Figure 3 First-order theoretical curvature K1.

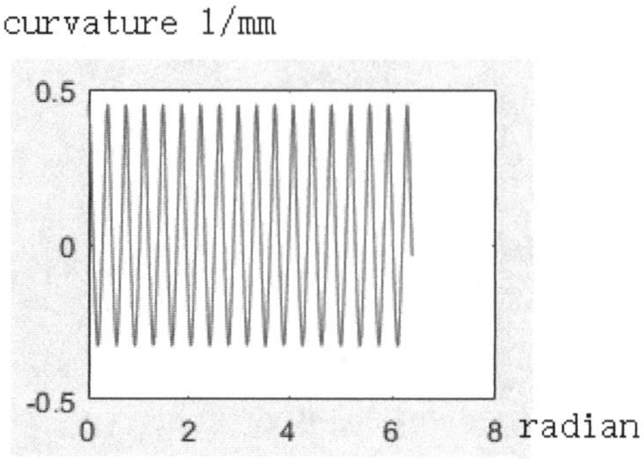

Figure 4 Second-order theoretical curvature K2.

3.2 Calculation of other parameters

The inner diameter RI of the movable rack should be greater than the length a-r of the long semi-axis of the working profile of the shock cam, i.e.

$$R_I > a - r \quad (4)$$

In order to avoid interference between the movable rack and the tooth tip of the internal gear, the outer diameter Ro of the movable rack should be smaller than the circular radius of the tooth tip of the internal gear. In addition, in order to ensure the continuity of transmission and prevent the movable rack pin from slipping out of the radial groove of the half disk during transmission, the outer diameter Ro of the movable rack should be larger than the theoretical long semi-axis A of the cam, as can be seen from the above figure.

$$a < R_I < b + r \qquad (5)$$

The radial groove width Lu should not be less than the diameter of the movable toothpin shaft.

The radial groove depth Ru should ensure that the movable teeth can always contact the shock cam during transmission, as shown in the above figure.

$$R_u < b - d_r/2 \qquad (6)$$

The value of pin column thickness dz should not be too small, otherwise the strength is not enough. When the inner and outer radii RI and Ro of the movable rack are taken as the minimum and maximum values respectively, the pin column thickness dz gets the maximum value, and the following formula is given.

$$d_{zmax} = 2r - (a - b) = 2r - e \quad (7)$$

The inner arc length S_I of the pin is the length of the arc connecting the inner side of the pin, which corresponds to the distance between two adjacent movable teeth. According to the geometric relationship shown in the above figure, the calculation formula of S_I can be deduced as follows.

$$S_I = 2R_I(\frac{\pi}{Z_a} - \sin^{-1}\frac{r}{R_I}) \quad (8)$$

In order to ensure the reliable transmission of the movable rack, the value of the arc length SI in the pin column should not be too small.

Cam transmission is a two-tooth difference transmission. When the transmission is relatively large, there are many movable teeth, and the radial dimension of the movable rack is required to be compact. The tooth extraction transmission structure can be adopted. That is to say, one tooth is removed from the adjacent movable teeth, and half of the movable teeth are kept to ensure that there is no motion interference between the remaining half of the movable teeth. Accordingly, according to the geometric relationship shown in the above figure, the calculation formula of the arc length SII in the connecting pin column after tooth extraction is as follows.

$$S_I = 2R_I(\frac{\pi}{Z_a} - \sin^{-1}\frac{r}{R_I} \quad (9)$$

The following is the changing trend of the arc length S_I and S_{II} in the connecting pin column with different values of transmission ratio I, theoretical long semi-axis A of shock cam, and movable tooth radius R. According to the above values of the theoretical long and short semi-shafts, combined with the transmission ratio, we select the appropriate movable tooth radius, and check the inner and outer diameters and thickness of the movable tooth frame, as shown in Table 2.

Table 2 Values of Gear Dimensions.

Variable symbol	Parameter	First-order value	Secondary value
a	semi-major axis	25.25	19.75
b	Short half shaft	24.25	18.75
e	Half axle difference	1	1
I	Single-stage transmission ratio	11	9.5
za	Number of movable teeth	22	19
zb	Number of internal teeth	20	17
r	Movable tooth radius	2	2
RI	Half-disk inner diameter	23.5	18
RO	Half-disk outer diameter	26	20.5
Ru	Radial groove depth	21.25	16.25
lh	Length of movable teeth	8	8
la	Pin spacing	4	4
dz	Pin thickness	2.5	2.5
sI	Internal arc length of pin column	2.63	1.86
sII	Internal arc length of tooth extraction pin column	9.27	7.73
dr	Movable pin diameter	4	4
Kmax	Maximum curvature	0.3698	0.4406
ρ	Minimum radius of curvature	2.70	2.27

Combined with the curve above, the circular arc lengths S_I and S_{II} in the half disk meet the requirements.

On this basis, the actual 3D verification is carried out, and Figure 5 shows the results of theoretical centerline generation.

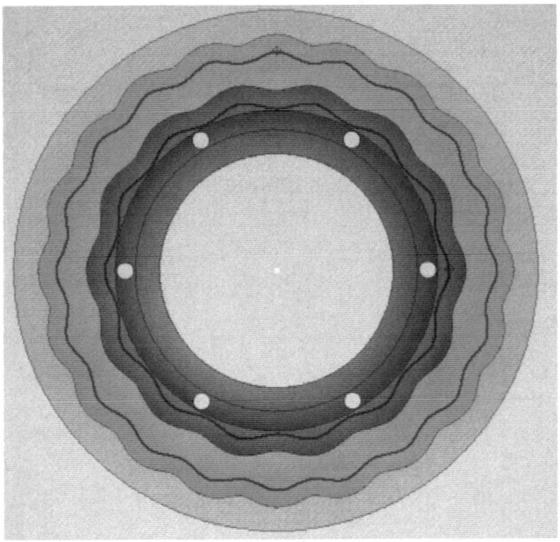

Figure 5 Theoretical centerline generation result.

The theoretical tooth profile equations of the central inner tooth and cam groove thus determined are as follows.

3.3 First stage transmission

Cam theoretical equation:

$$\rho = \frac{ab}{\sqrt{b^2 \cos^2\varphi + a^2 \sin^2\varphi}} \qquad \varphi \in [0, 2\pi) \qquad (10)$$

Derived from polar coordinate equation:

$$\rho_1 = (23.25) * (22.25)/((24.25 - 2)^\wedge 2 * \cos(a)^\wedge 2 + (25.25 - 2)^\wedge 2 * \sin(a)^\wedge 2)^\wedge 0.5 \quad T \in [0, 2\pi] \qquad (11)$$

Central wheel theoretical equation:

$$X1 = 25.25 * 24.25 * \cos(t)/(24.25^\wedge 2 * \cos(10 * t)^\wedge 2 + 25.25^\wedge 2 * \sin(10 * t)^\wedge 2)^\wedge 0.5 \qquad (12)$$

$$Y1 = 25.25 * 24.25 * \cos(t)/(24.25^\wedge 2 * \cos(10 * t)^\wedge 2 + 25.25^\wedge 2 * \sin(10 * t)^\wedge 2)^\wedge 0.5 \qquad (13)$$

3.4 Second stage transmission

Cam theoretical equation:

$$\rho = \frac{ab}{\sqrt{b^2 \cos^2\varphi + a^2 \sin^2\varphi}} \qquad \varphi \in [0, 2\pi) \qquad (14)$$

Derived from polar coordinate equation:

$$\rho_2 = (17.75) * (16.75)/((16.75)^\wedge 2 * \cos(a)^\wedge 2 + (17.75)^\wedge 2 * \sin(a)^\wedge 2)^\wedge 0.5 \quad T \in [0, 2\pi] \qquad (15)$$

Central wheel theoretical equation:

$$X2 = 19.75 * 18.75 * \cos(t)/(18.75^\wedge 2 * \cos(8.5 * t)^\wedge 2 + 19.75^\wedge 2 * \sin(8.5 * t)^\wedge 2)^\wedge 0.5 \qquad (16)$$

$$Y2 = 19.75 * 18.75 * \sin(t)/(18.75^\wedge 2 * \cos(8.5 * t)^\wedge 2 + 19.75^\wedge 2 * \sin(8.5 * t)^\wedge 2)^\wedge 0.5 \qquad (17)$$

4. Simulation verification

Through the hysteresis curve of the reducer (as shown in Figure 6 below), the backlash at 0 torque and the idle value at 3% rated torque are measured. The simulated hysteresis characteristic curves of Figures 7 and 8 are obtained[9-13].

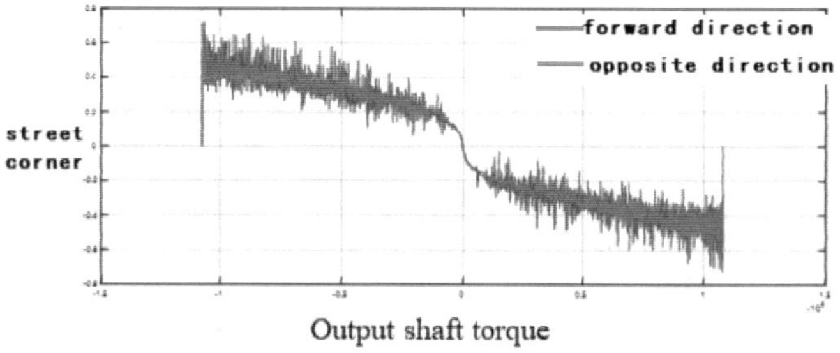

Figure 6 hysteretic characteristic diagram.

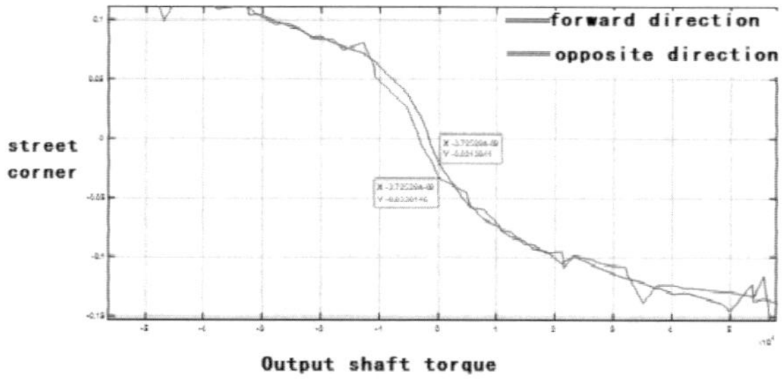

Figure 7 Return error measurement.

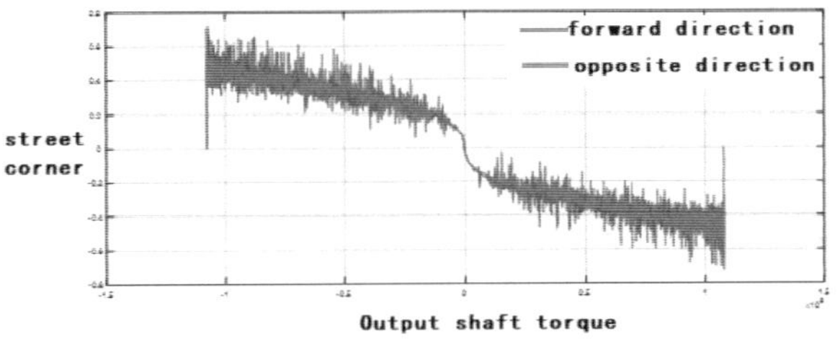

Figure 8 Idle distance measurement.

According to the requirements of GB/T37718-2019, when the input end of the reducer is fixed and the output end is gradually loaded to the rated torque of 1078000 Nmm (1078Nm), the difference between the rotation angles of the output end at 0 torque in two directions is measured by the results in Figure 6. At 0 torque, the rotation angle deviation of the output end is 0.0116, which meets the standard requirements.

According to the requirements of GB/T37718-2019, the input end of the reducer is fixed, the torque at the output end is changed from forward rotation to reverse rotation. The rotation angle difference is under 3% and -3% rated torque (32335.7Nmm). According to the measurement in Figure 7, the idle angle is 0.212, which meets the standard requirements.

5. Conclusion

In this paper, the double-click parallel shock wave cam reducer is taken as the research object. The size parameters of the reducer are obtained according to the working principle and two-stage theoretical curve of the two-stage parallel shock wave reducer. On this basis, the virtual simulation model is established by Solidworks. The reducer is simulated and evaluated based on MSC.adams. The hysteretic curve of the reducer is obtained. At the same time, the backlash measurement and idle distance measurement are carried out. According to the backlash and idle distance angle of the reducer, it is concluded that the reducer can meet the structural and performance requirements.

Acknowledgment

It is supported by the Science and Technology Research Program of Chongqing Municipal Education Commission (Grant No. KJZD-K202205201)".

References

[1] Li Yang. Design and experimental research of ground screeding construction robot. Beijing University of Civil Engineering and Architecture, 2021.
[2] Li Qinyang. Development and application of industrial robot deburring platform based on monocular vision. Xi'an University of Technology, 2021.
[3] Li Jianfeng, He Aiying, Dong Xinrui, et al. Geometric Design of Cam Shock Wave Rolling Movable Teeth Transmission, Journal of Mechanical Engineering, 2011.
[4] Wang Jianbin, Wan Jun, Yin Jishu. Virtual Design and Motion Simulation of Cycloidal Movable Teeth, Mechanical Transmission, 2009, 33(4): 45-47.
[5] Fu Qiang. Design and Implementation of Grabbing Robot System. Jilin University, 2021.
[6] Dong Xinrui, Li Jianfeng, Sui Dan, et al. Structure and tooth profile analysis of oscillating tooth drive with cam shock wave [J]. Mechanical transmission, 2007, 31(3)1-5.
[7] Han Liang. The Application of Robot in Construction Industry under Intelligent Manufacturing Era [J]. Building Economy, 2018, 39(3):23-27.
[8] Yan Jidong, Research on Cam Shock Wave Sinusoidal Movable Tooth Reducer [D]. Yanshan University, 2021.

[9] Yi Yali, Guo Zhenghui, Wei Rui, et al., Study on the design of movable teeth parameters of two-tooth difference pendulum based on dynamic characteristics. Vibration and impact, 2021, 40(11): 252-261.
[10] Li Jianfeng, Yu Yang, Lei Yu Zhang, et al. Tooth profile design and meshing force analysis of oscillating output movable tooth cam mechanism. Journal of Beijing institute of technology, 2018, 38(11): 1111-1119.
[11] Li Jianfeng, Dong Xinrui. Tooth profile modification method of internal gear in cam shock wave rolling movable teeth transmission [J. Journal of Beijing University of Technology, 2008(10): 1009-1014.
[12] Shen Haitao, An Zijun, Contact Stress Analysis and Optimization of Double Shock Sleeve Movable Tooth Transmission [Mechanical Strength, 2021, 43(04): 874-880.
[13] Liu Jun, An Zijun. Fuzzy reliability analysis of contact fatigue strength of double shock sleeve movable teeth transmission based on improved response surface method [J]. Mechanical strength, 2020, 42(06): 1362-1368.

Journal of Physics: Conference Series **2992** (2025) 012003 doi:10.1088/1742-6596/2992/1/012003

Reliability design of mechanical structures with small samples

[1st] **Shenye Zou and** [2nd] **He Zhao***

Dalian Polytechnic University, Dalain, China

*Corresponding Author: 842689576@qq.com

Abstract—In traditional mechanical structure reliability optimization design, the treatment of uncertain parameters relies on large sample data to obtain the probability density function or statistical characteristics, especially the first four moments. However, in actual engineering applications, the cost of obtaining data samples is high, and the number is limited, leading to the limitation of traditional methods in handling small sample problems, not only making it difficult to accurately represent uncertainty but also potentially leading to unreasonable optimization results and failure risks. Therefore, this paper proposes an improved method based on bootstrapping to expand the original small number of samples to a larger scale, so as to reconstruct the statistical characteristics of uncertain parameters in the case of data scarcity. Specifically, this paper first uses bootstrapping to perform self-sampling expansion of the sample data and calculates the first four central moments of the expanded sample to obtain information of the central moments of the uncertain parameters. On this basis, the probability density function is established using the maximum entropy principle, and the most consistent probability distribution model is generated by maximizing the information entropy, achieving reliability calculation. This method not only expands the applicability of reliability optimization design under small sample data conditions but also improves the accuracy of uncertainty description, providing a new method and idea for engineering structure reliability analysis.

1. Introduction

The reliability of mechanical structures has gained increasing significance, especially in complex environments and under variable loads. Reliability analysis is crucial for improving structural safety and extending service life. Traditional reliability assessment methods rely on large sample data to estimate failure probabilities and life distributions [1-2]. However, practical engineering constraints—such as limited test conditions, time, and costs—often result in insufficient data, particularly in extreme environments or specialized equipment assessments like life or non-failure tests [3]. To address challenges posed by small sample sizes, several methods have been developed.

Among these methods, bootstrapping has shown considerable potential. By resampling existing data, bootstrapping increases sample sizes, thereby enhancing the stability and accuracy of statistical inferences, making it particularly suitable for small sample reliability analysis [4]. Bootstrapping has been effectively applied in fields such as life assessment of systems without failure data and bearing reliability testing, improving both effectiveness and reliability. In recent years, it has garnered significant attention. Anonymous et al. [5] provided a comprehensive guide on its applications, detailing operational steps, use cases, and practical scenarios to facilitate efficient real-world implementations. The method's flexibility has proven valuable for conducting multiple comparison analyses. Wang H. et al. [6] proposed an improved bootstrapping method based on Radial Basis

Function (RBF) neural networks. By modifying the empirical distribution using exponential functions and combining it with RBF neural networks, this method generates extended samples, improving reliability estimates such as Mean Time Between Failures (MTBF), especially under limited data conditions. Shao Y. et al. [7] combined virtual sample augmentation with bootstrapping to evaluate the fatigue reliability of excavator components under extremely small sample conditions, achieving both interval and point estimates of life parameters. Zhao Yuan [8] employed B-spline functions to construct empirical distribution functions for life evaluation models, demonstrating their effectiveness. Additionally, Liu Jian [9] proposed expanding the sample generation range in bootstrapping, addressing the limitations of traditional methods. These advancements underscore bootstrapping's potential to improve reliability analysis in small sample scenarios.

Given the advantages of bootstrapping in small sample expansion, this paper adopts it as the foundation for sample expansion. By repeatedly resampling limited data, the bootstrap method generates numerous pseudo-samples, effectively increasing sample sizes while preserving the statistical characteristics of the original data. This approach addresses the instability in statistical inference caused by insufficient data, thereby improving the accuracy of reliability analysis under small sample conditions. After generating pseudo-samples, the first four central moments are calculated to characterize the distribution of uncertain parameters. Using these moments, the probability density function is inferred based on the maximum entropy principle, ensuring compliance with the moment constraints. This enables a quantitative evaluation of mechanical structure reliability. The proposed method maximizes the utility of limited data, reduces dependence on large samples, and offers a practical solution for conducting reliability analyses under small sample conditions.

2. Bootstrapping principle

The bootstrap method, introduced by Stanford Professor Efron, is a statistical inference technique. It involves a set of independent and identically distributed samples X=(X_1, X_2,..., X_n), each following a distribution F(x). A statistic R (X, F) is calculated from the sample and its distribution, allowing estimation of distribution characteristics based on observed samples. For instance, if θ represents a parameter in distribution F, the empirical distribution function of sample X estimates θ, with the estimation error computed as follows:

$$R_n = \hat{\theta}(F_n) - \hat{\theta}(F) \tag{1}$$

Drawing once from F_n generates a resampled set $X_1^* = (X_{11}^*, X_{12}^*, ..., X_{1n}^*)$ and compute $\theta\,(F)$ as a new estimate of d. Let F_n^* represent the empirical distribution function of the regenerated sample X^*. Then, the definition is:

$$R_n^* = \hat{\theta}(F_n^*) - \hat{\theta}(F_n) \tag{2}$$

Using the distribution property of R_n^*, the distribution property of R_n can be approximated, expressed as:

$$\theta\,(F) \approx \hat{\theta}(F_n) - R_n^* \tag{3}$$

Suppose the sample is repeatedly drawn from the bootstrap sample B times, and the regenerated sample matrix is $X_B^* = (X_{B1}^*, X_{B2}^*, ..., X_{Bn}^*)$, then a matrix containing B bootstrap samples can be constructed:

$$X_{\text{Bootstrap}} = [X_1^*, X_1^*, \cdots, X_B^*]^T = \begin{bmatrix} X_{11}^* & X_{12}^* & \cdots & X_{1n}^* \\ X_{21}^* & X_{22}^* & \cdots & X_{2n}^* \\ \vdots & \vdots & \vdots & \vdots \\ X_{B1}^* & X_{B2}^* & \cdots & X_{Bn}^* \end{bmatrix} \tag{4}$$

Therefore, the distribution characteristics or statistical properties of θ can be obtained from Equation (4), including mean, variance, covariance matrix, and third- and fourth-order central moments. The reliability or confidence level can be determined using the following equation:

$$R = \int_{g(X)>0} f_X(X)dX \tag{5}$$

Where $f_X(X)$ is the probability density function of $X = [X_1, X_2, \cdots X_n]^T$, representing the randomness of the structure. The function $g(X)$ defines the structural state, dividing failure and safety conditions as follows:

$$\left.\begin{array}{ll} g(X \le 0), & \text{Is the failure state;} \\ g(X \le 0), & \text{Is a safe state;} \end{array}\right\} \tag{6}$$

Denote the random variable as：

$$\boldsymbol{X} = \boldsymbol{X}_d + \varepsilon \boldsymbol{X}_p \tag{7}$$

Let the function $g(X)$ be expressed as：

$$g(\mathbf{X}) = g_d(\mathbf{X}) + \varepsilon g_p(\mathbf{X}) \tag{8}$$

Applying the mathematical expectation operation to the above expression, we can obtain：

$$\mathrm{E}(X) = \overline{X} = \mathrm{E}(X_d) + \varepsilon \mathrm{E}(X_p) = X_d \tag{9}$$

$$\begin{aligned} \mu_g &= \mathrm{E}[g(X)] = \overline{g}(X) \\ &= \mathrm{E}[g_d(X)] + \varepsilon \mathrm{E}[g_p(X)] \\ &= g_d(X) \end{aligned} \tag{10}$$

By analyzing the Kronecker product and related random variables, the first four moments are calculated.

$$\mathrm{Var}(X) = \mathrm{E}\{[X - \mathrm{E}(X)]^{[2]}\} = \varepsilon^2 [X_p^{[2]}] \tag{11}$$

$$\mathrm{C}_3(X) = \mathrm{E}\{[X - \mathrm{E}(X)]^{[3]}\} = \varepsilon^3 [X_p^{[3]}] \tag{12}$$

$$\mathrm{C}_4(X) = \mathrm{E}\{[X - \mathrm{E}(X)]^{[4]}\} = \varepsilon^4 [X_p^{[4]}] \tag{13}$$

Accordingly, the variance and higher-order moments of the function g(X) can be expressed as：

$$\begin{aligned} \mathrm{Var}[g(X)] &= \mathrm{E}\{[g(X) - \mathrm{E}(g(X))]^{[2]}\} \\ &= \varepsilon^2 \mathrm{E}\{[g_p(X)]^{[2]}\} \end{aligned} \tag{14}$$

$$\begin{aligned} \mathrm{C}_3[g(X)] &= \mathrm{E}\{[g(X) - \mathrm{E}(g(X))]^{[3]}\} \\ &= \varepsilon^3 \mathrm{E}\{[g_p(X)]^{[3]}\} \end{aligned} \tag{15}$$

$$\begin{aligned} \mathrm{C}_4[g(X)] &= \mathrm{E}\{[g(X) - \mathrm{E}(g(X))]^{[4]}\} \\ &= \varepsilon^4 \mathrm{E}\{[g_p(X)]^{[4]}\} \end{aligned} \tag{16}$$

Where, ⊗ denotes the Kronecker product, meaning: $(A_{p\times q} \otimes B)_{s\times r} = [a_{ij}B]_{ps\times qt}$。
When the random variable is small, $g_p(\boldsymbol{X})$ can be approximated by Taylor expansion, with:

$$g_p(\boldsymbol{X}) = \frac{\partial g_d(\boldsymbol{X})}{\partial \boldsymbol{X}^T} X_p \tag{17}$$

Substituting this into the correlation variance and moment expressions, we obtain:

$$\begin{aligned} \sigma_g^2 &= \mathrm{Var}[g(X)] = \varepsilon^2 \mathrm{E}\left[\left(\frac{\partial g_d(X)}{\partial \boldsymbol{X}^T}\right)^{[2]} \boldsymbol{X}_p^{[2]}\right] \\ &= \left[\left(\frac{\partial g_d(X)}{\partial \boldsymbol{X}^T}\right)^{[2]}\right] \mathrm{Var}(X) \end{aligned} \tag{18}$$

The higher-order moments can be expressed as:

$$\theta_g = \mathrm{C}_3[g(X)] = \varepsilon^3 \mathrm{E}\left[\left(\frac{\partial g_d(\boldsymbol{X})}{\partial \boldsymbol{X}^T}\right)^{[3]} \boldsymbol{X}_p^{[3]}\right] = \left[\left(\frac{\partial g_d(X)}{\partial \boldsymbol{X}^T}\right)^{[3]}\right] \mathrm{C}_3(\boldsymbol{X}) \tag{19}$$

$$\eta_g = \mathrm{C}_4[g(X)] = \varepsilon^4 \mathrm{E}\left[\left(\frac{\partial g_d(\boldsymbol{X})}{\partial \boldsymbol{X}^T}\right)^{[4]} \boldsymbol{X}_p^{[4]}\right] = \left[\left(\frac{\partial g_d(X)}{\partial \boldsymbol{X}^T}\right)^{[4]}\right] \mathrm{C}_4(\boldsymbol{X}) \tag{20}$$

Where Var(X), C$_3$(X), C$_4$(X) represent the variance, third-order central moment, and fourth-order central moment; σ_g^2, θ_g, η_g denote the variance, third moment, and fourth moment of $g(X)$.

3. Maximum entropy principle

The problem is transformed into an optimization problem by normalizing the functional function $Z = g(X)$ and using the first m order of origin moments as constraints:

$$\left.\begin{aligned} &\max H = -c\int_{-\infty}^{+\infty} f_Y(y)\ln f_Y(y)dy \\ &\mathrm{st.E}(Y^i) = \int_{-\infty}^{+\infty} y^i f_Y(y)dy, (i = 0,1,\cdots,m) \end{aligned}\right\} \tag{21}$$

Using the Lagrange multiplier method to solve this problem, the probability density function can be obtained in the form:

$$f_Y(y) = \exp\left(-\sum_{i=0}^{m} b_i y^i\right) \tag{22}$$

Here, $b_0, b_1, \cdots, b_m$ is a specific parameter, and the central moment of $Z = g(X)$ has the following relationship with the origin moment of Y:

$$\mathrm{E}(Y^i) = \frac{\mathrm{E}[(Z-\mu_g)^i]}{\sigma_g^i}, i = 0,1,\cdots,m \tag{23}$$

Substituting these conditions into the equation, the solution can be obtained, and the probability density function expression after $Z = g(X)$ standardization and the reliability of the structure can be obtained as follows:

$$R_{\mathrm{EM}} = \Pr(Z \geq 0) = \Pr\left(-\frac{\mu_Z}{\sigma_Z}\right) = \int_{-\frac{\mu_Z}{\sigma_Z}}^{+\infty} \exp\left(-\sum_{i=0}^{m} b_i y^i\right) dy \tag{24}$$

Where, R_{EM} represents the reliability calculated by the maximum entropy principle.

4. Verification of practical engineering applications

4.1 Reliability calculation of machine tool spindle

The spindle of a machine tool is a critical component, with its most vulnerable part defined by its inner diameter (d_1) and outer diameter (d_2). This part is prone to fatigue strength failure. Key factors influencing its performance include the bending moment (M), torque (T), and bending section coefficient (W).

$$W = \frac{\pi d_1^3}{32}(1-(\frac{d_2}{d_1})^4) = \frac{\pi(d_1^4 - d_2^4)}{32 d_1} \tag{25}$$

The function of its fatigue strength failure is:

$$g(X) = \sigma_{-1} - \sqrt{\sigma^2 + 4(\alpha\tau)^2} = \sigma_{-1} - \frac{32d\sqrt{M^2 + (\alpha\tau)^2}}{\pi(d_1^4 - d_2^4)} \tag{26}$$

Here, σ_{-1} is the bending fatigue limit of the spindle and α is the conversion coefficient. W_T is the torsional cross-section coefficient, and the torsional shear stress of the most dangerous part is τ, whose equation is:

$$\tau = \frac{T}{W_T} = \frac{T}{2W} \tag{27}$$

Due to the limitations of engineering conditions, the distribution type of T is unknown, and only ten sample data points are available (2652478, 3743129, 5478892, 4408383, 4052083, 4052072, 4157255, 3831879, 4082580, 4321639, 3632989).

By bootstrap, the torque can be calculated using this small sample data set. The first fourth-order center moment of T are calculated as: (4.037×10^6 N·m, 2.122×10^5 N·m, 4.656×10^{14} (N·m)^3, 6.290×10^{21} (N·m)^4).

Other random variables affecting the spindle's failure mode are normally distributed, with their statistical characteristics listed in Table 1:

Table 1. Parameter data.

Random variable	Mean value	Standard deviation
M(N·m)	5.4×10^6	2.7×10^6
d_1(mm)	70	0.35
d_2(mm)	50	0.25
σ_{-1} (MPa)	245	2.45

The reliability of the structure is calculated using the maximum entropy method:

R=0.9944

4.2 *Reliability calculation of hollow pressure bar*

Figure 1 [11] illustrates the force diagram of a simply supported hollow compression bar under axial load, with its failure mode described by a specific function.

$$g = \frac{\pi^2 E[(d+2t)^4 - d^4]}{64 l^2 dt} - \frac{F}{\pi t d} \tag{28}$$

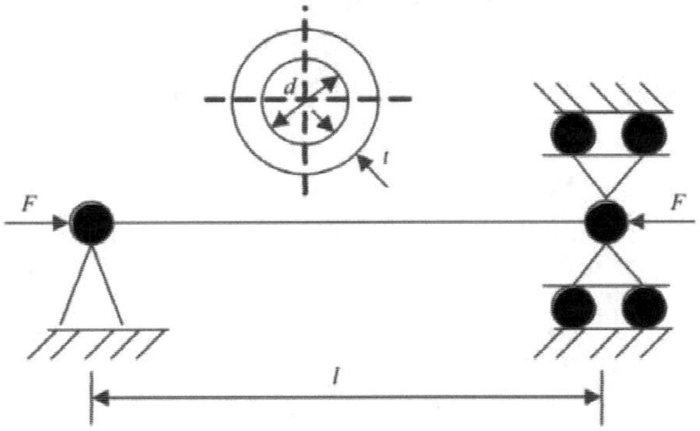

Figure 1. Hollow pressure bar.

Due to the limitations of engineering conditions, the distribution type of load F is unknown, and only ten sample data points are available as (96219, 91826, 87428, 187439, 97893, 154567, 180124, 106307, 91227, 95826). Using the bootstrap method, the first four central moments of F are calculated as: (1.188×10^5 kN, 1.171×10^4 kN, 5.064×10^{11} $(kN)^3$, 5.531×10^{16} $(kN)^4$), The other random variables follow a normal distribution, with their specific values shown in Table 2.

Table 2. Parameter data.

Random variable	Mean value	Standard deviation
Modulus of elasticity of material E /GPa	203.0	4.0
Material yield limit S /MPa	490.0	40.0
Middle diameter of section d / mm	42.5	0.2125
Thickness of wall t / mm	2.4	0.012
Length of rod l / mm	1090.0	5.45

The structure's reliability is again calculated using the maximum entropy method:

$$R=0.9737$$

5. Conclusions

The limitation of traditional mechanical structure reliability optimization design in handling small sample problems can be extended by the bootstrap method. Using the maximum entropy principle to establish central moments effectively addresses uncertainty under data scarcity. The study shows that this proposed method not only improves the accuracy of statistical properties for uncertain parameters but also significantly enhances the applicability and robustness of reliability optimization design under small sample conditions, providing new ideas for engineering practice. Future studies could further optimize the sample expansion strategy, improve the representativeness and diversity of the extended data, and explore more efficient probabilistic model-building methods by combining other data-driven approaches or machine-learning techniques. As data acquisition methods improve and intelligent optimization algorithms develop, the potential of this method in cross-domain engineering applications will be more widely explored and validated.

References

[1] Xie N, Liu J K, Ma L F, et al. Reliability analysis method of slope stability in permafrost area under small sample condition [J]. Engineering Mechanics. (2008). (07):166-171.

[2] FAN Lei. Non-failure Small Sample Reliability Evaluation Based on Bootstrap Method [J]. Microcomputer Information, 2008, (33):209-211.

[3] Wang Guo-dong, Niu Zhan-wen, Qu Liang, et al. Life Data Analysis Based on Improved Two-stage Method and Bootstrap Method[J]. Journal of Systems Engineering, 2016, 31(05):710-718.

[4] Wang Ruixiang, Xu Lingtian, Chen Xiaoyang, et al. Reliability Assessment of Bearing with Small Sample Non-Failure Life Test Data [J]. Journal of Aerospace Power,2021,36(11): 240-2409.

[5] Anonymous. Bootstrap Methods: A Guide for Practitioners and Researchers (2nd ed.) [J]. Technometrics,2008,50(4):550.

[6] Wang H, Liu H, Shao S. Improved Bootstrap Method Based on RBF Neural Network for Reliability Assessment[J]. Applied Sciences,2024,14(7):

[7] Shao Y, Lu P, Wang B, et al. Fatigue reliability assessment of small sample excavator working devices based on Bootstrap method[J]. Fracture and Structural Integrity,2019,13(48):757 -767.
[8] Zhao Yuan, Yang Lin. Life evaluation model of small sample based on Bootstrap theory [J]. Journal of Beijing University of Aeronautics and Astronsutics, 2012,48(1):106-112.
[9] Liu Jian, Wu Yi, Tan Lu. Improvement of Bootstrap sampling for Bootstrap method [J]. Mathematical Theory and Applications,2006(1):69-72.
[10] LiSha Z, Hao L, YiMin Z. A system reliability estimation method by the fourth moment saddle point approximation and copula functions[J]. Quality and Reliability Engineering International, 2021, 37(6):2950-2969.
[11] Han Wen-qin, ZHOU Jin-yu, SUN Kui-zhou. A Copula analysis method for mechanical structure reliability related to failure mode [J]. China Mechanical Engineering, 2011, 22 (03):278-282.

Study of application of genetic aggregation in optimizing the dimensions of EDM pillar

Hui Li [1], * Jianyong Liu [1], Dianxin Li [1], Jianrong Yu [1], Wenming Zhang [2] and Dan Liu [2]

[1] Department of Mechanical Engineering, Beijing Institute of Petrochemical Technology, Beijing 102617, China
[2] AECC Shenyang Liming Aero-Engine Co., Ltd., Shenyang 110043, China

*Corresponding author's e-mail: 0020200028@bipt.edu.cn

Abstract. A multi-objective optimization method based on a combination agent model is proposed to address the problem of insufficient stiffness of a typical electric discharge machine pillar. Firstly, the main size parameters of the pillar were selected as the design variables, and static and dynamic performance indicators such as maximum deformation were selected as the target variables. The Optimal Space Filling method was used for experimental design. Then, a multiple surrogate model construction method based on penalty prediction score and genetic algorithm was proposed, and a response surface model between the target response of the pillar and the design variables was built based on this method. Finally, a multi-objective optimization calculation was conducted on the performance response of the pillar using the screening method on the constructed response surface model, and the final optimization design scheme was determined through comparative analysis. The results showed that without increasing mass, the maximum deformation of the optimized pillar was reduced by 13.36%, and the 1st order frequency was increased by 11.73%. The dynamic and static performance of the pilar was significantly improved.

1. Introduction

With the rapid development of aviation, aerospace, automobile, molding, and other industries, modern machine tool design is moving toward the high-end, intelligent, and green direction. The EDM micro hole machining tool is one of the core equipment to realize the localization of air film hole machining of aviation blades in China. The pillar, as the key component and guiding element of the machine tool, directly affects the machining accuracy of the micro-hole and is one of the key components that determine the performance of the machine tool [1-3]. Many scholars have carried out extensive research on the performance optimization of the pillar of the machine tool. Wang et al. [4] used the Kriging function to fit the response surface to optimize the key indicators of the pillar of a small drilling machine. Under the condition that the total deformation is unchanged, the 1st-order natural frequency of the pillar was increased by 16% and the mass was reduced by 5.3%. Combined with bionic design, Li et al. [5] optimized the internal complex structure of the pillar of the gear grinding machine with a heavy spherical worm wheel and achieved the best lightweight and size. Liu et al. [6] identified the pillar as the key structural part affecting the dynamic performance of the whole machine through dynamic analysis, took the pillar structure as the optimization goal, and carried out size optimization so that the 1st order natural frequency of the whole machine was increased by more than 10%.

In the current research, a single-agent model is often used to construct the response surface. Although the selected model has a suitable fitting effect for its specific problem, the selection often depends on experience, and the results are uncertain. Moreover, when the target variables are increased, a single agent model makes it difficult to achieve a good fitting effect on all the target responses. In response to such problems, You et al. have proposed a combination construction method [7-8]. The research shows that the combined agent model has better robustness, higher precision, and more accurate prediction performance than the single model [9-11]. Wang et al. have applied the combined agent model in various optimization, shape construction, and parameter identification and achieved good results [12-13], but the application of the combined agent model to the optimization of machine tool pillars is rarely studied.

Taking the pillar of a typical EDM small hole machine as the research object, the Optimal Space Filling method was used to carry out the experimental design. The fitting effect of a typical single agent model was evaluated, and a combination agent model construction method based on penalty prediction score and genetic algorithm was proposed to construct a response surface model between the design variable and the target variable. On the constructed response surface model, the multi-objective optimization was carried out by using the screening method, and the dynamic and static performance of the optimized pillars was verified.

2. Design of experiment

2.1. Introduction to ultra-precision EDM small hole machine

An important application of ultra-precision EDM small-hole machines is the processing of small holes with a large depth-to-diameter ratio of aerospace engines, such as engine injectors, and engine precision spiral groove, which is characterized by small size, special materials, and high precision requirements. Such as the aerospace engine injector, a titanium alloy material injector in every 2 tiny holes to form an attitude control engine “impact pair”, requiring dozens of “impact pair” along the spray disk circumference uniform distribution, all “impact pair” center line intersection point gathered to the center intersection point to ensure better impact effect. Another example is the hole plate of a new model of aerospace engine, the hole diameter series is 0.1 mm, 0.11 mm, and 0.12 mm, the depth to diameter ratio is >15, and the aperture consistency deviation: ±0.003 mm (>10 holes). This kind of hole has very high requirements for perpendicularity and cylindricity.

Figure 1 is an ultra-precision EDM small hole machine, which is mainly composed of a bed, X-axis, pillar, processing head, Y-axis, and working table. The pillar is made of mineral material marble, and the machine tool structure schematic and physical picture are shown in Figure 1.

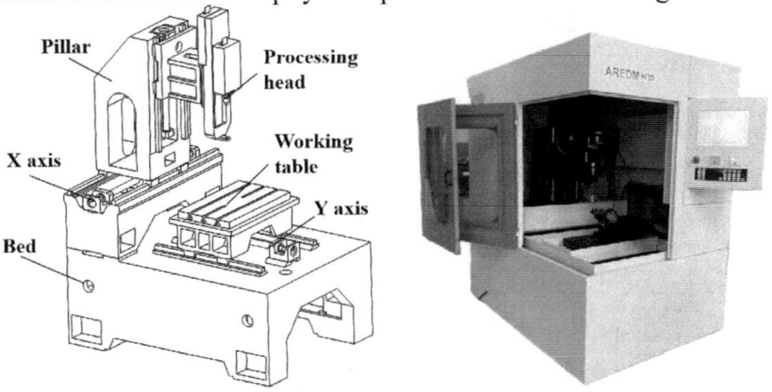

Figure 1. Sketch and picture of the machine tool.

The structural characteristic of the machine tool is that the gravity of the spindle structure will cause deformation of the supporting part of the machining head and the Z-pillar so that the axis of the machining head will produce a small tilt angle, and then affect the precision of the machining hole. Therefore, it is necessary to further optimize the stiffness of the pillar. Considering that the main size of the pillar design is determined according to experience, there is a certain optimization space for the size of the pillar.

2.2. Finite element simulation analysis of the pillar

Static and modal finite element simulation was carried out on the pillar, and material parameters were set as: marble mass density 2, 600 kg/m^{-3}, elastic modulus 30 GPa, Poisson ratio 0.18. We set the boundary conditions as follows: the base of the pillar is set as a fixed contact, and the spindle and machining head are equivalent to 1, 500 N remote force in the static analysis, which acts on the center of the guideway surface. We export the deformation and stress nephogram of static analysis and the 1st order vibration mode nephogram of modal analysis, as shown in Figure 2.

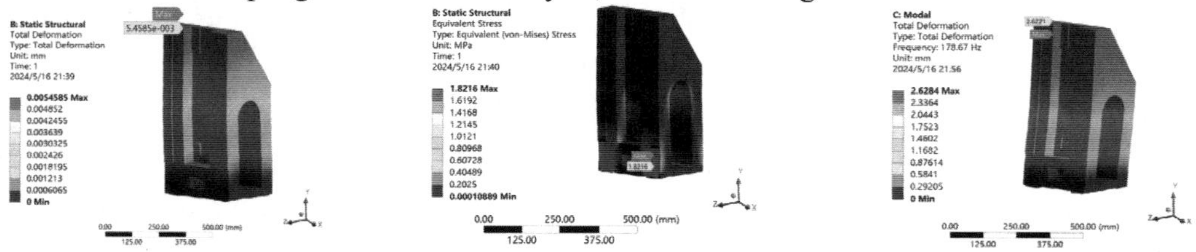

(a) Deformation nephogram (b) Stress nephogram (c) First-order mode nephogram

Figure 2. Simulation nephograms of the initial design.

Figure 2 (a) shows that the overall deformation of the pillar is distributed in layers from top to bottom. The closer the distance to the upper end is, the larger the deformation is. The maximum deformation is 0.0054585 mm. Figure 2 (b) shows that the overall stress distribution is relatively uniform, and the stress at the shape transition and near the bottom is larger, with the maximum stress being 1.8216 MPa. Figure 2 (c) shows that the 1st order vibration of the pillar is mainly the forward and backward swing of the vertical guide rail surface, and the 1st order natural frequency is 178.67 Hz.

2.3. Selection of design variables and target variables

In the optimization, the static and dynamic stiffness and vibration resistance of the pillar are considered as a priority, so the maximum deformation, 1st order natural frequency, maximum stress, and pillar mass are selected as the optimization objectives, denoted as δ, f, σ, and M respectively. Several main sizes in the pillar design (Figure 3) are selected as design variables, and their initial values and value ranges are given according to design requirements, as shown in Table 1.

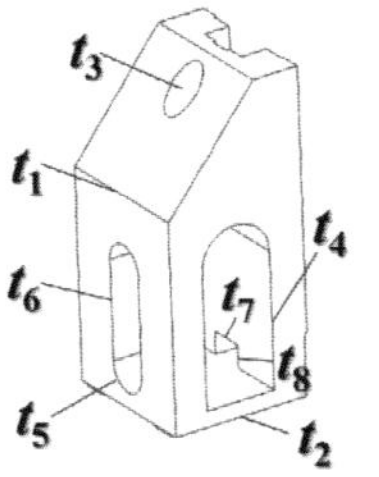

Figure 3. General view of design variables side-by-side to save space.

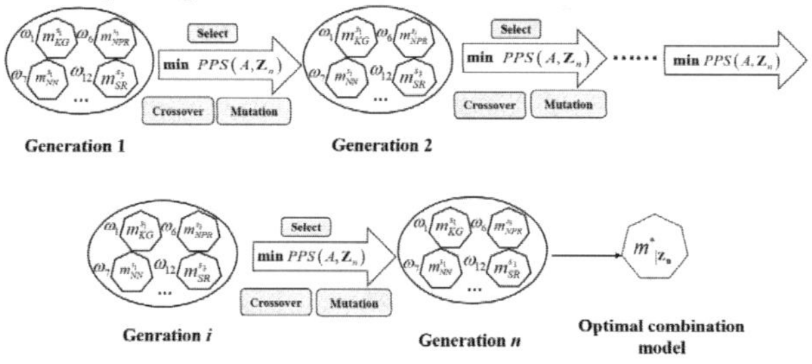

Figure 4. Schematic diagram of genetic algorithm iteration process.

Table 1. The design variables.

Variables	Name	Initial/mm	Minimum/mm	Maximum/mm
t_1	Pillar length	330	310	350
t_2	Pillar width	310	300	330
t_3	Middle excavation diameter	80	70	110
t_4	Side cutting height	300	240	350
t_5	Straight groove cutting radius	50	35	70

t_6	Straight groove cutting length	220	180	280
t_7	Bottom window width	80	75	200
t_8	Bottom window height	50	45	100

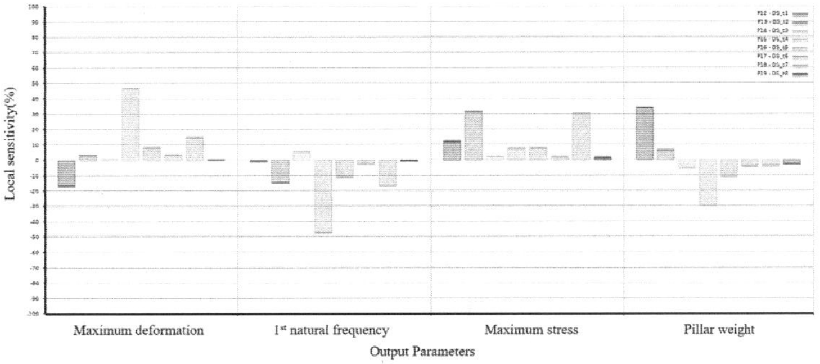

Figure 5. Local sensitivity analysis.

2.4. Generation of parameter space

We conducted a local sensitivity analysis of input parameters on output parameters, and the result is shown in Figure 5. As shown in Figure 5, all 8 parameters have a strong correlation with performance response, among which t_4 has the strongest correlation, followed by t_1 and t_2.

By defining the global variable of the relevant design size in SolidWorks and adding DS before the variable name, such as "DS_t_1", it can be automatically recognized by ANSYS and realize the parameter passing between SolidWorks and ANSYS. The 8-dimension parameters are set as variables, imported into ANSYS Workbench, and the 8-dimension variables identified are set as input parameters. The finite element simulation of the initial model was carried out. The maximum deformation and maximum stress were extracted from the results of static analysis, and the 1st order of natural frequency in modal analysis and the mass of the pillar were set as output parameters.

Sample point selection is a key issue in response surface technology. Optimal Space Filling (OSF) is an improvement of the Latin Hypercube sampling method. The design points obtained are more distant and more evenly distributed, which is suitable for more complex response surface algorithms. Therefore, the OSF method is adopted in this paper for experimental design, and 81 sample points are generated. The parameter space is shown in Table 2.

3. Response surface model construction of the pillar

3.1. Response surface construction of single agent model

Table 2. Some sample points.

No.	Design variables/mm								Target variables			
	t_1	t_2	t_3	t_4	t_5	t_6	t_7	t_8	δ/mm	f/Hz	σ/MPa	M/kg
1	317.17	325.44	79.08	312.35	42.41	229.45	117.78	52.17	0.0062	170.4	2.55	256.83
2	341.36	319.26	86.12	344.45	63.52	266.12	180	76.24	0.0072	158	2.68	249.82
3	322.6	326.92	87.97	341.98	46.86	248.34	191.86	67.6	0.0076	157.6	2.83	243.75
4	332.97	321.98	84.63	258.03	59.08	182.78	83.71	68.21	0.0049	185.25	2.37	286.79
5	344.33	312.84	77.23	296.3	69.82	226.12	174.08	73.15	0.006	168.68	2.48	270.25
6	310.75	320	93.52	302.47	42.78	201.67	169.63	69.45	0.0064	173.11	1.99	246.72
7	324.08	323.21	72.41	308.65	67.97	218.34	166.67	95.38	0.0066	162.83	2.56	250.02
8	341.86	317.04	90.93	250.62	45	239.45	132.6	54.63	0.0046	190.89	2.14	296.31
						......						
78	336.92	317.78	70.56	306.18	48.71	202.78	141.49	50.93	0.0057	172.17	2.14	276.18
79	333.46	321.49	74.63	316.05	40.93	203.89	197.78	83.65	0.0065	164.43	3	267.4
80	349.76	320.99	93.89	303.71	69.45	231.67	113.34	77.47	0.0057	170.09	2.55	274.62
81	344.82	327.17	73.15	286.42	66.86	206.12	129.63	60.19	0.0055	169.48	2.61	287.26

The size optimization of the pillar needs to call the maximum deformation, 1st order natural frequency, maximum stress, and mass under different size parameters. The process involves 3D modeling, mesh division, finite element analysis, and other steps, and the direct call to solve the optimization cost is quite high. Therefore, the response surface modeling is introduced to approximate the implicit relationship between the structural dimension parameters and the performance response, and then the approximate model can be directly invoked to obtain the optimal structural dimension parameters in the optimization solution.

The response surface model, also known as the meta-model or agent model, is often used to replace expensive numerical calculations and is a common model construction method in engineering problems. Common response surface types include Full 2nd-order Polynomials (FSP), Kriging (KRG), Radial Basis Function (RBF), Support vector regression model (SVR), Non-Parametric Regression (NPR), Neural Network (NN), etc.

In the problem studied in this paper, several representative agent models are selected: FSP, KRG, NPR, and NN, and the response surface is constructed respectively. Among them, the FSP model is simple in construction, small in calculation, and high in linear fitting precision; the KRG model is high in local fitting precision, and NPR and NN models have good fitting effects in highly nonlinear response. The response surface fitting results are shown in Table 3. Among them, R^2 is the determination coefficient, of which the best value is 1, and *RMSE* is the root mean square error, of which the best value is 0, and both are commonly used model performance evaluation indexes. It should be noted that the Kriging method is an interpolation-type response surface, and its determination coefficient and RMS error are always the best, which cannot objectively reflect the quality of the response surface, so verification points are added to evaluate.

Table 3. Evaluation of the single model fitting effect.

	Neural network		Non-parametric regression		Full 2nd-order polynomials		Kriging		
	R^2	*RMSE*	R^2	*RMSE*	R^2	*RMSE*	R^2	*RMSE*	*RMSE*-VP
δ	0.99852	2.7314e-5	0.99840	2.8401e-5	0.99969	e-5	1	e-12	1.0817e-4
f	0.99552	**0.54130**	0.99856	**0.30697**	0.99954	**0.17308**	1	e-8	**1.28680**
σ	**0.84190**	0.09455	0.99762	0.01161	**0.88928**	0.07509	1	e-9	**0.75685**
M	0.99851	**0.64056**	0.99870	**0.59809**	1	0.01171	1	e-9	0.06173

As can be seen from Table 3, the fitting accuracy of the four models cannot meet the requirements. The coefficient of determination of the neural network and the complete quadratic polynomial model is less than 0.9 on the maximum stress of the target variable, which does not meet the engineering requirements; The root-mean-square error of the non-parametric regression model is larger in 1st order frequency and mass, which indicates that the fitness degree is low in these two target variables. Because the Kriging model is an interpolation method, the coefficient of determination and root mean square error are the best values, but the results of the verification points are larger on the 1st order frequency and maximum stress, indicating that the model has a poor prediction effect on these two target variables.

It can be seen that the fitting quality of the single agent model is unstable when there are many output variables, and it takes loads of time to select the type and adjust the settings when facing new problems. Therefore, a combination agent model construction method based on penalty prediction score and genetic algorithm is proposed in this paper, which can automatically select, configure, and generate the response surface type or combination that is most suitable for each target variable in the problem. This model construction process and method are introduced in the following section.

3.2. Combination agent model building method

A combined agent model is a linear combination of single agent models, and its general form is:

Journal of Physics: Conference Series **2992** (2025) 012004 doi:10.1088/1742-6596/2992/1/012004

$$\hat{y}_A(x) = \sum_{i=1}^{p} \omega_i \hat{y}_i(x) \tag{1}$$

$$\sum_{i=1}^{p} \omega_i = 1 \tag{2}$$

where $\hat{y}_A(x)$ is the predicted value of the combined model, ω_i is the weight coefficient of the ith response surface model, $\hat{y}_i(x)$ is the predicted value of the ith response surface model, and ω_i must satisfy Equation (2).

The key to constructing the combined agent model lies in the definition of error metric and the choice of weight factor calculation method. Many scholars have conducted extensive research on the definition and selection of error measurement. Salem and Tomaso [11] proposed a general evaluation method-Penalized Predictive Score (PPS), which includes internal accuracy, prediction performance, and roughness penalty. With good adaptability and robustness, PPS can be applied to any type of agent model, the calculation is simple, estimation is unbiased, and overfitting can be prevented. PPS is selected as the error metric in this paper, and its formula is as follows.

$$PPS(m, \mathbf{Z_n}) = \alpha\hat{\Re}_{l_2,\mathbf{Z_n}}(m) + \beta\Re_{k-CV}(m) + \gamma\Re_{LRM}(m) \tag{3}$$

where $\hat{\Re}_{l_2,\mathbf{Z_n}}(m)$, $\Re_{k-CV}(m)$, and $\Re_{LRM}(m)$ are the internal accuracy, K-fold cross-validation error, and penalty function of model m respectively; α, β, and γ are the weights, usually $\alpha=2\beta$ and $\beta=2\gamma$ in practice.

At present, there are two kinds of calculation methods for weight factors: first is error minimization, and second is weight factor iterative. The precision and robustness of the first kind are poor, and the method is easy to fall into the problem of local optimization. The second type is to judge whether the convergence criterion is met after each weight factor is obtained. If it is not met, it will continue to iterate until the convergence criterion is met, which can achieve a good balance between model accuracy and modeling time.

In this paper, a genetic algorithm is selected to compute the weight factor iteratively. In this genetic algorithm, each different agent model is the individual, and all the individuals constitute the population. The specific setting of the agent model is the genetic information of the individual, and the PPS value of each model is the adaptability function of the evaluation model. Through the operation of selection, cross-over, and mutation on the current population, a new generation of population is generated, and the population gradually evolves to the state containing the optimal solution. Its iterative process is shown in Figure 4.

In each iteration, a new combination is generated, the weight factor of which is obtained by solving the PPS minimization problem of the current combination model in each iteration [11], and the solution is defined in Equation (4).

$$\mathbf{W} = \frac{\mathbf{R}^{-1}}{\mathbf{1}^{\mathrm{T}}\mathbf{R}^{-1}\mathbf{1}} \tag{4}$$

where $\mathbf{R} = \alpha\mathbf{E} + \beta\mathbf{C} + \gamma\mathbf{K}$, $\mathbf{W}$ is the matrix of weights, and $\mathbf{E}$, $\mathbf{C}$, and $\mathbf{K}$ is the matrix of internal accuracy, cross-validation error, and roughness penalty.

The mutation and cross-over operations are performed on the settings of the agent model. We perform the mutation operation by modifying the settings of the agent model. There are two types of cross-over operations, in which the first is to exchange partial settings between two agent models of the same type, and the second is to generate a combination of equal weights of two different types of agent models. New models or combinations are generated through mutation and cross-over operations to maintain the diversity of the population. The iteration stop criterion is to not exceed the maximum generations set. The general flow chart of model construction is shown in Figure 6.

Table 4. Error results of multiple models.

Target variables	Fitness		Crossover error	
	R^2	*RMSE*	R^2	*RMSE*
δ	1	e-11	0.99943	e-5
f	1	0.00026	0.99930	0.05858
σ	1	e-8	0.99716	0.07577
M	1	0.00116	1	0.03629

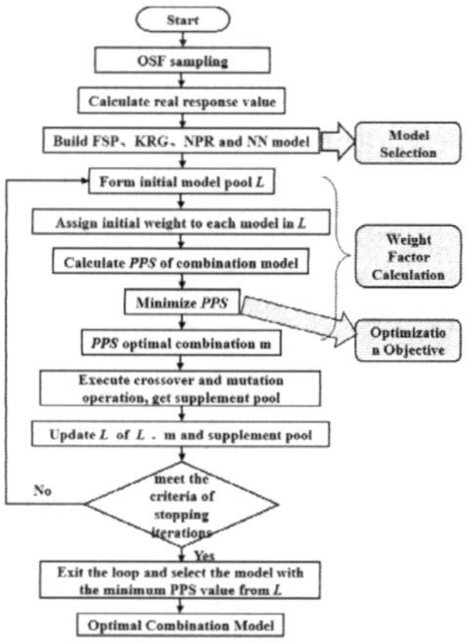

Figure 6. Process diagram of constructing the multiple surrogate model.

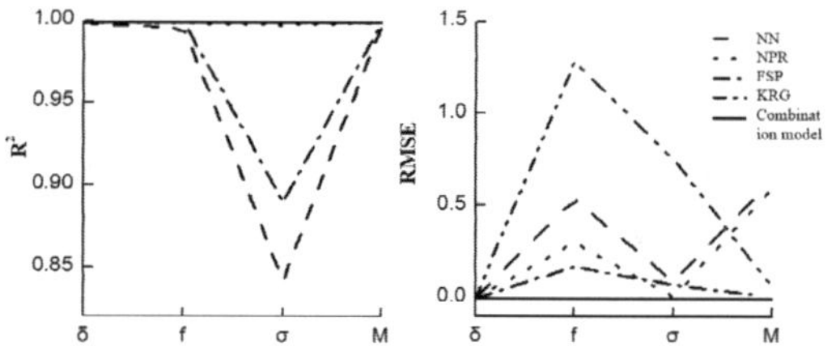

Figure 7. Precision comparison of single agent model and combined model (R^2 and *RMSE*).

3.3. Response surface construction of combination agent model

The response surface model between the performance responses δ, f, σ, and M of the pillar of the EDM small hole machine and the design variables t_1~t_8 was established by using the method proposed in the former section. The error values and cross-validation errors on the four target responses were calculated, as shown in Table 4.

As shown in Figure 7, the R^2 and *RMSE* of the four target variables are all very close to the optimal values, which indicates that the model has good fitting accuracy. In the cross-validation error, the coefficient of determination of the target variable is 0.99943, 0.99930, 0.99716, and 1 respectively, which meets the general requirement of no less than 0.9 in engineering. Meanwhile, the root mean square error is about 0, indicating that the model has good prediction performance.

4. Pillar optimization design

4.1. Multi-objective optimization based on screening method

After the response surface is constructed, a multi-objective optimization mathematical model of the pillar is established with 8 dimensions as design variables and maximum deformation, 1st order natural frequency, maximum stress, and pillar weight as objective functions, as shown in Equations (5)-(7):

Objective function:
$$\min \delta(\mathbf{t}), \max f(\mathbf{t}), \max \sigma(\mathbf{t}), \min M(\mathbf{t}) \quad (5)$$

Design variables:
$$\mathbf{t} = [t_1, t_2, t_3, t_4, t_5, t_6, t_7, t_8]^{\mathrm{T}} \quad (6)$$

Constraints:
$$\begin{cases} \delta(t) \le \delta^{'}(t) \\ f(t) \ge f^{'}(t) \\ \sigma(t) \ge \sigma^{'}(t) \\ M(t) \le M^{'}(t) \\ \alpha_i \le t_i \le \beta_i (i = 1, 2, ..., 8) \end{cases} \quad (7)$$

where t_i is the design variable, α_i is the lower limit constraint of the design variable, β_i is the upper limit constraint of the design variable, $\delta'(t), f'(t), \sigma'(t)$, and $M'(t)$ are respectively the initial maximum deformation, initial 1st order natural frequency, initial maximum stress, and initial mass of the pillar.

The screening method is used to search and calculate the optimization target set. The Pareto optimal solution set of each target variable is obtained, and 3 recommended schemes are given from it, as shown in Table 5.

Table 6 shows the effect comparison of the three recommended design schemes. It can be seen that: Recommended Design 1 has the best optimization effect on the target response δ and f, but the lightening effect is the worst; Recommended Design 2 has the best effect on the target response σ, but the maximum deformation optimization effect is average and the 1st order natural frequency optimization effect is the

Table 5. Recommendation designs.

Design variables	Initials	Candidate 1	Candidate 2	Candidate 3	Rounding of Candidate 1
t_1	330	316.55	320	310.51	317
t_2	310	312.88	315.56	310.42	313
t_3	80	99.81	97.14	99.59	100
t_4	300	250.46	250.07	250.60	250
t_5	50	39.65	59.37	61.95	40
t_6	220	270.49	240.19	268.47	270
t_7	80	81.83	102.41	93.42	82
t_8	50	52.98	77.38	59.63	53

Table 6. Comparison of the 3 recommendation designs.

Target variables	δ/mm		f/Hz		σ/MPa		M/kg	
	Value	%	Value	%	Value	%	Value	%
Initial design	0.005459	/	178.67	/	1.8216	/	266.69	/
Candidate 1	0.004908	**-12.43%**	199.37	**11.21%**	2.2519	23.62%	265.14	-0.58%
Candidate 2	0.005037	-10.13%	191.71	6.94%	2.3667	**29.92%**	263.46	-1.21%
Candidate 3	0.005287	-5.67%	195.07	8.81%	2.2335	22.61%	247.62	**-7.14%**

worst; Recommended design 3 has the best lightweight effect, and the 1st order natural frequency is also improved a lot, but the maximum deformation optimization effect is the worst.

Considering that the ultimate purpose of our optimization is to improve stiffness and anti-vibration performance, the target response δ and f are of the highest importance, followed by M and σ. According to this design principle, recommended Design 1 is selected as the final scheme and the results are rounded, as shown in Table 5.

4.2. Verification of optimization results

Since the optimization scheme is obtained by response surface calculation, it must be verified by actual simulation calculation. The three-dimensional model of pillar optimization design was imported into ANSYS Workbench and transformed into a finite element model. Statics and modal analysis were performed according to the same settings in the former section. The response values of the four target variables were obtained, as shown in Table 7. Compared with the predicted value of the response surface and the real response value calculated by the simulation, the deviation of the four target variables is within 4%, which meets the design requirements.

Table 7. Comparison of the simulation result and response surface result.

Target response	Candidate 1	Simulation	Deviation
Maximum deformation/mm	0.0049081	0.0047292	3.8%
First-order natural frequency/Hz	199.37	199.62	0.13%
Maximum stress/MPa	2.2519	2.1956	2.5%
Pillar weight/kg	265.14	264.81	0.12%

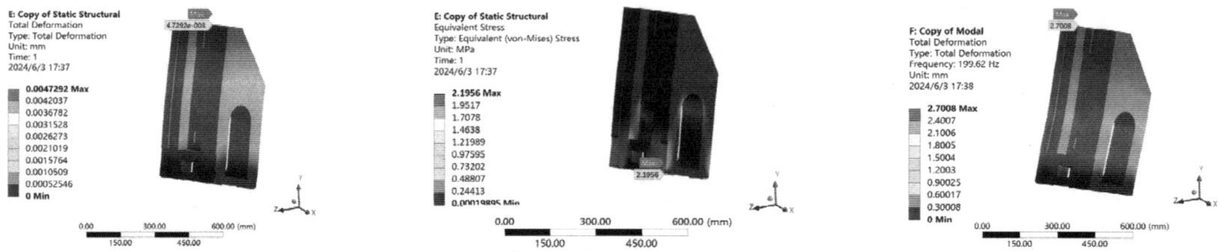

(a) Deformation nephogram (b) Stress nephogram (c) First-order mode nephogram

Figure 8. Simulation nephograms of the optimal design.

Table 8. Comparison of the initial decision and optimal design.

	Maximum deformation/mm	First-order natural frequency/Hz	Maximum stress/MPa	Pillar weight/kg
Initials	0.0054585	178.67	1.8216	266.69
Optimals	0.0047292	199.62	2.1956	264.81
Percentage	-13.36%	11.73%	20.53%	-0.7%

The deformation and stress nephogram of static analysis and the 1st order mode nephogram of modal analysis are exported, as shown in Figure 8. Figure 8 (a) shows that the maximum deformation is reduced to 0.0047292 mm after optimization, compared with the initial design of 0.0054585 mm; Figure 8 (b) shows that the stress increased to 2.1956 MPa after optimization; Figure 8 (c) shows that compared with the initial design of 178.67 Hz, the 1st order natural frequency after optimization increased to 199.62 Hz. Compared with the initial design (Table 8), under the premise of no increase in mass, the maximum deformation of the pillar is reduced by 13.36%, and the 1st order natural frequency is increased by 11.73%. The optimization effect is remarkable.

5. Conclusions

The main size of the pillar was selected as the design variables, and the maximum deformation, 1st order natural frequency, maximum stress, and mass were selected as the target variables. The test scheme was designed by the Optimal Space Filling method. In the construction of the response surface, a combination agent model construction method based on penalty prediction score and genetic algorithm is proposed. The response surface model between the target response of the pillar and the design variable is constructed by using this method. Compared with the single agent model, the model constructed by the method proposed in this paper has higher fitting accuracy and more accurate prediction. The coefficient of determination and root-mean-square errors on all 4 target variables meet the requirements, and the cross-validation errors are very close to the ideal value.

The constructed response surface model is used for multi-objective optimization calculation, three recommended schemes are selected from the Pareto optimal solution set that balances each objective, and the final optimization scheme is determined according to the principle of optimal design. The optimized design scheme is verified by simulation calculation. The results show that the maximum deformation of the pillar after the optimized design is reduced by 13.36%, and the 1st order natural frequency is increased by 11.73%, which indicates that the static and dynamic performance of the pillar is significantly improved. At the same time, the method proposed in this paper also provides a reference for the response surface construction in the machine tool pillar optimization.

Acknowledgments
This work was sponsored by the National Key Research and Development Program of China (Grant No. 2023YFB4605900), Beijing Nova Program (No. 20240484626), and BIPT Zhiyuan Research Fund (No. 2023003).

References

[1] LI Shifeng, HUANG Kang, MA Husheng, et al. Research progress on the design and manufacturing technologies of film cooling holes for aero-engine turbine blades [J]. Journal of Engineering for Thermal Energy and Power, 2022, 37 (9): 1-11.
[2] Harane P. P., Unune R. D., Ahmed R., et al. Multi-objective optimization for electric discharge drilling of waspaloy: A comparative analysis of NSGA-II, MOGA, MOGWO, and MOPSO [J]. Alexandria Engineering Journal, 2024, 991-16.
[3] XING Jun. The structure design of machine tool and study on tool electrode wear for EDM & ECM hybrid machining of micro-hole [D]. Nanjing University of Aeronautics and Astronautics, 2016.
[4] WANG Dongcheng, WANG Yu, LIU Shulian, et al. Optimization of the column structure of a small drilling machine based on reverse engineering [J]. Manufacturing Technology & Machine Tool, 2024, (03): 94-101.
[5] LI Yupeng, BA Chunlai, and LIU Laichao. Comprehensive Optimization of Heavy Machine Columns Using Structural Bionics [J]. CHINA MECHANICAL ENGINEERING, 2019, 30 (13): 1, 621-1, 625+1, 637.
[6] LIU Chengying, TAN Feng, WANG Liping, et al. Research on Optimization of Column Structure Design for Dynamic Performance of Machine Tool [J]. JOURNAL OF MECHANICAL ENGINEERING, 2016, 52 (03): 161-168.
[7] YOU Xiongxiong and NIU Zhanwen. Ensemble surrogate assisted evolutionary algorithm for complex system many-objective optimization [J]. Computer Integrated Manufacturing Systems, 2024, 30 (04): 1, 201-1, 212.
[8] LI Zhihua, ZENG Huiyi, NIE Chao, et al. Optimal Iterative Weight Factors Method for Constructing Ensemble of Surrogate Model [J]. Transactions of the Chinese Society for Agricultural Machinery, 2016, 47 (07): 391-397.
[9] Viana C. A. F., Haftka T. R., and Steffen V. Multiple surrogates: how cross-validation errors can help us to obtain the best predictor [J]. Structural and Multidisciplinary Optimization, 2009, 39 (4): 439-457.
[10] Acar E. Various approaches for constructing an ensemble of metamodels using local measures [J]. Structural and Multidisciplinary Optimization, 2010, 42 (6): 879-896.
[11] Salem B. M. and Tomaso L. Automatic selection for general surrogate models [J]. Structural and Multidisciplinary Optimization, 2018, 58 (2): 719-734.
[12] Wang Z., Almeida Jr J. H. S., Ashok A., et al. Lightweight design of variable-angle filament-wound cylinders combining Kriging-based metamodels with particle swarm optimization [J]. Structural and Multidisciplinary Optimization, 2022, 65 (5): 140.
[13] Zhang X., Wang D., Kong D., et al. The anti-fatigue lightweight design of heavy tractor frame based on a modified decision method [J]. Structural and Multidisciplinary Optimization, 2022, 65 (10): 280.

Experimental Research on ultrasonic-assisted direct writing metal additive manufacturing technique

Guofang Hu*, Ning Wang and Yanzhen Zhang

College of Mechanical and Electronic Engineering, China University of Petroleum (East China), Qingdao, China

* Corresponding author: 13963760980@163.com

Abstract—While high-energy beam-based laser additive manufacturing technology boasts high formability efficiency, its large equipment size and high manufacturing cost limit its application in certain scenarios. In contrast, the direct writing metal additive manufacturing approach showcases attributes of a compact heating apparatus, superior energy utilization, and the utilization of metal wire as raw material, culminating in notable reductions in manufacturing costs. However, conventional direct writing technology often faces issues of nozzle clogging, which adversely impacts the quality of the formed structures. This paper introduces an ultrasonic-assisted direct writing metal additive manufacturing (UADWMAM) technique that addresses the nozzle clogging problem by applying ultrasonic vibrations, thereby enhancing the stability of molten metal extrusion. Further investigations were conducted through single-pass forming experiments, examining the influence of process parameters, such as writing speed and heating temperature, on writing formability. The findings from this study have the potential to advance the development of direct writing metal additive manufacturing technology.

1. Introduction

Metal additive manufacturing, commonly referred to as metal 3D printing, is a transformative process enabling the rapid fabrication of intricate high-performance metal components devoid of traditional molds [1, 2]. This technique boasts attributes such as expedited production cycles, full density constructs, and near-net shaping, rendering it applicable across diverse domains including aerospace, medical, automotive, and rail transportation industries [3-5].

In pursuit of cost reduction and broadened applicability, researchers have proposed the extrusion-based metal direct writing additive manufacturing technique. This approach offers the advantages of compact heating apparatus, heightened energy utilization efficiency, and reduced raw material costs, rendering it particularly suitable for the production of small-scale components. The metal direct writing method operates on the principle of continuous extrusion deposition from liquid metal materials [6, 7]. Filament-based extrusion in metal direct writing technology bears similarity to plastic Fused Deposition Modeling (FDM) techniques [8-10]. In this approach, a metal wire is fed into a heated nozzle, wherein the molten metal is propelled through the nozzle by the advancing unmelted portion of the wire, facilitating deposition-based shaping. This method not only boasts a straightforward operational principle but also allows for concurrent feeding and direct writing, while obviating the need for post-processing steps like sintering [11, 12].

In order to address the nozzle clogging issue and enhance control over the efflux of molten metal, ultrasonic vibration is applied to the nozzle to facilitate the flow of molten metal. In light of the aforementioned challenges, an ultrasonically assisted direct writing metal additive manufacturing technique is proposed herein, aiming to realize stable operations and cost-effective manufacturing within the realm of metal additive manufacturing.

In order to conduct a more comprehensive investigation into the ultrasonic-assisted direct writing metal additive manufacturing technique, the present study has established a prototype of an ultrasonic-assisted direct writing metal additive manufacturing system. Subsequently, a series of systematic experiments involving process parameter optimization and temperature simulations were conducted, with the overarching aim of enhancing the fabrication quality of the produced components.

2. Experimental method

The ultrasonic-assisted direct-writing metal additive manufacturing prototype built by self-modification is shown in Figure 1. The prototype mainly includes the motion system and control system of a commercial printer, ultrasonic-assisted metal direct writing nozzle, wire feeder, ultrasonic generator, heating module and other parts. The ultrasonic-assisted metal direct writing nozzle is the key to the whole device. The piezoelectric ultrasonic transducer has a natural frequency of 40 kHz and a power of 60 W. The metal direct writing nozzle uses a 304 stainless steel tube with an inner diameter of 1.8 mm and an outer diameter of 2 mm. The heating module adopts a self-developed miniature electromagnetic induction heating device, and the heating temperature can reach up to 1000°C. The technical exploratory material of the experiment is Sn63Pb37 alloy, which is a wire with a diameter of 1.5 mm.

During the experimental procedure, the direct writing nozzle was subjected to induction heating through an induction heating coil, while the metal wire was introduced into the direct writing nozzle via a wire feeding mechanism for subsequent heating. Upon the melting of the metal wire, the molten metal within the nozzle emerged in a continuous liquid flow manner, driven by the propulsive force of the molten metal and influenced by gravity. The motion execution module adhered to a predetermined printing program, affecting the direct writing process. Concurrently, the piezoelectric ultrasonic device commenced its operation, facilitating the smooth outflow of the molten metal.

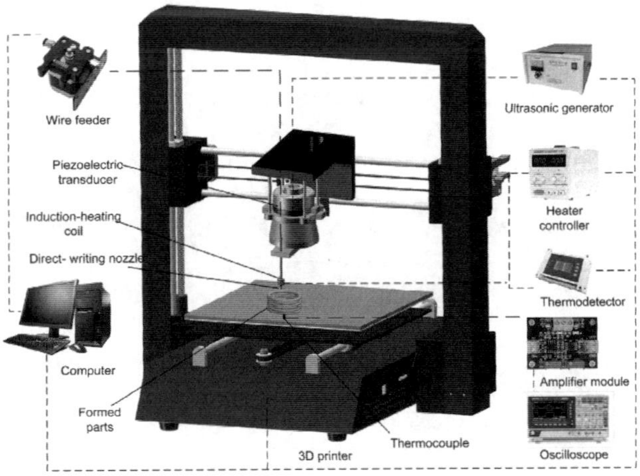

Figure 1. Schematic diagram of ultrasonic-assisted direct writing metal additive manufacturing prototype.

3. Results and discussions

Figure 2 shows the results of the direct writing single-pass forming experiment. The nozzle moving speed is 6.7 mm/s. In Figure 2(d) I, the outcome of a single-pass direct writing experiment without ultrasonic assistance is portrayed. It is discernible that during the linear direct writing process,

Journal of Physics: Conference Series **2992** (2025) 012005 doi:10.1088/1742-6596/2992/1/012005

irregularities in the outflow of molten metal occur due to nozzle obstructions, resulting in discontinuous flow and accumulation phenomena, ultimately inducing defects in the formed structure and compromising surface smoothness (as depicted in Figure 2(a)). Moreover, in cases of severe nozzle blockage, a constriction-induced flow disruption is observed, leading to discontinuity in the formed object and, in extreme instances, even fracture (as exemplified in Figure 2(c)). As depicted in Figure 2(d) II, the outcomes of a single-pass direct writing experiment with the application of ultrasonic assistance are presented. A comparative analysis of the results obtained without such assistance reveals a notably uniform formation of the tracks and an enhanced quality of the formed surface (as evidenced in Figure 2(b)). This observation underscores that the introduction of ultrasonic vibration-assisted direct writing serves to facilitate a smoother and more stable outflow of molten metal, thereby ameliorating the challenges posed by nozzle blockages and subsequently enhancing the quality of the formed structures.

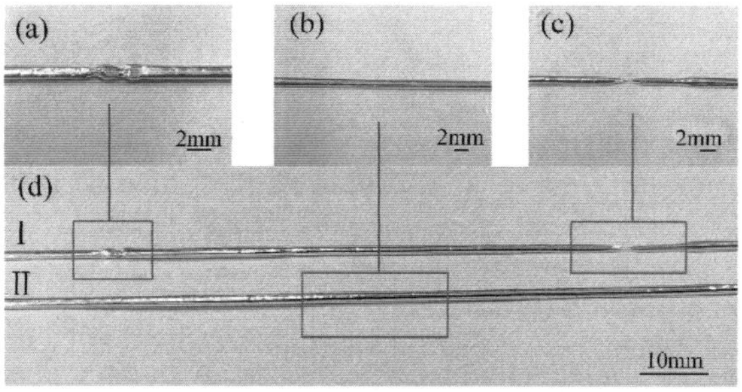

Figure 2. Experimental results of single-pass forming with or without ultrasonic-assisted direct writing.

The molten metal swiftly solidifies upon egress from the nozzle, effecting a rapid single-pass formation upon the substrate. Within the context of direct writing additive manufacturing, all three-dimensional entities are constructed through the sequential deposition and stacking of individual single-pass layers. In the present study's direct writing methodology, the principal process parameters comprise the writing speed, the nozzle-to-substrate distance, and the heating power. Direct writing speed affects the amount of molten metal deposition per unit time, the nozzle-to-substrate distance affects the drag form of molten metal during direct writing, and the heating power affects the forming temperature of molten metal. The nozzle-to-substrate distance is set to 1 mm and heating power to 36 W. From Figure 3, it is evident that an increase in the writing speed leads to a gradual reduction in the dimensions of the single-pass metallic lines, as indicated by decreasing line width, reduced line height, and diminishing cross-sectional area. When the writing speed is too low, under the condition of a constant wire feed rate, the molten metal extruding from the nozzle tends to accumulate and, due to limitations imposed by the nozzle-to-substrate distance, might fail to fully discharge. Subsequent to the influence of the molten liquid thrust, splattering occurs upon the substrate's surface, preventing successful formation. Maintaining a consistent volumetric discharge, different single-pass metallic line widths and formative precision can be attained by modulating distinct writing speed.

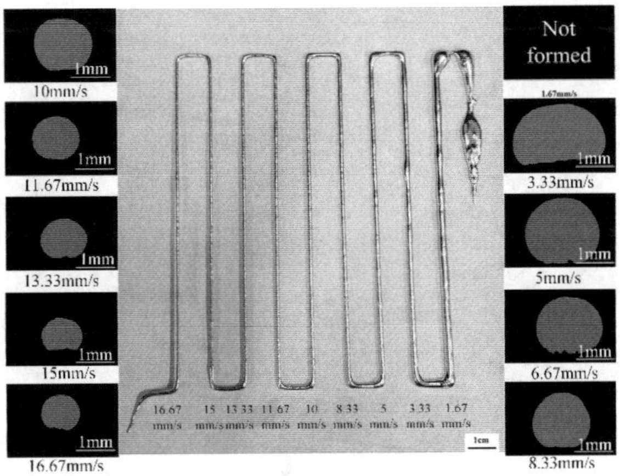

Figure 3. Effect of writing speed on single pass forming quality.

Upon exiting the nozzle, the molten metal undergoes a dual constraint exerted by both the nozzle and the substrate. The distance between the nozzle orifice and the substrate surface significantly influences the cross-sectional dimensions and morphology of single-pass formations. The direct write speed is set to 6.67 mm/s and the heating power to 36 W. As depicted in Figure 4, when the distance between the nozzle orifice and the substrate is minimal and the wire feed rate remains constant, the molten metal tends to engulf the entire nozzle. In this case, due to the excessively small separation distance, the nozzle becomes submerged within the flowing molten metal, constituting a coating-like direct writing state. Such reduced distances result in inadequate single-pass formation heights, coupled with an excessive accumulation of molten metal that leads to splattering and compromised formative success. Conversely, an appropriately set distance facilitates a dragging direct writing state, wherein the nozzle's movement imparts a dragging effect on the flowing molten metal. Under stable operating conditions, the molten metal uniformly accumulates and forms layers within the drag-induced nozzle configuration. However, when the distance between the nozzle orifice and the substrate is excessively large, the dragging effect on the molten metal becomes overly pronounced. Subject to surface tension, this condition reduces the contact area between the molten metal and the substrate, consequently impacting the interlayer adhesion quality during three-dimensional construct formation. Additionally, an analysis of the cross-sectional profile of single-pass formations reveals that, apart from instances where distances are too small, variations in line width remain relatively modest with increasing nozzle-to-substrate distances. In contrast, line height gradually increases, while the contact area between the metallic line and the substrate diminishes. Investigating the nozzle-to-substrate distance offers valuable guidance for probing the interlayer bonding in three-dimensional construct formation experiments.

Journal of Physics: Conference Series **2992** (2025) 012005 doi:10.1088/1742-6596/2992/1/012005

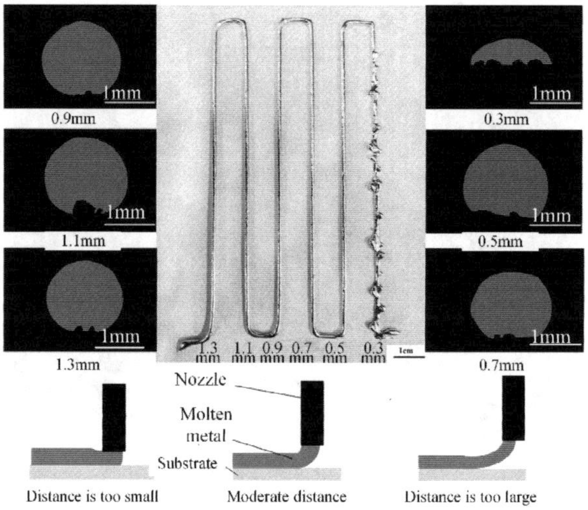

Figure 4. The effect of the distance between the nozzle and the substrate on the quality of single-pass forming.

The heating power governs the temperature of the extruded molten metal from the nozzle. Diverse temperatures induce varying states in the extruded molten metal. The direct write speed is set to 6.67 mm/s and nozzle-to-substrate distance to 1 mm. When the temperature of the extruded molten metal from the nozzle is in proximity to or below its melting point, as depicted in Figure 5, effective single-pass formation becomes challenging, resulting in inadequate bonding with the substrate. Under identical process parameters, elevated temperatures diminish the surface tension of the metal and concurrently reduce its viscosity. This diminished viscosity minimizes the adhesive friction upon nozzle exit, rendering the metal more prone to easy extrusion, consequently leading to intermittent flow phenomena under constant wire feed rates. Excessive extrusion temperatures hinder rapid cooling and deposition of the metal upon outflow. Subsequent cooling upon the substrate induces an increased surface tension that prompts metal coalescence, thereby impeding successful formation. Thus, the regulation of applied heating power to govern the temperature of the extruded molten metal is of paramount significance for both single-pass formation and subsequent three-dimensional additive manufacturing.

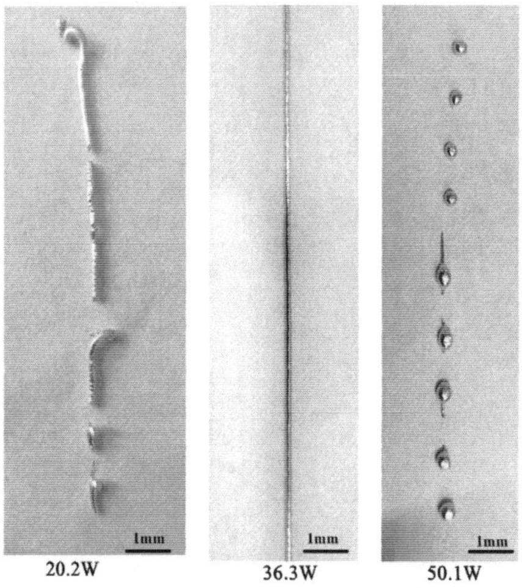

Figure 5. Effect of heating power on single-pass forming quality.

The interaction of different process parameters needs to be further investigated by orthogonal experiments, which will not be discussed further. Following optimization design, the ultrasonically-assisted molten metal direct writing apparatus demonstrates prolonged and stable operational performance with uniform liquid extrusion. It affords the capability to fabricate three-dimensional entities of varying bonding strengths, as exemplified in Figure 6, tailored to specific processing requirements. Through the adjustment of process parameters, diverse shapes of articles can be formed, characterized by high-quality formative outcomes. This system is adept at producing structures with inclined angles and smoothly transitioning radius variations, such as the waist structure. Additionally, it enables the continuous deposition-forming process, thus substantiating the device's capacity for continuous shaping. The direct writing process remains seamless, ensuring superior formative quality.

Figure 6. Direct writing multi-layer 3D molded parts.

4. Conclusion

In order to address certain challenges posed by existing metal additive manufacturing techniques, this study introduces an ultrasonically-assisted molten metal direct writing technology. It encompasses the design of an ultrasonically-assisted molten metal direct writing apparatus, investigates the role of solving blockage on the direct writing nozzle under ultrasonic assistance, conducts explorations into process parameters within the context of single-pass formation experiments, and culminates with three-dimensional solid formation experiments. These experiments delve into the temperature variations during the forming process under ultrasonic assistance, ultimately employing this technology to produce well-constructed articles with superior formative quality.

Single-pass formation experiments were conducted to comparatively assess the effects of ultrasonic assistance on the process, revealing that the application of ultrasonic vibration yields uniformly shaped and high-quality single-pass formations. The influence of parameters such as direct writing velocity, distance between the nozzle aperture and substrate, and temperature of the extruded molten metal on the quality of single-pass formation was explored, thereby offering parameter references for subsequent three-dimensional solid formation experiments. By adjusting process parameters, well-constructed articles with superior formative quality were achieved, thereby substantiating the formative capacity of ultrasonically-assisted direct metal additive manufacturing technology.

Acknowledgment

This work was financially supported by the National Natural Science Foundation of China (Grant No. 52075548), the Taishan Scholar Program of Shandong Province (Grant No. tsqn201909068).

Journal of Physics: Conference Series **2992** (2025) 012005 doi:10.1088/1742-6596/2992/1/012005

References

[1] Laleh, Majid, et al. "Heat treatment for metal additive manufacturing." Progress in Materials Science 133 (2023): 101051.
[2] H. Yi, Z. Wang, H. Cao, M. Liu, J. Li, Metal droplet deposition: from foundation to engineering manufacturing, Advanced Engineering Materials, 25 (2023) 2201003.
[3] B. Blakey-Milner, P. Gradl, G. Snedden, M. Brooks, J. Pitot, E. Lopez, M. Leary, F. Berto, A. du Plessis, Metal additive manufacturing in aerospace: A review, Materials & Design 209 (2021).
[4] Sarkar, Devashish, Angshuman Kapil, and Abhay Sharma. "Advances in computational modeling for laser powder bed fusion additive manufacturing: A comprehensive review of finite element techniques and strategies." Additive Manufacturing (2024): 104157.
[5] Kapil, Angshuman, et al. "Exploring impact, spreading, and bonding dynamics in molten metal deposition for novel drop-on-demand printing." Materials & Design 238 (2024): 112633.
[6] Meng, Meng, et al. "3D printing metal implants in orthopedic surgery: Methods, applications and future prospects." Journal of Orthopaedic Translation 42 (2023): 94-112.
[7] T.V. Neumann, M.D. Dickey, Liquid Metal Direct Write and 3D Printing: A Review, Advanced Materials Technologies 5(9) (2020)
[8] S.C. Altıparmak, V.A. Yardley, Z. Shi, J. Lin, Extrusion-based additive manufacturing technologies: State of the art and future perspectives, Journal of Manufacturing Processes 83 (2022) 607-636.
[9] Hu, Guofang, et al. "Ultrasonic-assisted direct writing metal additive manufacturing technique." Journal of Materials Processing Technology 312 (2023): 117830.
[10] M. Ma, Z. Hu, K. Zhang, Z. Wang, H. Zhang, A metal additive manufacturing methodology: Pneumatic extruding direct-writing deposition, Additive Manufacturing 46 (2021).
[11] Herrada-Manchón, Helena, Manuel Alejandro Fernández, and Enrique Aguilar. "Essential guide to hydrogel rheology in extrusion 3D printing: how to measure it and why it matters?." Gels 9.7 (2023): 517.
[12] H. Ramazani, A. Kami, Metal FDM, a new extrusion-based additive manufacturing technology for manufacturing of metallic parts: a review, Progress in Additive Manufacturing (2022).

Simulation analysis and optimization design of forklift AGV fork rack

Yafeng Jiang [1], Shuaimei Lian [2*] and Senyao Wu [1]

[1] School of Intelligent Manufacturing and Information Engineering, Sichuan Technology & Business College, Chengdu 611830, China
[2] School of Intelligent Manufacturing, Chengdu Technological University, Chengdu, China

*Corresponding author's e-mail: 317760287@qq.com

Abstract: Forklift AGV is a widely used intelligent warehousing and logistics equipment, and is one of the key equipment for promoting the construction of intelligent workshops and factories. As one of the key load-bearing components of a forklift-type AGV, the fork frame affects its working ability. This article analyzes the structure and stress situation of a forklift-type AGV fork frame, obtains the force diagram and dangerous section of the fork frame, and verifies the strength of the dangerous section according to classical mechanical methods. At the same time, a three-dimensional model of the fork frame was established and subjected to static analysis by using Ansys Workbench software. The results showed that the theoretical calculation results were consistent with the simulation analysis results, verifying the correctness of the calculation results. In addition, the structure was optimized with quality objectives, reducing the weight of the fork by 11.5%. Simulation analysis was conducted again on the optimized fork, and the results showed that the fork structure design can meet the requirements for use.

1. Introduction

Forklift AGV is a typical intelligent warehousing logistics equipment, with a high degree of intelligence, strong carrying capacity, high calibration, flexible movement, and other characteristics, which is one of the key equipment of modern factory automation logistics systems. It is widely used in factory workshop material handling, and warehouse palletizing, and can effectively improve the production efficiency and intelligent production level of enterprises [1]. As the core component of forklift AGV, the working device of the forklift is mainly composed of fork, fork rack, door frame, hoisting mechanism, tilting mechanism, and so on. The mechanical properties of each structure are directly related to whether the forklift can work normally or not[2,3]. At present, the research on the working device of forklift is mainly focused on the key components such as the gantry, fork, and chassis, mainly using simulation software for static analysis and dynamic analysis, optimizing design according to specific working conditions, and achieving good results. Wei et al. [4] designed a new type of forklift door frame hoist structure. Sun [5] re-optimized the cross-section of the portal frame steel and the production process of the gantry steel based on the lightweight design goal. Guo [6] carried out the finite element static analysis and modal analysis of the forklift door frame system under multiple working conditions and optimized its structure. Li et al. designed a new type of forklift AGV structure, used ABAQUS software to carry out static analysis and modal analysis of the forklift AGV fork and the door frame of the forward mechanism, and verified its reliability [7]. Lemerle et al. [8] established the multi-body dynamics model of the forklift, studied its lateral stability during turning, and obtained the relationship between forklift stability and related parameters. Chen and Yan [9] established the dynamic model of forklift and analyzed the influence of forklift load, lifting speed, and road slope on the stability of forklift based on ADAMS software.

To sum up, the researchers have done a lot of work on the working device of the forklift, but it does not involve the research of the transition connecting device-fork between the fork and the door frame. A fork is a device used to install a fork or other working equipment, also known as a sliding frame, which drives the fork and the goods carried up and down along the door frame under the pull of the chain to complete the lifting of the goods. In this paper, the fork of a forklift AGV is calculated and analyzed, and the finite element analysis software is used for simulation.

2. Structural analysis and theoretical calculation

2.1 Structural analysis of fork

The fork structure studied in this paper is a plate, which is cut by a 32 mm steel plate, and the rear support is welded by a 20 mm steel plate and 10 mm triangular steel bar. The material is Q235A, and the structure and size are shown in Figure 1. According to the analysis of the motion characteristics of the fork, it completes the lifting movement in the gantry, which is similar to the trolley with vertical motion. Structurally, it is composed of the front frame and the rear roller frame with rollers, the front and rear parts are made into a whole, the forks or fixtures are installed on the front frame, and the rear roller frame interacts with the wing plate of the door frame through the roller wheel, so that the whole fork frame moves up and down under the action of chain tension along the door frame [5].

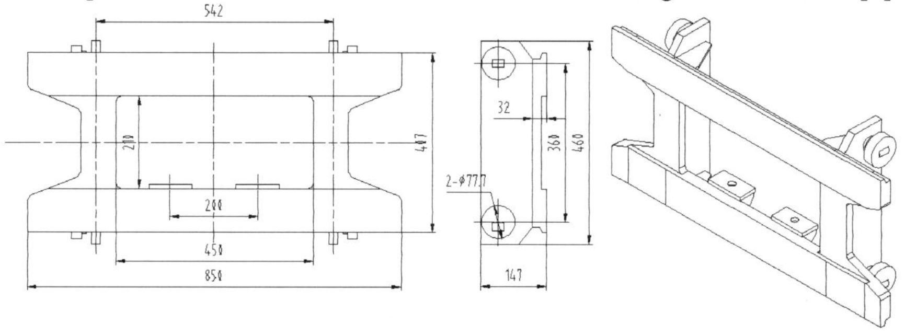

Figure 1 Fork frame structure and size.

2.2 Strength calculation of fork

The force of the fork is very complex, which involves the calculation of the statically indeterminate frame. The process is tedious and the workload is heavy, so the fork is usually simplified. It is usually simplified to a cantilever simply supported beam, and the force model, calculation diagram, and bending moment diagram shown in Figure 2 can be obtained after simplification. F is the acting force of the fork on the frame, and its magnitude is 10, 052 N; P is the calculated load, and its size is 7, 644 N.

IOP Publishing
doi:10.1088/1742-6596/2992/1/012006

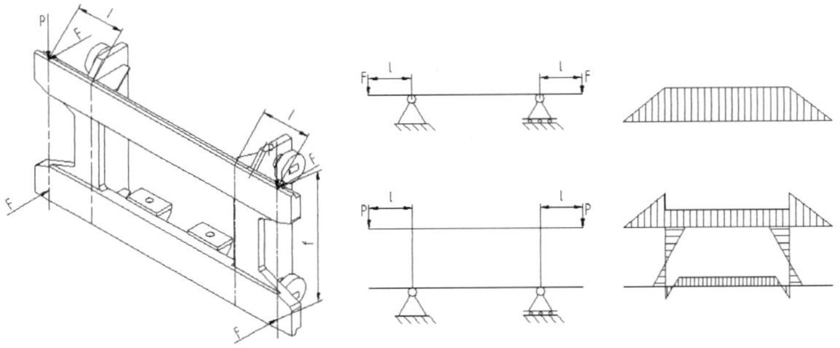

Figure 2 Fork force diagram.

As shown in Figure 2, if the force F of the fork on the frame is perpendicular to the frame plane, the upper and lower beams can be simplified to a double cantilever simply supported beam with the roller frame as the fulcrum, and the action point of the force F is on the upper and lower edges of the crossbeam. It is subject to the combined action of bending and torque at the same time, and under the action of external load P in the frame plane, it forms a symmetrical cubic statically indeterminate frame [10]. It can be seen from the bending moment diagram that the bending moment of the root of the cantilever of the crossbeam is the largest under the action of F and P at the same time, so the section in this position is dangerous. The strength of the fork should be designed and checked according to this section.

Bending moment in the frame plane:

$$M_x = Pl = 7644 \times 45 = 343980 N.mm \tag{1}$$

Bending moment perpendicular to the frame plane:

$$M_y = Fl = \frac{P(c+a+{}^{b}\!/_{2})l}{f} = \frac{7644(450+60+{}^{32}\!/_{2}) \times 45}{400} = 452334 N.mm \tag{2}$$

The torque of the section:

$$M_T = \frac{Fh}{2} = \frac{10052 \times 86}{2} = 432236 N.mm \tag{3}$$

where P is the calculated load; l is the cantilever length of 45 mm; c is the load center distance of 450 mm; a is the fork thickness of 60 mm; b is the cross-beam section thickness of 32 mm; h is the calculated height of cross beam section of 86 mm; F is the distance between the upper and lower beams of the frame of 400 mm.

Considering the weakening effect of the fork positioning groove on the upper beam, from the point of view of safety and simplified calculation, the cross-section is calculated directly as the rectangular section, and the stress at A and B can be verified, as shown in Figure 3.

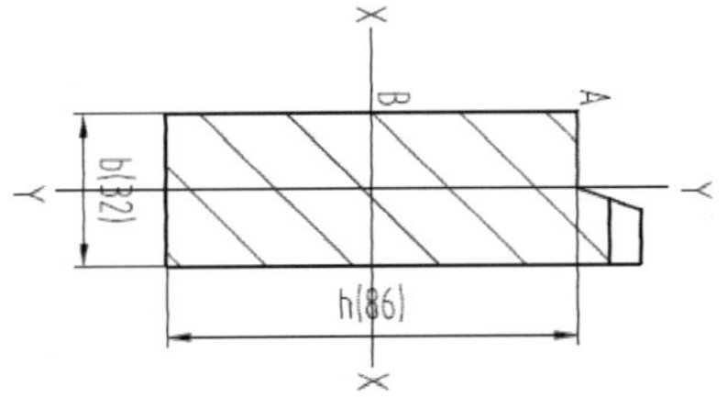

Figure 3 The dangerous section.

Stress at Point A: $$\sigma_A = \frac{M_x}{W_x} + \frac{M_y}{W_y} = \frac{Pl}{bh^2/6} + \frac{Fl}{b^2h/6} \tag{4}$$

If you bring in the data, you can get: $\sigma_A = \frac{343980}{32\times 86^2/6} + \frac{452334}{32^2 86/6} \approx 40MPa$

The shear stress is the highest at Point B: $\tau_{\max} = \frac{M_T}{\alpha b^2 h} = \frac{432236}{3\times 32^2 \times 86} = 1.6MPa$ (5)

where α is the aspect ratio coefficient of the rectangular section (h/b); Point B is a two-way stress state. According to the fourth strength theory:

$$\sigma_b = \sqrt{\left(\frac{M_y}{W_y}\right)^2 + 3\tau_{\max}} = \sqrt{\left(\frac{452334}{32^2\times 86/6}\right)^2 + 3\times 1.6^2} = 31MPa \quad (6)$$

The fork is welded with a Q235A steel plate, and the above calculation meets the requirements if you choose the steel [σ=] 110 MPa.

3. Finite element static analysis and optimization of fork frame

3.1 Finite element static analysis

We import the fork model into Ansys Workbench and set the material properties of the fork. The material of the fork is Q235A, and the specific parameters are shown in Table 1.

Table 1 Fork frame materials and features.

Part name	Materials	Elastic modulus (GPa)	Poisson's ratio	Density (Kg/m^3)	Yield limit (MPa)	Strength limit (MPa)
Fork rack	Q235A	206	0.3	7, 850	235	400

Definition of elements and meshing [11, 12]: Element selection should be comprehensively considered according to the type of structure, shape characteristics, stress and deformation characteristics, precision requirements, and hardware conditions. As the three-dimensional model has been established, the three-dimensional solid element is selected for meshing and analysis in this paper. For 3D entities, Ansys Workbench software has five meshing methods, namely Automatic, Tetrahedrons, Hex Dominant, Sweep, and Multi-Zone. The automatic meshing method is selected here. According to the structural characteristics of the fork and the performance of the computer, the meshing parameters are set as the default values of the system. The structure includes 8, 422 nodes and 4, 130 3D solid elements.The grid division of the fork frame is shown in Figure 4.

The application of constraint and load: Through the structural design and force analysis of the fork, we can know that when the fork is carrying goods, the tension of the moving end of the chain is balanced with the weight, and the degree of freedom of the roller in the Z direction is limited. The whole fork can be regarded as fixed in the roller, and all its degrees of freedom should be limited, that is, Fixed Support constraints should be added to the end face of the roller. According to the real working conditions, the concentrated force 10, 052 N is applied in the Z direction where the upper surface of the fork is in contact with the fork, 7, 644 N is applied in the calculated load in the Y direction, and 10, 052 N is applied in the Z direction where the lower part is in contact with the fork. The specific constraints and the application of loads are shown in Figure 5.

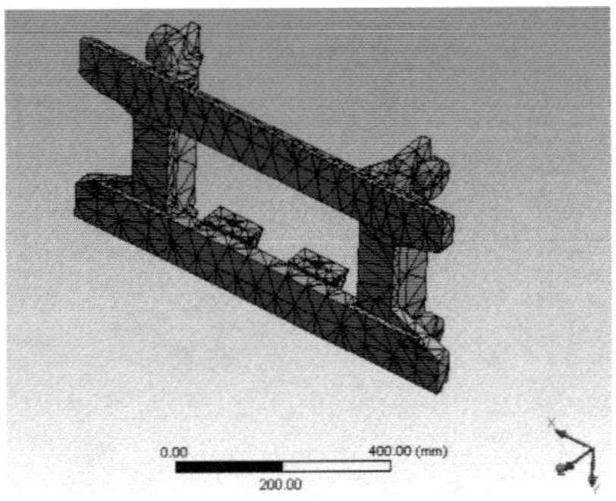

Figure 4 Fork frame meshing diagram.

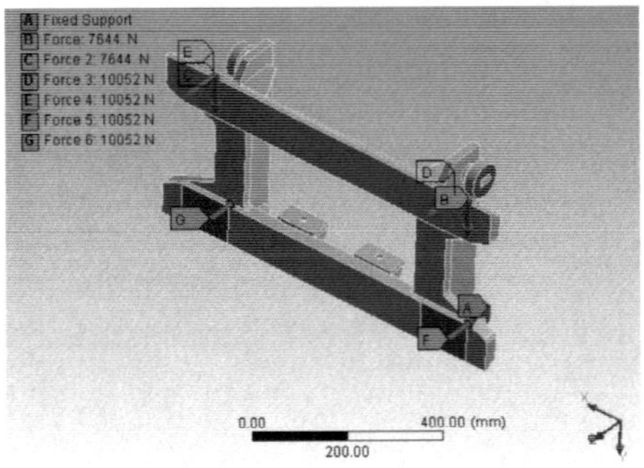

Figure 5 Load and constraints.

Results post-processing and analysis: After completing the above steps, we enter the post-processing stage of the results, and we are required to calculate the equivalent stress and total deformation according to the analysis, as shown in Figures 6 and 7.

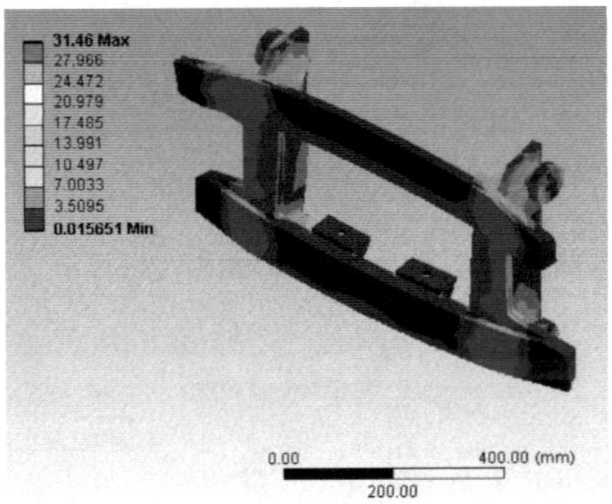

Figure 6 Fork frame equivalent stress nephogram.

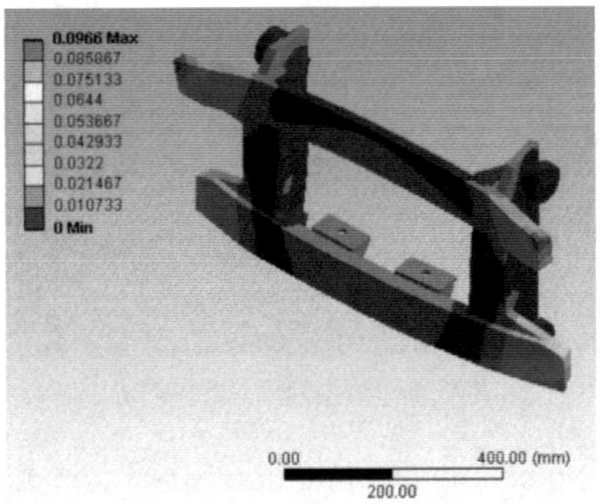

Figure 7 Fork frame total deformation nephogram.

From Figure 6, the maximum stress of the fork frame under load occurs at the root of the cantilever of the fork frame, and the stress distribution is consistent with the theoretical analysis. In addition, it can be seen from the figure that the maximum equivalent stress value is 31.46 Mpa, which is close to the theoretical calculation value of 31 Mpa. From Figure 7, the upper part of the fork frame panel deforms forward and the lower part deforms backward under load, and the deformation trend is consistent with the actual force situation. The maximum deformation occurs at the left and right ends of the upper cantilever, with a maximum deformation of 0.0966 mm. Therefore, the entire analysis result has a certain reference value in practical engineering applications, and the designed structure meets the requirements of the working conditions.

3.2 Structural optimization design and analysis

From the calculation and simulation analysis results, the maximum equivalent stress value of the fork frame is 31 Mpa, which is much lower than the allowable stress of the material, 110 Mpa. The original design of the fork frame has a large safety factor and can be further optimized to reduce the overall weight. Further analysis from the stress cloud map and deformation cloud map shows that the equivalent stress value of the rear bracket is less than 3.5 Mpa, which has a large safety factor. Therefore, this article takes the rear bracket as the optimization object, and the area near the roller is the main stress area. Therefore, the material near the roller remains unchanged, and the material far away from the roller area is removed. The optimized structure is shown in Figure 8. We perform static analysis on the optimized fork frame, and the results are shown in Figures 9 and 10.

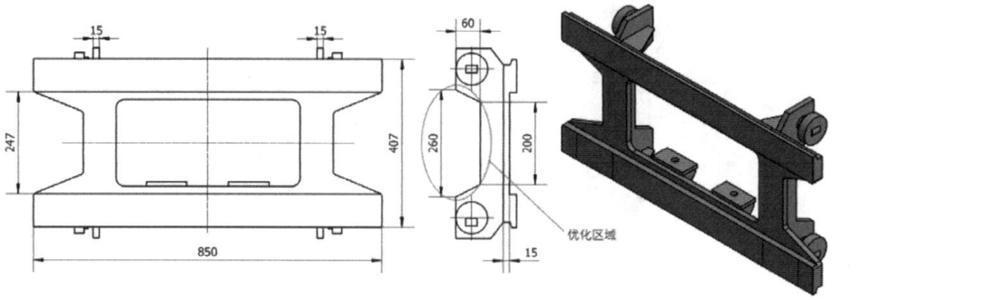

Figure 8 The optimized fork frame structure.

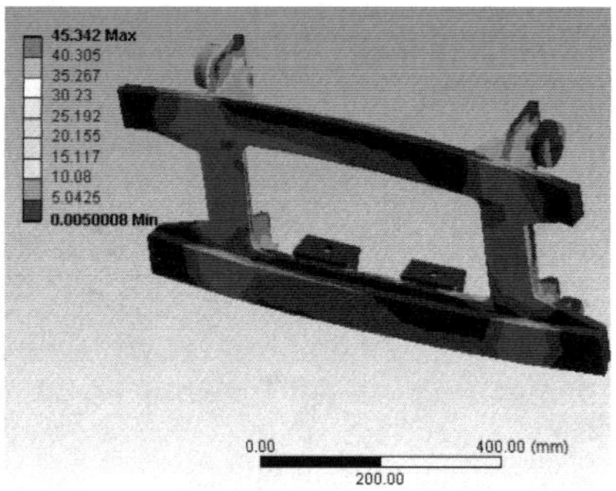

Figure 9 Fork frame equivalent stress nephogram.

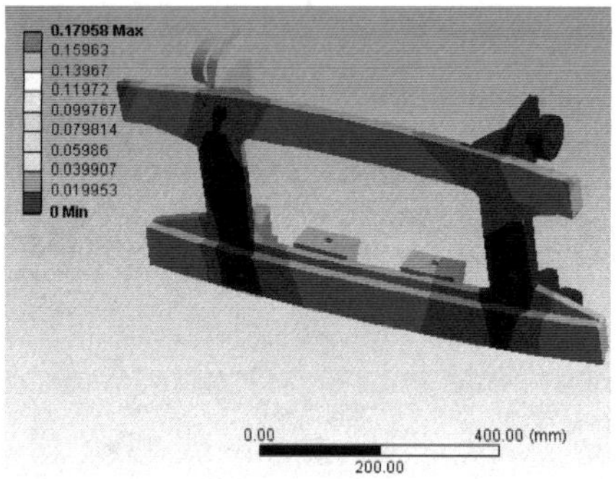

Figure 10 Fork frame total deformation nephogram.

From Figures 9 and 10, the optimized fork has a maximum equivalent stress of 45.342 MPa and a maximum deformation of approximately 0.18 mm. Both the strength and stiffness meet the requirements, and there is a significant safety factor. The weight of the optimized fork has been reduced from 61 kg to 54 kg. The comparison of stress, deformation, and weight changes before and after optimization is shown in Table 2.

Table 2 Stress, deformation, and quality before and after optimization.

Parameter	Equivalent stress value (Mpa)	Maximum deformation (mm)	Fork weight (Kg)
Before optimization	31.46	0.0966	61
After optimization	45.342	0.17958	54

4. Conclusion

The force of a forklift-type AGV fork is analyzed and calculated, and the maximum stress values of Point A and Point B of the panel section of the fork are 40 MPa and 31 MPa, respectively. The simulation analysis by Ansys Workbench verifies the results of theoretical calculation and analysis. The results show that the fork can meet the working requirements. At the same time, according to the analysis results, the safety coefficient of the fork is relatively large. Through the optimization of the support behind the fork, the weight of the fork is reduced from 61 kg to 54 kg, and the quality is reduced by 11.5%. The analysis method and process in this paper provide a certain reference for the design optimization of the forklift, and the analysis results have a certain guiding significance for the design of the forklift AGV.

Journal of Physics: Conference Series **2992** (2025) 012006 doi:10.1088/1742-6596/2992/1/012006

Acknowledgment
The paper is supported by the Sichuan Technology & Business College campus-level project funding (Project number: 2024CC05).

References

[1] Eui-Jung Jung, Jae Yeon Choi, and Sung Ho Hong. "Localization for an unmanned forklift in a refrigerated warehouse". 2016 13th International Conference on Ubiquitous Robots and Ambient Intelligence (URAI).
[2] Chu Weijie. Optimization Design of AGV Forklift Structure [D]. Southeast University, 2020. DOI: 10.27014/d.cnki.gdnau.2020.003394.
[3] Liu Meihong. Finite Element Analysis of the Overall Design and Key Structures of a Certain 2T Electric Forklift [D]. Chang'an University, 2016. https://www.cnki.net.
[4] Wei Pengfei, Gao Youshan, Jia Tingchen, et al. Research on the Lifting Mechanism and Safety Protection of New Forklift Doors [J]. Journal of Taiyuan University of Science and Technology, 2018, 39 (06): 467-472. https://www.cnki.net.
[5] Sun Changyuan. Optimization Design and Verification of Forklift Door Frame Structure [D]. Dalian University of Technology, 2013. https://www.cnki.net.
[6] Guo Jiliang. Analysis and Optimization Design of Forklift Door Frame Structure Based on Finite Element Method [D]. Dalian Jiaotong University, 2021. https://www.cnki.net.
[7] Li Yang, Cai Wenwei, and Li Weiguang. Structural Design and Simulation Analysis of Forklift AGV [J]. Mechanical and Electrical Engineering Technology, 2021, 50 (12): 16-22. https://www.cnki.net.
[8] Lemerle P., Höppner O., and Rebelle J. Dynamic stability of forklift trucks in cornering situations: parametrical analysis using a driving simulator [J]. Vehicle system dynamics, 2011, 49 (10): 1, 673-1, 693.
[9] Chen Huang and Yan Shirong. Stability Analysis of Forklifts Based on ADAMS [J]. Mechanical Manufacturing and Automation, 2015, 44 (04): 108-111. https://www.cnki.net.
[10] Yang Longwu. Research on Mechanical Characteristics and Optimization Design of Forklift Door Frame [D]. South China University, 2014. https://www.cnki.net.
[11] Guo Jie. Finite Element Analysis and Experimental Study on Key Structural Components of Forklifts [D]. Shandong University of Technology, 2012. https://www.cnki.net.
[12] Loli. Simulation Study on Forklift Working Device [D]. Shandong University of Technology, 2015. https://www.cnki.net.

Computer simulation of the auxiliary feed mechanism and opening closing boxes in a ginning machine

Shuai Cui[1], Xuan Sun[1], Quancheng Dong[1*], Haiqiang Gao[2], Yue Zhang[1] and Huaikun Li[2]

[1] School of Mechanical Engineering, University of Jinan, Jinan 250022, China;
[2] Shandong Swan Cotton Industrial Machinery Stock Co., LTD, Jinan 250032, China

[*]Corresponding author's e-mail: me_dongqc@ujn.edu.cn

Abstract: The ginning process significantly impacts the quality of cotton lint. However, an accurate mathematical model to describe the relationship between ginning process parameters and cotton lint quality is still unavailable, which hinders the optimization of the ginning process. To improve the production efficiency of ginning machines, this study uses simulation experiments to analyze the force and deformation of seed cotton rolls within the ginning machine's working chamber. By examining the maximum stress and maximum strain values of the seed cotton rolls during the simulation, the study evaluates the stability and processing efficiency of the ginning process. Experimental results indicate that when the ejector speed is 300 r/min and the adjustable chamber angle is 7°, the ginning process is stable and the machine operates efficiently.

1. Introduction

The ginning process is a crucial stage in cotton processing and plays an essential role in determining the quality of the cotton. The saw-tooth gin is widely used in ginning operations. The front and middleboxes of the gin, along with the auxiliary feed mechanisms within them, significantly influence ginning efficiency. Currently, the adjustment of the auxiliary feed speed and the opening and closing angles of the boxes relies on manual control based on the experience of workers. This method severely restricts the fine-tuning of the ginning process, preventing real-time dynamic adjustments based on cotton properties and adversely affecting both the quality and efficiency of the ginning process.

To design a ginning machine model with high operational efficiency, superior cotton quality, and strong adaptability, while maximizing both cotton quality and production efficiency, this study conducts an in-depth analysis of the mechanical properties of seed cotton and their variations over time. The mechanical properties tested vary according to different application objectives [1-2]. The pulling action of the saw blades on seed cotton bales can damage cotton fibers. This interaction is closely linked to the adjustment of the process parameters. Existing research has further validated this relationship, focusing on the impacts of key factors such as moisture content, ginning speed, seed cotton bale density, and feeding rate on the ginning outcome [3-6]. In recent years, numerous researchers have conducted extensive simulations of the ginning process, demonstrating that simulation can effectively replicate and analyze the ginning process [7-10].

This study applies finite element analysis for the first time to simulate the auxiliary feed mechanism and the opening and closing boxes in a cotton ginning machine. The simulation focuses on the stress states of seed cotton rolls under varying operating conditions of the auxiliary feed mechanism and the opening-closing boxes. By analyzing stress and deformation, the study evaluates how these parameters affect the quality and efficiency of ginning. This method provides reliable analytical results in the absence of real-time data, effectively predicts and enhances ginning quality and efficiency, reduces experimental costs, and offers data support and theoretical guidance for process improvements.

2. Factors influencing ginning performance in the work box

The raw seed cotton, upon entering the ginning machine's working box, is rolled into a uniform cylindrical structure through mechanical agitation and compaction. This rolled form is referred to as the seed cotton roll. The auxiliary feeding device is installed at the center of the working box to provide rotational assistance to the seed cotton bales, ensuring that they rotate within the box at an appropriate speed and condition. The structure of the auxiliary feeding device is shown in Figure 1. The cross-sectional view of the auxiliary feeding device blade is shown in Figure 2. The working box of the ginning machine not only serves as the primary location for the formation of the seed cotton roll but also has an adjustable function. It can regulate the pressure angle and control the working point by rotating around the axis. The rotational speed of the auxiliary feeding device and the opening and closing angle of the boxes are key process parameters of the ginning machine's working box and have a significant impact on the formation and processing quality of the seed cotton bales.

Figure 1 Auxiliary feeding mechanism.

Figure 2 Auxiliary blade profile.

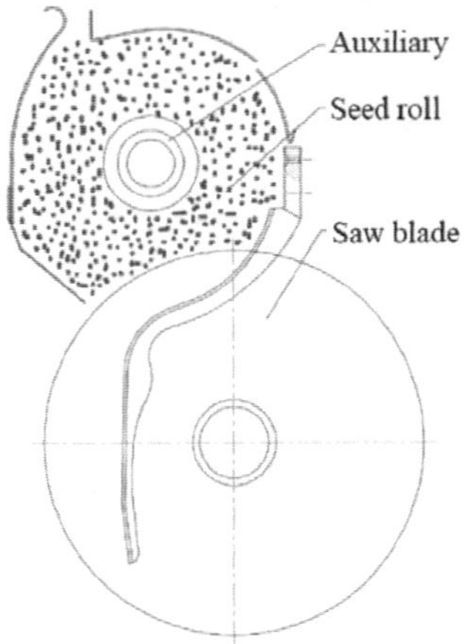

Figure 3 Schematic diagram of the ginning machine.

This study investigates the impact of process parameters in the ginning machine working box on seed cotton rolls, identifying the auxiliary feed speed and the opening-closing box angle as the primary factors influencing product quality and production efficiency. The focus is on the specific mechanisms by which variations in individual parameters affect the processing quality and efficiency in the saw gin working box. During mathematical modeling, auxiliary feed speed and box angle are assumed to be the main influencing factors, while external conditions such as temperature and humidity are considered secondary and thus disregarded. Although the model may exhibit some deviations due to dynamic changes in actual production, the in-depth analysis of the effects of individual parameters provides a theoretical foundation and practical guidance for optimizing ginning machine process parameters.

In this study, based on the actual structural parameters of the ginning machine, a model of the seed cotton roll was created using 3D software. A schematic diagram of the ginning machine's working setup is shown in Figure 3. In the diagram, the saw blades rotate around their center, while the seed cotton roll and ejector rotate around the center of the ejector. For large deformation calculations, the SOLID185 element was selected, as it is capable of performing various types of nonlinear analyses.

Journal of Physics: Conference Series **2992** (2025) 012007 doi:10.1088/1742-6596/2992/1/012007

The elastic modulus of the saw blade and auxiliary is E1 = 2.1e11, Poisson's ratio is 0.3046, and density is DENS1 = 7850 kg/m³; the elastic modulus of the cotton roll is E2 = 2.4e9, Poisson's ratio is 0.40, and density is DENS2 = 400 kg/m³. The viscoelastic characteristics of the cotton roll are described using the Prony series, with parameters in the Prony series set as i1 = 3.53, a1 = 0.123, i2 = 924.09, and a2 = 0.877 [8]. The models for the ejector, seed cotton roll, and saw blades were all meshed using tetrahedral elements. Figure 4 shows the mesh division of the auxiliary equipment and the seed cotton roll, while Figure 5 illustrates the mesh division of the saw blade and the seed cotton roll, including a magnified view of the contact points. In setting the boundary conditions, the auxiliary saw blade, and seed cotton roll is defined to rotate around their axes, with the rotational velocity of the auxiliary and saw blade specified. The stiffness matrix is formulated in an asymmetric manner, constraining the five degrees of freedom of the auxiliary and saw blade, allowing rotation only along the Z-axis.

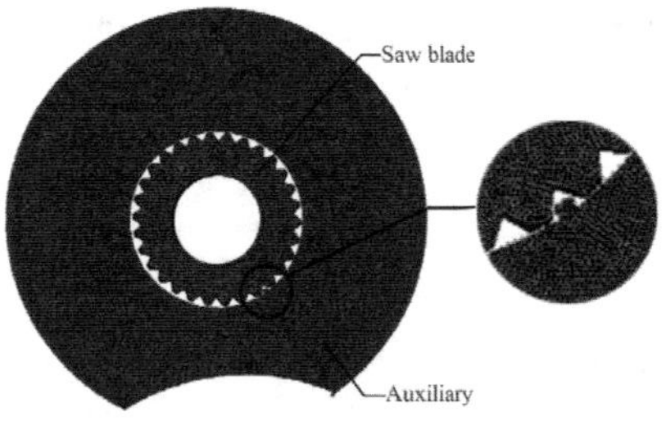

Figure 4 Mesh division diagram of the auxiliary and seed cotton roll.

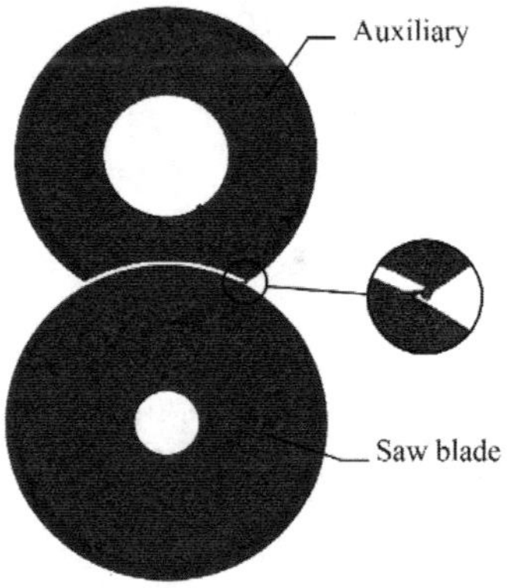

Figure 5 Mesh division diagram of the saw blade and seed cotton roll.

The simulation adopts a nonlinear analysis to examine the stress and strain during the interaction between seed cotton rolls and auxiliary equipment or saw blades. The stress and strain experienced by seed cotton rolls directly determine their mechanical behavior during processing, significantly influencing cotton quality and production efficiency. The rapid growth of stress and strain generates sufficient force in a short period to complete the separation of cotton fibers and seeds, accelerating the processing pace. Conversely, slow growth of stress and strain causes the seed cotton roll to stay too long in the work area, increasing unnecessary friction losses and reducing efficiency. Optimizing the auxiliary equipment speed and box opening angle ensures an ideal stress and strain distribution, improving cotton quality while enhancing production efficiency and reducing processing time.

2.1. Impact of ejector speed on ginning production
To investigate the impact of auxiliary feed speed on ginning machine production quality and efficiency, multiple auxiliary feed speeds were set in the simulation software based on practical conditions. The considered auxiliary feed speeds were 240 r/min, 270 r/min, 300 r/min, 330 r/min, and 360 r/min. Through simulation analysis, stress and strain contour maps at different speeds were obtained, and the variation of cotton roll stress and strain values over time under each speed condition was analyzed.

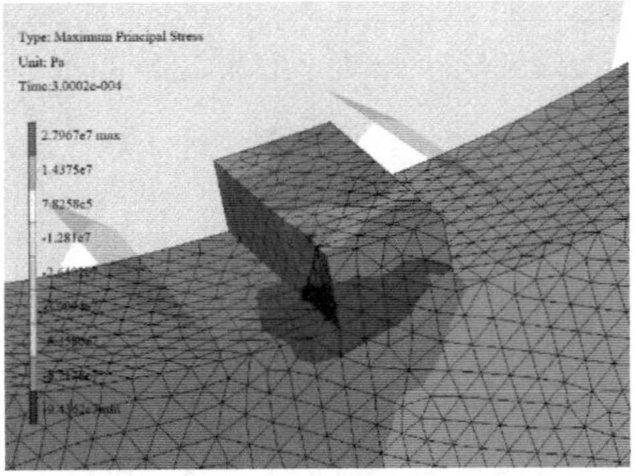

Figure 6 Local stress contour map of seed cotton roll and saw blade under the auxiliary speed.

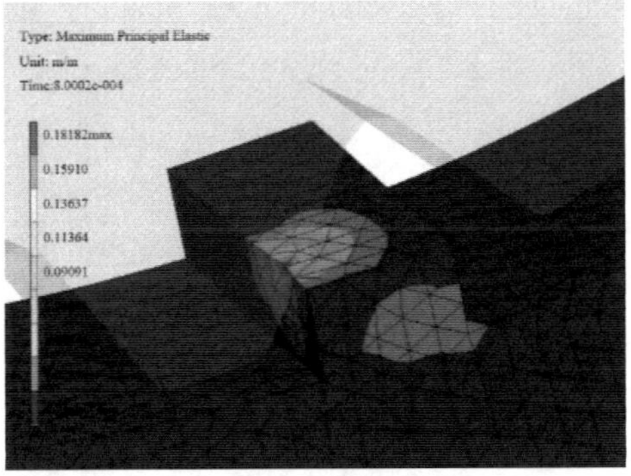

Figure 7 Local strain contour map of seed cotton roll and saw blade under the auxiliary speed.

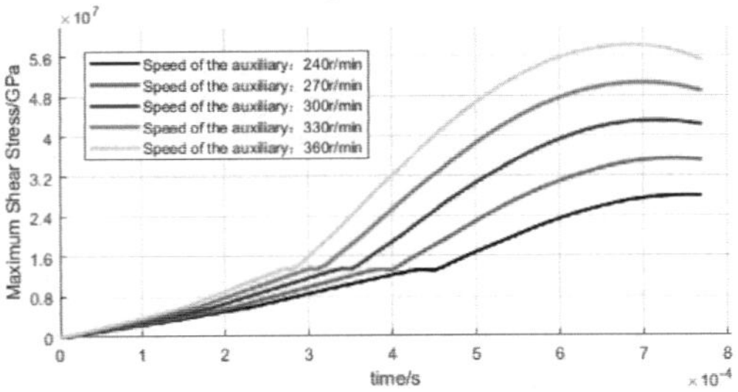

Figure 8 Time Response Chart of Stress in Cotton Roll under Different Auxiliary Speeds.

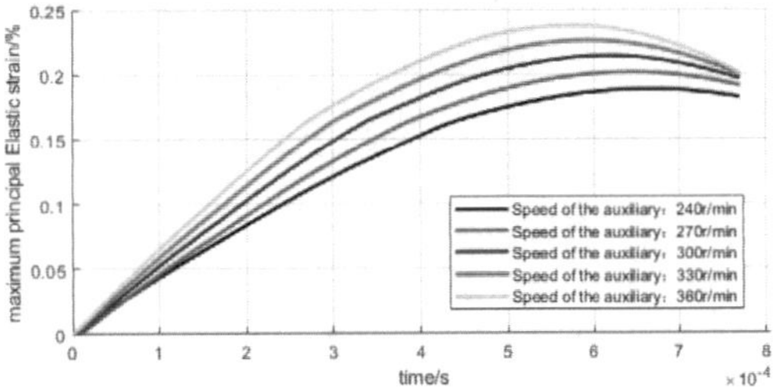

Figure 9 Time Response Chart of Strain in Cotton Roll under Different Auxiliary Speeds.

Figures 6 and 7 show the stress-strain contour maps of the cotton roll under an auxiliary speed of 240 r/min. This paper only presents the stress-strain contour maps at an auxiliary speed of 240 r/min, while the contour maps for other speeds are not shown to avoid repetition and simplify the content. Figures 6 and 7 show that the auxiliary, through rotation, drives the seed cotton roll, inducing internal stress and strain responses. As the auxiliary rotates, the seed cotton roll experiences deformation due to the influence of centrifugal force, causing stress within the fibers. As the seed cotton roll rotates, the internal fibers undergo an alternating strain process of tension and compression, ultimately forming a more uniform stress distribution. Figures 8 and 9 show the line charts of stress and strain variation over time at different auxiliary speeds. From the data in Figures 8 and 9, it can be observed that under different ejector speeds, the overall trend of each curve is generally consistent. All curves show a gradual increase in stress and strain at the initial stage. The growth rate of these quantities significantly accelerates as the ejector speed increases. After approximately 6×10^{-4} seconds, although the growth rate of stress slows down, it continues to increase, while the strain begins to decrease.

At higher auxiliary feed speeds, the rapid increase in stress and strain enhances fiber separation efficiency. High speeds generate greater shear forces in a short period, enabling faster fiber separation. However, this rapid stress and strain variation may lead to process instability, increased mechanical load and wear, and significantly higher energy consumption. At lower auxiliary feed speeds, stress and strain changes are more gradual, ensuring a stable process suitable for scenarios requiring consistency and low energy consumption. While separation efficiency at low speeds may not match that of high speeds, the economic and stable operation makes it preferable when extreme separation speed is unnecessary. Considering the trends in stress and strain variations and their effects on ginning performance, selecting 300 r/min as the auxiliary feed speed is a reasonable compromise. This speed produces sufficient stress and strain within a short time, improving fiber separation efficiency while maintaining process stability and avoiding the instability and excessive energy consumption associated with high speeds. This choice optimizes separation efficiency, effectively controls energy usage, and ensures operational stability, contributing to the economic and sustainable performance of the ginning process.

2.2. Analysis of the effect of opening and closing box angle on ginning production

The opening angles of the front, middle, and rear boxes in the ginning machine are set within the range of 4° to 8°. To study the impact of the box opening angle on ginning yield, this study sets the opening angles at 4°, 5°, 6°, 7°, and 8°, and establishes corresponding models.

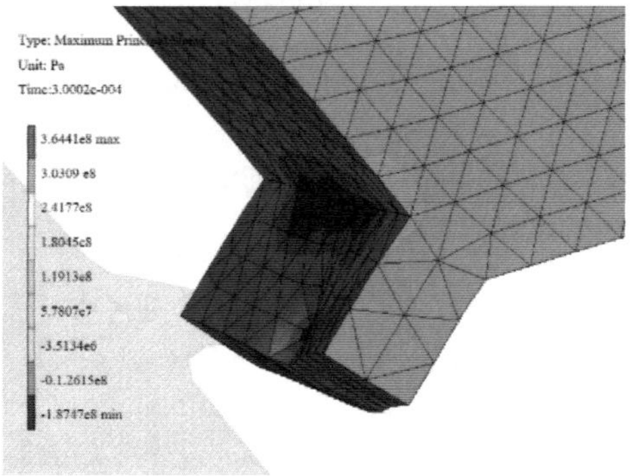

Figure 10 Local stress contour map of seed cotton roll and saw blade under opening angle.

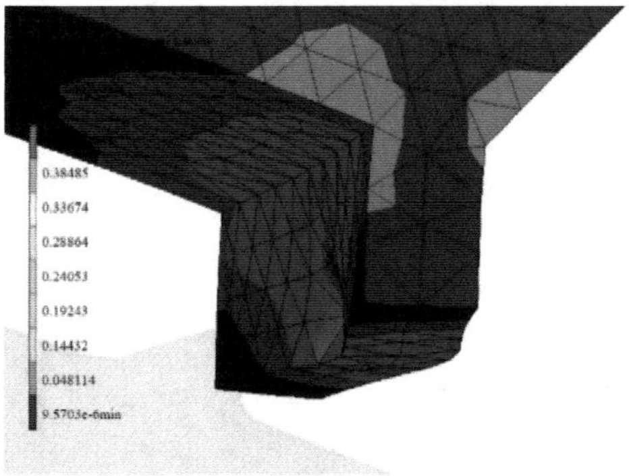

Figure 11 Local strain contour map of seed cotton roll and saw blade under opening angle.

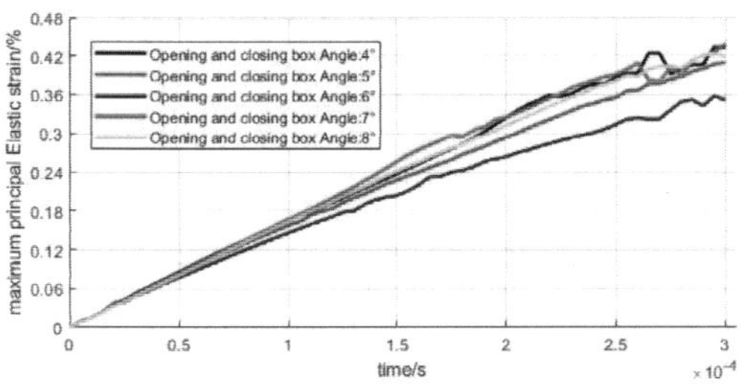

Figure 12 Time response chart of stress in cotton roll under different opening and closing box angles.

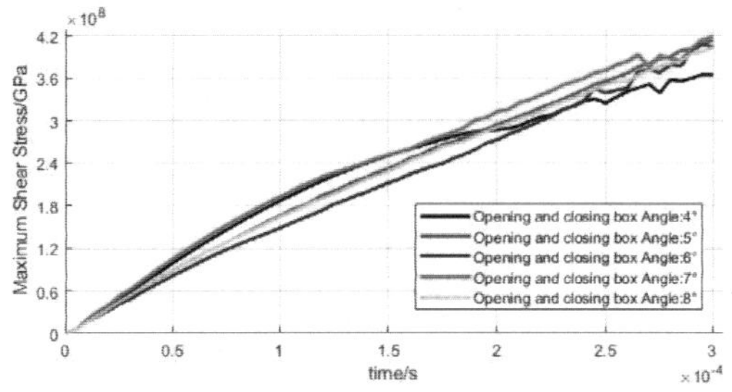

Figure 13 Time response chart of strain in cotton roll under different opening and closing box angles.

In ANSYS, the saw blade speed was set to 640 r/min for simulation analysis. The simulation results provided stress and strain contour maps at different box opening angles. The variation of cotton roll stress and strain values over time under each opening angle condition was analyzed.

Figures 10 and 11 are local magnified views of the stress-strain contour maps of the cotton roll at an opening and closing box angle of 4°. From Figures 10 and 11, it can be observed that the interaction between the saw blade and the seed cotton roll induces stress and strain responses within the roll, characterized by significant stress concentration in the contact area between the saw blade and the seed cotton roll. Figures 12 and 13 show the line charts of stress and strain variation over time at different opening and closing box angles. Figures 12 and 13 further illustrate that under different open-box angles, the stress and strain variations exhibit a gradually increasing trend over time. Specifically, as the opening angle increases, both strain and stress continue to grow, though at varying rates. Notably, when the opening angle is 4°, strain increases most slowly. In contrast, at an opening angle of 7°, stress increases the most rapidly.

The simulation results indicate that when the opening and closing box angle is 7°, stress, and strain increase rapidly, which aligns with the characteristics of significant mechanical forces in the actual ginning process. At this angle, the speed of fiber separation is significantly accelerated. The efficiency of fiber separation during compression and shear is greatly improved. The larger mechanical forces effectively facilitate cotton processing, not only shortening processing time but also maximizing ginning efficiency. This is especially beneficial when processing cotton with high impurity content or strong adhesiveness, as it enables faster separation. By dynamically monitoring stress and strain changes through simulation, the impact of different opening and closing box angles on cotton separation can be reflected in real time, providing scientific support for optimizing process parameters. This is particularly helpful in complex production environments, where it aids in better adjusting equipment settings to improve overall efficiency. Combining simulation results with actual production needs, it can be concluded that an opening and closing box angle of 7° is the optimal state for the highest ginning efficiency. This finding offers theoretical support for optimizing parameters in practical production, enhancing both production efficiency and product quality. In practice, the opening and closing box angle can be dynamically adjusted according to the cotton's characteristics, such as impurity content and adhesiveness, to further optimize the ginning process.

3. Conclusion

This study conducts a preliminary investigation into the ginning process in the ginning machine working box, using finite element analysis to systematically analyze the impact of auxiliary feed speed and opening-closing box angle. By establishing simulation models for the auxiliary feed box, seed cotton rolls, and sawteeth, and designing experiments with varying feed speeds and opening-closing box angles, the main factors influencing the ginning process are explored. The following conclusions are drawn: At an auxiliary feed speed of 300 r/min, sufficient stress and strain are generated in a short time, improving fiber separation efficiency, while avoiding the instability, excessive mechanical load, and high energy consumption associated with higher speeds. Therefore, 300 r/min is a reasonable choice that balances economy, efficiency, and stability. At an opening-closing box angle of 7°, the system's stress and strain increase most rapidly, significantly enhancing the fiber separation rate during compression and shear, thereby shortening processing time and improving production efficiency. Under the condition of seed cotton roll density of 400 kg/m^3, selecting an auxiliary feed speed of 300 r/min and an opening-closing box angle of 7° not only effectively improves ginning efficiency and optimizes fiber separation but also achieves the best balance between machine stability and energy consumption. In conclusion, the results of this study provide valuable references for the production operations of ginning machines, helping operators better adjust process parameters to achieve higher efficiency and better cotton quality in actual production. These findings offer a scientific basis for the economic viability and sustainable development of the ginning process and lay a solid foundation for further research and the promotion of related technologies.

Acknowledgments

This work was financially supported by Agricultural Machinery R&D Manufacturing Promotion and Application Integration Project of Xinjiang Autonomous Region (YTHSD2022-11) fund.

References

[1] Liu Q T, et al. Study progress on mechanics properties of crop stalks [J]. Transaction of the CSAM, 2007(7): 172-176. https://www.cabidigitallibrary.org/doi/full/10.5555/20073263736
[2] Pandey C P S. Mechanical properties of sorghum stalk in relation to quasi-static deformation [J]. J Agric Engng Res, 1999, 73(2): 199-206. https://doi.org/10.1006/jaer.1999.0406
[3] Rodgers, et al. Laboratory Ginning and Blending Impacts on Cotton Fiber Micronaire Measurements [J]. AATCC Journal of Research, 2015, 2(4):1-7. https://journals.sagepub.com/doi/abs/10.14504/ajr.2.4.1
[4] Lang X X, et al. Analysis of Factors Affecting the Yield of Saw Gin [J]. *China Cotton Processing*, 2020, (03): 4-6. https://doi.org/10.3969/j.issn.1003-0662.2020.03.002.

[5] Wang X T, et al. Experimental Optimization of Process Parameters for Saw Gin [J]. *China Cotton Processing*, 2023, (06): 4-7. https://doi.org/10.3969/j.issn.1003-0662.2023.06.001.
[6] Wang D, Qiu Y P, and Wang J. Finite Element Analysis of Cotton Ginning State Based on ANSYS [J]. *Textile Research Journal*, 2019, 89(11): 2142–2153. https://doi.org/10.1177/0040517518786274
[7] Hu W, et al. Construction and Finite Element Simulation of the Saw Type Gin Stand Using Fractal Theory [J]. *Textile Research Journal*, 2020, 90(23–24): 2755–2768. https://doi.org/10.1177/0040517520929358
[8] Ge Y, et al. Ginning Speed Prediction Method Based on ANSYS and BP Neural Network [J]. *Journal of Donghua University (Natural Science Edition)*, 2017, 43(06): 886-892. https://doi.org/10.3969/j.issn.1671-0444.2017.06.017
[9] Yang J G, et al. Inversion of Textile Production Process Parameters Based on Genetic Neural Network Algorithm [J]. *Journal of Donghua University (Natural Science Edition)*, 2016, 42(2): 234-241. https://doi.org/10.3969/j.issn.1671-0444.2016.02.013
[10] Li Y, et al. BP Neural Network-Based Prediction Model for the Quality Indicators of Mechanically Harvested Cotton [J]. *Cotton Textile Technology*, 2015, 43(1): 17-20. https://doi.org/10.3969/j.issn.1001-7415.2015.01.005

Journal of Physics: Conference Series **2992** (2025) 012008 doi:10.1088/1742-6596/2992/1/012008

Design and application of four-wheel steering system for integration of theory and practice teaching

En Wei [1], Jinggong Wei [2*] and Dongfang Zhang [3]

[1] Tianjin University of Technology and Education, Tianjin, China
[2] Tianjin University of Technology and Education, Tianjin, China
[3] Beijing Pangxin Changda Technology Co., Ltd., Tianjin, China

[*] Corresponding author e-mail: weijinggong@tute.edu.com

Abstract—Aiming at the problems of backward technical performance and poor visual presentation of the teaching equipment of the "integration of theory and practice" course in the chassis steering department of auto repair major in vocational colleges, a new energy wire-controlled four-wheel steering car which can be used for steering demonstration in the classroom and other limited venues is designed and developed to enhance the intuitionistic teaching, improve the teaching quality and help talent training. This paper covers the general layout, steering motor selection, steering ladder design check, vehicle modeling, motion simulation, real car performance evaluation, and so on. The relationship of wheel Angle, steering ladder parameter, Ackermann rate, and minimum turning radius are preset, checked, and verified by means of calculation and diagram. The CATIA mechanical design and electronic prototype module are used to model and assemble the vehicle. The motion simulation of the steering system and shock absorber system is carried out to check the motion condition, verify the minimum clearance, and prevent interference. Finally, the whole vehicle trial production was carried out, and the measured Ackermann rate of the actual vehicle was 72.2%, with a minimum turning radius of 2.1 m, which meets the design requirements. The formed design loop ensures the design quality of the product and verifies the reliability of the design method.

1. Introduction

According to the statistics of the China Association of Automobile Manufacturers, the share of new energy vehicles in the entire automobile market exceeds 30%, which has become the main development direction of the automobile industry [1]. With the vigorous development of new energy vehicles, major automobile manufacturers have taken "chassis intelligence" as the development goal of their products. Geely Auto and Hella plan to jointly develop a steer-by-wire mechanism, which is expected to be mass-produced in 2026; Great Wall Motors' "Smart drive-by-wire chassis" and Tesla's electric multi-purpose truck are both planned to be equipped with drive-by-wire steering systems [2]. Through research on high steering performance at home and abroad, it is planned to combine steering by wire and four-wheel steering to improve the flexibility and stability of new energy vehicles [3].

In order to follow the development of the market, we promote technological upgrading and promote industrial upgrading, middle. Higher vocational colleges have fully opened the application and maintenance of new energy vehicles. Due to the short construction time and many new technologies, there are some problems in the "integration of science and practice" teaching links of the chassis, such as less teaching equipment and poor operation demonstration. Even some colleges still use traditional automotive anatomy-based chassis steering system teaching aids, which cannot be used for on-site steering demonstrations in the classroom. Students cannot intuitively see vehicle concepts such as minimum turning radius and Ackerman angle. The equipment is seriously lagging behind current development trends, unable to demonstrate cutting-edge technologies such as steer-by-wire and four-wheel steering, which is not conducive to talent cultivation. In order to keep teaching up with the development trend of new technologies, enhance the intuitive teaching of "integration of theory and practice" for the chassis of new energy vehicles, and better cultivate students' practical ability, a new energy-driven four-wheel steering car with a turning radius of no more than 2.4 m that can be used for steering demonstration in the classroom is designed and developed. This car can enable students to intuitively experience concepts such as minimum turning radius and Ackermann angle and gain a deeper understanding of cutting-edge technologies such as steer-by-wire and four-wheel steering. This design can thus play a role in meeting teachers' daily teaching needs, promoting students' understanding of knowledge, improving their practical ability, and facilitating innovative development.

2. General layout and parameter presets

The new energy by-wire four-wheel steering trolley (hereinafter referred to as the "trolley") combines the by-wire steering and four-wheel steering on the base four of the new energy power chassis. The front and rear axles of the car are equipped with electric by-wire steering devices. The ECU receives the steering signal and outputs the angle signal through the signal receiver, and controls the steering motor to drive the rack and pinion mechanism to drive the steering tie rod to move left and right to achieve four-wheel steering. The effect of car construction is shown in Figure 1.

Figure 1 New energy wire control four-wheel steering trolley structure.

2.1 General layout scheme

2.1.1 Power part. The car adopts the rear-rear drive arrangement. The lithium-ion battery is selected as the power source of the whole vehicle [4]. The three-phase permanent magnet synchronous drive motor is selected as the power output device. The car maximizes the use of high and low-voltage exclusive accessories for new energy vehicles and controls the high-voltage on-off process in strict accordance with the control logic of new energy vehicles. The power control unit PEU is composed of four parts: drive motor controller MCU, vehicle controller VCU, DC-DC, and high voltage distribution. It uses the CAN bus to communicate with the outside world. The control unit collects the data information of the power battery, such as temperature, single voltage, total voltage, and current, and displays it in the instrument, which is convenient for users to check the system status. Power battery has a nominal voltage 64 V and a rated current 80 A. Three-phase permanent magnet synchronous motor has rated voltage 64 V and rated power 3 kW. The power battery, power control unit, and drive motor are water-resistant closed treatments. The car only carries out steering teaching demonstrations and can run for more than 5 hours.

2.1.2 Chassis. The car adopts a hydraulic disc brake system controlled by wire. All four wheels are equipped with disc brakes to ensure a good braking effect. In order to meet the requirements of lightweight and reduce the manufacturing cost of the vehicle, the car simplifies the shock absorption system as much as possible. There is a double fork arm suspension between the steering knuckle and the frame. When the wheel jumps up and down, the upper and lower fork arm drives the shock absorber to absorb energy and ensure the normal work of the parts carried by the car. The wheels are made of 235/30-10 tires. The frame is made of 40×40 square steel.

2.1.3 Body parts. The trolley adopts a non-bearing body. The body truss is welded by round pipe steel and connected to the frame by bolts. The left and right sides of the car body are designed with one removable steel plate, which is convenient for students to observe the movement and working conditions of the vehicle chassis braking system, steering system, and shock absorption system. The rear board of the car is provided with an emergency stop switch, a charging jack, a one-button start button, a charging indicator, a working indicator, etc. The front panel is equipped with illuminated headlights, which are powered by a 12 V battery. The car body ensures the intuitiveness of teaching to the maximum extent while taking into account integrity, aesthetics, and economy.

2.2 Parameter presets
The car makes the following presets according to the use requirements. The detailed data are shown in Table 1.

Table.1 Parameter preset table.

Parameter	Numerical value
Kingpin center distance/mm	780
Front and rear wheelbase/mm	1050
Left and right wheelbase/mm	1120
Tire pressure/Mpa	0.25
Total mass/kg	230
Front axle load/N	1000
Rear axle load/N	1300
Kingpin internal inclination Angle/°	6
Minimum ground clearance/mm	90

3. Design essentials of four-wheel steering system
The general design flow of the automotive steering system is shown in Figure 2. In the early stage of this design process, scheme selection and layout presetting are carried out. User needs are reasonably input into the system design according to the actual working environment. Referring to the materials of conventional automotive steering system components for material selection can shorten the design cycle and facilitate later processing and manufacturing. In the parameter calculation process, the calculation accuracy can be guaranteed by pre-setting and verifying based on empirical formulas. By establishing a numerical model for motion simulation, we analyze the rationality of production and processing, and avoid interference in secondary processing.

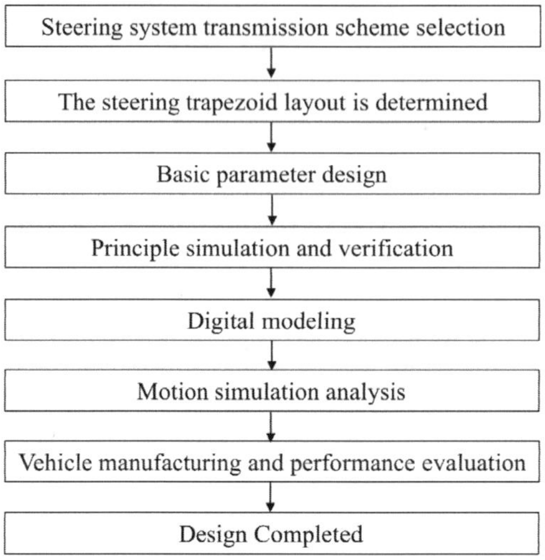

Figure 2 Steering system design flow char.

3.1 Scheme selection

Steering systems are divided into mechanical steering systems and power steering systems according to different driving energy sources. The power steering system includes an electronic hydraulic power steering system, electric power steering system, wire control steering system, and so on. The steering-by-wire system mechanically decouples the steering disc from the steering wheel [5] and uses the angle sensor to receive the steering signal and control the steering motor to drive the axial movement of the steering bar to realize the wheel steering. Compared with the traditional steering system, the drive-by-wire steering system has the advantages of fewer mechanical parts, a flexible system layout, and space-saving. The car sends the steering signal through the remote control. The angle sensor transmits the steering signal to the ECU and converts it into the angle signal to control the steering motor. Since this product does not have the function of human driving, it has not been designed for steering disc assembly, road sense analog motor, and redundant system.

The traditional front-wheel steering system has the characteristics of a simple structure and convenient layout, but the steering response is slow and the steering radius is large when the speed is low. Due to the limited teaching space, four-wheel steering technology is used to reduce the steering radius. The steering mechanism is provided on the front and rear axles of the trolley. The front and rear steering motors drive the front and rear steering rods to move axially. The transverse rod is connected with the steering knuckle arm through the universal ball head, so as to drive the wheel to achieve steering. The trolley adopts the broken steering trapezoid structure, the front axle steering trapezoid and steering gear are rear, and the rear axle steering trapezoid and steering gear are front. The steering system components are made of Q235 steel, which has good economy, weldability, and certain strength, and is widely used in vehicle chassis structures.

3.2 Calculation of steering resistance moment

The steering resistance moment of the steering shaft is related to many factors such as front and rear axle load, tire size structure, front and rear wheel positioning parameters, driving speed, road adhesion coefficient, etc. It is quite difficult to accurately calculate the steering resistance moment [6]. In the practice of automobile chassis design, the resistance moment of stationary steering is usually replaced by the resistance moment of stationary steering. The steering resistance moment is calculated according to the empirical Formula (1) summarized from the results of the in-situ steering test. The car only turns during the driving process and does not do 360° in-situ steering design.

$$M_w = \frac{\mu}{3}\sqrt{\frac{G_t^3}{p}} \qquad (1)$$

In the above equation, Mw is the steering resistance moment; Gt is the steering shaft load; p is for tire pressure; μ is the coefficient of sliding friction between the tire and the ground, generally μ=0.7. By calculation, the steering resistance moment of the front axle is 14.7 N·m. The steering resistance moment of the rear axle is 21.87 N·m. The steering resistance moment of the rear axle is greater than that of the front axle, so the preset steering resistance moment is 21.87 N·m.

3.3 Selection of steering motor

The power source of the steering motor is the DC 12 V vehicle battery. When the steering motor is working, the maximum steering resistance moment should be overcome to drive the rack and pinion mechanism in the steering gear. Formula (2) is the calculation formula of the output torque of the motor. Te is the output torque of the motor; Rack and pinion transmission efficiency ηgr =0.9; k is the safety factor, generally let k =1.5; Mw is the steering resistance moment. By calculation, the output torque of the motor is 36.4N ·m.

$$T_e = \frac{kM_w}{\eta_{gr}} \qquad (2)$$

The design speed of the steering gear rack is 70 mm/s, the gear speed is 320°/s, and the calculated motor power is about 205 W. Common DC 12 V automotive steering motors on the market have 170 W, 220 W, 360 W, etc., and 220 W steering motors are selected here.

3.4 Presets of steering mechanism

3.4.1 Ackermann Angle presets. For the design of the steering system, in addition to meeting the maneuverability and handling stability, it is necessary to ensure the minimum slip rate of the internal and external steering wheels to reduce the wear of parts. When the car makes a turn, the track circumference formed by the inner and outer wheels is different. If the inner and outer steering wheels turn at the same angle, the wheels will slip, which is easy to cause tire wear and component deformation [7]. Only with proper design can all wheels make pure rolling without sliding on different track circles around one instantaneous center as much as possible [8]. This ideal inner and outer front wheel angle relation is called the Ackermann steering geometry relation. When the car is used for teaching demonstration, the speed is low and the tire side stiffness is large, and the influence of tire side stiffness on the angle is ignored. The car is designed for four-wheel steering. The steering must meet the Ackermann steering relation of four-wheel steering. Formulas (3), (4), (5) and (6) are the Ackermann steering relation of four-wheel steering. In order to take into account the steering stability and maneuverability of the car, the maximum limit angle of the inner wheel is designed to be 45°.

$$\theta_1 = \theta_3 \qquad (3)$$

$$\theta_2 = \theta_4 \qquad (4)$$

$$\cot\theta_2 - \cot\theta_1 = \frac{2K}{L} \qquad (5)$$

$$\cot\theta_4 - \cot\theta_3 = \frac{2K}{L} \qquad (6)$$

In the formula, θ1, θ2, θ3, and θ4 are respectively the corners of the left front wheel, right front wheel, left rear wheel, and right rear wheel. K is the center distance of the left and right kingpins; L is the wheelbase. When the inner steering wheel rotates 5°, the theoretical angle relation curve of the inner and outer steering wheels can be obtained, as shown in Figure 3.

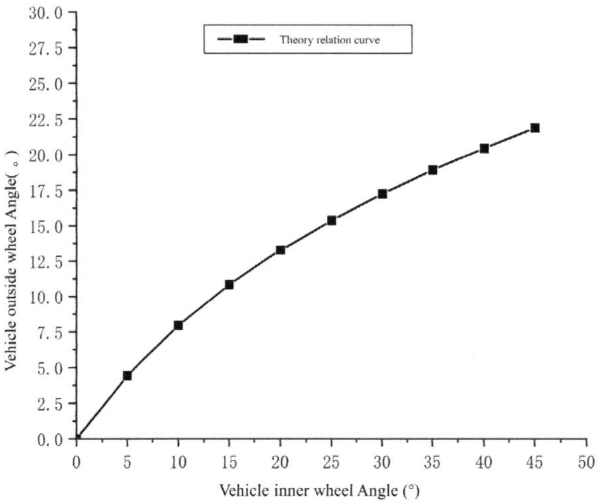

Figure 3 Theoretical curve of angle relation between inner and outer wheels.

3.4.2 Preset minimum turning radius. When the inner wheel turns to the limit position, the track circle radius of the center of the outer steering wheel rolling on the supporting plane is called the minimum turning radius when the car is steering at the lowest stable speed. Figure 4 shows the minimum turning radius. When the inner wheel turns to the design limit corner position, the outer wheel is 21.88° according to the Ackerman steering relation, the instantaneous center of the vehicle steering is point P, and the theoretical minimum turning radius, that is, the distance between line segments UP, is 1.6m. According to the Design Code for Primary and Secondary Schools (GB50099-2011), the width of the middle school classroom should not be less than 4.8m. The car can be turned around.

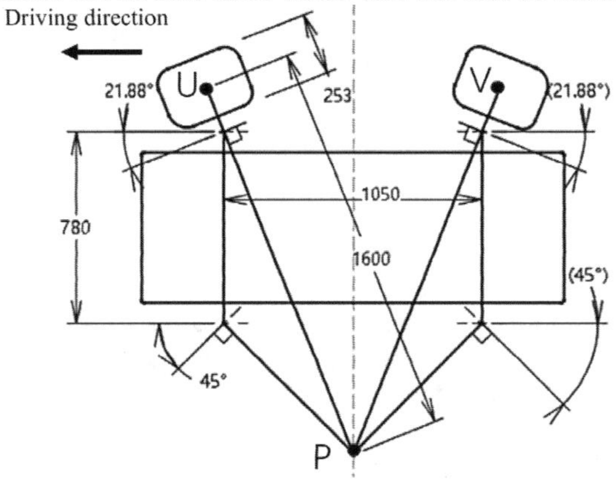

Figure 4 Minimum turning radius diagram.

3.4.3 Preset disconnect point of steering tie rod. The trolley steering trapezoid is designed as a broken steering trapezoid. The main advantage is that when one side of the wheel beats up and down due to the uneven road surface, the other side of the wheel can still maintain a normal driving state, without interference from the other side of the wheel. The position of the breakpoint of the tie rod is related to the position of the independent suspension. The position of the hinged point between the double-arm independent suspension and the body should be preset first. Then the position of the breaking point of the steering tie rod should be determined according to the "three-center theorem".

Taking the left front wheel of the car as an example, the design steps are as follows: ① According to the selected tire type and the minimum ground clearance of the body, we determine the hinge center of the lower cross arm of the suspension and the frame and the kingpin. ② The kingpin internal

IOP Publishing
doi:10.1088/1742-6596/2992/1/012008

inclination angle and kingpin tow distance are preset to determine the hinge center of the suspension cross-arm, the frame, and the kingpin. ③ According to the size of the pre-selected steering knuckle, we judge the height of the rack from the ground, and the height of the center point between the tie rod and the steering knuckleball pin. ④ We apply the graphic method based on the "three-center theorem" to judge the breaking point of the tie rod under the condition of no beating wheel. Figure 5 shows the process of parametrically solving the breakpoint position using the CATIA sketch editor.

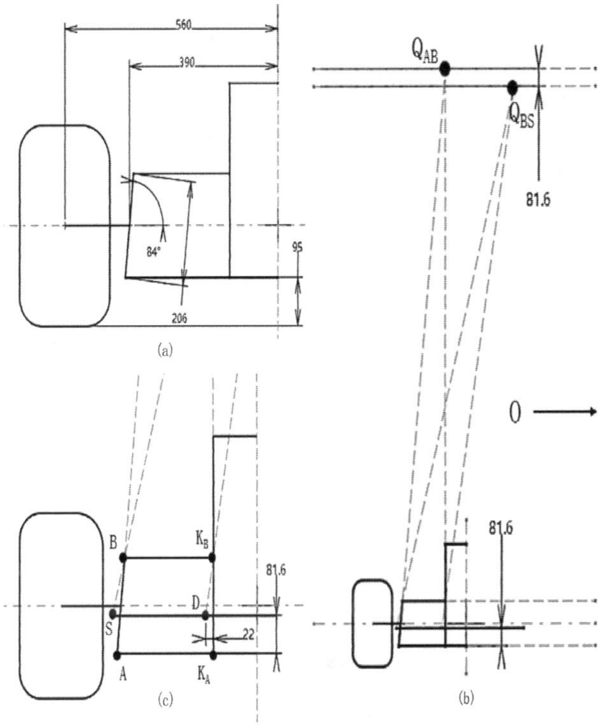

Figure 5 Breakpoint solution procedure.

In the figure, points A and B are the endpoints of the main pin, point S is the center of the steering tie rod and the steering knuckleball pin, points KB and KA are the hinge points of the double-arm suspension and the frame, and point D is the breaking point of the tie rod.

3.4.4 Trapezoidal mechanism presupposition and check calculation. In order to ensure that the inner and outer wheel angles of the car keep the Ackermann steering relationship to the maximum extent, the trapezoidal steering mechanism is preset and checked. The trapezoidal arm length and bottom angle are calculated according to the empirical formula in the preset section. The inner and outer wheel angle angles are solved graphically in the checking section. The relationship curves are drawn.

3.4.4.1 Default. The size of the planned layout space and the axial force of the steering tie rod are the main factors affecting the length of the trapezoid arm. The axial force of the tie rod is inversely proportional to the length of the trapezoidal arm. Too small an arm's length makes it difficult to increase the axial force of the tie rod. According to the empirical Formulas (7) and (8), the preset value of the trapeze arm length m is 93.6mm. The preset value of the base angle γ is 55.6°.

$$m = (0.11 \sim 0.15)K \qquad (7)$$

$$\gamma = \arctan\frac{\frac{L}{2}}{\frac{K}{2} - b} \qquad (8)$$

where K is the center distance between the left and right kingpins; L is the wheelbase; b is the distance between the ground projection points at the upper and lower ends of the main pin.

3.4.4.2 Check the calculation. Different steering trapezoidal structures can be obtained by presetting different base angles and arm lengths. When the steering trapezoid size is initially determined, the angle relation should be checked to form the curve of the angle relation between the inside and outside wheel. The curve is compared with the theoretical curve of the angle relation between the inside and outside wheel to make the two as consistent as possible. The specific verification methods are divided into calculation methods and graphic methods. In order to make the process more intuitive, the graphic solution method is taken as an example to demonstrate. The graphic process is shown in Figure 6.

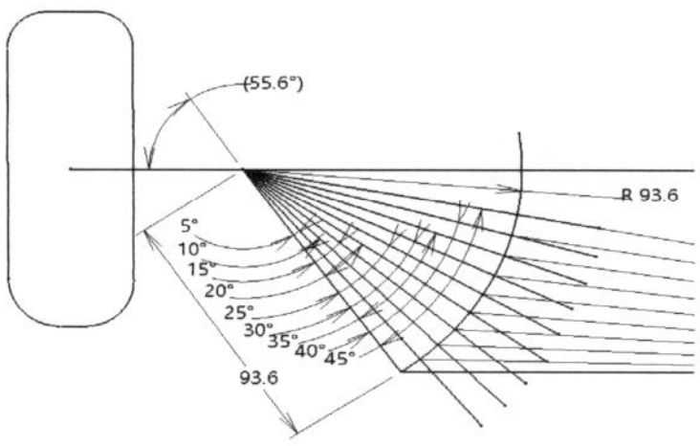

a Process of determining the inner wheel steering angle

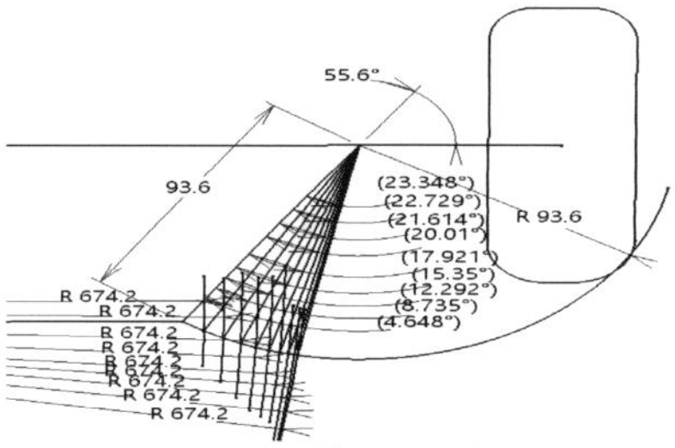

b Corner corresponding to the outer wheel

Figure 6 The process of checking the length of the bottom angle arm of the trapezoid mechanism.

The diagram process is as follows: ① We take respectively the left and right kingpin as the center of the circle, with the trapezoidal arm length as the radius of two upper arcs. ② We take the left wheel as the inner steering wheel, and draw the ray intersecting the left upper arc every 5°. ③ The intersection point of the left ray and the upper arc is the center of the circle. The length of the bottom side of the trapezoid is the radius of the arc. ④ We connect the right kingpin and the upper arc intersection points respectively, and the measured sngle between the line and the original trapezoid arm is the corresponding outer wheel angle. According to the angle of rotation of the inner and outer wheels, the relation curve of rotation of the inner and outer wheels is drawn to verify the calculation, as shown in Figure 8. According to Formula (9), R represents the Ackermann rate, γ_2 and γ_1 represent the actual inner wheel angle and actual outer wheel angle, θ_2 and θ_1 represent the theoretical inner wheel angle and actual inner wheel angle, then the limit position angle is brought in. The Ackermann rate is calculated to be 93.6%. The presupposition is reasonable.

$$\mathrm{R} = \left(\frac{\gamma_2 - \gamma_1}{\theta_2 - \theta_1}\right) \times 100\% \qquad (9)$$

3.5 Check of steering mechanism

3.5.1 Model construction. The steering mechanism is sketched using the CATIA sketch editor. When drawing, the parameters such as kingpin position, breakpoint position, trapezoid bottom angle, and trapezoid arm length are considered comprehensively. The influencing factors such as chassis space size and processing technology are taken into account. The building points for the attention of the model are as follows: ① The trapezoidal mechanism itself is fully constrained by each other. ② We constrain the kingpin point position with the coordinate system. ③ We add the distance constraint at the middle point of the trapezoid bottom edge, and simulate the movement process of the steering rack by changing the distance. (4) The inner and outer wheel angles are densified for reference to facilitate data observation and recording. Figure 7 shows the effect of building the simulation model of the steering ladder mechanism.

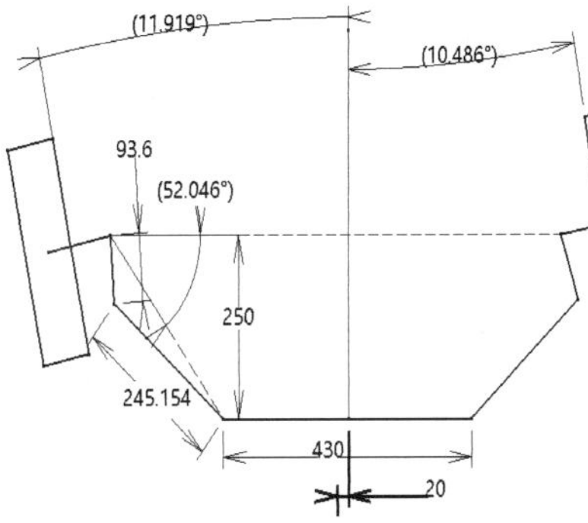

Figure 7 Steering ladder simulation model building effect.

3.5.2 Measurement of data. The corresponding inner and outer wheel angles are obtained by changing the drive length. The resulting inner and outer wheel angle relationship curves are drawn in Figure 8. The Ackermann rate of the simulation model is 82.7%. The minimum turning radius is 1.87m. The car steering meets the design requirements.

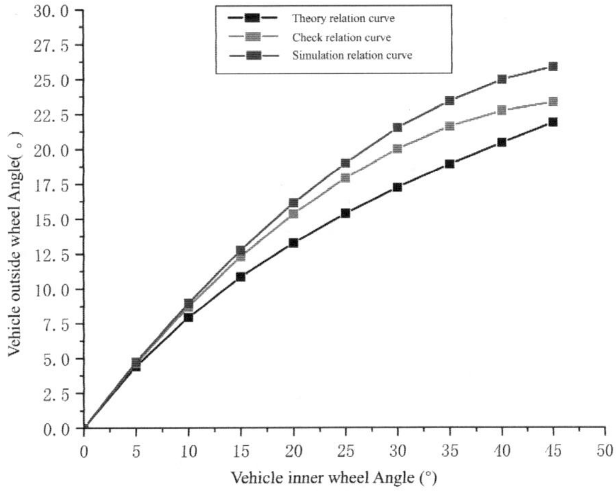

Figure 8 Theory, check calculation and simulation of the relationship between the angle of the inner and outer wheels.

3.5.3 Error analysis. Ackermann's steering relation is an ideal inner and outer steering relation, which is difficult to realize in actual production. The average passenger car Ackermann rate is about 60%. In the trolley steering ladder design work, the Ackermann rate obtained by theoretical calculation, presetting, checking calculation and simulation is different, the main reasons are as follows: (1) There are errors between theoretical calculation and geometric figure. The geometric figure cannot accurately express the relationship of theoretical calculation. (2) There is an error between the geometric figure and the rod design. The movement trajectory of the rod cannot be completely coincident with the movement trajectory of the geometric figure. (3) The size of the chassis space is limited. It is affected by design factors such as wheel base, wheelbase, and drive assembly layout during the design process, which cannot be completely designed according to the theoretical value and the checked calculation value. ④ It is necessary to comprehensively consider the relationship between the disconnect point position, the length of the tie rod, the length of the steering knuckler arm, etc., to ensure that all lengths are reasonable, and should not only meet a certain length and make the rest of the length close to the extreme.

4. DMU motion simulation analysis

CATIA is a CAD/CAE/CAM integrated software developed by the French company Dassault System, which is the mainstream design software in the world. Since its release, it has been widely used in the automobile manufacturing industry. Its surface design, structural design, parametric design, vehicle assembly, and motion simulation capabilities occupy a leading position in the automobile manufacturing industry [9]. The electronic prototype module in CATIA, namely the Digital Mock-Up module (DMU module), can simulate the motion process of the product, check the correctness and safety of the product assembly, and provide visual analysis of the product in different assembly states. With the application of the DMU module, users can conduct motion simulation, collision detection, gap analysis, etc., so as to avoid interference in the actual assembly process, ensure design quality, shorten the manufacturing cycle, and improve production efficiency [10].

4.1 Digital model establishment and preinstallation

The parts of the frame, suspension, upper U-arm, lower A-arm, steering knuckle, steering bar, wheel, brake, and so on are established respectively. When establishing the chassis parts' digital model, the axis of the rotating parts should be coaxial and the intersection of the axis should be copoint. When the established part is added to the product file, attention should be paid to adding the part to be simulated under the first level of the product to avoid the situation that constraints cannot be added in the constraint addition link. After the parts are added, the parts and components are preassembled by moving and intelligent moving commands to reduce the difficulty of adding moving pairs. Figure 9 shows the effect diagram of the car chassis modeling pre-assembly.

Figure 9 Chassis modeling preassembly effect.

4.2 Motion analysis and addition of motion pairs

The motion simulation analysis of the steering system and shock absorber system of the car is emphasized. The motion interference of parts that occur in the process of wheel beating and swing is analyzed. When using the DMU moving mechanism table to add constraints, there is a process of motion analysis, constraint point, line and surface drawing, degree of freedom calculation, etc.

The wheel and the steering knuckle remain stationary, using a rigid combination. The upper and lower kingpin ball pins and the steering cross tie rod ball pins are fixed on the steering knuckle through fasteners, using a rigid combination. When the wheel does the steering movement, the inner and outer steering wheels swing left and right are under the tension and thrust of the steering rod, and the steering knuckle and the steering rod are a U-shaped combination. The steering rack is simplified to a straight line fixed on the frame, the connection point between the rack and the tie rod is regarded as a point. The relative movement between the tie rod and the rack is realized by combining the point-curve and adding the driving length.

When the wheel jumps due to the unsmoothness of the road surface, the ball pin on the steering knuckle and the U-arm ball head are spherically combined. The U-arm and the frame are rotating combined in a swinging state. The upper end of the U-shaped arm is connected to the shock absorber system. The lower shock absorber barrel is combined with the U-shaped arm. The upper and lower shock absorber barrel is combined with a cylinder to simulate the tensioning process of the shock absorber. The upper shock absorber barrel is combined with a cylinder of the frame. The steering knuckle-down ball pin is spherical combined with the A-arm ball head. The A-arm is rotary combined with the frame. To simulate wheel runout, a drive angle is added to the rotation combination of the A-arm and the frame.

In the process of adding motion pairs, in order to conveniently select the points, lines, and planes to be constrained, relevant parts can be activated to draw the central rotation points, axes, and reference planes. The drawn elements are all parameterless elements. The parts can be colored for easy observation. After adding all the motion pairs, the degree of freedom of the mechanical device is 0, and the table pops up for simulation. Taking the left front wheel of the car as an example, the addition of main motion pairs is shown in Table 2.

Table.2 The main movement pair of the left front wheel of the car is added.

Part 1	Part 2	Kinematics Joints
The upper ring of the shock absorber	Upper frame hinge	Cylindrical Joints
Upper shock absorber cylinder	Lower shock absorber cylinder	Cylindrical Joints
The lower ring of the shock absorber	U-shaped arm	Revolute Joints
U-shaped arm	Mid frame hinge	Revolute Joints
U-shaped arm ball head	The ball pin on the steering knuckle	Spherical Joints
Middle ball pin of steering knuckle	Steering tie rod ball joint	Universal Joints
Steering rack	Steering tie rod	Point Curve Joint
Steering joint ball pin	A-type arm ball head	Spherical Joints
A-arm	Under frame hinge	Revolute Joints

4.3 Driver add and edit simulation

In order to ensure good mobility and maneuverability of the car, the design tire runout height is ±90 mm, and the driving angle is ±9° on the lower A-arm rotating pair after calculation. The limit angle of the inner tire is 45°. The driving length of 94 mm is calculated on the point-curve combined motion pair. By using the editing simulation command, the key frame of the wheel swing and jump process is inserted at the same time. The projection step is adjusted to 0.01. Finally, playback is generated. Driver adds and edit simulation effects are shown in Figure 10.

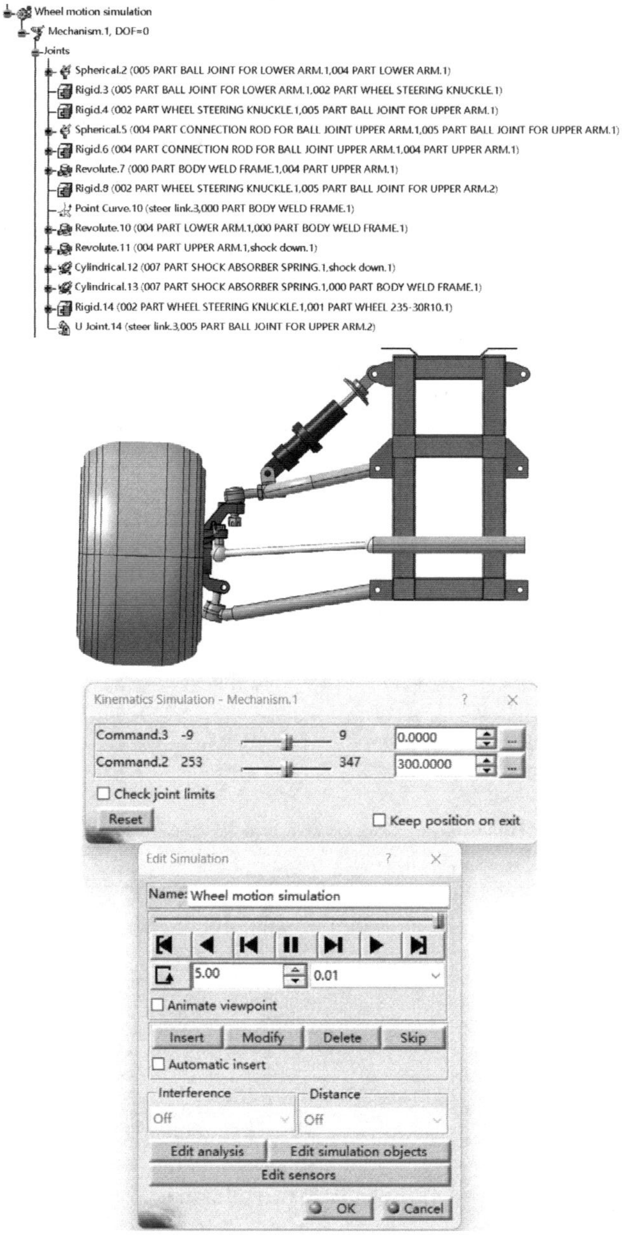

Figure 10 Drive add and edit simulation effects.

4.4 Analysis of envelope and distance region

On the basis of generating and reproducing the mechanical device, the key moving parts are analyzed by sweeping envelope and distance region. The sweeping envelope command and distance region analysis command can visualize the sweeping volume, movement trajectory, and relative distance of the mechanical device in the process of motion, which is convenient for interference inspection and assembly size verification. The following takes the left front wheel of the car as an example to analyze and check.

4.4.1 Envelope analysis. The steps for generating and analyzing the bounding and swinging sweep envelope of tires and steering tie rods are as follows: ① We apply the sweep envelope command, select the tire and steering tie rod for the product to be swept, and select other relative moving parts for reference. ② We set the filtering accuracy reasonably. If the setting is too large to affect the scanning precision, or too small to affect the calculation efficiency, we usually set to 0.2mm. ③ We form a preview and save the envelope as a ".cgr" file. ④ We switch back to the assembly table and insert the envelope in the original part position. (5) We observe and analyze the envelope continuity, path smoothness, integrity, volume change rate, surface quality, and edge feature retention, and determine whether interference occurs. The effect of sweeping the envelope during the movement of the wheel and steering tie rod is shown in Figure 11. The envelope is complete and good without interference, which meets the design requirements.

Figure 11 Wheel, steering tie rod sweep envelope body effect picture.

4.4.2 Distance region analysis. The parts that are prone to collision and interference are selected for distance region analysis. The steps are as follows: ① We adjust the left front wheel to the swinging and jumping extreme motion position by simulating the command. ② We use the range area analysis command to select the parts to be measured respectively. ③ The measured distance is compared with the design distance to complete the check. The measured minimum clearance between the wheel and the body is 13.53mm. The minimum clearance between the wheel and the steering tie rod is 32.37mm. No interference occurs, which meets the design requirements. The effect of distance region analysis is shown in Figure 12.

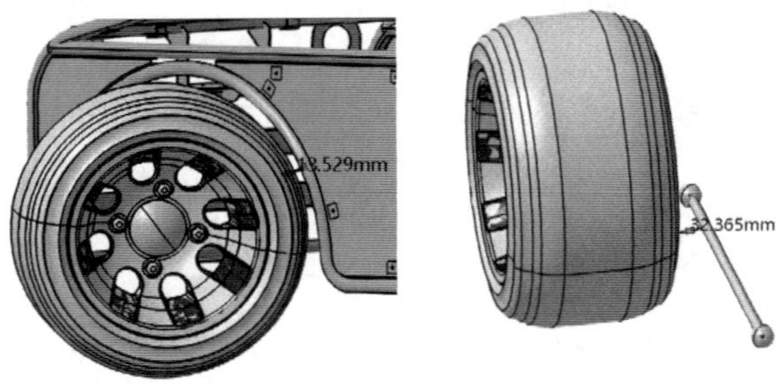

Figure 12 Wheel, steering tie rod distance analysis effect.

5. Whole vehicle test

The trial production effect of the sample car is shown in Figure 13.

Figure 13 Prototype production effect drawing.

5.1 Real car Ackermann rate assessment

We check the tire pressure and choose a flat road for testing. In the course of steering, the protractor is used to measure and record the degree of wheel angle inside and outside. We make the inner steering wheel value every 5° and measure the corresponding outer wheel angle degree. The angle relationship curves of the inside and outside wheels of the real vehicle are drawn in the angle relationship diagram, as shown in Figure 14. The calculated real Ackermann rate is 72.2%.

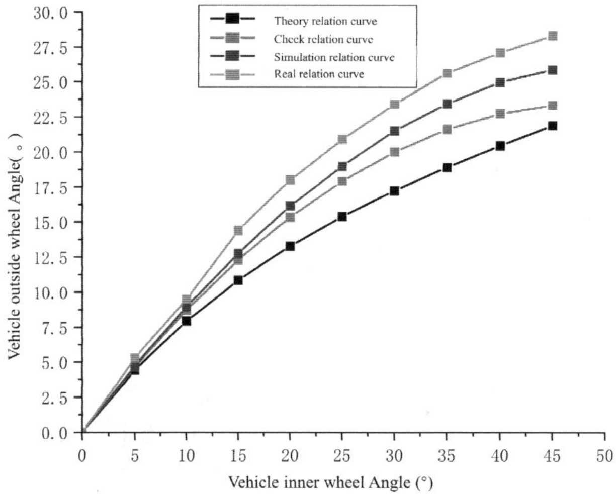

Figure 14 Theory, check calculation, simulation, real interior and exterior wheel Angle relationship diagram.

5.2 Real car turning radius evaluation

In the process of steering, when the wheel turns to the limit angle position, we mark the position of the front, rear, and outer wheels, keep the limit angle unchanged, and drive the car to continue driving. When the horizontal line of the front and the initial horizontal line of the rear coincide, we mark the position of the outer wheel again. The straight line connected to the position of the outer wheel is the minimum turning diameter, as shown in Figure 15. The measured minimum turning radius is 2.1m, which meets the design requirements.

Figure 15 Minimum turning radius measurement procedure.

6. Conclusions

In order to meet the teaching needs of "integration of science and practice" in middle and higher vocational colleges, a new energy drive-by-wire four-wheel steering car was developed. This paper focuses on the design process of the four-wheel steering system of the car and makes presets for the four-wheel angle relationship, steering trapezoid structure, turning radius, and other parameters. Through checking and checking, it is obtained that the design Ackermann rate of the car is 82.7% and the design turning radius is 1.87m. CATIA digital assembly table is used to conduct motion simulation analysis of the steering system and shock absorber system of the trolley, generate an envelope of key parts such as tires and tie rods, and check the clearance to ensure normal movement of the mechanism without interference. Through the trial production of the whole vehicle, the measured Ackermann rate of the real vehicle is 72.2%, and the turning radius is 2.1m, which conforms to the design objective, and verifies the reliability of the design method. This small car will be applied to vocational college training teaching. This design method will provide new ideas for the teaching of automotive design discipline.

References

[1] WANG S Y. Review on the Development Status of New Energy Vehicles [J]. Internal Combustion Engine and Parts, 2024(05): 135-137.
[2] Wang G C, Gao S Q. Research Review of Automotive Steering-by-wire System [J]. Automotive Digest, 2024(03): 9-20.
[3] Li B, Bai Y F, Li S S, et al. Research on Design and Control Method of four-wheel Omnidirectional Steerable by Wire System [J]. Mechanical Engineer, 2022(12): 143-148+152.
[4] AN F Q, ZHAO H L, CHENG Z, et al. Development status and research progress of power battery for pure electric vehicles [J]. Chinese Journal of Engineering, 2019, 41(1): 22-42.
[5] Zou L Y, Zhu Z M, Chen X X. Analysis on Key Technologies of Automobile Steering-by-wire [J]. Internal Combustion Engine and Accessories, 2024(04): 106-108.
[6] WANG W Y. Automobile Design [M]. Beijing: China Machine Press, 2011: 232-233.
[7] Wang R C, Yin Z K, Liu Z S, et al. Optimization Design and Analysis of Traction Steering Mechanism of Loading Transfer Vehicle [J]. Journal of Mechanical Transmission, 2021, 45(05): 42-49.
[8] Li P, Yang X H, Fang Q F, et al. Steering Ladder Optimization Design of Four-wheel Steering Vehicle Based on Ackermann Theory [J]. Beijing Automobile, 2021(03): 16-20.
[9] Tang M D, Suo C, Liu M G. Application of CATIA in the Design of Aircraft Iron Bird Platform [J]. Public Standardization, 2024(08): 146-148.
[10] Zhao Y, Chen Y, Lv Y, et al. Automotive Steering System and Its Controller Design for Intelligent Vehicles [J]. SAE International Journal of Sustainable Transportation, Energy, Environment, & Policy, 2023, 4(13-04-02-0011): 177-188.

Optimization design of wind turbine blade structure

Yuhua Hao*and Xiujuan Zhang

Dalian Jiaotong University, Dalian, Liaoning, 116028, China

*Corresponding author's e-mail: 1799215415@qq.com

Abstract: With the intensification of the global energy crisis and the increasing severity of environmental pollution, the development and utilization of new and renewable energy sources have become an inevitable trend. Wind energy, with its advantages of being locally available, widely distributed, pollution-free, clean, and renewable, has become an important direction for the development of new energy. The blades of wind turbines are the most critical components of wind power generation systems. The airfoil and structural type of the blades directly affect the power and performance of the wind power generation devices. Therefore, optimizing the structure of wind turbine blades to reduce deformation, improve the utilization of wind energy, and extend service life has significant practical importance, leading to the proposal of optimized design for wind turbine blades.

1. Introduction

Faced with the intensification of the global energy crisis and the increasing severity of environmental pollution, the development and utilization of new and renewable energy sources have become an inevitable trend. Among the development of various renewable energy technologies, wind power generation technology stands out as particularly important. The blade is a crucial component of wind turbines. Kim et al. [1] optimized the cross-sectional dimensions of wind turbine blades through fluid dynamics analysis. Nong [2], based on the drag reduction theory of convex hull non-smooth surfaces, used numerical simulation methods to analyze the drag reduction performance of non-smooth blades under different wind speeds and attack angles and selected the variant convex hull structure with the best drag reduction performance. Ali et al. [3] introduced an optimization method for the design of uniform blade structures used in vertical-axis wind turbines. The blade cost accounts for 20% of the total turbine cost, with inertia loads being the dominant design loads. The aim is to optimize weight and cost while maintaining structural integrity. Jeon et al. [4] developed a fully coupled simulation model to analyze the unsteady aerodynamic behavior of wind turbines. Herrmann and Bangga [5] used genetic algorithms to optimize the DU00-W-401 airfoil to enhance its aerodynamic performance. Wang et al. [6], based on the fundamental information of existing blades, have carried out the design and structural simplification of flexible blades and established a fluid-structure interaction response analysis framework for flexible blades. Zhang [7] used three-dimensional modeling and CFX fluid-structure coupling technology to analyze the pressure loads of horizontal axis wind turbines. Tang [8] adopted an improved Mix-up data augmentation method to address the issue of class imbalance in wind turbine SCADA data, achieving early warning of icing faults in wind turbine blades. Xu [9] from Lanzhou University of Technology carried out design optimization of the wind turbine blades for a 1.5MW horizontal axis wind turbine. Chen et al. [10], through the study of the wake flow field of horizontal axis wind turbine blades, found that optimizing the tip airfoil layout can improve aerodynamic performance.

In this paper, the optimized blade structure is analyzed using ANSYS software to verify the reliability of the blades. These studies indicate that the optimized blade design will not experience resonance phenomena and can effectively improve the utilization rate of wind energy, which is of great significance in addressing the energy crisis.

2. Finite element analysis (FEA)

2.1 Finite element method

With the development of computer technology, the Finite Element Method (FEM) has been increasingly applied in the structural analysis of blades. Its powerful modeling and structural analysis capabilities are suitable for the analysis of stress, deformation, frequency, buckling, fatigue, and root strength of blades, providing the details needed for the final blade design.

2.2 Modal analysis

Modal analysis is the most fundamental aspect of all dynamic analyses, primarily used to determine the vibration characteristics—natural frequencies and mode shapes—of designed structures or machine components, allowing the structural design to avoid resonance or vibrancy at specific frequencies. Additionally, modal analysis can serve as a starting point for other dynamic analysis problems, such as transient dynamic analysis, harmonic response analysis, and spectral analysis.

Natural frequency analysis is typically conducted using the Finite Element Method. Inertia forces are forces related to the mass and acceleration of an object. The total inertia force acting on an object is:

$$\{F(t)\} = \sum_{e=1}^{n0}\{F(t)\}_1^e = -\sum_{e=1}^{ne}[M]^e\xi\ddot{u}\{(t)\}^e = -[M]\{\ddot{u}(t)\} \tag{1}$$

$$[M] = \sum_{e=1}^{n_0}[M]^e = \sum_{e=1}^{n_0}\iiint \rho[N]^T[7V]\,dV \tag{2}$$

$[M]^e$——Module quality matrix

$\{F\ (t)\}_1^e$——Inertial forces of the ut

n_0——Total number of units of vibrating structures

The total damping acting on the object is:

$$\{F\{t\}\}_d = \sum_{e=1}^{n_0}\{F(t)\}_d^e = -\sum_{e=1}^{n_0}[C]^e\{\dot{u}(t)\}^e = -[C]\{u(t)\} \tag{3}$$

$$[C] = \sum_{e=1}^{n_0}[C]^e = \sum_{e=1}^{n_0}\iiint \mu[N]^T[N]d_v \tag{4}$$

The total elastic force acting on an object:

$$\{F(t)\}_e = \sum_{e=1}^{n_0}\{F(t)\}^e = -\sum_{e=1}^{n_0}[k]^n\{u(t)\}\frac{e}{0} - [k]\{u(t)\} \tag{5}$$

$$[k] = \sum_{e=1}^{n_0}[k]^e = \sum_{e=i}^{n_0}\iiint [B^e]^T[D][B^e]\,dv \tag{6}$$

From D'Alembert's principle, the forces acting on an object at any instant of time form an equilibrium system of forces, then:

$$[M]\{\ddot{u}(t)\} + [C]\{\dot{u}(t)\} + [k]\{u(t)\} = \{F(t)\} \tag{7}$$

For undamped free vibration, both the damping term and the external load term are zero, so Equation (7) becomes:

$$[M]\{\ddot{u}(t)\} + [k]\{u(t)\} = 0 \tag{8}$$

We let the solution of Equation (8) be:

$$\{u(t)\} = \{X\}\sin wt \tag{9}$$

{X}——The column vector of amplitudes of the displacement {u(t)}
ω——Natural oscillation rate
Bringing Equation (9) into Equation (8) yields:

$$([k]-\omega^2\{M\})\{X\}=0 \tag{10}$$

Let ω^2 = λ, then we have:

$$([k]-\lambda\{M\})\{X\}=0 \tag{11}$$

For Equation (10) to yield a non-zero solution, it must be treated as a chi-square equation. The determinant of the matrix ([k]-λ{M} should be zero, i.e.

$$\det([k]-\lambda\{X\})=0 \tag{12}$$

From this, n generalized eigenvalues λ_i=(i=1, 2...n) can be determined and the nth order intrinsic frequency of the structure can be expressed as:

$$w_i=\sqrt{\lambda_i}$$

3. Building 3D models

3.1 Model building

In this paper, the NACA64 series airfoil model is selected for optimization, and Solidworks 3D modeling software is appliedto model the optimized wind turbine blade. In this study, Glass Fiber fiber-reinforced plastic (GFRP) is selected as the material for wind turbine blades. GFRP is well known for its low density, which makes the material not only lightweight but also strong. In addition, GFRP's wind pressure resistance meets or exceeds international standards. It is due to GFRP's superior performance in terms of strength and quality that it is widely used in contemporary wind turbine blade manufacturing. The elastic modulus E of the material is 17.6 GPa, the Poisson's ratio v is 0.17, and the tensile strength is approximately between 350 MPa and 450 MPa.

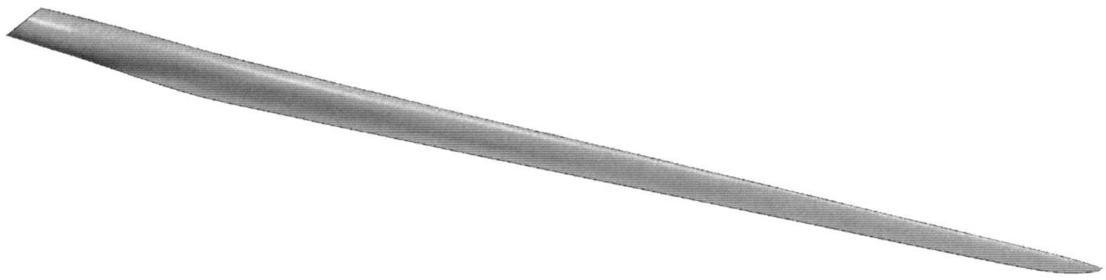

Figure 1. 3D model of wind turbine blades.

3.2 Meshing

The accuracy of simulation analysis relies on the computation of grid nodes. The quality of the grid directly impacts the simulation outcomes and computation time. For complex simulation models, generating high-quality grids is extremely time-consuming and error-prone. To enhance the precision and efficiency of the calculations, this paper employs local mesh refinement, particularly on the surface of the blade, utilizing hexahedral meshing of higher quality on the blade surface to ensure the accuracy of the calculations. The workflow within the software includes operations such as generating surface meshes, describing geometric structures, adding boundary layers, and generating volume meshes, with a total of 341, 091 nodes and 296, 045 grids. The final finite element mesh model is shown in Figure 2.

Figure 2. 3D Modeling of mesh generation.

4. Modal analysis solution

When wind turbine blades resonate, they are subjected to severe vibration damage. Therefore, it is necessary to analyze the blades to determine their frequencies and vibration modes under various conditions. Initially, full constraints are applied to the entire surface at the east end of the blade, with the tip being free. A load of 0.01 MPa is applied at the blade tip. Under the influence of a strong wind load at a wind speed of 20 m/s, without considering the effects of material temperature, we can observe from Figure 3 and Figure 4 that the overall deformation and equivalent elastic strain of the blade are both within the reasonable range.

Figure 3. Overall deformation.

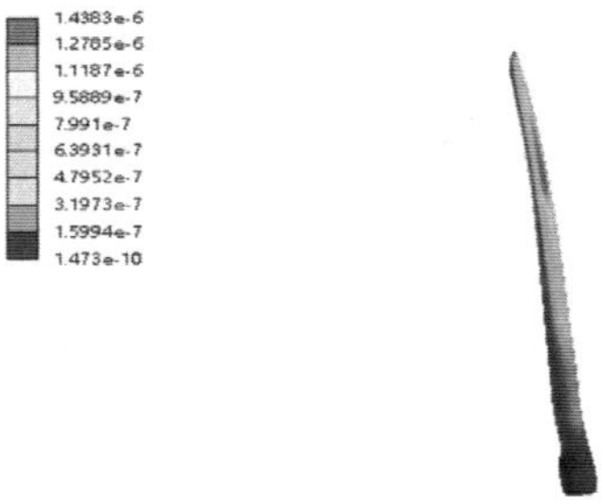

Figure 4. Equivalent Elastic Strain.

Through finite element analysis, different loads are applied to extract the first eight vibration modes. By solving each mode, the first eight vibration mode shapes of the blade under loads of 0.01 MPa and 0.05 MPa are obtained as follows.

The following figure shows the first eight mode shapes under a load of 0.01 MPa:

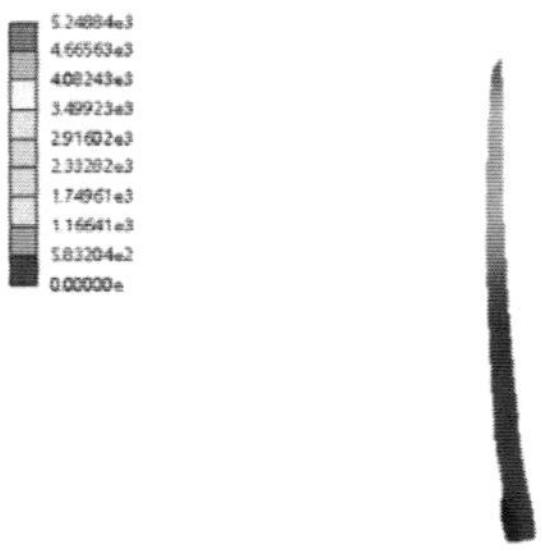

Figure 5. first-order mode.

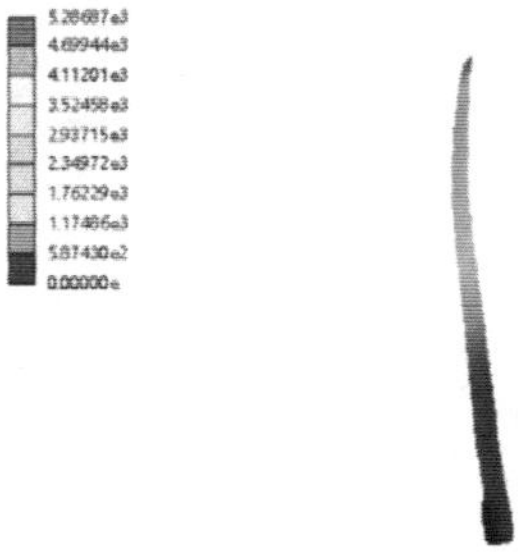

Figure 6. Second-order mode.

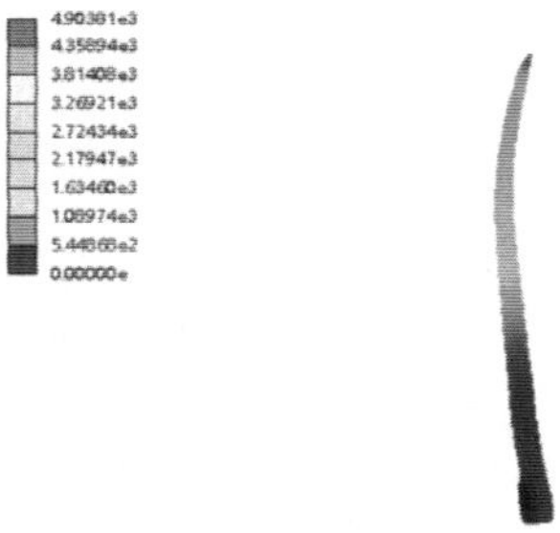

Figure 7. Third-order mode.

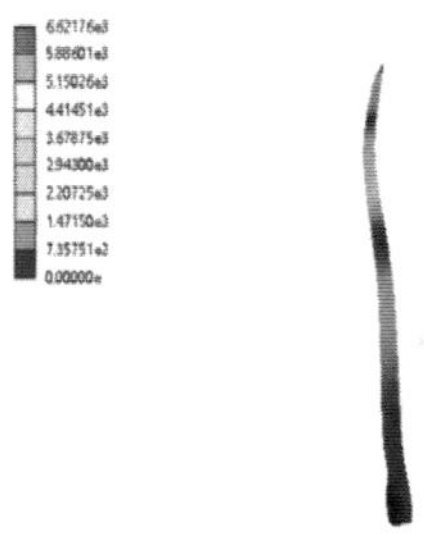

Figure 8. Fourth-order mode.

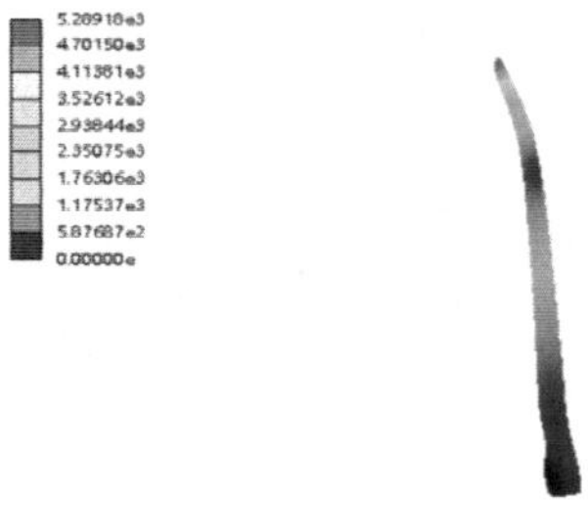

Figure 9. Fifth-order mode.

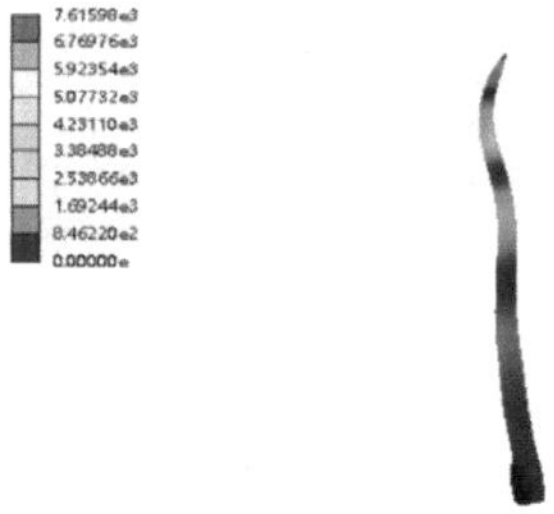

Figure 10. Sixth-order modes.

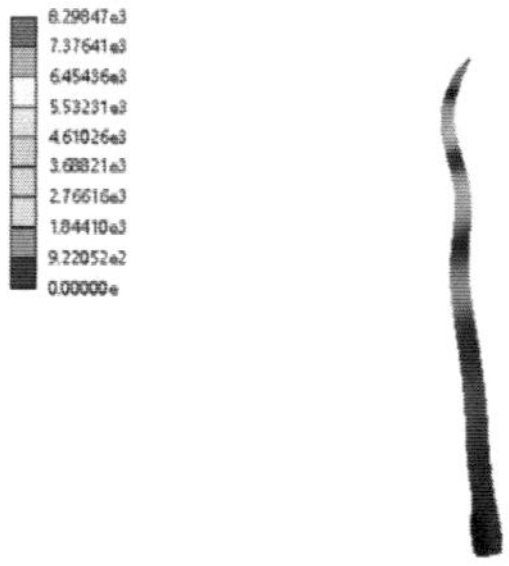

Figure 11 seventh-order mode.

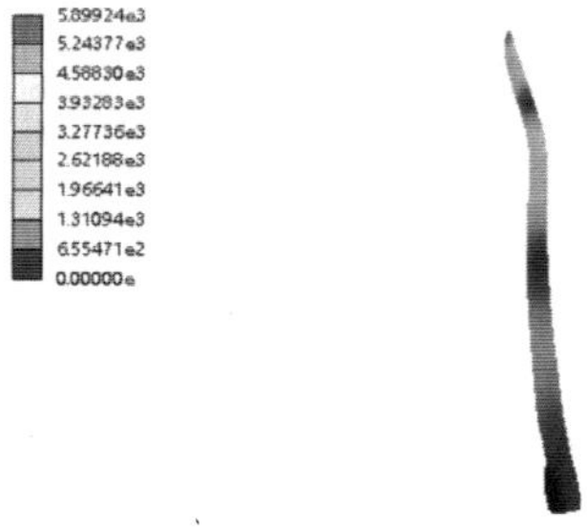

Figure 12. eighth-order mode.

Due to the subtle changes, only the first six Mode Shapes Under A Load of 0.05 MPa are presented below.

Figure 13. First-order mode.

Figure 15. Second-order mode.

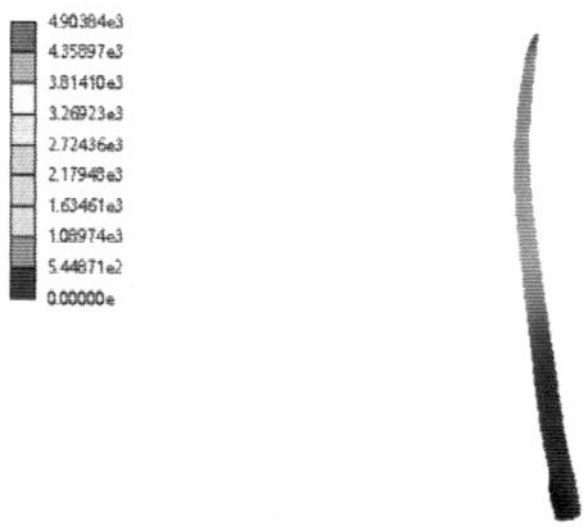

Figure 16. Third-order mode.

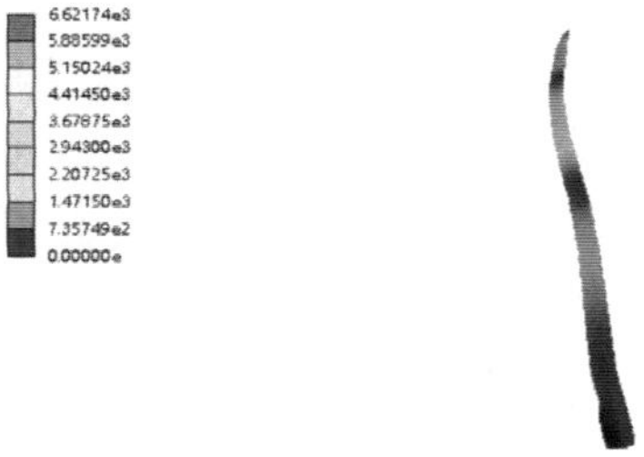

Figure 17. Fourth-order mode.

Figure 18. Fifth-order mode.

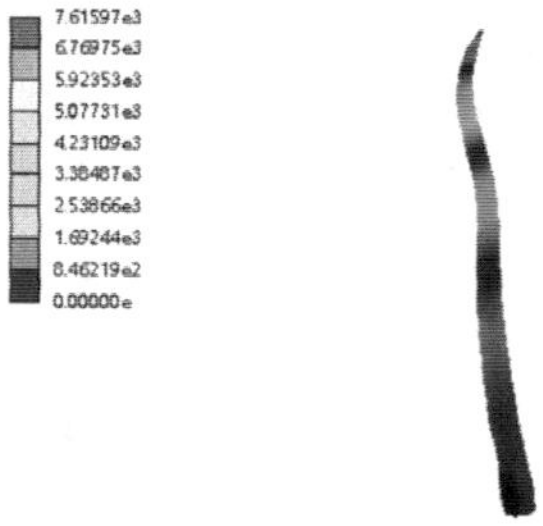

Figure 19. Sixth-order modes.

When in the case of pre-stress constraints, the rated speed of the blade is 12.5 r/min. The following table shows a comparison of some of the working conditions of the two comparisons comparison.

Table 1. Results of modal analysis conclusions.

Serial number	prestressed	unprestressed	Vibration pattern
1	3.5241	2.4532	stagger
2	5.1452	3.4658	stagger
3	8.1954	6.2983	stagger
4	11.2985	9.2643	Stagger、oscillaion

5. Conclusion

This project is dedicated to analyzing the first eight modes of the wind turbine blade, as the operating conditions of the blade are primarily reflected in these lower-order modes. Comparing Figures 5 to 19, it can be observed that the main modes manifest as both stagger and oscillation, with some modes encompassing both stagger and oscillation, indicating a coupling of the two forms. Through modal analysis of the wind turbine blades under different loads and with or without pre-stressed conditions, the data in Table 1 confirm that resonance will not occur in these wind turbine blades during operation, thus meeting the requirements of the design specifications.

References

[1] Dong-Hoon Kim, O-Kaung Lim, Eun-Ho Choi, Yoojeong Noh. Optimization of 5-MW wind turbine blade using fluid structure interaction analysis [J]. Journal of Mechanical Science and Technology, 2017, 31(2)

[2] Nong Yuchang. Research on the Structural Optimization of Wind Turbine Blades for Low Wind Speed in Complex Terrain [D]. Guizhou University, 2022. DOI: 10.27047/d.cnki.ggudu.2022.001112.

[3] Geneid Ahmed Ali, Atia Mostafa R. A., and Badawy Ahmed. "Multi-objective optimization of vertical-axis wind turbine's blade structure using genetic algorithm." Journal of Engineering and Applied Science 69.1(2022):

[4] Minu Jeon, Seungmin Lee, Soogab Lee. Unsteady aerodynamics of offshore floating wind turbines in platform pitching motion using vortex lattice method [J]. Renewable Energy, 2014, 65

[5] Jordis Herrmann, Galih Bangga. Multi-objective optimization of a thick blade root airfoil to improve the energy production of large wind turbines [J]. Journal of Renewable and Sustainable Energy, 2019, 11(4)

[6] Wang Xiaozhe, Wang Yibo, Wu Kun, & Wan Zhiqiang. Structural optimization of flexible blades for vertical axis wind turbines. Journal of Beijing University of Aeronautics and Astronautics, 1-13.doi:10.13700/j.bh.1001-5965.2024.0444.

[7] Zhang Zhiyang. Research on the wake effect of horizontal axis wind turbine [D]. Shihezi University, 2020.

[8] Tang Wenyang. Research on Structural Health Monitoring of Offshore Wind Turbine Blades [D]. Yancheng Institute of Technology, 2023. DOI: 10.44381/d.cnki.gycit.2023.000061.

[9] Zhu Xu. Optimized design and finite element analysis of large wind turbine blades [D]. Lanzhou University of Science and Technology, 2013.

[10] Chen Kai, Zhang Zhenyu, WANG Tongguang, HUANG Huixiu. CFD-based study of power enhancement of horizontal-axis wind turbine blade tiplets [J]. Journal of Solar Energy, 2021.

Journal of Physics: Conference Series **2992** (2025) 012010 doi:10.1088/1742-6596/2992/1/012010

Research on the modeling of thin-walled casing bolted flange connections considering nonlinear effects

Yansheng Wang *[1] and Yuefang Wang [2]

[1] College of Mechanics and Aerospace Engineering, Dalian University of Technology, China
[2] College of Mechanics and Aerospace Engineering, Dalian University of Technology, China

*Corresponding author's e-mail: 2440937392@qq.com

Abstract: The bolted flange connection structure in thin-walled casings is widely used in mechanical assemblies. However, finite element simulation of bolted connections requires the inclusion of a large number of nonlinear contacts, leading to long computational times and convergence difficulties. Therefore, it is meaningful to develop an equivalent model that is computationally efficient and easy to converge. One approach is to use a spring-damping element to model the contact interface. The most commonly used model for the spring-damping element is the Iwan model, which is a commonly used method for simulating nonlinear and hysteresis behavior. Firstly, the cyclic loading of the contact model is investigated to explore the nonlinear characteristics of the interface. The tangential sliding hysteresis curve of the interface under cyclic loading is obtained. The Iwan parameters are then obtained by discretizing the hysteresis curve. A spring-damping element equivalent modeling method for bolted flange connections is proposed. Secondly, the effectiveness of the equivalent model's hysteresis behavior is verified under different loads, with a correlation error of less than 1% and an area error of the hysteresis curve of less than 9%. Finally, a modal analysis of the equivalent model is performed. The results show that the equivalent model has high computational efficiency and can be used for equivalent modeling of thin-walled casing bolted connections.

1. Introduction

Bolted flange connection structures are widely used in mechanical assemblies, and bolted contacts represent a form of nonlinear contact that results in energy dissipation and damping within the structure [1]. Although linearization methods are frequently employed when modeling and analyzing nonlinear connecting structures, these methods do not accurately describe the nonlinear behavior of bolted connection structures nor do they fully capture the nonlinear response phenomena of bolted structures [2]. Therefore, to conduct a more precise analysis and computation of bolted connection structures, it is necessary to establish a model that can parameterize and describe the nonlinear characteristics of bolted structures [3]. Iwan [4] proposed a method for modeling the hysteretic characteristics of bolted connection structures using a continuous Iwan model. Zhu et al. [5] constructed six-parameter Iwan model parameters based on experimental identification. Their results indicated that the cyclic energy dissipation of the model decreases as the bolt diameter increases. Mao et al. [6] introduced an equivalent model using spring elements to simulate threaded connections, employing the super-element method to extract the stiffness matrix for determining spring stiffness. Their findings showed that this equivalent model significantly enhances the modeling and computational efficiency of the threaded connections while accurately reflecting their mechanical characteristics. Yang et al. [7], based on experimental data, applied the Iwan model theory to construct a nonlinear numerical dynamic model and solved it using the fourth-order Runge-Kutta method. Their results indicated that the model effectively characterizes the modal characteristics of bolted connection structures. Chen et al. [8] proposed substituting solid thin layers with virtual thin layers for the equivalent modeling of engine casing bolt connections. The comparison of modal shapes demonstrated that the virtual thin layer element method effectively simulates bolted connections and successfully avoids axial position deviations in the engine casing caused by thin layer thickness. Liu et al. [9] established an improved equivalent model for the connecting layer of bolted connection beams, showing that this model can predict the nonlinear behavior of bolted connection assembly structures. You et al. [10] proposed a simplified modeling method for bolted connections using a combination of beam and spring elements, employing dynamic coupling constraints and an equivalent spring connection model for bolted contacts.

This paper presents a finite element modeling of the bolted flange interface in thin-walled casing structures and proposes an equivalent modeling method using spring-damping elements for the bolted interface. The most commonly used model for the spring-damping element is the Iwan model. To investigate the hysteresis characteristics of the restoring force and displacement of the bolted interface, a modified discrete Iwan model is adopted, and the hysteresis curve is discretized to obtain the parameters of the Iwan model. Subsequently, an equivalent model of the bolted flange interface is established in ANSYS software. Compared to the bolt contact model, the equivalent model improves computational efficiency while ensuring key details. It can be used for equivalent modeling of the bolted flange connection in thin-walled casing structures.

2. Theoretical foundation

2.1. Jenkins element

The Jenkins unit is an idealized elastoplastic unit consisting of a linear spring with stiffness k connected in parallel with a slider with a critical slip force f^*, as shown in Figure 1.

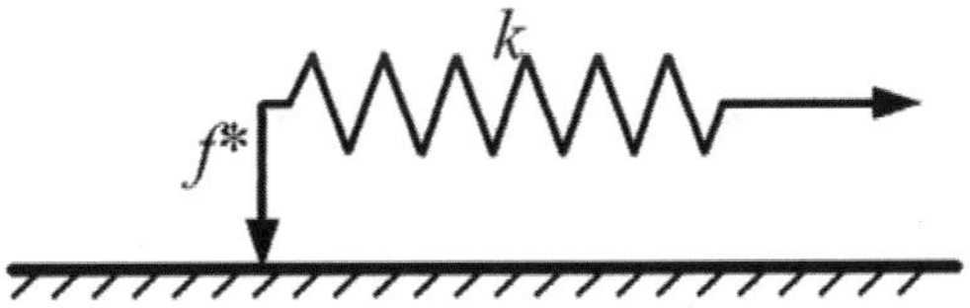

Figure 1. Jenkins element.

The loading end is referred to as the active end, and the end where the slider is located is referred to as the sliding end. When F is less than the critical slip force f^*, the spring deforms, the slider position remains unchanged, the external force F and the active segment displacement exhibit a linear relationship, with stiffness equal to the stiffness of the linear spring k, and the unit's restoring force being equal to F; when F reaches the critical slip force f^*, the spring deformation no longer increases, and the maximum deformation is f^*/k, denoted as the unit's yield displacement x^*, and the slider begins to slide, with the unit's restoring force reaching its maximum value f^*. Based on the above analysis, the restoring force-displacement relationship of the Jenkins unit can be expressed by Equation (1). According to the above analysis, the restoring force of a single Jenkins unit is given by:

$$f=\begin{cases}kx, -x^{*}\le x\le x^{*}\\ f^{*}, x<-x^{*} \text{ or } x\ge x^{*}\end{cases} \tag{1}$$

When deriving the relationship between the restoring force and displacement of the Jenkins unit, the inertia of the slider and spring are neglected, and the maximum restoring force of the spring is equal to the critical slip force of the slider. When the restoring force of the spring is less than the critical slip force of the slider, the slider is in a stationary state, and only the spring stretches; when the restoring force of the spring is equal to the critical slip force of the slider, the spring no longer stretches, and the slider is in a sliding state.

2.2. Discrete Iwan model

The discrete Iwan model is composed of N Jenkins units and a spring-damping unit connected in parallel, as shown in Figure 2.

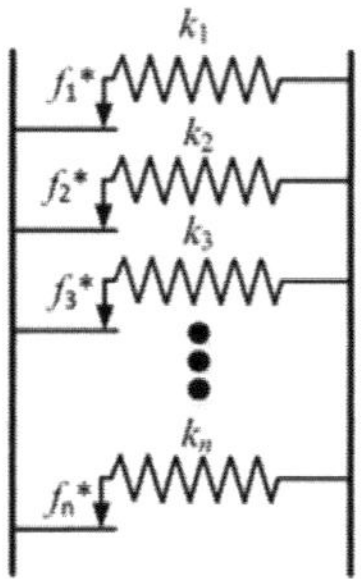

Figure 2. Discrete iwan model schematic.

In a system of N Jenkins units with spring stiffness values k_1, k_2..., and k_N, we assume that the yield displacement of the units satisfies Equation (2).

$$x_1^{*}\le x_2^{*}\le..\le x_{N-1}^{*}\le x_N^{*} \tag{2}$$

When an external force F is applied to the active end of the discrete Iwan model, the N Jenkins units and the spring-damping unit experience the same displacement. The sum of their restoring forces constitutes the restoring force of the discrete Iwan model. Under static cyclic loading conditions, when the free end of the discrete Iwan model moves from its initial position to a displacement magnitude of A and then reverses the load, the relationship between the restoring force and displacement is given by Equation (3). In this equation, the first term represents the restoring force provided by the Jenkins units that yielded during the initial loading phase and also yielded during the reverse loading phase. The second term accounts for the restoring force from those Jenkins units that yielded during the initial loading phase but did not yield during the reverse loading phase. The third term corresponds to the restoring force from Jenkins units that did not yield in either the initial or the reverse loading phases. The third term represents the restoring force provided by the spring units.

$$f(x)=\sum_{i=1}^{l} f_i^{*}+\sum_{i=l+1}^{p}\left[k(x-A)+f_i^{*}\right]+\sum_{i=p+1}^{N} k_i x+k_a x, -A\le x\le A \tag{3}$$

For the discrete Iwan model consisting of 3 Jenkins units and 1 linear spring unit, its restored force-displacement curve under cyclic loading is shown in Figure 3.

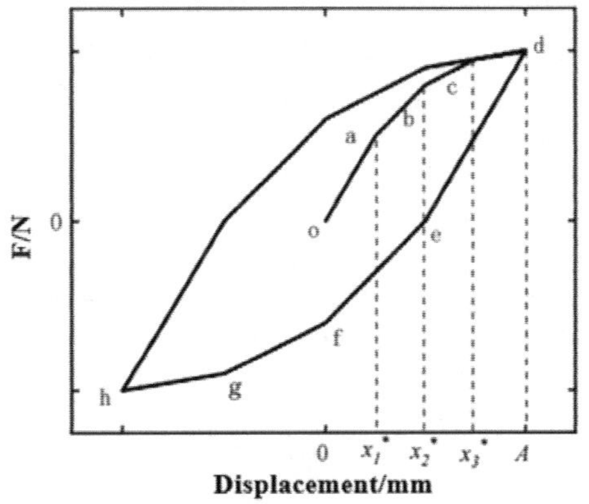

Figure 3. The hysteresis curve of the discrete Iwan model.

At the initial loading stage, Points a, b, and c, the 3 Jenkins units reach their yield point in sequence, and the model stiffness decreases. The stiffness in the cd segment is equal to that of the linear spring.

3. Contact model simulation

The research object of this study is the bolted connection structure of a thin-walled casing (millimeters). As shown in Figure 4, the structure mainly consists of upper and lower casings, upper and lower flanges, and 24 M6 bolt components. To simplify the modeling, the screw is represented as a standard cylinder, neglecting the threads inside the nut. The nut and screw are modeled as continuous components.

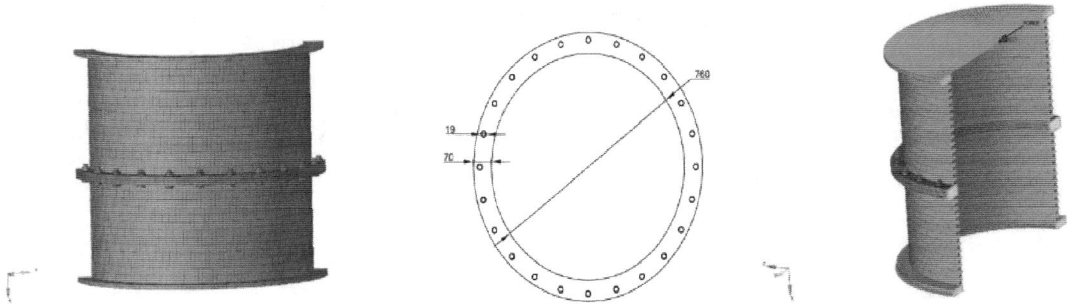

Figure 4. (a) Finite element model (b) Bolt connection interface geometric model (c) Boundary condition.

Table 1. Material mechanical performance parameters.

Material	Young's modulus (Mpa)	Poisson's ratio	Density (tone/ mm^3)
LY12	7.1E4	0.33	2.7E-9
45#steel	2.1E5	0.3	7.9E-9

The materials used for the housing and flange are aluminum alloy LY12, while the bolt and nut are made of 45# steel. The main mechanical performance parameters of the material used for the bolt-flange connection structure are shown in Table 1.

The unit types of the flange and bolt-nut parts are both SOLID185 elements. The shape of the elements is hexahedral. Frictionless contact is set between the screw and the flange hole. Frictional contact is used between the upper and lower casings and the flange, between the upper and lower flanges, between the nut and the upper flange, and between the bolt cap and the lower flange. Since the static friction coefficient between aluminum and steel under unlubricated conditions is 0.17, the friction coefficient for the contacts between the parts is set to 0.17. The target face elements use TARGET170 elements, and the contact face elements use CONTACT173 elements, with a total of 173088 mesh elements.

Due to the symmetry of the model, to reduce computational workload, a half-structure is used for analysis. In the half-structure, constraints that comply with symmetry conditions are applied at the disconnected surface. As shown in Figure 4(c), the nodes at the section discontinuity are constrained in three degrees of freedom: Z-axis translation, X-axis rotation, and Y-axis rotation. A Mass21 element is utilized to establish a mass point at the center of the upper casing surface. Fixed constraints are applied to the nodes on the lower casing surface, and an RBE3 element is used to couple this mass point with the upper surface nodes of the upper casing. The model includes bolts with a diameter of 6mm. According to HB6586-92, for a bolt with a nominal diameter of 6mm and a strength grade of 590MPa, the recommended torque value for tightening is 370±40N.cm. The recommended preload force for an M6 bolt is 2570±277N.cm. In the numerical simulation, the bolt preload force is set to 2700N.

Under the specified boundary conditions and with a bolt preload of 2700N applied, a periodic varying external force F is applied at the mass point established at the center of the upper casing, as shown in Equation (4).

$$F = 10000 \times \sin(50 \times \pi \times \mathrm{t}) \tag{4}$$

The external force F is a time-varying harmonic load with an amplitude of 10000N and a period of 0.04s. The positive value of the external force F is parallel to the Y-axis, which is the contact surface with the flange. The analysis duration is 0.16s, corresponding to four load cycles, with a substep duration of 0.0002s. The contact model takes 5 hours and 12 minutes for computation, while the equivalent model takes 46 minutes. The computational efficiency has improved by 80%.

Under the same pre-tension force of bolts (2700N) and friction coefficient (0.17), different external forces F (4kN, 7kN, 8kN, 10kN) are applied at the center mass point on the upper casing surface for analysis. The displacement along the Y-direction caused by the external force F during the loading is extracted and shown in Figure 5.

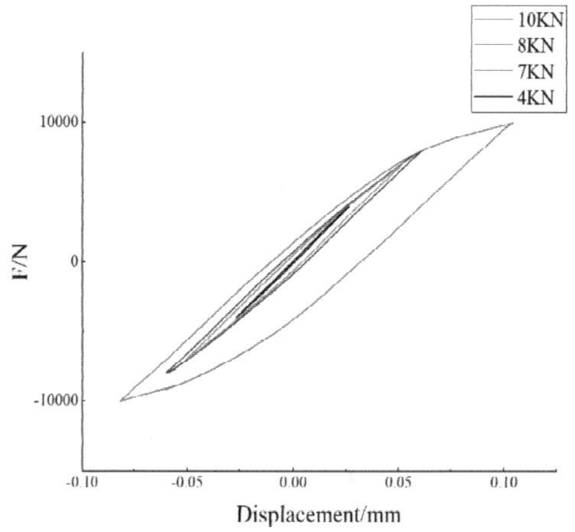

Figure 5. Hysteresis curve under different load amplitudes.

Combined with micro and macro slips on the surface, tangential slip will develop, leading to hysteresis characteristics. The area inside the hysteresis curve represents the energy dissipated on the contact surface within one loading cycle. From Figure 5, it can be seen that when the external force amplitude is small, the hysteresis curve is basically linear, and the contact surface is in a state of micro-slip. This is because the loading amplitude is small, resulting in only a small amount of macro slipping. As the external force amplitude increases, the area of the hysteresis curve also increases. Therefore, the energy dissipated by the bolt flange contact surface in one loading cycle becomes larger. When the bolt flange contact surface is subjected to a larger load, more macro slipping will occur, resulting in greater energy dissipation. These observations are consistent with the phenomena of micro and macro slips.

4. Equivalence model simulation

According to the method in Section 2.2, the discrete Iwan model has a clear physical meaning and can be used to simulate nonlinear and hysteretic characteristics. The discrete Iwan model can be implemented in Ansys using the COMBIN40 element, COMBIN14 element, and concentrated mass element MASS21.

As shown in Figure 6(a), the COMBIN40 element is a spring-slider, linear spring, and damper connected in parallel. The COMBIN40 element requires 6 input parameters, which are the first spring stiffness $\boldsymbol{K_1}$, critical sliding force FSLIDE, second spring stiffness $\boldsymbol{K_2}$, mass M, gap GAP, and damping C. The COMBIN40 element has an ideal elastoplastic constitutive relationship, making it suitable for replacing the Jenkins units in the discrete Iwan model. By setting the 6 input parameters to $\boldsymbol{K_1}$=$\boldsymbol{K_i}$, FSLIDE=$\boldsymbol{f_i^*}$, K2, M, GAP, and C all equal to 0, the COMBIN40 element can be used to simulate the nonlinear behavior of the discrete Iwan model.

The COMBIN14 element, as shown in Figure 6(b), is a spring-slider and damper connected in parallel, which can be used to implement the linear spring in the discrete Iwan model. The input parameters are set to K= and C=0.

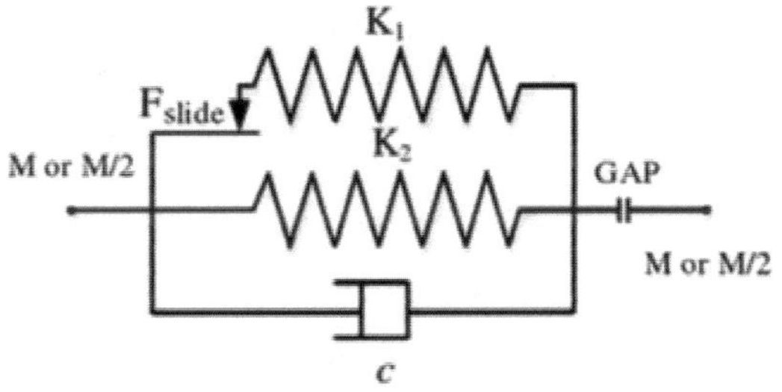

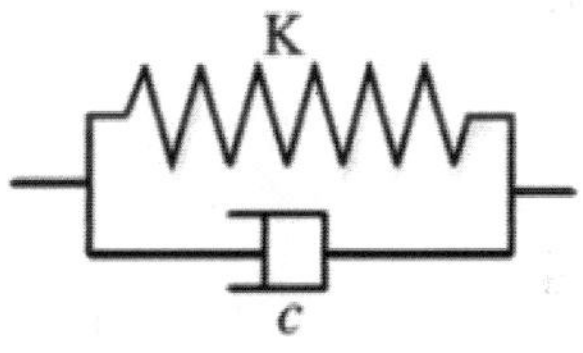

Figure 6. COMBIN40 element and COMBIN14 element.

The discrete Iwan model is a commonly used method to simulate nonlinearity and hysteresis characteristics. However, when establishing a model using finite element software Ansys, connector elements can be used as substitutes for ideal elastic-plastic elements in the discrete Iwan model. This simplifies the complexity of model solving. When setting up the finite element model, mass points can be established at the bolt-hole centers of the upper and lower flanges, using the MASS21 element type to represent these mass points. To simulate the interaction between the bolt and the flange, the RBE3 element can be used to couple the mass points with the nodes around the flange holes. Through the RBE3 element, the loads applied to the mass points can be distributed to the nodes around the flange holes.

For the bolt part, the COMBIN40 and COMBIN14 elements can be used to connect the mass points established in the bolt holes of the upper and lower flanges. This can replace the bolt part in the equivalent model. The contact model is a 3D model, so in the equivalent model, connector elements need to be established in three dimensions, including two tangential dimensions and one normal dimension. By using the tangential and normal stiffness of the connector elements, the tangential and normal stiffness of the bolt-flange contact surface can be approximately simulated. In the equivalent model, there is no need to perform contact calculations between the bolt and the flange, greatly reducing the solving complexity. This approach can effectively simulate the interaction between the bolt and the flange. The local schematic diagram of the equivalent model is shown in Figure 7(a).

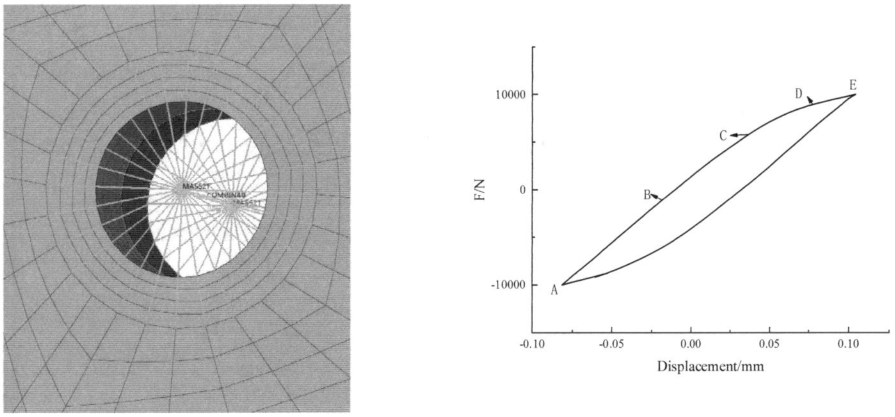

(a) (b)

Figure 7. (a) Equivalent model local diagram, (b) Discretization diagram of the hysteresis curve.

In order to obtain the parameters of the equivalent model of the spring damping unit, the hysteresis curve of the contact model between the surfaces in the bolted flange connection under tangential load can be discretized. The discretized hysteresis curve can be approximated by multiple linear segments, including 3 Jenkins units and a linear spring. In this study, the hysteresis curve for the condition of a pre-tension force of 2700N and a load of 10kN (F=10kN) is discretized for the upper part, as shown in Figure 7(b).

Firstly, the linear stage of microslip in the hysteresis curve needs to be determined. This stage corresponds to the AB segment in the hysteresis curve, assuming no yielding occurs in this stage. The stiffness of segment AB, (k_{AB}), can be represented by the sum of the stiffness of three Jenkins units. The segment DE in the hysteresis curve corresponds to the stage where all Jenkins units have yielded, and its stiffness, (k_{DE}), can be represented as the stiffness of the linear spring, (k_a), in the discrete Iwan model. The remaining curve parts can be approximated by three linear segments, B, C, and D representing the yield points of the 1st, 2nd, and 3rd Jenkins units, respectively. The difference in slope values before and after these points represents the stiffness of the corresponding Jenkins unit. Moreover, half of the displacement difference of points B, C, and D relative to point A represents the yield displacement of the corresponding points. The critical slip force F^* can be obtained by the product of the corresponding yield displacement x^* and stiffness k. The formulas for calculating the parameters of the three Jenkins units and the linear spring are as follows:

$$k_{AB} = \sum_{i=1}^{3} k_i + k_a, k_{BC} = \sum_{i=2}^{3} k_i + k_a, k_{CD} = k_3 + k_a, k_a = k_{DE} \tag{5}$$

$$k_1 = k_{AB} - k_{BC}, x_1^* = \frac{1}{2}(x_B - x_A), F_1^* = k_1 x_1^* \tag{6}$$

$$k_2 = k_{BC} - k_{CD}, x_2^* = \frac{1}{2}(x_C - x_B), F_2^* = k_2 x_2^* \tag{7}$$

$$k_3 = k_{CD} - k_{DE}, x_3^* = \frac{1}{2}(x_D - x_E), F_3^* = k_3 x_3^* \tag{8}$$

According to the coordinates of the points on the hysteresis curve of the contact model, the discretization can be performed using the above formula to obtain the parameters of three Jenkins elements and one linear spring in the discrete Iwan model. These parameters can be defined in the finite element software Ansys using the COMBIN40 and COMBIN14 elements, as shown in Table 2.

Table 2. Parameters for COMBIN40 and COMBIN14 elements.

Element	Parameters		
	x^* (mm)	k (N/mm)	F^* (N)
COMBIN40(1)	0.0264	9378	247.58
COMBIN40(2)	0.0528	4673	2469.68
COMBIN40(3)	0.0793	44155	3501.49
COMBIN14	/	40684	/

5. Verification of static and dynamic characteristics of the equivalent model

5.1. Verification of static hysteresis characteristics

Based on the hysteresis curve of the contact model for bolted flange connections, we can discretize and identify parameters to obtain the parameters of the connecting unit. In Ansys, an equivalent model of the spring element for the contact interface is established, and the relevant parameters of the connecting unit are defined. Then, the equivalent model is simulated and analyzed by setting the same boundary conditions as the contact model and applying different tangential loads (4KN, 7KN, 8KN, 10KN). The hysteresis curves of the equivalent model and the contact model are compared to observe the differences between them, as shown in Figure 8.

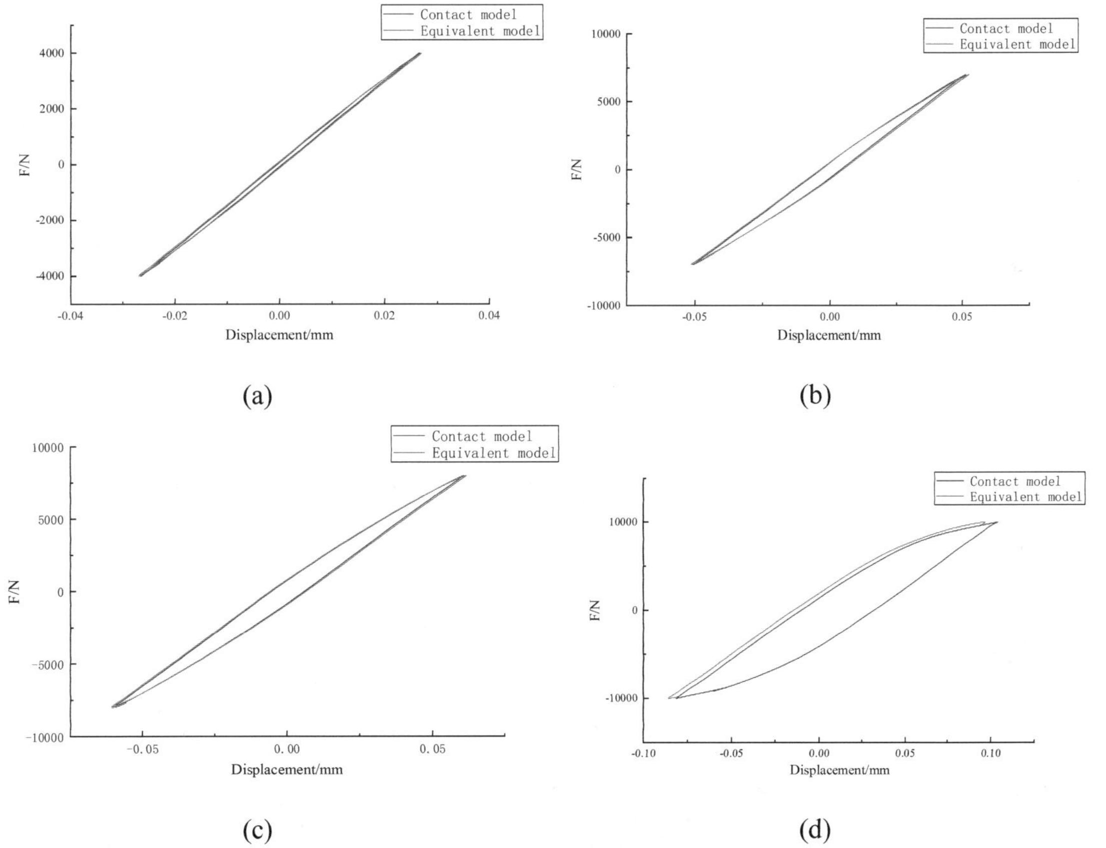

Figure 8. Comparison of hysteresis curves between the equivalent model and the contact model. (a) F=4KN, (b) F=7KN, (c) F=8KN, (d) F=10KN.

The area of the hysteresis curve of the bolted flange connection joint under tangential load can be used to reflect the energy dissipation characteristics of the joint surface. In order to quantitatively evaluate the error between the hysteresis curves of the equivalent model and the contact model, a metric can be defined, which is the ratio of the area of the hysteresis curve of the equivalent model to the area of the hysteresis curve of the contact model, denoted as S_{12}. This metric can be used to measure the difference between the hysteresis curves of the equivalent model and the contact model. If this ratio is close to 1, it indicates that the hysteresis curve area of the equivalent model is similar to that of the contact model, indicating similar energy dissipation characteristics. However, if this ratio is close to 0 or very small, it indicates that the equivalent model cannot accurately simulate the energy dissipation characteristics of the contact model. To quantitatively measure the degree of correlation between the hysteresis curves of the equivalent model and the contact model, the Pearson correlation coefficient, denoted as $r < x_1, x_2 >$, can be used to calculate their correlation. The formula for calculating the correlation coefficient is as follows:

$$S_{12} = \frac{A_1}{A_2}, r < x_1, x_2 >= \frac{x_1 \cdot x_2}{| x_1 || x_2 |} \tag{9}$$

where is the displacement value vector of the equivalent model and is the displacement value vector of the contact model. The area ratio and the correlation coefficient for the four operating conditions with load amplitudes of 4KN, 7KN, 8KN, and 10KN are shown in Figure 9.

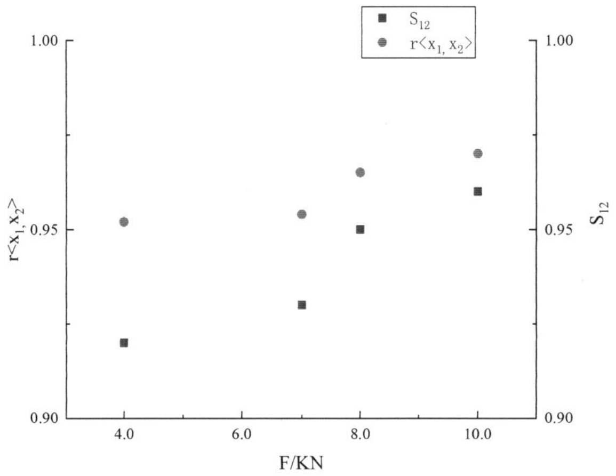

Figure 9. Correlation and Area Ratio between Contact Model and Equivalent Mode.

From the graph, it can be observed that there is a high degree of correlation between the hysteresis curves of the equivalent model and the contact model. The correlation coefficient is greater than 0.95 and less than 1, while the area ratio is greater than 0.9 and less than 1. This indicates that the equivalent model can effectively replace the contact model and demonstrates reliability in solving the nonlinearity and hysteresis characteristics of tangential slip in flange joint surfaces.

5.2. Verification of dynamic hysteresis characteristics

We compute the first 10 natural frequencies of the Iwan model under free boundary conditions. Figure 10 shows some low-order mode shapes of the Iwan model, which are consistent with the corresponding mode shapes of the contact model. Table 3 lists the first 10 natural frequencies of the Iwan model, along with the natural frequencies of the contact model and the calculated errors.

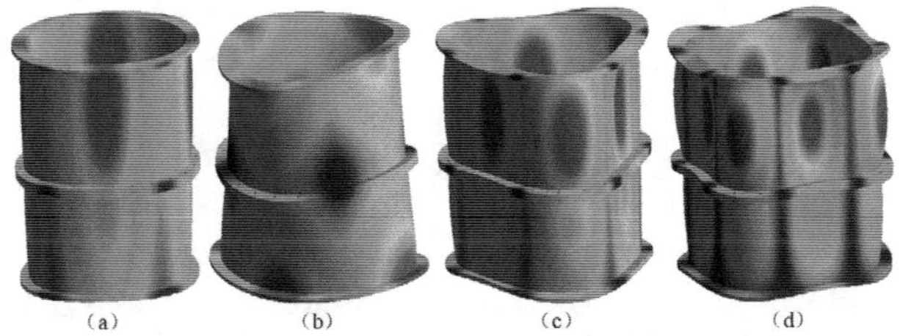

Figure 10. Low-order mode shapes of the equivalent model.

Table 3. The first 10 natural frequencies are equivalent to the model.

Mode Order	*Equivalent model (Hz)*	*Contact model (Hz)*	*Error (%)*
1	170.52	164.76	3.5
2	170.52	164.77	3.5
3	190.54	184.81	3.1
4	190.54	184.81	3.1
5	457.46	427.53	7.0
6	457.46	427.58	7.0
7	530.49	483.14	9.8
8	530.49	483.14	9.8
9	841.71	739.64	13.8
10	841.71	739.67	13.8

According to the calculation results, the relative error of the first 4 natural frequencies of the Iwan model compared to the contact model is within 4%, and the maximum relative error of the first 10 natural frequencies is 13.8%. The main reason for the poor accuracy of the Iwan model in predicting natural frequencies is that it only considers the normal and tangential characteristics of the connection structure and neglects the torsional behavior. This leads to significant errors in predicting the torsional natural frequencies. In the calculation of natural frequencies, the contact model takes 6 hours and 20 minutes, while the equivalent model takes 1 hour. The computational efficiency has improved by 80%.

6. Conclusions

(1) Under the same preload and friction coefficient, we found that the correlation error between the equivalent model and the contact model is less than 1% for different loading conditions. The area error of the hysteresis curve is also below 9%. This indicates that the equivalent model performs well in simulating the nonlinear hysteresis behavior of the interface.

(2) The modal analysis of the equivalent model shows that some low-order vibration modes are consistent with the corresponding vibration modes of the contact model, but the maximum error of the first 10 natural frequencies relative to the contact model is 13.8%. This is because the equivalent model did not consider the structural twist, which leads to a relatively large error in the twist's natural frequencies. The equivalent of twists still needs to be improved.

(3) Compared to the contact model, the equivalent model greatly reduces the number of elements, reducing the number of contact pairs by 100%. The reduction in the number of elements and contact pairs improves computational efficiency.

Acknowledgments

The authors thank the National Science and Technology Major Project (Grant 2019-IV-0019-0087) for financial support.

References

[1] Li Ling, Cai Anjiang, Ruan Xiaoguang. (2015) Equivalent linear method for hysteresis nonlinear of bolted joints. Journal of Vibration Engineering. 28: 560–566. https://link.cnki.net/doi/10.16385/j.cnki.issn.1004-4523.2015.04.008.

[2] Cao Junyi, Liu Qinghua, Hong Jun. (2021) Overview of micro friction to macro dynamics for bolted connections. China Mechanical Engineering. 32: 1261-1273. https://link.cnki.net/urlid/42.1294.TH.20201112.0922.002.

[3] Ma Shuangchao, Zang Chaoping, Lan Haibo. (2013) Dynamic model updating of an aero-engine casing. Journal of Aerospace Power. 28:878-884. https://link.cnki.net/doi/10.13224/j.cnki.jasp.2013.04.024.

[4] Iwan W. D. (1966) A distributed-element model for hysteresis and its steady-state dynamic response. Journal of Applied Mechanics. 33:893-900. https://doi.org/10.1115/1.3625199.

[5] Zhu Mingyang, Miao Huihui, Li Bing. (2022) Nonlinear Contact Modeling of Screws Based on the Iwan Model and Parameter Identification Science. Technology and Engineering. 22: 14729-14735. http://www.stae.com.cn/jsygc/article/pdf/2202332.

[6] Mao Fanghai, Tang Kaifei, Wang Delun. (2022) Equivalent Model of Screw Connections Based on Finite Element Spring Units. Mechanical Engineering and Automation. 5: 48-50.https://kns.cnki.net/kcms2/article/abstract?v=62vjN2oCPVYN-m6V29ox9NId9tAYWk3yT.

[7] Yang Haoran, An Lulíng, Zou Liyang. (2023). Dynamic Modeling Method for Screw Connections Considering Sliding. Journal of Nanjing University of Aeronautics and Astronautics. 5:924-932. doi: 10.16356/j.1005-2615.2023.05.019.

[8] Chen Yun, Zang Zhaoping, Yang Zhiqiang. (2022). Virtual Thin Layer Element Modeling Method for Machine Housing Screw Connections. Mechanical Manufacturing and Automation. 3:117-119+129. Doi: 10.19344/j.cnki.issn1671-5276.2022.03.029.

[9] Liu Pengtao, Guan Tianchi. (2023) Nonlinear Modeling and Parameter Identification of Single Bolt Connection Layer in Beams. Vibration and Impact. 42:190-197.doi:10.13465/j.cnki.jvs. 1000-3835.

[10] You Jian, Jiang Haocheng, Shi Chaoming (2023) Application of Beam-Spring Unit in Simplified Modeling of Internal Hexalobular Screw. Journal of Ningbo University (Science and Engineering Edition).4:112-120. Doi: 10.20098/j.cnki.1001-5132.2022.1129.

Journal of Physics: Conference Series **2992** (2025) 012011 doi:10.1088/1742-6596/2992/1/012011

Optimization design of asymmetric fork structure for shifting mechanism based on topology method

Zhonglin Jiang[1], Yong Li[2,*], Mingge Wu[3] and Jie Ren[4]

[1,2,4]College of Mechanical and Electrical Engineering, Wenzhou University, Wenzhou, China

[3]RUI'AN Graduate college, Wenzhou University, Wenzhou, China

[*]Corresponding author's e-mail: lyre@wzu.edu.cn

Abstract—The reliability of the shifting mechanism is crucial for the comfort and stability of heavy-duty commercial vehicles. How to reduce the local deformation and the maximum stress through structural optimization is the key to improving the stability and service life of heavy-duty commercial vehicle shifting devices. To improve the fracture problem of shift forks, a topology optimization method based on finite element analysis was applied. The finite element analysis results serve as constraints for topology design, optimizing areas that are difficult to manufacture and produce to produce better structures. This topology optimization is mainly based on three judgment objectives, namely maximum stress, maximum displacement, and cycle life. Firstly, the optimized structure reduces stress in three directions, effectively reducing stress concentration. Secondly, the maximum displacement in the three directions decreased by 0.765 mm, effectively reducing the deformation gap between the two arms. Finally, the design achieved a weight reduction of 0.32 kg and a volume reduction of 15%, which to some extent achieved the goal of being lightweight. The increase in the number of cycles also extends the service life of the shifting mechanism, ensuring its durability.

1. Introduction

With the continuous development of the transportation industry, the demand for heavy commercial vehicles in various industries is also increasing. Therefore, people have higher expectations for their economic efficiency and safety. The transmission is an important component of commercial vehicles, which has a significant impact on the comfort and safety of the vehicle. The shifting mechanism is an indispensable part of the transmission, used for actions such as shifting and engagement. As the actuating element of the shifting mechanism, the shift fork is subjected to various external forces during the shifting process, such as gear meshing impact and friction, making it one of the most susceptible components to fatigue failure. Due to the specialized use of commercial vehicles for transporting heavy goods and their large weight, their structures bear considerable loads, which may lead to component fatigue damage. In the entire shifting system, the shifting fork serves as the executing element, and its performance directly affects the safety and comfort of the vehicle during operation. To extend the service life of the actuating element, reduce the possibility of fatigue damage, and improve the shifting performance of the transmission, the optimization design of the shift fork structure has become particularly important.

With the advancement of technology, researchers are increasingly using finite element analysis software to guide the design process. At present, the shifting mechanism of high-performance vehicles is a popular research topic. Numerous scholars both domestically and internationally have researched the lifespan and optimization of gear-shifting mechanisms. For example, Li et al. optimized the size parameters of the shift fork for small passenger cars and compared the results of finite element analysis, making some progress[1]. Jia et al. analyzed the fatigue life of heavy-duty commercial vehicle shift forks and improved their service life to some extent by changing the material[2]. Ankur Varma et al. studied the stiffness and shifting mass of shift forks and found that the vibration of manual shift levers and transmission gears increased with the contact between the fork and the engagement sleeve, thus optimizing the shift fork[3]. In addition, topology optimization methods have been widely applied in the design and production of automotive-related components. For example, Liu et al. studied the lower arm of automobile suspension and optimized the topology design of the lower arm using topology optimization methods, achieving the goal of being lightweight while meeting design requirements[4]. In summary, research on shifting mechanisms at home and abroad mainly focuses on the static strength analysis of shifting forks and the optimization of the size of individual shifting forks. Most of the research objects are shifting forks for small passenger cars. There is relatively little research on the shifting mechanism of heavy-duty commercial vehicles. In addition, according to their shape, forks can be divided into symmetrical forks and asymmetrical forks. Due to its structural characteristics, there are differences in stiffness and displacement deformation between the two arms of the asymmetric fork, which can easily lead to fatigue damage and other failure phenomena. At present, there is relatively little research on the stiffness and displacement deformation of asymmetric fork arms in heavy-duty commercial vehicles[5].

This article aims to reduce stress concentration, stiffness differences between the two arms, and displacement deformation through topology optimization techniques[6]. Using the optimization results from the conceptual stage, apply the reverse design method to the shift fork to obtain an optimized structure. Eliminating unnecessary materials and reducing the weight of shift forks have achieved the goal of being lightweight to a certain extent. Even with lightweight processing, this structure can still meet the optimization goal of reducing the stiffness and displacement deformation gap between the two arms of the asymmetric shift fork. This can not only reduce stress concentration but also improve its service life.

2. Structure and model

2.1 The structure of the shifting mechanism

The research object of this article is the actuator in the shifting mechanism, namely the asymmetric shifting fork. Figure 1 shows the assembly diagram of the shifting mechanism, which mainly consists of the shifting fork of the main and front boxes, the shifting cylinder of the main and front boxes, the shifting shaft, and other components. The main function is to control the actuator through the switch of the solenoid valve group to complete the gear selection and shifting operation after receiving the shift command issued by the control system ECU[7].

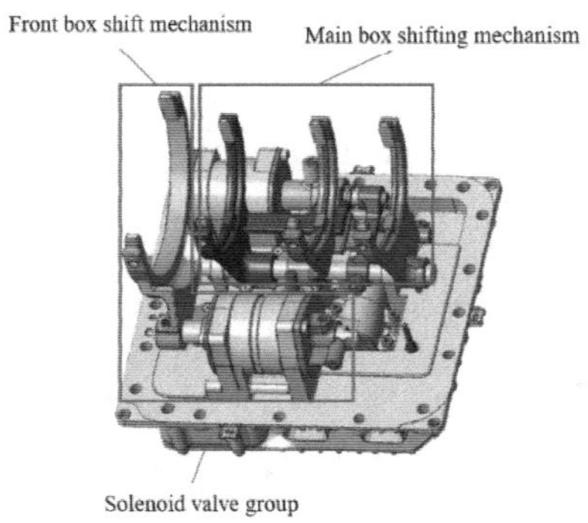

Figure 1. Structural diagram of gear shifting device.

2.2 *Model analysis before optimization*

Considering the limited space of asymmetric forks in topology optimization, they were re-modeled before finite element analysis to increase the optimization area. In order to provide constraints for topology optimization, this paper analyzed the model before optimization, using the transient dynamics method to apply an initial velocity of 400 mm/s to the impact force loaded on the shift fork, which is more in line with actual working conditions. The shifting force is 1750 N, which is the experimental load provided by the enterprise. The finite element model is shown in Figure 2.

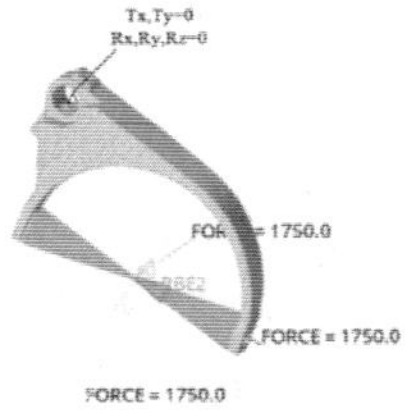

Figure 2. Finite element model of the shift fork.

The displacement deformation and stress in three directions are shown in Figure 3.

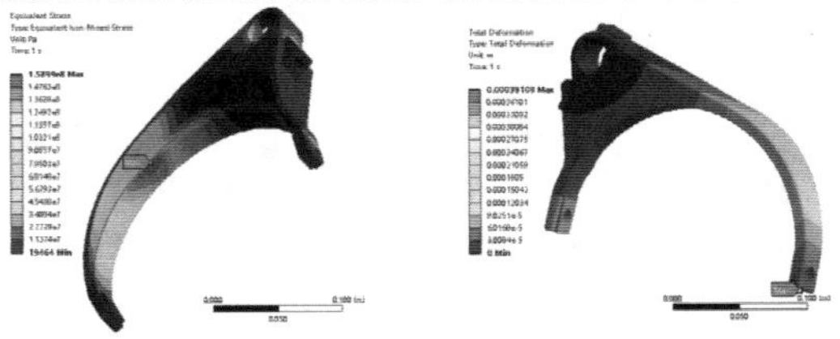

(a)

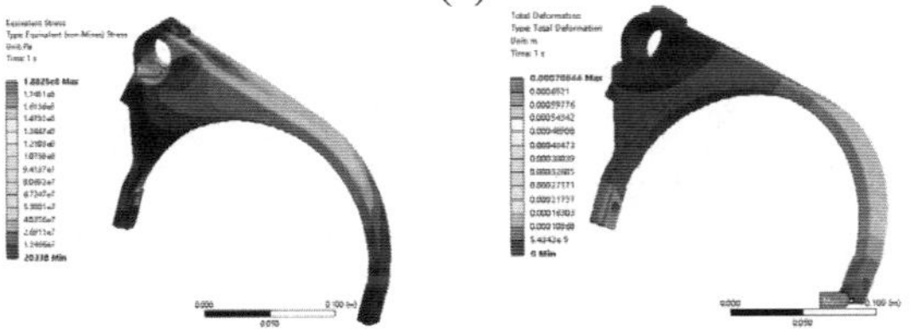

(b)

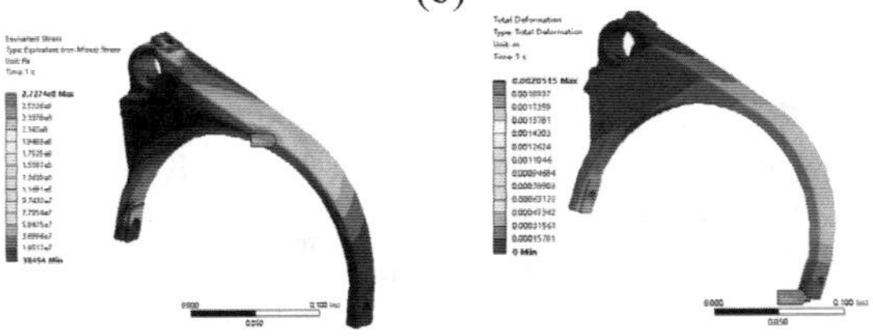

(c)

Figure 3. Maximum displacement and stress in the X、Y and Z directions.

3. Determine the topology method

Topology optimization in the design phase refers to finding the optimal material distribution or force transmission path within a given design space to achieve maximum stiffness, minimum mass, or more stable structure while satisfying various performance conditions.

3.1 *Three elements of topology optimization*

The three elements of optimal design are as follows[8]:

$$Minimize: f(x) = f(x_1, x_2, \dots, x_n) \tag{1}$$

$$Subject: c_i(x) \leq 0, i = 1, \dots, m \tag{2}$$

$$h_x(x) \leq 0, k = 1, \ldots, m \quad (3)$$
$$X = \{x \in R^n | c_i(x) \leq 0, i = 1, \ldots, m\} \quad (4)$$

Where $X = (x_1, x_2, \ldots, x_n)$ is design variables, such as structure size, shape, etc; $f(x)$ is design objective; $c_i(x)$and $h_x(x)$ are Constraint condition.

3.2 Determine optimization space

According to the assembly of the fork with other parts in the shift device and the fork model before optimization, the optimization space of the structure was determined, as shown in Figure 4. Area 1 is connected with the shifting shaft, which is the constraint of the model and belongs to the non-design area. Area 3 is assembled by slider and synchronizer as the loading part of the model, which also belongs to the non-design area. Therefore, only region 2 is the design space for topology optimization.

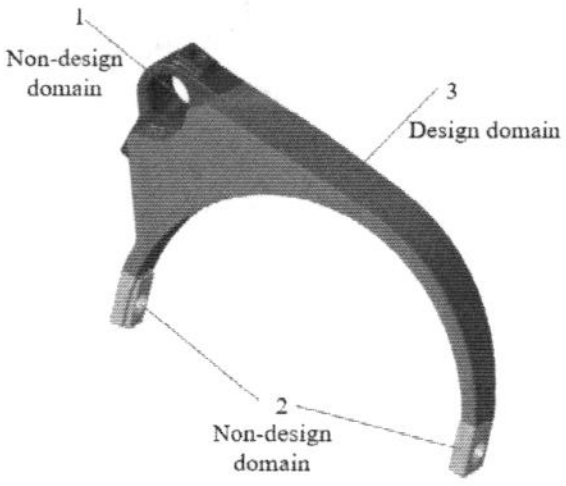

Figure 4. Determination of optimization area.

3.3 Determine the optimization constraints

The object optimized in this paper is an asymmetrical shifting fork, the stiffness of its long and short arms is quite different, and it is easy to have stress concentration and shifting of the joint sleeve during shifting, which leads to the failure to complete shifting and unshifting normally. Therefore, it needs to be constrained by displacement and deformation. In addition, considering the feasibility of processing and manufacturing, it is also necessary to optimize the manufacturing process constraints[9].

Further, organize the displacement and deformation results from the previous section into Table 1. From the table, it can be seen that whether it is the long arm or the short arm, the deformation of the asymmetric fork in the Z direction is greater than in other directions, and there is a large stiffness gap between the two arms, which is prone to fracture during operation. To reduce the failure caused by the stiffness gap, appropriate deformation constraints are applied to the Z-direction of the topology optimization model:$Z_1 < 1.2mm$, $0.8mm < Z_2 < 1.0mm$. The deformation in the Y and Z directions is relatively small, but deformation constraints still need to be applied. The deformation constraints in these two directions have been appropriately relaxed: $X_1 = X_2 < 0.2mm$, $Y_1 = Y_2 < 0.3mm$. In addition, considering the potential adverse effects on subsequent manufacturing of cavities, trusses, and other structures during topology optimization, this article also sets constraints on the unidirectional draft for the fork model[10].

Table 1. Displacement deformation results.

Deformable fork arm	Direction	Deflection(mm)
Long fork arm	X+	0.391
	Y+	0.707
	Z+	2.052
Short fork arm	X+	0.211
	Y+	0.435
	Z+	0.947

3.4 Determine optimization objectives

The paper takes the minimum flexibility of the asymmetric fork, that is, the maximum stiffness, as the objective function z, and the mathematical model is as follows[11]:

$$min(z) = F^T U = U^T K U = \sum_{i=1}^{n} E_i(p_i) u_i^T k_i u_i \quad (5)$$

$$s.t. K = \frac{V_{(\rho)}}{V_0} \quad (6)$$

$$V = \sum V_i \rho_i \leq f V_0 = V^* \quad (7)$$

where z is the objective function; ρ_i is a design variable; ρ is relative density; F is a structural load; U is a global displacement vector; K is the global stiffness matrix; E_i is the Young's modulus of the unit material; $E_i(p_i)$ is a penalty function; k_i is the stiffness matrix of the unit; V_i and V^* are respectively the initial structure volume and the optimized volume.

4. Results analysis and comparison

4.1 Analysis of topology optimization results

After defining the optimization space, constraints, and optimization objectives, use the Optimal solver to solve topology optimization. After 38 iterations, the simulation result reached the minimum value, that is, the stiffness reached the maximum value, and the result tended to converge[12].

After each iteration, a relative density cloud map will be generated at the current iteration step, with values ranging from 0 to 1 for relative density. The closer the relative density is to 1, the more important the material is; Otherwise, you can consider deleting it. This article compares two cases with relative densities of 0 and 0.3, namely the density cloud map without material removal and the relatively complete density cloud map retained after material removal, as shown in Figure 5.

According to the results, it can be seen that the materials at locations 1, 2, and 3 have little effect on the stiffness of the asymmetric fork and can be appropriately removed, but the materials at these three locations cannot be completely removed. This section is divided into the main stress areas of the fork except for the two arms. If the material in this area is completely removed, the stiffness will decrease. The stress will concentrate on two steel bars, and the stiffness cannot be guaranteed, which is prone to fracture. Therefore, the material here can be removed appropriately, but not completely. The structures at positions 4 and 5 are reinforced structures generated under draft constraints, indicating that they are the main force-bearing area in the middle of the fork and need to be appropriately preserved and modified. The material at position 6 is the area on both sides of the fork. If the material in this area is removed, the two arms of the fork will be too thin to achieve this optimization goal. Therefore, the materials in this place need to be preserved. In summary, when the relative density is 0.3, the structure is relatively complete and the material distribution is uniform. Therefore, the topology optimization result with a relative density of 0.3 is chosen as the final result of the conceptual stage.

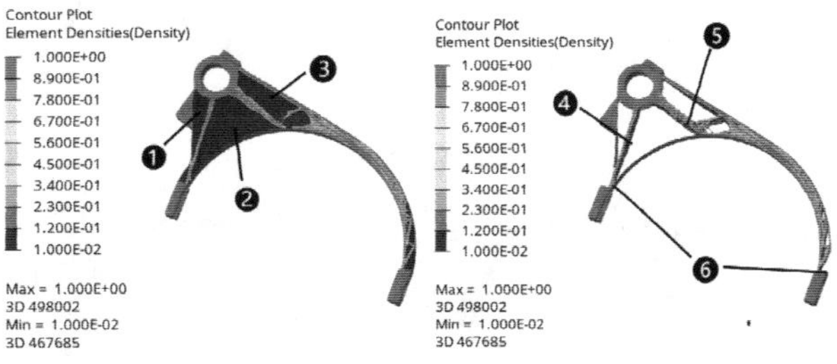

Figure 5. Topology optimization results.

4.2 Optimization simulation results verification

Based on the topology optimization results during the conceptual design phase, the original fork was reverse-engineered using 3D software, and the final product structure is shown in Figure 6.

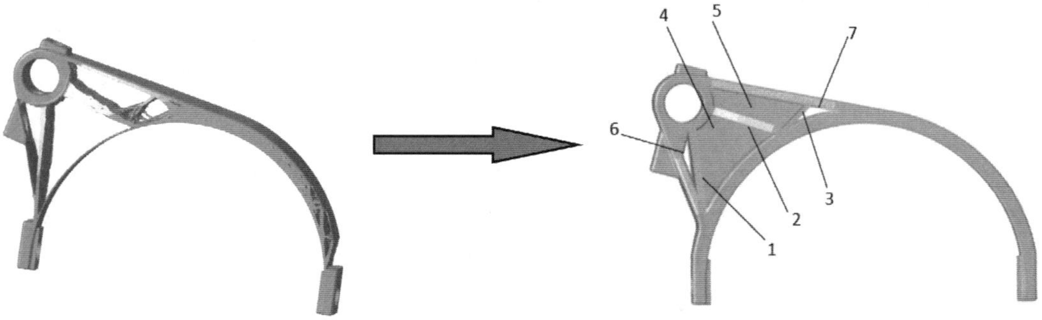

Figure 6. Optimized fork structure.

Specifically, due to the limitations of Optistruct optimization on the drawing process, the positions of the two steel bars are displayed as 1 and 2, respectively. The two steel bars have the same size and an angle of about 60 degrees between them, which improves their stability. According to the optimization results, a transverse reinforcement bar has been added. The materials at locations 4 and 5 have been properly stored, and based on lightweight design, efforts have been made to avoid stress concentration, which is beneficial for reducing the possibility of fatigue damage. The materials at locations 6 and 7 had little impact on the reliability of the structure, so they were removed and the entire structure was reduced in weight. The weight of the original fork was 1.77 kg. After optimization and weight reduction, the mass is 1.45 kg, a decrease of 0.32 kg, and the volume is 86% of the original fork.

After topology optimization and reverse design of the asymmetric fork structure, the optimization design of the structure was completed. The purpose of optimization is to reduce the deformation gap and possible stress concentration between the long and short arms of the asymmetric fork, but whether the redesigned structure meets the requirements needs to be verified through simulation. The optimized fork was reanalyzed using the finite element method without changing the boundary conditions and parameter settings. By analyzing the displacement, deformation, and maximum stress of the fork under different working conditions, it can be determined whether the new structure meets the requirements. The results of maximum stress and displacement deformation are shown in a, b, and c of Figure 7.

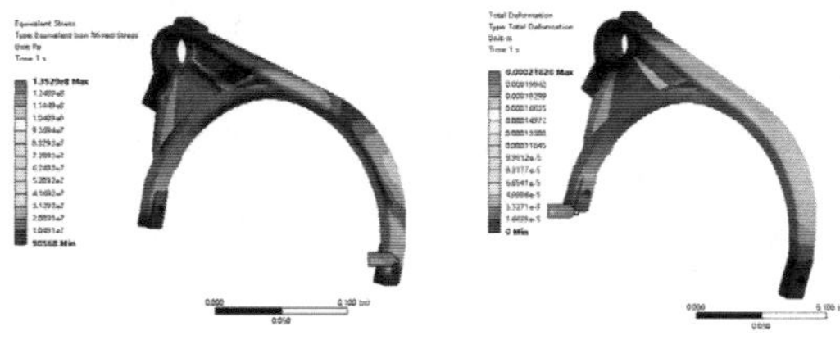

(a)

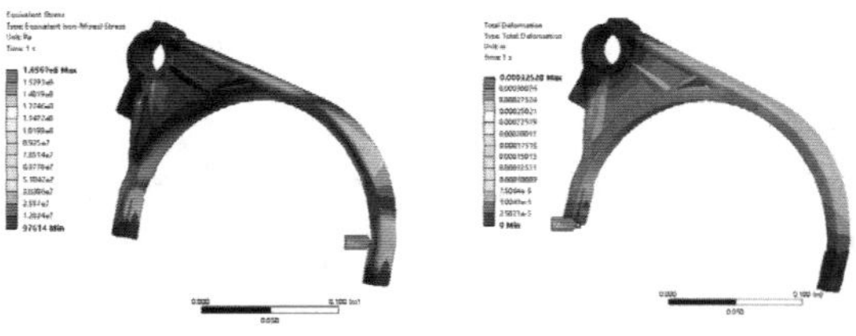

(b)

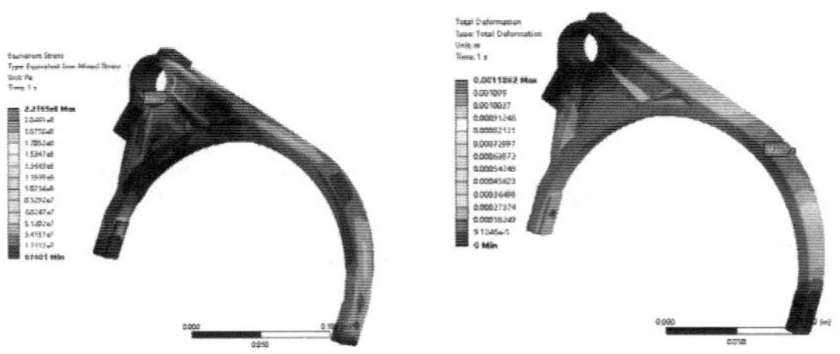

(c)

Figure 7. Optimized displacement and stress in three directions.

Fatigue analysis of the optimized fork is carried out by fatigue analysis software[13]. The finite element results obtained from the transient dynamic analysis were taken as the pre-input conditions, and the working life of the optimized fork was simulated, as shown in Figure 8. It can be seen from the figure that the shortest point lifetime under the transient dynamic condition is second. Most of the short-life points are concentrated in the position of the fork reinforcement and the long arm straddle, which is close to the actual position of fatigue fracture. It can also be seen from the average biaxial ratio cloud image that the ratio between the long arm and the vertical reinforcement is close to -1, which is the part prone to fatigue damage, which is consistent with the actual working condition and can provide certain reference value for engineering practice.

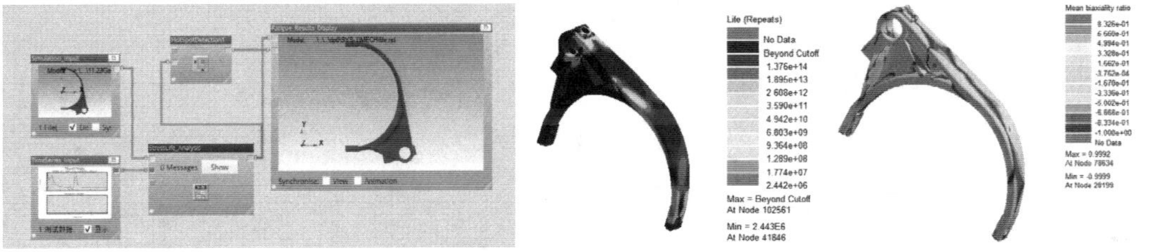

Figure 8. Optimized fork lifetime number and mean biaxial ratio cloud map

4.3 Comparison of optimization results

The above simulation results are compared with the data before optimization and sorted into a table, as shown in Table 2. It can be seen that the maximum displacement and deformation in the three directions are reduced, especially the maximum deformation in the Z-direction along the shifting direction is reduced by 0.865mm, and the optimization effect is the most obvious. In addition, the maximum stress in the three directions is also significantly reduced, which better reduces the possibility of stress concentration phenomenon. The cycle life is also obtained, so the structure is reasonable and can meet the design requirements.

Table 2. Performance comparison before and after optimization.

	Before optimization			post-optimization		
item	**Maximum displacement deformation (mm)**	**Maximum stress (MPa)**	**Cycle number**	**Maximum displacement deformation (mm)**	**Maximum stress (MPa)**	**Cycle number**
1	0.391	159.99		0.216	135.29	
2	0.707	188.25	1.760×10^{6}	0.326	165.67	2.442×10^{6}
3	2.052	272.74		1.187	221.65	

5. Conclusion

This article takes the shift fork of a certain enterprise as an example to re-model the shift fork, expand the optimization space, and conduct preliminary simulation analysis on it. In the conceptual design phase, the preliminary simulation results are used as inputs for topology optimization, and corresponding constraints and optimization objectives are set. According to actual manufacturing requirements, reverse engineer the topology model and optimize the shift fork structure. The optimized shift fork has replaced the original single reinforcement with a three-rib structure, with two symmetrical reinforcement ribs designed at an angle of approximately 60 degrees. The weaker reinforcement ribs extend from the shift shaft to the fork feet, enhancing its structural stability. After optimization, the maximum stress of the fork is 221.65MPa, and it can be seen from the stress cloud map that the possibility of stress concentration is effectively reduced. Secondly, the maximum displacement and deformation of the long arm of the optimized fork is 1.187mm, which is reduced by 42.3% compared with that before optimization. The maximum displacement and deformation of the short arm is 0.73mm, which is 23% less than before. And the stiffness and deformation of the two arms are close, which shortens the gap between the two arms. The mass is also reduced from the original 1.45kg to 1.13kg, and the volume is 86% of that before optimization, and the overall system has reached the goal of lightweight to a certain extent. Finally, under the transient dynamic conditions, the minimum point lifetime is also increased to 2.442 × 10^6 times, which has a relatively obvious improvement. The topology optimization method adopted in this paper is very helpful in the initial design stage of products, saving time and cost, and has reference value for the subsequent design and manufacturing process of enterprises.

Acknowledgment

This work was financially supported by the key scientific and technological innovation projects in Wenzhou, China (ZG2022024).

References

[1] Wenlin L. Optimization and Improvement of Key Components of internal shift mechanism of dual-clutch transmission. M. D. Dissertation. Chongqing University of Technology. Chongqing, China, 2020.

[2] Jia Y., Chen, Y. (2014). Fatigue Life Analysis of Transmission Shift Fork. Automotive Engineer,2014:08,48-50.

[3] Jaideep S., Rohit K., Ankur V., Ayush S. and Mangesh Dhane. Weight Reduce on of Shifter Forks using Steel Inserts. C. International Commercial Vehicle Engineering Congress, 2013. 44-49.

[4] Wu N., Liu C., Yang Z. A. and Lian, Z. L. Structural Design of Special Vehicle Components Based on Topology Optimization. J. International Journal of Automotive Technology, 2023. 24,2,541-549.

[5] Xi Li. Finite element dynamic analysis and fork structure optimization of single cone lock ring synchronizer. M. D. Dissertation. Huazhong University of Science and Technology. Wuhan, China,2020.

[6] Gui, W. Q. Transmission Fork Strength Simulation and Optimization Based on ANSYS. J. Journal of Jiamusi University: Natural Science Edition, 2018: 03,387-389.

[7] Xie, C. Research on lightweight optimization design method of multi-material cab structure for a commercial vehicle. Ph. D. Jilin University. Jilin, China,2022.

[8] S. Yuvaraja., G. Arunkumar., B. Venu Sai. and P. Rajan. V. Dhinakaran. Design and Development of a Compliant Clutch Fork using Topology Optimization. J. International Journal of Innovative Technology and Exploring Engineering (IJITEE), 2019: 3625-3629.

[9] Lim, S., Misawa, R., Furuta, K., Maruyama, S., Izui, K. and Nishiwaki, S. Weight reduction design of multi-material vehicle components using level set-based topology optimization, 2022: 22,4,659-674.

[10] Zhu, H., Zhang, H., and Xia. L. Topology Optimization in Aircraft and Aerospace Structures Design. J. Arch Computat Methods Eng, 2016: 23,595-622.

[11] Park, S. W., Choi, J. H., and Lee. B. C. Multi-objective Optimization of An Automotive Body Component With Fiber-reinforced Composites. J Structural and Multidisciplinary Optimization, 2018: 58,2203-2217.
[12] Saputro, B. A., Ubaidillah, Triono, D. A., Pratama, D.R., Cahyono, S. I. and Imaduddin, F. Static load simulation of steering knuckle for a formula student race car. AIP Conf. Proc, 2018: 1931,030049.
[13] Yoo, S., Doh, J., Lim, J., Kang, O., Lee, J. and Kang, K. Topologically optimized shape of CFRP front lower control ARM. Int. J. Automotive Technology, 2017: 18, 4,625-630.

Autofrettage design and study of ultrahigh-pressure emergency pressure relief valves

Yuehao Gao[1st], Xuesheng Wang[2nd]*, Qinzhu Chen[3rd]

Work unit: East China University of Science and Technology School of Mechanical and Power Engineering, Xuhui District, Shanghai, China

*Corresponding author's e-mail: wangxs@ecust.edu.cn

Abstract: The problems encountered in the development of ultra-high-pressure valves were analyzed, and the valve body of an ultra-high-pressure emergency relief valve was chosen as the research object, which was subjected to finite element static analysis to find out the location of the stress concentration of the valve body structure. Based on the Bauzinger effect of the material, the autofrettage process of the valve body is simulated and analyzed, and the stress of the valve body during the autofrettage process is analyzed, which shows that the autofrettage can improve the distribution of the stress of the valve body from the perspective of the internal stress of the valve body. By comparing the stress distribution cloud diagrams of the autofrettage treated and non-autofrettage valve bodies under the working conditions, the stress distribution of the former is more uniform, and the stress maximum value is smaller, which changes the state of the maximum stress on the inner surface.

1. Introduction

In recent years, with the development of industry, the application of ultrahigh-pressure valves has become more and more extensive, and they are mainly used in the production of polyethylene, isostatic pressure treatment, artificial crystal, powder metallurgy, food pressure processing, water jet cutting, test equipment and other fields. Ultra-high-pressure valves are the control unit in the ultra-high-pressure circuit, and the stability of the valve directly affects the reliability of the whole system [1,2]. Ultra-high pressure valves are used in many conditions and very harsh conditions, such as ultra-high pressure (≥ 100 MPa), high temperature, medium corrosion, cyclic load, shock load, etc. These harsh conditions require very high performance of valves.

In the process of opening and closing the ultra-high-pressure valve, the internal pressure will fluctuate, and this pressure generated by the fluctuation of ultra-high-pressure valves will form an impact on the impact of the valve, resulting in fatigue damage to the valve. At the same time, in the petrochemical industry, the corrosiveness of the fluid medium will further accelerate the damage of the valve, especially the valve body damage. At present, ultrahigh-pressure valve bodies at home and abroad mainly use high-strength materials, increase the thickness of the ultrahigh-pressure valve body and use better processing technology to improve the bearing capacity of the ultrahigh-pressure valve body, but these methods increase the cost of the valve body, and do not provide a proportional increase in the service life of the valve body [3-7].

Compared with the above methods to improve the strength of the valve body, the self-reinforcement treatment of the valve body can better and more effectively improve the service life of the valve body. Self-enhancement treatment, also known as prestressing treatment, is that essentially in the valve body before service, by adding pressure inside the body, the valve body plastic deformation of the inner wall layer produces residual stress, and the residual stress can effectively offset part of the stress generated by the operating pressure, so as to achieve the purpose of improving the valve body stress to improve the service life of the valve body [8,9].

This paper takes the ultra-high pressure emergency relief valve in the polyethylene plant as the research object, simulates the strength of the valve body using CFD technology, carries out the self-enhancement treatment of the valve body, and compares to get the strength of the valve body after the self-enhancement treatment relative to the previous obvious enhancement.

2. Valve body strength analysis

2.1 Computational model

Figure 1 shows the valve body of an ultra-high-pressure emergency relief valve in a polyethylene plant built with Solidworks software. It is a simplified model of parts of the actual valve body that do not have an impact on the results of our study, such as the bolt holes for the connecting flanges and the threads used to connect the gland.

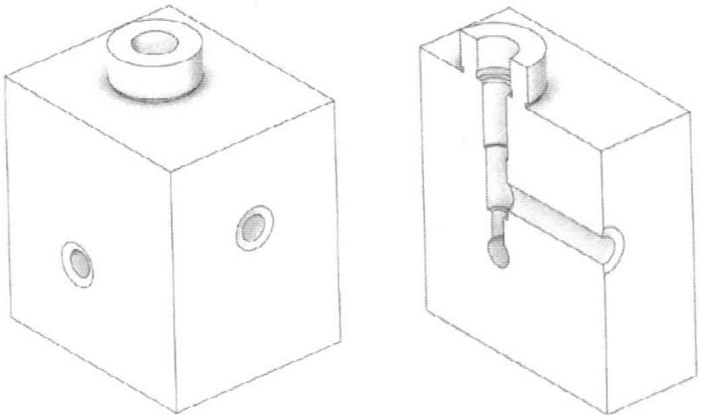

Figure 1. Three-dimensional solid model of the valve body.

The valve body is made of 35CrNi3MoVR alloy steel with a yield limit of $\sigma s = 960$ MPa, a tensile limit of $\sigma b = 1070$ MPa, a modulus of elasticity of $E = 2.06 \times 105$ MPa, and a thermal conductivity of 20.1 W/(m·°C).

2.2 Mesh Setup and Boundary Conditions

In the finite element analysis, due to the symmetry of the UHP valve body model and load constraints, half of the valve body structure is taken as the model for the finite element analysis calculations in order to save computer computational resources. When meshing the model, the internal pressure surface meshing is dense, the external mesh is sparse, and the structure at the coherent line of the valve body is meshed finely. The meshing results are shown in Figure 2. The average quality of meshing is 0.85773, the number of meshes is 1377788, and the number of mesh nodes is 2127426.

Figure 2. Finite element model diagram.

The ultrahigh-pressure valve body was statically analyzed by applying a pressure load of 360 MPa and a temperature load of 350°C on the internal pressure surface of the body, and by applying a symmetric load on the symmetric surface.

2.3 Valve body static analysis
When the pressure inside the pipeline is higher than the rated pressure, the Ultra-high-pressure emergency relief valve opens to relieve the pressure. Under the condition of pressure load of 360 MPa and temperature load of 350°C, the stress cloud of the valve body is shown in Figure 3.

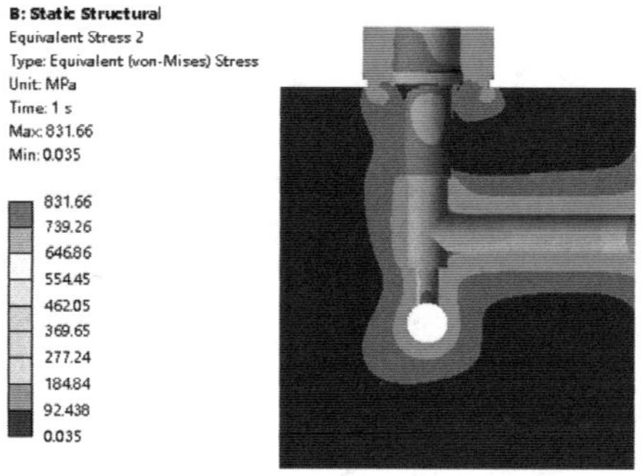

Figure 3. Stress cloud of valve body in open state.

Through the static analysis, it can be seen that under the working conditions of 360 MPa, 350°C, the stress level of the valve body is mostly under 600 MPa, which is less than the yield limit of the valve material at 350°C, 758 MPa. However, there is a concentration of stress near the flow path of the valve body, which is mainly distributed in the corners of the coherent line of the inner surface of the flow path. The maximum stress value is 831.66 MPa, and the stress here exceeds the yield limit of 758 MPa at the working temperature. Therefore, in order to improve the stress distribution of the stress concentration inside the valve body, high pressure self-reinforcement treatment is carried out for the valve body, and the pressure load is added to the valve body inside.

3. Valve body autofrettage analysis

3.1 The principle of autofrettage
Before the valve body is formally put into service, high pressure is applied to the inside of the valve body, making the inside of the valve body undergo plastic deformation, and then a certain amount of residual stress will be generated on the inner wall of the valve body. When the valve body works, the residual stress can offset part of the stress generated by the working pressure, and the valve body stress distribution has been improved.

To carry out the autofrettage treatment, you need to know the maximum autofrettage pressure of the material, and to calculate the maximum autofrettage pressure, you need to know the reverse yield stress of the material, which is determined by the Bauchinger coefficient of the material. The formula for calculating the reverse yield stress is as follows:

$$\sigma_S' = \sigma_S \times BEF$$

where BEF is the Bauzinger factor; σ_S is the material yield limit; σ_S' is the reverse yield stress.

35CrNi3MoVR yield limit at room temperature is 960 MPa, and the material does not have a standard Bauer Singer coefficient. From the relevant research, it can be known the Bauer Singer coefficient can be taken as 0.86 due to the absence of experimental data [10,11], so the material's reverse yield limit stress is 825.6 MPa.

3.2 Valve body autofrettage analysis
There is unloading of autofrettage pre-treatment using stepwise loading. The stepwise loading is used

to understand the process of autofrettage pretreatment of the valve body, and ultimately the monotonic loading of 360 MPa is used to understand the stress distribution condition of the valve body after autofrettage and then loaded again for use. In order to compare the autofrettage effect under different loading conditions, three different maximum loading loads ("different" refers to that the maximum value of the loading load is different) were selected for calculation, respectively, to calculate the residual stresses under different loading conditions as well as the size of the stress distribution in the final work. The loading conditions are shown in Figure 4.

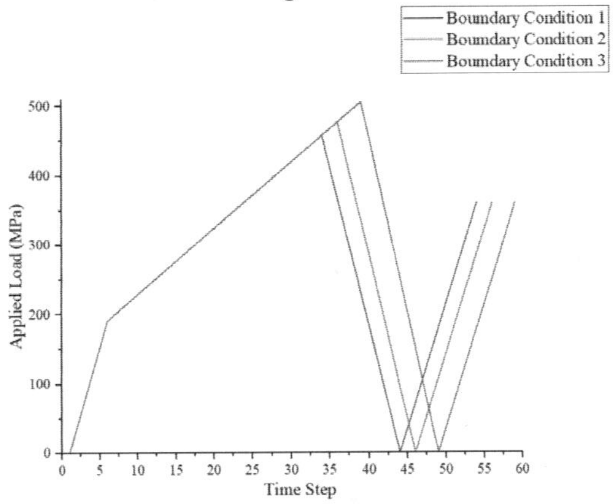

Figure 4. Plot of load loading scenarios.

As can be seen from Figure 4, the common denominator of using three different loading conditions is the preloading to a load of 190.5 MPa at which the yield point occurs, which is obtained by continuous trial loading until initial local yielding occurs. Then the load amplitude of the initial yield limit of 5% of the load amplitude gradually increased until the load to the autofrettage pressure, and then the load amplitude of 10% of the autofrettage pressure gradually unloaded to 0. At this time, the residual stress of the valve body should not be greater than the material's reverse yield limit stress. Then the working load of 360 MPa is applied to the valve body to obtain the stress state of the valve body after the autofrettage treatment.

In the process of autofrettage, the internal stress of the valve body reaches the initial yield limit of the internal pressure, which can be obtained as 190.5 MPa by finite element analysis, i.e., under this pressure, the internal stress of the valve body reaches the material's yield limit of 758 MPa at 350°C. The stress cloud is shown in Figure 5.

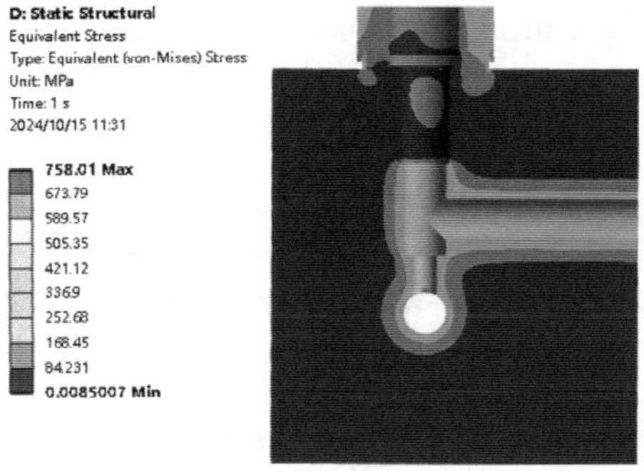

Figure 5. Stress cloud inside the valve body at 190.5 MPa.

Subsequently, the loading amplitude of 5% of the internal pressure at the initial yield limit was continued to each of the three different automatic wheel pressures, and then the 10% amplitude autofrettage pressure was unloaded to 0. At this time, the stress distribution of the body of the valve body for the residual stresses continues to load to the working pressure of 360 MPa after the autofrettage treatment to improve the surface of the valve stress distribution. The valve body surface at this point in the stress concentration has been well improved. The stress concentration condition of the valve body has been improved.

The residual stresses of the valve body under different autofrettage pressures and the stress magnitude of the valve body in the final operating state are shown in Table 1, as shown by finite element analysis. With different autofrettage pressures, the residual stresses are different, and thus the stress state of the final valve body will also change.

Table 1. Table of valve body stresses corresponding to different autofrettage pressures.

Autofrettage pressure /MPa	Residual stress /MPa	Maximum final valve body stress /MPa
457.2	677.62	782.41
476.25	742.88	744.85
504.825	840.46	707.35

The specific variation diagram is shown in Figure 6.

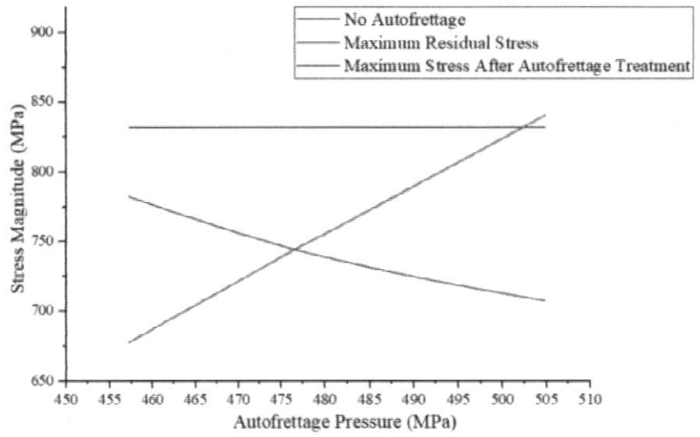

Figure 6. Residual stresses at different autofrettage pressures and the maximum stress magnitude after the autofrettage treatment.

From Table 1 and Figure 6, it can be seen that the maximum residual stress of the valve body increases with the increase of the autofrettage pressure, while the maximum stress of the valve body after autofrettage treatment at the working pressure of 360 MPa decreases with the increasing of the maximum autofrettage pressure. Finally, the stress of the valve body after autofrettage treatment under working conditions is within the yield limit of the material.

However, according to Table 1, when the autofrettage pressure is 504.825 MPa, the maximum residual stress of the valve body is 840.46 MPa, which is beyond the material's reverse yield limit stress of 825.6 MPa. Therefore, it can be seen that the maximum autofrettage pressure is between 476.25 MPa and 504.825 MPa, and the maximum residual stress is 823.95 MPa, which is near the material's reverse yield limit stress of 825.6 MPa, so the maximum autofrettage pressure is 500 MPa. The maximum residual stress of the valve body is 823.95 MPa, which is close to the reverse yield limit stress of the material 825.6 MPa, so the maximum autofrettage pressure of the valve body is near 500 MPa.

At an autofrettage pressure of 500 MPa, the stress cloud of the valve body at operating pressure is shown in Figure 7.

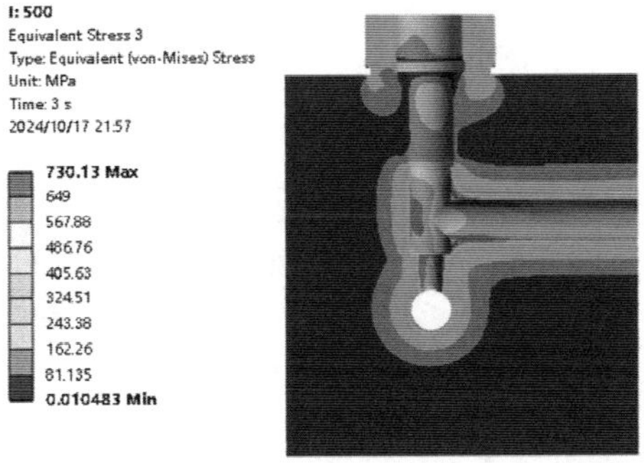

Figure 7. Stress cloud of the valve body in the working condition under an autofrettage pressure of 500 MPa.

Comparing Figure 3 and Figure 7, the maximum stress value of the valve body without autofrettage treatment is 831.66 MPa under working pressure, and the maximum point is located in the coherent line of the inner surface of the valve; the maximum stress value of the valve body after autofrettage treatment is 730.13 MPa under working pressure, the stress distribution is more uniform, and the maximum point is located in the inner wall of the valve body, and the maximum stress value is reduced by 12.2%. The stress concentration on the inner surface of the valve body has been improved.

4. Conclusions

(1) The stress level of the valve body under normal operating conditions of 350°C and 360 MPa is mostly below 600 MPa, which is less than the yield limit of the valve body material at 350°C, 758 MPa. However, there is a concentration of stress at the position of the coherent line of the runner on the inner surface of the valve body, and the maximal stress on the inner surface of the valve is 831.66 MPa, which is over the yield limit, so it is necessary to carry out the autofrettage treatment on the valve body to make the stress distribution on the inner surface of the valve body more uniform. Self-reinforcement treatment of the valve body is needed to make the stress distribution on the inner surface more uniform.

(2) Through the finite element analysis, it can be seen that after the autofrettage treatment of the inner wall of the valve body to form a strong residual stress, the distribution of stress under the operating conditions is more uniform, the maximum stress has been reduced, and the inner surface of the valve body of the worse working conditions in a lower stress state. With the increase of autofrettage pressure, the maximum residual stress of the valve body increases, and the maximum stress on the inner surface of the valve body decreases. When the maximum autofrettage pressure is 500 MPa, the maximum stress on the inner surface of the valve body decreases to 730.13 MPa, i.e. the autofrettage treatment effectively improves the stress state of the inner surface of the valve body.

References

[1] Zhu R.L. (2024) Theoretical study and design method discussion of pressure vessel autofrettage technology based on Mises yield condition. Chemical Equipment Technology, 45(03):28-35.
[2] Ma X. (2016) Research on the best autofrettage pressure of ultra-high pressure valve body. Key Engineering Materials, 667: 524-529.
[3] Shufen R, Dixit U S. (2021) Autofrettage: from development of guns to strengthening of pressure vessels. Cham: Springer International Publishing, 143-160.
[4] GENG S., WANG W., MING Y., et al. (2023) Research progress of ultra-high pressure valves. Fluid Machinery, 51(10):33-42.
[5] Yan T. (2024) Analysis of high temperature creep strength of high pressure steam valves. Modern Industrial Economy and Informatization, 14(05):256-258+262.

[6] Beune A, Kuerten J G M, Van Heumen M P C. (2012) CFD analysis with fluid–structure interaction of opening high-pressure safety valves. Computers & Fluids, 64: 108-116.
[7] Jin H, Zheng Z, Ou G, et al. (2015) Failure analysis of a high pressure differential regulating valve in coal liquefaction. Engineering Failure Analysis, 55: 115-130.
[8] Avitzur B. (1994) Autofrettage—Stress distribution under load and retained stresses after depressurization. International journal of pressure vessels and piping, 57(3): 271-287.
[9] CHEN B.L, TAN G.Y. (2023) Design and finite element analysis of ball valve. China Machinery, (24):7-10+15.
[10] Gao J. Translation. [Japanese] (1983) Guidelines for the design of ultra-high pressure cylindrical vessels. Equipment Design and Technology Center of the Ministry of Chemical Industry. Shanghai.
[11] YU Y.S. (2022) Measurement research on the behavior of metal sheet bauching based on pre-strain-three-point bending. Jilin University:1-90.

Journal of Physics: Conference Series **2992** (2025) 012013 doi:10.1088/1742-6596/2992/1/012013

Topology optimization-based lightweight design of the winch bracket for charged fire-connecting robots

Decheng Meng [1], Xuan Sun [1*] and Bo Chen [2]

[1] School of Mechanical Engineering, University of Jinan, Jinan, China
[2] Linyi Lingong Intelligent Information Technology Co., Ltd., Linyi, China

[*] Corresponding author's e-mail: me_sunx@ujn.edu.cn

Abstract: The charged fire-fighting robot is a high-altitude operating robot, minimizing its weight while ensuring structural strength is an important part of the design goal. Topology optimization is a mathematical method to perform structural optimization, which can help designers optimize the design of products in the product development process in order to reduce the weight of the product, improve the performance of the product, and reduce the cost. In this paper, Inspire software is applied to carry out topology optimization design with a winch bracket as an example, which improves the bracket frame structure, effectively solves the material redundancy problem of the winch bracket, and reduces the maximum Mises equivalent force at the stress concentration.

1. Introduction

With the continuous development of science and technology, the material needs of the people are increasing. At the same time, the requirements for the reliability of electricity consumption are also gradually increasing. In order to minimize power outages caused by accidents and maintenance, and to improve the reliability of electricity consumption, distribution network power-carrying operation technology has been proposed and applied to the actual distribution network. The traditional charged fire-connecting relies on operators wearing insulated shielding clothing using insulated bucket arm trucks or climbing tools and insulated operating rods. The traditional method of charged fire-connecting has problems such as high labor intensity, high danger, high cost, and long time, so the charged fire-connecting robot came into being [1]. The charged fire-connecting is suspended on the 10KV cable line through the hook, sling, and winch on the line to carry out the work. As high-altitude work equipment, in order to ensure strength and life, the lightweight of the components of the charged fire-connecting robot is one of the important design objectives [2]. This paper takes the winch bracket as an example and carries out structural analysis and optimization based on the principle of topology optimization.

Topology optimization is a mathematical method to optimize the material distribution in a specified region according to the given load cases, constraints, and material performance parameters. Through topology optimization, it is possible to remove excess materials and realize the light weight of the structure under the premise of ensuring the structural strength of the charged fire-connecting robot. It is able to make a more reasonable layout for the distribution of materials, improve the stress situation of components, and extend the service life.

Journal of Physics: Conference Series **2992** (2025) 012013 doi:10.1088/1742-6596/2992/1/012013

2. Pre-processing for topology optimization of winch mounts

2.1. Model construction

This paper takes the winch bracket as the object of topology optimization design. Figure 1(a) shows its 3D model, in order to ensure the support of the bracket to the winch parts, the overall outline (dark part) is retained and set as the non-design space, and the light color part is divided into the design space. In Figure 1(b), the winch bracket is connected to the robot body by bolts and to the winch parts by pins.

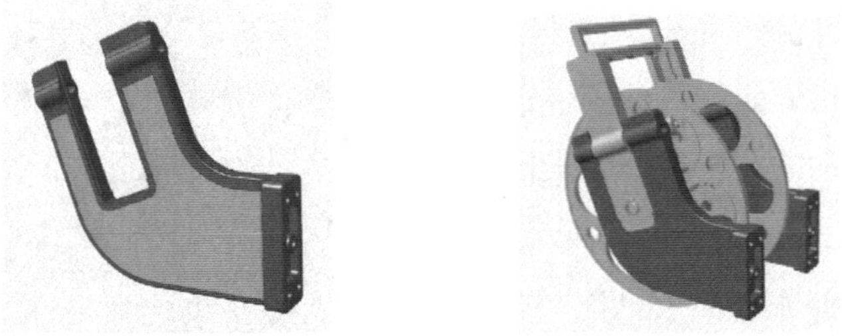

(a) Winch bracket (b) Synthesis

Figure 1. Winch bracket 3D model.

In charged fire-connecting operations, the robot works in a high-voltage charged environment. Good electrical insulation is the key to ensuring the safety of the operation and the normal operation of the robot. Compared with steel, aluminum alloy, and other metal materials, first of all, nylon material has high volume resistance and breakdown voltage, which can effectively prevent current leakage, preventing the robot from leaking electricity and causing safety accidents. Secondly, the lower density of nylon material is conducive to reducing the overall weight of the charged fire-connecting robot, making it more flexible and convenient to move on the overhead distribution line, reducing the load impact on the line. At the same time, it is also easy for the operator to carry and install the robot.

In summary, the winch bracket material selection is nylon, and the winch bracket weight is 738g. The material properties are shown in Table 1.

Table 1. Material properties of winch mounts.

Materials	Densities /(kg·m^{-3})	Young's modulus /MPa	Poisson's ratio	Yield strength /MPa	Tensile strength /MPa
Nylon	1.23	$2.91*10^3$	0.41	75	70

2.2. Load analysis

When working, the charged fire-connecting robot rises to a position close to the main line, through the symmetrical two groups of winches in the same direction or reverse rotation, so that the walking wheel mounted on the main line, the analysis can be seen. In the process of mounting the line, the robot deflected to the maximum angle of the winches components of the load is maximum. The force is shown in Figure 2.

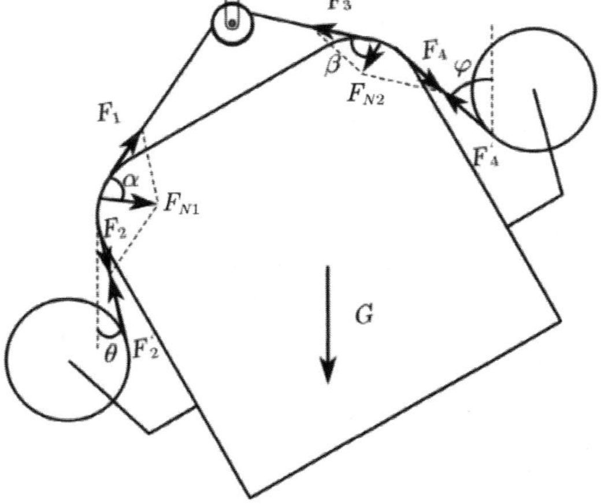

Figure 2. Force diagram of the charged fire-connecting robot.

Neglecting the weight of the sling, the tension on the sling can be viewed as an action and reaction force according to Newton's third law.

For two slings, we get:

$$F_1=F_2=F_2^{'}$$
$$F_3=F_4=F_4^{'}$$

The combined force acting on the frame from the bend in the sling is:

$$F_{S1}=2F_1\times\cos\alpha$$
$$F_{S2}=2F_3\times\cos\beta$$

For the robot column force balance equations, we get:

$$\begin{cases} F_2^{'}\times\cos\theta+F_4^{'}\times\cos\varphi=F_{N1}\times\sin\left(\frac{\pi}{2}-\alpha-\theta\right)+F_{S2}\times\sin\left(\frac{\pi}{2}-\beta+\varphi\right)+G \\ F_{S1}\times\cos\left(\frac{\pi}{2}-\alpha-\theta\right)=F_{S2}\times\cos\left(\frac{\pi}{2}-\beta+\varphi\right)+F_2^{'}\times\sin\theta+F_4^{'}\times\sin\varphi \end{cases}$$

When the overall deflection angle of the robot is 20°, the walking wheels can be extended to cross the main line at this point:

$$\theta=9.3°,\ \alpha=65.5°,\ \beta=82.6°,\ \varphi=30.7°$$

Substituting the above parameters into each equation gives:

$$F_1\approx281\text{N},\ F_3\approx253\text{N}$$

For the two winch brackets, the torque generated by F_1 and F_3 is transmitted to the motor through the gears and finally acts with the motor bracket, so it does not need to be considered. The decomposition of F_1 and F_3 along the direction of relative motion and its normal direction is shown in Figure 3.

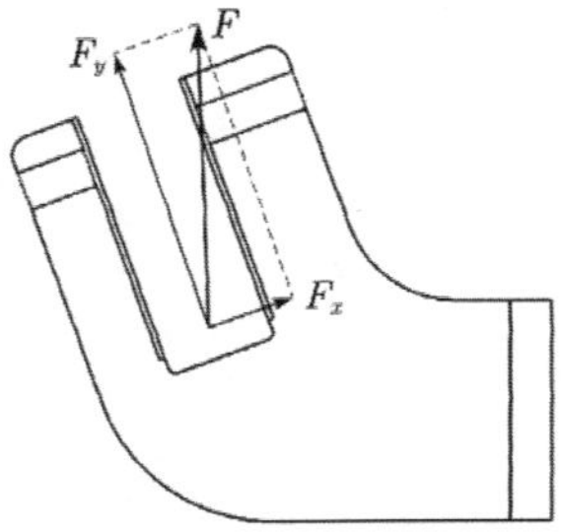

Figure 3. Winch bracket force diagram.

From the above analysis, it can be seen that the left winch is subjected to a greater load than the right winch, so the left winch is selected for analysis:

$$\begin{cases} F_x = F \times \cos\omega \\ F_y = F \times \sin\omega \end{cases}$$

We substitute F=281N and ω=59.3° to obtain:

$$F_x \approx 144\text{N},\ F_y \approx 242\text{N}$$

2.3. Load and boundary condition setting

Fixed constraints are applied to the five bolt holes on the side, 144N load is applied perpendicular to the direction of relative motion, and 121N load is applied to each of the two pinholes, with the direction outward along the direction of relative motion; the contact between the design space and the non-design space is set as the default (bound contact); the cell size is set as 3 mm; and the shape control scheme is symmetric control and bi-directional pullout. The loads and boundary conditions of the model are shown in Figure 4.

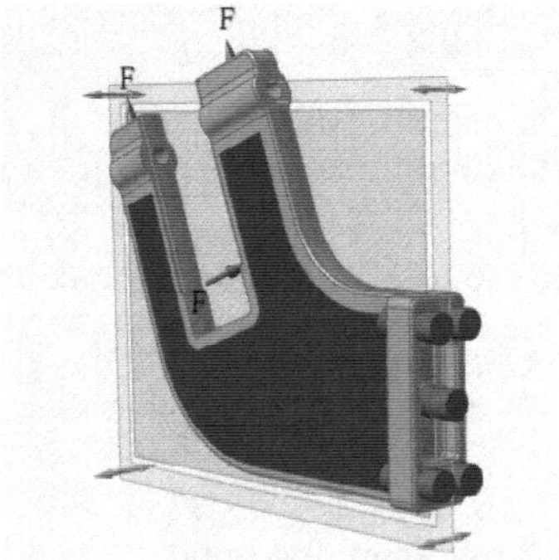

Figure 4. Loads and Boundary Conditions.

2.4. Initial finite element analysis

The finite element analysis results of the winch bracket are shown in Figure 5, from which it can be seen that the overall stress level is low, and the stress concentration phenomenon occurs at the position of the shape mutation, with the maximum Mises equivalent stress of 9.11MPa, but it is still small compared to the yield strength of nylon material of 75MPa. It can be seen that the original structure has a large optimization space.

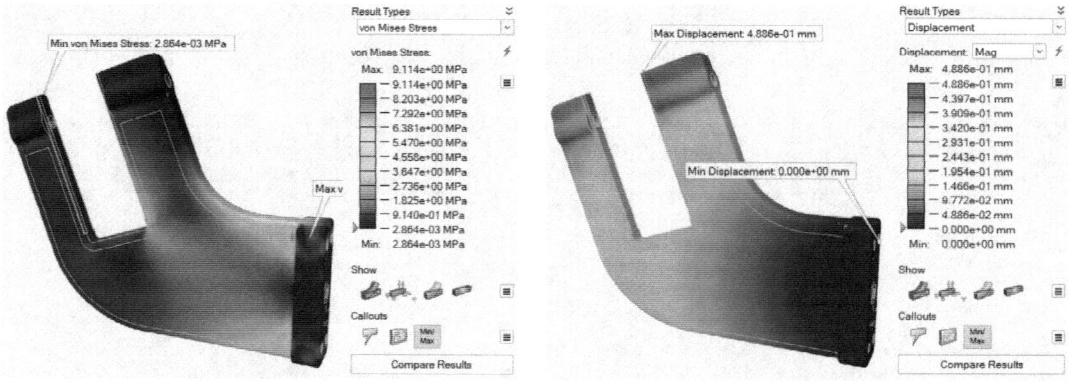

(a) Mises equivalent force (b) Displacement

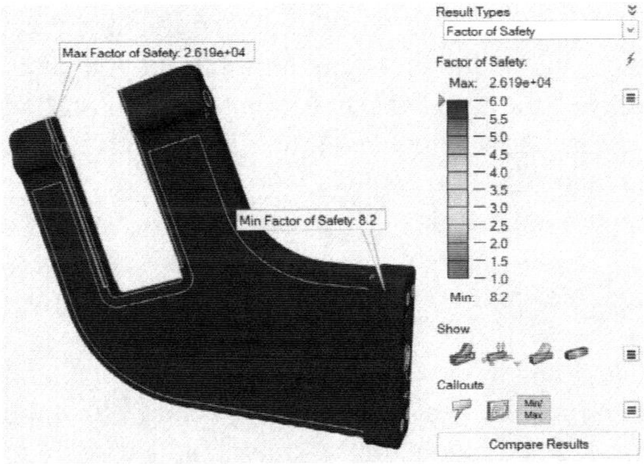

(c) Safety factor

Figure 5. Initial model static strength analysis results.

3. Topology optimization

3.1. Principles of topology optimization

Topology optimization is a mathematical method to optimize the material distribution in a specified region according to the given load cases, constraints, and material performance parameters, which is a kind of structural optimization [3]. Generally, for an engineering structural optimization problem, the optimization process is shown in Figure 6:

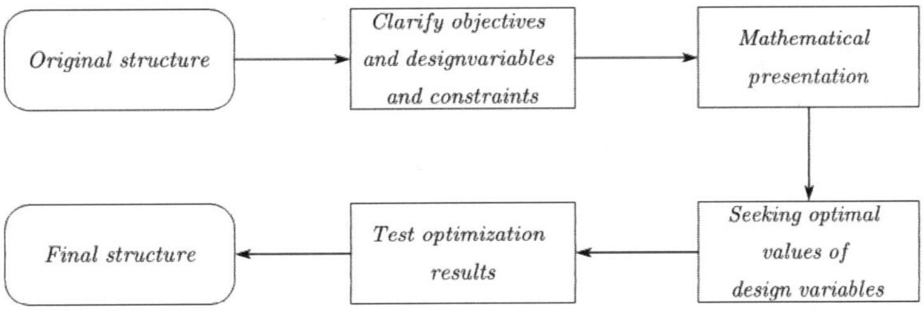

Figure 6. Structural optimization process.

3.1.1. Clarify objectives, design variables, constraints. Currently, the most widely used topology optimization method is the variable density method [4]. The core of the variable density method is to optimize the density of each unit in the structure, so as to obtain the optimal solution of the overall structure. In topology optimization problems, the minimum structural flexibility is often taken as the objective function. Topology optimization relies on the finite element method. For a finite element mesh, the cell stiffness array is [5]:

$$K^{e}=\int_{V_e} \mathrm{B}^{\mathrm{T}}\mathrm{DB}\,\mathrm{d}V$$

$$\mathrm{D}=\frac{\mathrm{E}_0}{1-v_0^2}\begin{bmatrix} 1 & v_0 & 0 \\ v_0 & 1 & 0 \\ 0 & 0 & \frac{1-v_0}{2} \end{bmatrix}$$

where B is the strain matrix; and D is the material elasticity matrix.

It is assumed that the modulus of elasticity of each cell is different so that the modulus of elasticity of each cell can be used as a design variable [6]. However, in general, it is necessary to select the material first and then perform the topology optimization. In the variable density method, we let:

$$E_0(x)=\rho(x)^p E$$

where $E_0(x)$ is the Young's modulus of the xth cell used to calculate the stiffness array; $\rho(x)$ is the "density" of the xth cell, which is the design variable in the structural optimization problem, and takes the value range $0<\rho(x)<1$; E is the Young's modulus of the material itself; p is the density coefficient, which is used to adjust the intermediate density cell, so that $\rho(x)^p$ is converging to 0 or 1, with 0 being the absence of the material and 1 being the full material, and taking the value range $p>1$ full of material, and the value range $p>1$.

Therefore, for an n×m finite element mesh, it has n×m design variables, i.e., the density of these n×m meshes.

The variable density method of topology optimization often uses the volume fraction as a constraint function [7].

3.1.2. Mathematical presentation. The mathematical model of the topology optimization variable density method is:

$$Find\ \rho=(\rho_1,\rho_2,\ldots,\rho_{m\times n})$$

$$\min C(\rho)=U^TKU=\sum_{e=1}^{m\times n}E_e(\rho_e)u_e^Tk_0u_e$$

$$\text{s.t.}\begin{cases}\dfrac{V(\rho)}{V^*}\leq f\\ 0<\rho_{\min}<\rho<1\end{cases}$$

where ρ_i is the design variable; $C(\rho)$ is the structural flexibility function; flexibility is inversely proportional to stiffness, the smaller the flexibility, the larger the stiffness; k_0 is the stiffness matrix; u_e is the displacement column vector; $V(\rho)$ is the volume of the design space after optimization; V^* is the volume of the design space at the time of the initial design; f is the fraction of the volume of the design space; and $\rho_{\min}$ is the minimum value of the relative density [8].

3.1.3. Finding optimal values of design variables and testing optimization results. For general structural problems, the optimal value of the design variables is the value of the design variables when all possible damage modes occur at the same time. Then the optimal structure is obtained [9]. The commonly used method for solving the optimal value is the iterative method, in which the values of the design variables, the upper and lower bounds, the gradient of the objective function, the value of the constraint function, and the gradient are substituted into the solver. Iterative calculations can be carried out to obtain the optimal value of the design variables.

Finally, the optimization results are re-imported into the finite element software for re-analysis to check whether the stresses and displacements meet the design requirements.

3.2. Optimization results

The optimization objective is set to maximize stiffness, the mass objective is set to 30%, and the thickness constraints are a minimum of 5mm and a maximum of 10mm. The results of topology optimization are shown in Figure 7, from which it can be found that the topology optimization removes most of the regions with low stress, and only retains the regions that generate stress concentration and some of the necessary connections, making the structure more compact and reasonable.

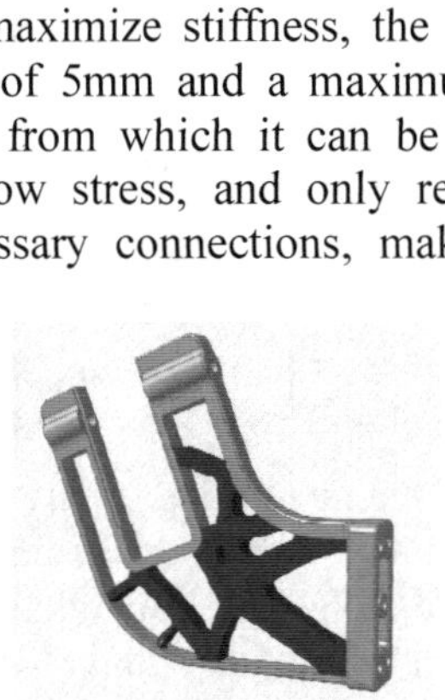

Figure 7. Topology optimization results.

Journal of Physics: Conference Series **2992** (2025) 012013 doi:10.1088/1742-6596/2992/1/012013

3.3. Model reconstruction

The topology-optimized model has rougher edges, which are not easy to manufacture for machining and 3D printing, so the model needs to be reconstructed [10]. PolyNURBS auto-fitting is a featured function of Inspire software that creates smooth, tight surface models. In the PolyNURBS module, the number of polyNURBS faces was set to 1000, the curvature was set to 50%, and the number of iterations was 40. The reconstruction of the model in Figure 4 was carried out to obtain a stent model with a smooth and smooth surface. The results are shown in Figure 8. Using the software to analyze the weight, the mass of the original stent is about 0.113 kg. The mass of the reconstructed model is about 0.071 kg, which is about 37.2% reduction in mass.

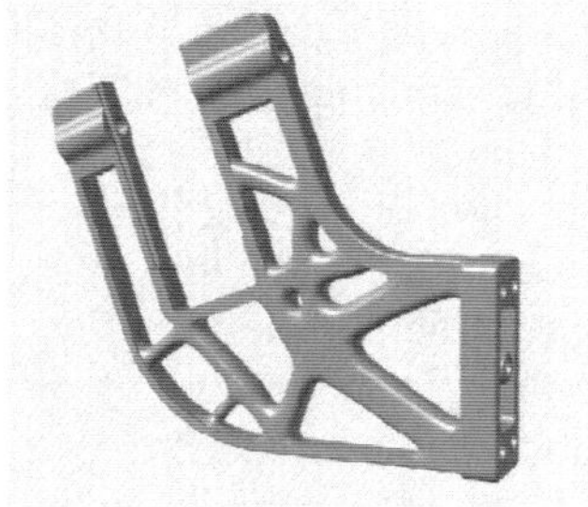

Figure 8. Reconstructed model.

3.4. Reconstruction model finite element analysis

Using Boolean operations to merge the models and redefine the same loads and boundary conditions as the original structure, the static strength analysis was performed on the model of Figure 8. The results are shown in Figure 9.

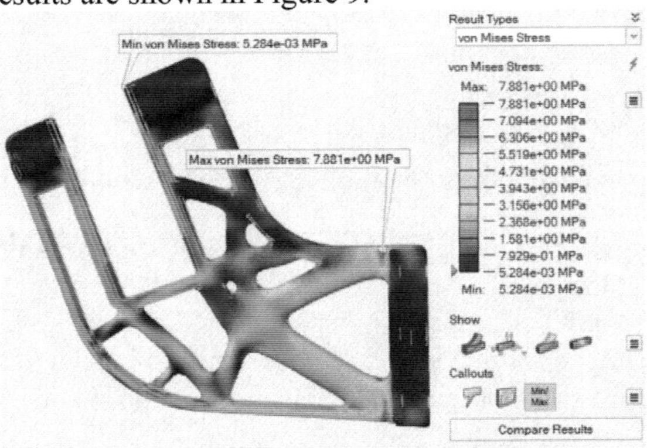

(a) Mises equivalent force

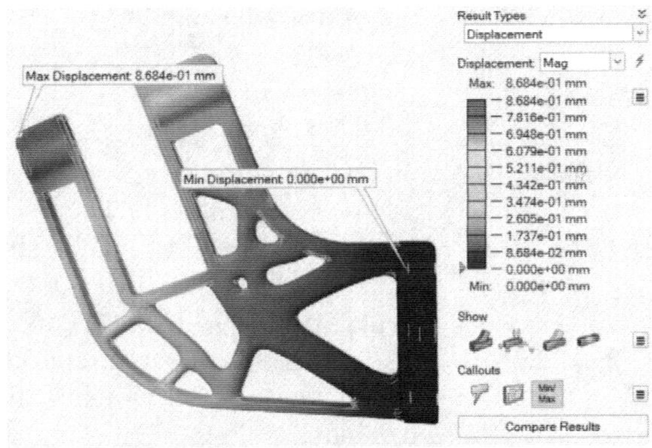

(b) Displacement

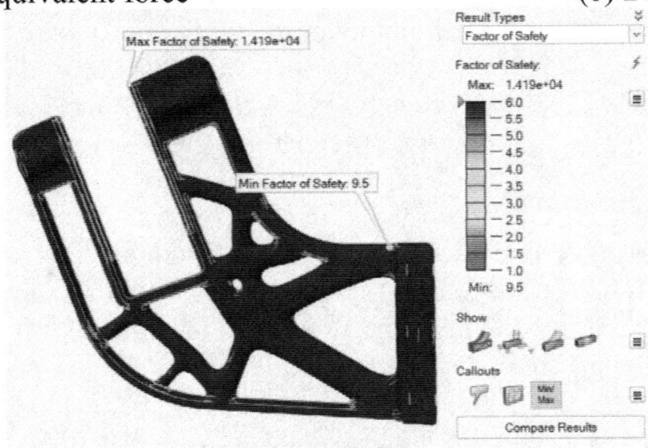

(c) Safety factor

Figure 9. Static strength analysis results of the optimized model.

As can be seen in Figure 9, the maximum displacement of the winch bracket after geometric reconfiguration by PolyNURBS fitting is 0.868 mm, the minimum safety factor is 9.5, and the maximum Mises equivalent force is 7.9 MPa, which meets the strength design requirements of the lightweight target.

4. Conclusion

According to the finite element analysis results of the initial and optimized winch bracket, the following conclusions can be drawn: 1) The overall mass of the optimized winch bracket decreases significantly. Its maximum displacement increases, but it is still less than 1 mm, which is in line with the design requirements. Reduced mass makes it easier to transport the fire-fighting robot to the job site, allowing it to better face complex terrain and line conditions. 2) After the reconstruction of the structure of the winch bracket using the PolyNURBS auto-fitting, the maximum Mises equivalent stress decreases, which indicates that its mechanical performance improved. The stress distribution is more uniform and reasonable, which can effectively improve the bearing capacity and anti-fatigue performance of the key parts, reduce the structural deformation and damage caused by uneven stress during the operation, and improve the reliability and safety of the operation.

At present, the lightweight of the robot in this project is only for the external frame. Future research will focus on developing more advanced internal actuation tools, including the crimp joints of hydraulic pliers and the rotating components of wire strippers. The goal is to reduce the weight of the charged fire-connecting robot and enhance the force distribution across its various parts, ensuring it meets the practical requirements of its intended use.

Acknowledgements

Thanks to the Key R&D Program of Shandong Province (China, Project Number: 2022TSGC2376, 2023TSGC0738, 2023CXPT047) for funding this study.

References

[1] Liu, B. Y., Zheng, S. Y., Zhao, H. (2020) Research on live line operation and maintenance of 10 kV distribution network overhead line. Electric Power Equipment Management, 10: 75-76+89. https://c.aiiz.cn/qkdbqy

[2] Jiang, P., Wu, G. P., Yang, S. (2021) The structure design and implementation of tree pruning robot for 220kV live line. Machinery Design & Manufacture, 08: 152-156. DOI: 10.19356/j.cnki.1001-3997.2021.08.036.

[3] Lei, L. M., 2017. Analysis of structural topology optimization method. In: 2017 Aviation Intelligent Equipment and Test Technology Annual Conference. WuHan. pp. 83-86+89. DOI: 10.26914/c.cnkihy.2017.004271.

[4] Ji, L. F., Mi, Y. H., Wu, M. Q., Chen, G. H., Su, Y. (2024) Lightweight design of automotive shock absorber structures based on inspire software. Automobile Applied Technology, 49(13): 43-48.DOI: 10.16638/j.cnki.1671-7988.2024.013.009.

[5] Guo, Y., Parker, R. G. (2012) Stiffness matrix calculation of rolling element bearings using a finite element/contact mechanics model. Mechanism and Machine Theory, 51(5): 32–45. https://doi.org/10.1016/j.mechmachtheory.2011.12.006

[6] Zhang, D. H., Lu, X. Y. (2017) The theory and method of structural topology optimization design. Chemical Engineering & Equipment, 01: 155-158. https://c.aiiz.cn/UwwYqy

[7] Yao, Z. K., Gong, P., Ji, S. D. (2024) Lightweight design of multi-habitat unmanned aerial vehicle's fuselage based on topology optimization. Journal of Machine Design, 41(09): 58-65. DOI: 10.13841/j.cnki.jxsj.2024.09.003.

[8] Bendsøe, M. P., Sigmund, O. (1999) Material interpolation schemes in topology optimization. Archive of Applied Mechanics, 69: 635-654. https://doi.org/10.1007/s004190050248

[9] Meng, P., Liu, M., Yu, M. (2022) Optimal suspension points based on synchronous failure principle for a new double hoisting reclaiming boom. Mechanics in Engineering, 44(04): 884-889. https://c.aiiz.cn/azbBeT

[10] Xiang, G., Zhu, Y., Xu, R. (2022) Lightweight topology optimization design of arc support basedon inspire software. Automobile Applied Technology, 47(23): 83-87. DOI: 10.16638/j.cnki.1671-7988.2022.023.015.

Journal of Physics: Conference Series **2992** (2025) 012014 doi:10.1088/1742-6596/2992/1/012014

The design and analysis of an industrial five-axis laser elbow cutting machine

1st Wentao Yang, 2nd Junwei Liu, 3rd Tao Fan and 4th Yong Wang

School of Mechanical Engineering, Jiangsu University of Technology, Changzhou, China

* Corresponding author: 1010188314@qq.com

Abstract: In the field of industrial manufacturing, metal elbows have garnered significant attention due to their extensive applications. With the advancement of technology, traditional cutting methods struggle to meet the precision and efficiency demands of modern industry. Laser-cutting technology, known for its high precision and efficiency, has become the mainstream method for processing metal elbows. This paper introduces the working principle and system components of laser cutting, with a focus on the design of a processing system based on a horizontal laser cutting machine. The CNC system adopts multi-axis synergistic control and realizes precise positioning of the workpiece and laser cutting through servo motors, which can meet the demand for high-efficiency cutting of metal elbow.

1. Introduction

In today's industrial manufacturing sector, metal elbows have garnered significant attention due to their wide range of applications. From aerospace and automotive manufacturing to petrochemical industries, metal elbows serve as critical structural and functional components, playing an essential role. With continuous technological advancements, traditional methods for cutting metal elbows have struggled to meet the increasing demands for precision and efficiency [1-3]. Laser cutting technology, with its superior cutting accuracy, high processing speed, excellent material adaptability, and strong automation control capabilities, has gradually become the mainstream method for processing metal elbows [4-5].

Due to its high energy and high coherence, laser technology can be used for drilling and cutting industrial components. A horizontal laser beam emitted by the laser passes through an expander lens and focusing lens for shaping, producing a high-energy laser beam that is concentrated into an extremely small spot at the focal point [6-7]. The energy density at this spot is incredibly high, instantly bringing the material to its ignition point. Through height control of the cutting head, the laser irradiates the material, causing it to melt and evaporate instantly, forming a hole [8]. Assisted by gases such as air or oxygen, the molten metal is blown away. As the focal point moves continuously, a slit of approximately 0.1 mm is formed in the material, achieving the purpose of cutting [9-10].

Modern laser cutting technology has evolved into a comprehensive system, equipped not only with high-power, high-quality mode lasers but also with high-precision worktables driven by ball screws. By utilizing multi-coordinate linkage control through a computer, various complex spatial shapes can be cut [11-12]. The metal elbow laser cutting system is primarily composed of two parts: the laser and the laser cutting machine, the latter of which includes the laser head, control system, workpiece clamping system, workpiece movement system, and various auxiliary peripheral equipment [13].

2. Main structure design

In the preliminary design of laser processing for elbows, several technical approaches are considered based on different processing needs and scales:

1) Wire cutting

The workpiece remains stationary, but this method has a slow cutting speed and takes a long time, making it unsuitable for large-scale production.

2) Vertical laser cutting machine with multi-angle and multi-size devices

While this method allows processing, it suffers from poor flexibility, frequent fixture changes, and low efficiency, which makes it challenging to meet the demands of mass production for elbows.

3) Horizontal laser cutting machine with clamping device

In this approach, the workpiece is fixed while the laser head moves and adjusts its path to complete the cutting. This method offers high efficiency and strong flexibility, making it ideal for large-scale production.

4) Rotary clamping device with a vertical laser machine

This approach uses relative motion cutting and is advantageous for handling complex workpieces, though it is not as efficient as horizontal cutting.

The horizontal laser cutting technology was chosen. This method achieves precise cutting by clamping the workpiece and moving the laser head, offering significant advantages in both processing efficiency and accuracy. Compared to other methods, the horizontal laser machine is better suited for large-scale production, as it minimizes downtime and enhances overall production efficiency.

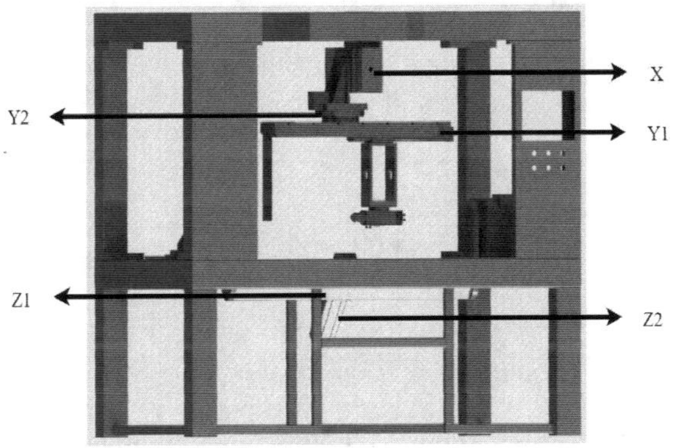

Figure 1. The cutting machine's main mechanical structure.

In the design of the laser elbow cutter in Figure 1, the main part is five axes (X-axis, Y1-axis, Y2-axis, Z1-axis, Z2-axis), each of these axes is responsible for a different function, and through the synergy of the work, to achieve high-precision laser cutting of the elbow.

2.1 X-axis—laser rotation and movement axis

The main function of the X-axis is to adjust the position of the laser head to ensure that the laser head always stays in the same straight line with the center of the circle of the workpiece, to reduce the substantial displacement of the Y1-axis during the cutting process, and to improve the overall cutting accuracy. The X-axis controls the horizontal movement of the laser head through the servo drive and precisely locates the initial position of the laser head to align it with the center of the circle of the workpiece. By reducing the displacement requirement of the Y1 axis, the X-axis significantly reduces the positioning error during the cutting process and automatically adjusts the position of the laser head according to PLC commands at different cutting stages.

Journal of Physics: Conference Series **2992** (2025) 012014 doi:10.1088/1742-6596/2992/1/012014

2.2 Y1-axis—laser head following moving axis

The main function of the Y1 axis is to monitor the distance between the laser head and the surface of the elbow in real-time by means of a capacitive height adjuster and make dynamic adjustments by means of a servo drive to ensure that the laser head maintains the ideal cutting distance from the surface of the workpiece. This axis not only copes with irregularities in the surface of the workpiece but also adjusts the position of the laser head in real-time during the cutting process to ensure that the laser focal point is always in the optimal working position. The precise control of the Y1 axis avoids inconsistencies in cutting quality due to changes in distance.

2.3 Y2-axis--laser rotation cutting axis

The Y2 axis is responsible for controlling the laser head to rotate 360° around the workpiece for cutting, thus enabling wrap-around cutting operations on elbows. By controlling the rotation of the laser head, the Y2 axis is able to adapt to the complex geometry of the workpiece, ensuring continuity and accuracy of the cut. This axis adjusts the rotation angle and speed in real-time during the cutting process according to the shape of the workpiece and the processing requirements to meet different cutting path requirements and maintains stable operation of the laser head throughout the cutting process to avoid interruptions.

2.4 Z1-axis--workpiece aotation and lifting axis

The function of the Z1 axis is to control the rotation and lifting of the workpiece so that it can be cut at different angles. This axis adjusts the position of the workpiece according to the bending angle of the workpiece so that the laser head can contact the surface of the workpiece from the optimal angle to adapt to the needs of different cutting angles. The Z1 axis can dynamically adjust the height of the workpiece to ensure that the cutting of different parts can be carried out at the appropriate height and angle.

2.5 Z2-axis—workpiece movement axis

The Z2 axis controls the cutting path and feed of the workpiece through precise linear movement to ensure high accuracy in cutting operations. This axis is capable of adjusting the workpiece's movement speed and position to control the depth and quality of the cut according to the cutting requirements. The Z2 axis works in concert with the other axes to ensure precise movement of the workpiece along a predetermined path as the laser head rotates and the workpiece is lifted to achieve synchronized multi-axis motion. This coordinated motion greatly improves the efficiency of cutting while ensuring the precision of the workpiece.

3. Hardware system construction

The control system architecture of the laser elbow-cutting machine is shown in Figure 2. The system is divided into two major parts: the upper computer and the lower computer, which are responsible for the processing of non-real-time tasks and real-time tasks respectively. The main functions of the upper computer include data acquisition and display, motion control command generation and transmission, etc., relying on Kunlun Tongtai HMI (Human Machine Interface) to realize the operation and monitoring. The lower computer runs on Huichuan H5U PLC, which is mainly responsible for receiving commands from the upper computer and carrying out real-time task processing, such as servo motor drive control, laser power adjustment, etc. Specifically, the upper computer receives commands from the H5U PLC through the HMI (Human Machine Interface).

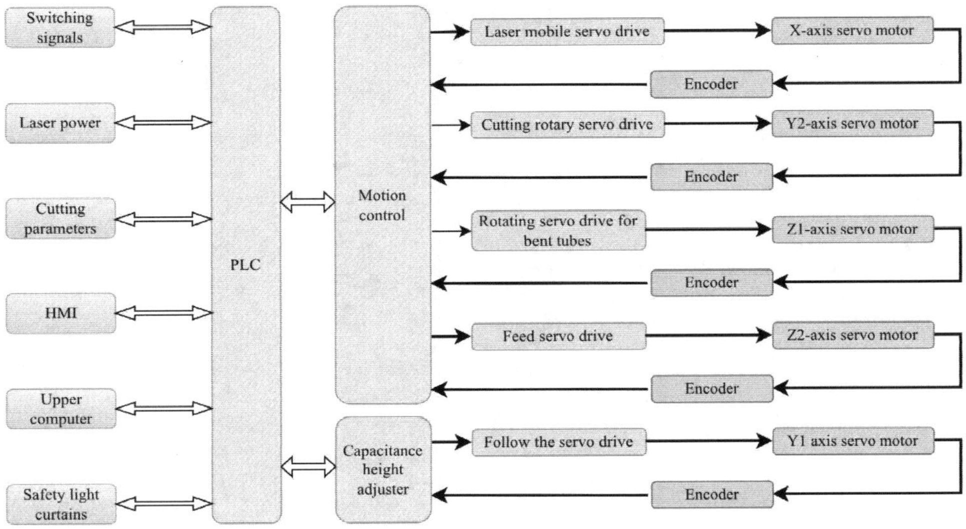

Figure 2. Hardware system structure diagram of laser elbow cutting machine.

Specifically, the upper computer receives and analyzes the relevant cutting parameters inputted by the operator through HMI. The data refresh cycle of HMI is 100ms, which ensures the real-time and interactivity of the system. After the cutting parameters are processed, the valid motion commands and laser commands are stored in the command queue. These instructions are then transferred to the lower computer for further execution.

In the lower computer, the real-time tasks run on the PLC, which has an execution cycle of 4ms to ensure the high efficiency of the system response. The PLC generates the position information of each servo axis through the motion control module and encoder to drive the servo motors of each axis to complete the precise movement of the cutting head. The servo system adopts the Huichuan SV660 series drive, which is connected to the servo motor for controlling the precise movement of the X, Y, and Z axes. These motion data are fed back through encoders to ensure high-precision control.

The height adjustment module is an important part of the system, which uses a Panasonic MBDKT2510CA1 model motor and servo drive combined with a CHC-1000L capacitive height adjustment system to dynamically adjust the distance between the laser head and the workpiece through real-time detection of the height of the surface of the workpiece, so as to ensure the stability of the cutting quality. This module effectively compensates the height error in high-speed movement, which further improves the processing accuracy and surface quality.

The whole system ensures precise control of multi-axis servo motors and efficient laser cutting through the cooperative work of Huichuan H5U PLC with servo drive and capacitor height adjuster. In addition, the safety module also integrates a safety light curtain, which can immediately stop the equipment operation when detecting the operator entering a dangerous area, thus guaranteeing the safety of the operator.

4. Follower mechanism design

The follower system of the laser elbow cutter is a key component to ensure efficient and accurate cutting. This system integrates the CHC-1000L height adjuster mainframe, preamplifier, laser head, servo motor, servo driver, and rail module to form an efficient control architecture, as shown in Figure 3. The CHC-1000L height adjuster host is responsible for the precise adjustment of laser power, which can monitor and stabilize the laser energy output during the cutting process in real-time to ensure the consistency of cutting quality. The introduction of a preamplifier enhances signal strength and reduces noise interference, making the laser-cutting process more efficient.

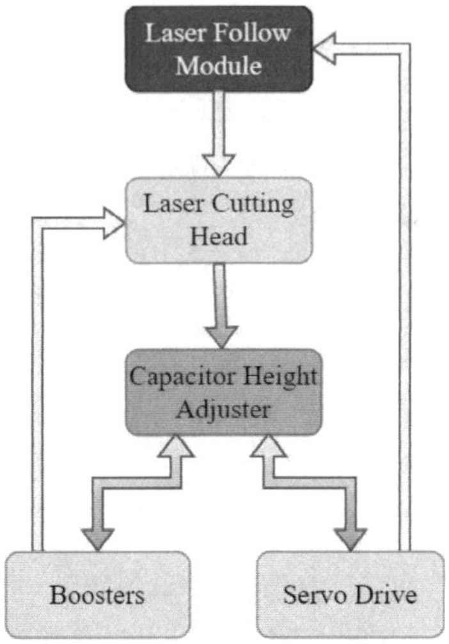

Figure 3. Schematic diagram of laser following system.

The laser head, on the other hand, integrates a high-precision optical system that tracks the position of the workpiece in real-time and dynamically adjusts the cutting parameters according to the feedback to adapt to the cutting needs of different materials and thicknesses. The combination of servo motor and servo drive provides excellent motion control capability, responding quickly to changes in the cutting path and optimizing cutting efficiency. Meanwhile, the guide rail module ensures stable movement of the laser head during the cutting process.

5. Cooperative motion control

In the control system of the laser elbow-cutting machine, the operator first inputs the geometric parameters of the elbow and the relevant process parameters for laser cutting through the HMI (Human Machine Interface). Subsequently, the elbow is fixed on the fixture and the PLC automatically controls the motion of each axis according to the input parameters to ensure efficient and accurate cutting.

First, the PLC controls the X-axis to move the laser head to the initial position corresponding to the elbow to ensure that the laser head is aligned with the center of the workpiece circle. Next, the Z1 axis rotates and lifts the workpiece to the desired cutting angle, ensuring that the laser head can contact the workpiece at the optimal angle. At the same time, the Z2 axis controls the linear feed of the workpiece, moving it precisely to the position where it needs to be cut in preparation for subsequent cutting operations.

After the positioning and adjustment of the workpiece and the laser head are completed, the Y2 axis, as shown in Figure 4, begins to rotate 360° around the workpiece for cutting. In the process of continuous cutting, the Y2 axis can not only realize forward rotating cutting but also reverse rotating to cut the next part. This alternating forward and reverse cutting can reduce the pause in the cutting process and improve the efficiency and continuity of cutting.

Figure 4. Laser rotary cutting elbow.

This multi-axis coordinated motion design ensures a high degree of coordination between the laser head and the workpiece throughout the entire process. The PLC realizes high-precision cutting of the elbow by accurately controlling the motion of each axis, adapting to complex geometries and diversified cutting needs. The coordinated work of the X-axis, Y1-axis, Y2-axis, Z1-axis, and Z2-axis at different stages not only ensures the cutting precision but also significantly improves overall processing efficiency.

6. Cutting test

The effect of laser cutting parameters on the cutting accuracy and surface roughness of the elbow is explored through experiments. Then the optimal combination of parameters is determined to improve the processing quality and efficiency of the laser elbow-cutting machine. This experiment uses the cutting parameters in Table 1. The experiment uses the laser elbow cutting machine designed in this paper, equipped with a fiber laser generator. The power range is set between 6000W and 10000W. The experimental object is a stainless steel elbow with a diameter of 40mm and a bending radius of 76mm. In order to ensure the accuracy of the experimental results, a high-resolution optical microscope and a surface roughness meter were used to evaluate the edge morphology and surface quality after cutting, respectively.

Table 1. Laser cutting parameter setting.

Cutting parameters	Digital
Bending radius	76mm
Elbow radius	40mm
Initial angle	0°
Bend thickness	1mm
Output power	6000W~10000w
Cutting rotation angle	370°
Cutting gas	Helium
Cutting speed	150°/s~250°/s

This experiment is based on the single-variable method, which gradually adjusts key parameters such as cutting power and speed to observe their effects on cutting quality. In the cutting power experiment, the cutting speed was kept fixed at 200°/s, the laser power was set in the interval of 6000W to 10000W, and the laser power was gradually adjusted upward to observe the changes in cutting accuracy and surface roughness under different powers. Subsequently, in the cutting speed experiment, the power was kept fixed at 9000W. The cutting speed was gradually increased in the range of 150°/s to 250°/s to analyze the cutting effect under different speeds.

Journal of Physics: Conference Series **2992** (2025) 012014 doi:10.1088/1742-6596/2992/1/012014

The microscopic morphology of the cutting edge was observed using an optical microscope after the cutting was completed. The finish of the cut surface was measured with the help of a surface roughness meter. The cutting effect is shown in Figure 5:

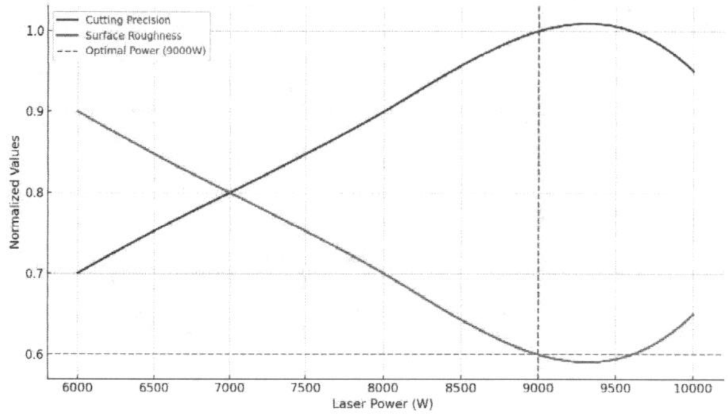

Figure 5. Flight trajectory comparison.

The results show that the increase in laser power has a significant positive effect on the quality of the bend cut. With the increase in power, the cutting accuracy steadily increases, while the surface roughness gradually decreases. When the power reaches 9000W, the cutting accuracy and surface quality have reached the optimal state. Although further increasing the laser power can accelerate the processing speed, too high a power may cause the intensification of the thermal effect on the surface of the material, thus affecting the cutting accuracy and surface finish.

In terms of cutting speed, the increase in speed helps to improve the cutting quality within a certain range. When the cutting speed reaches 230°/s, the cutting accuracy and surface roughness are in an optimal state. Then after the speed continues to increase to 250°/s, the cutting quality begins to decline slightly, and the surface roughness increases. Too high cutting speed may lead to insufficient contact time between the laser and the material, affecting the cutting effect.

For a wall thickness of 1mm, bend diameter of 40mm, and bending radius of 76mm for a stainless steel bend, the optimal combination of cutting parameters includes a laser power of 9000W and a cutting speed of 230°/s. Under this parameter, the cutting efficiency and quality have reached the ideal state. Both ensure efficient processing speed and also ensure the stability of the cutting accuracy and surface quality.

7. Conclusions

This paper designs an industrial five-axis laser elbow-cutting machine tool. Metal elbows are widely used in aerospace, automobile manufacturing, and petrochemical industries, etc. Traditional cutting methods are difficult to meet the needs of modern industry for precision and efficiency. Laser-cutting technology has gradually become the mainstream method due to its high precision and efficiency. This paper discusses the principle of laser cutting and its system components and introduces a multi-axis cooperative control system based on a horizontal laser cutting machine.

The design uses five axes (X, Y1, Y2, Z1, Z2). Each axis realizes precise positioning and cutting through servo motors, which meets the demand for efficient cutting of metal elbows. It completed the construction of the hardware system of the laser elbow cutting machine, including the control architecture of the upper computer and the lower computer, and its cooperative work.

Finally, the laser elbow-cutting machine was subjected to cutting experiments to explore the influence of laser cutting parameters on the cutting accuracy and surface roughness of the elbow. The optimal combination of cutting parameters was determined to improve the processing quality and efficiency. It is initially verified that horizontal laser cutting technology can significantly improve cutting accuracy and efficiency while meeting the demands of mass production.

References

[1] Alsaadawy, M, Dewidar, M., Said, A., et al, "A comprehensive review of studying the influence of laser cutting parameters on surface and kerf quality of metals," Int J Adv Manuf Technol, vol. 130, pp. 1039–1074, 2024.
[2] Han, Bing, "Editorial for the Topic on Advanced Laser Fabrication Technologies for Cross-Field Applications," Micromachines, 15, 2024.
[3] Villumsen S, Joergensen S N, Kristiansen M, "Flexible laser metal cutting: An introduction to the robocut laser cutting technique," Towards New Frontiers. Springer International Publishing, 217-228, 2014.
[4] Gautam G D, Mishra D R, "Pulsed Nd: YAG laser cutting: accuracy improvement and parametric influences," Accuracy enhancement technologies for micromachining processes, 109-119, 2020.
[5] Alsaadawy M, Dewidar M, Said A, et al, "Investigation of the Effect of Laser Cutting Parameters on Surface and Kerf Quality of Thick Ti–6Al–4V Alloy Sheets," Arabian Journal for Science and Engineering, 1-18 2024.
[6] Alsaadawy, M, Dewidar, M., Said, A., et al, "A comprehensive review of studying the influence of laser cutting parameters on surface and kerf quality of metals," Int J Adv Manuf Technol 130, 1039–1074, 2024.
[7] Wandera C, Kujanpää V, Salminen A, "Laser power requirement for cutting thick-section steel and effects of processing parameters on mild steel cut quality," Proceedings of the Institution of Mechanical Engineers, Part B: Journal of Engineering Manufacture, 225(5): 651-661 2011.
[8] Advances in laser materials processing: technology, research and applications, Woodhead Publishing, 2017.
[9] Ma Z C, Liu X Q, Han B, "Editorial for the Topic on Advanced Laser Fabrication Technologies for Cross-Field Applications," Micromachines, 15(9), 1138, 2024.
[10] Pocorni J, Powell J, Frostevarg J, et al, "Investigation of the piercing process in laser cutting of stainless steel," Journal of laser applications, 29(2), 2017.
[11] Riveiro A, Quintero F, Boutinguiza M, et al, "Laser cutting: A review on the influence of assist gas," Materials, 12(1), 157, 2019.
[12] Shin J S, Song K H, Oh S Y, et al, "Laser cutting studies on 10–60 mm thick stainless steels with a short focus head for nuclear decommissioning," Optics & Laser Technology, 169, 110-121, 2024.
[13] Al Habsi G H R R K H, Rameshkumar G R, "Design and fabrication of 3-axis computer numerical control (CNC) laser cutter," International Journal of Multidisciplinary Sciences and Engineering, 7(5), 7-16, 2016.

Two-dimensional ultrasonic vibration assisted WEDM technology and simulation analysis

1st Minggang Xu, 2nd Zuhuang Liu*, 3rd Shengzhi Guan and 4th Zhen Wang

Department of Mechanical and Materials Engineering, North China University of Technology, Beijing, China

* Corresponding author: 603020942@qq.com

Abstract: To investigate the impact of two-dimensional ultrasonic vibration loading on the electrode wire in WEDM, a two-dimensional ultrasonic vibration scheme was first designed. The amplitude modulation rod was also designed accordingly. The vibration transmission mechanism of the two-dimensional ultrasonic vibration was then analyzed, focusing on the tool head's motion trajectory and the vibration transfer in the electrode wire. Finally, a simulation analysis of the electrode wire vibration was conducted. By comparing the two-dimensional and one-dimensional vibrations on the electrode wire, the results showed that: 1) Near the excitation point, the vibration amplitude decays over time; however, it exhibits better stability compared to regions farther from the excitation point. At positions farther from the excitation point (75mm and 90mm), the overall amplitude changes little, but there are more significant variations in certain intervals; 2) Throughout the entire stage, no significant amplitude decay was observed. The overall amplitude at positions farther from the excitation point (75mm and 90mm) remained relatively stable, showing better stability, and 3) during the vibration transmission process, coupling effects were observed. Although a 10N tension force was applied in the simulation, it was insufficient to dissipate the vibration at both ends, and the vibration continued to propagate.

1. Introduction

WEDM has the advantages of being unaffected by the material strength, toughness, and hardness of the workpiece, as well as being a non-contact processing method. It is widely used in fields such as defense, healthcare, chemicals, and instrumentation. However, WEDM also has some drawbacks, such as relatively low processing efficiency and the potential for wire breakage during the process. To address these issues, researchers have explored the addition of ultrasonic vibration during the cutting process and have achieved positive results. By applying ultrasonic vibration to the electrode wire, the circulation state of the working fluid and the discharge condition between the electrodes can be improved, which enhances the processing efficiency of WEDM and reduces the wire breakage rate. This approach is particularly significant for processing thick workpieces and certain difficult-to-machine materials, such as hard alloys and titanium alloys.

Zhang [1] conducted a study on the machining mechanism and process principles of one-dimensional ultrasonic vibration-assisted slow-wire WEDM composite machining. Xu [2] introduced ultrasonic vibration into gas medium electrical discharge machining. Dong [3] applied ultrasonic vibration to electrical discharge surface strengthening. Chang [4] applied ultrasonic vibration to electrical discharge milling (EDM) and achieved good machining results. Zhang [5] developed a composite machining technology integrating rotation, ultrasonic vibration, and micro-electrical discharge, which demonstrated good results and high efficiency for machining ultra-hard materials. Wang et al. [6] conducted four sets of comparative experiments to study the effects of ultrasonic vibration on the surface roughness and surface morphology of TC4 titanium alloy during high-speed wire WEDM. Gao and Liu [7] applied ultrasonic vibration to the workpieces processed by a micro-electrical discharge perforating machine. Hoang and Yang [8] introduced an ultrasonic vibration-assisted machining method in electrical discharge micro-wire cutting and compared the machining performance of applying ultrasonic vibration to the electrode wire versus applying it to the workpiece. Shabgard and Alenabi [9] applied ultrasonic vibration to the tool electrode in electrical discharge forming for the machining of Ti6Al4V. Wang et al. [10] proposed a novel composite material processing technique—vertical ultrasonic vibration-assisted low-speed wire electrical discharge machining (WEDM-LS)—to improve the performance of traditional WEDM-LS.

From the above research findings, it can be observed that although there is extensive research on introducing ultrasonic vibration into electrical discharge machining (EDM), studies on incorporating ultrasonic vibration into WEDM are relatively scarce. Current research primarily focuses on one-dimensional ultrasonic vibration and experimental studies of related processes, with limited work on two-dimensional studies, which mostly concentrate on applying vibration to the workpiece. Therefore, this paper primarily investigates the effects of applying two-dimensional ultrasonic vibration to the electrode wire in the machining process. The research will use finite element simulation to analyze the vibration behavior of the electrode wire under the influence of two-dimensional ultrasonic vibration.

2. I.Design of the two-dimensional ultrasonic vibration system

2.1 Scheme design

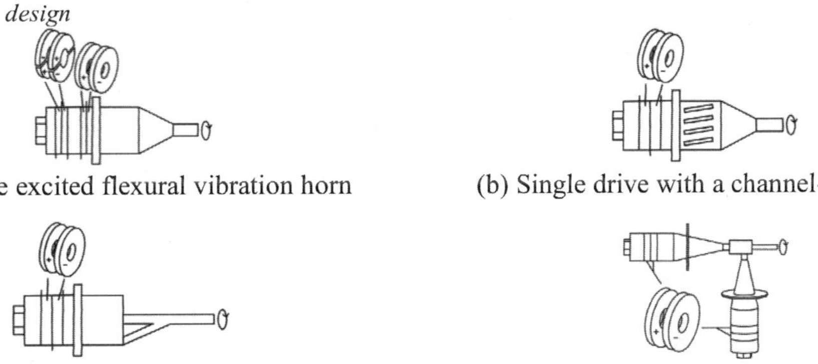

(a) Double excited flexural vibration horn

(b) Single drive with a channel-type horn

(c) Single excitation with mode converter type horn

(d) Double excitation one-dimensional longitudinal vibration scheme

Figure 1 Scheme Design.

As shown in Figure 1, b and c are single-drive transducers, while a and d are dual-drive transducers. The transducer generates two-dimensional ultrasonic vibration by combining bending vibration and longitudinal vibration. This vibration scheme features a compact structure and high energy transfer efficiency. Additionally, the two-dimensional ultrasonic vibration trajectory can be controlled by adjusting the longitudinal and bending excitation signals of the transducer. However, it is difficult to achieve accurate output.

b converts longitudinal vibration into two-dimensional ultrasonic vibration through a modal converter, featuring a compact structure. However, the trajectory is not adjustable and is difficult to express in a formulaic form.

c generates two-dimensional ultrasonic vibration by decomposing longitudinal vibration through an amplitude rod slot, also with a compact structure. The trajectory is not adjustable and cannot be presented in a formulaic form.

d combines two longitudinal vibrations to produce two-dimensional ultrasonic vibrations. The output trajectory can be expressed in a formula and is adjustable.

Fig. 2 is a schematic diagram of two-dimensional ultrasonic vibration-assisted WEDM processing.

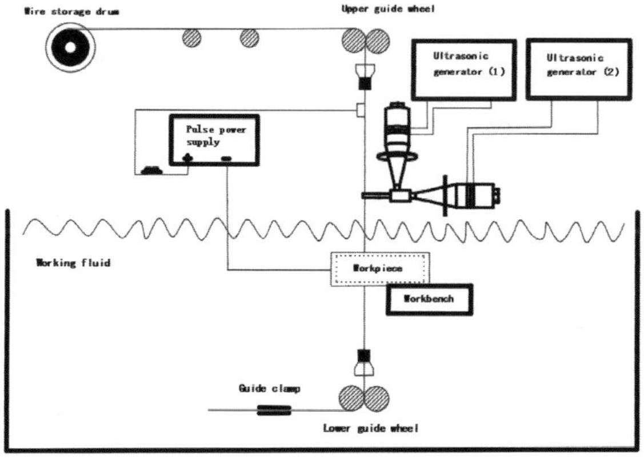

Figure 2 Schematic diagram of two-dimensional ultrasonic vibration-assisted WEDM.

2.2 Design of the amplitude rod

The design of the amplitude rod requires solving complex transcendental equations. Even for amplitude rods with simple shapes, solving the transcendental equations is relatively difficult. For the composite amplitude rod with a conical transition section designed in this paper, the problem becomes even more challenging. To simplify the design calculation, three coefficients defined by Ruan Shixun of Guangxi University are used: 1. Half-wavelength resonance length coefficient $H = \frac{2L}{\lambda}$; 2. Displacement node length coefficient $G = \frac{2x_0}{\lambda}$; 3. Strain peak length coefficient $E = \frac{2x_M}{\lambda}$: x_M represents the distance from the strain peak to the large end.

The material selected for the amplitude rod is 45# steel, mainly because this material has a relatively low resonance energy loss coefficient and good acoustical elasticity. Its performance parameters are as follows: density $\rho = 7.9\ g/cm^3$, elastic modulus $E = 20920\ kg/mm^2$, sound velocity $c = 5.16 \times 10^3\ m/s$, tensile strength $\sigma = 61\ kg/mm^2$, the frequency of the transducer $f = 20kHz$, diameter $\phi = 50mm$, and input amplitude $a_1 = 3\mu m$. The required output amplitude at the end of the amplitude rod is $a_2 = 15\mu m$. The following parameters are then determined for the uniform amplitude rod: the maximum particle velocity that can be achieved $v = 598.5\ cm/s$, the wavelength in the amplitude rod $\lambda = 258$ mm, the shape factor of the amplitude rod $\varphi = 0.315$, and the amplification factor of the amplitude rod $M_p = 5$.

Referring to the design tables compiled by Ruan Shixun using a computer, and through a table lookup, the stepped composite amplitude rod with a conical transition section ($kl_1 = kl_3 = 0.2.5$) closely matches the above design requirements. The values obtained from the calculation table are as follows: $H = 0.7806, G = 0.2285, M = 5.25, \varphi = 1.54$.

The transition section length $l_2 = 100mm$, the length of the two cylindrical sections $l_1 = l_3 = 29$ mm, the total length of the amplitude rod $L = I_1 + I_2 + I_3 = 158$ mm, the diameter of the small end cylinder of the amplitude rod $D_2 = 14.39mm$, and the displacement node $X_0 = 29.48$ mm are obtained. The dimensions of each section are 29mm for both cylindrical sections, 100mm for the variable cross-section section in the middle, 29.5mm for the nodes, and a total length of 158mm.

The modal analysis of the horn is shown in Figure 3.

Journal of Physics: Conference Series **2992** (2025) 012015 doi:10.1088/1742-6596/2992/1/012015

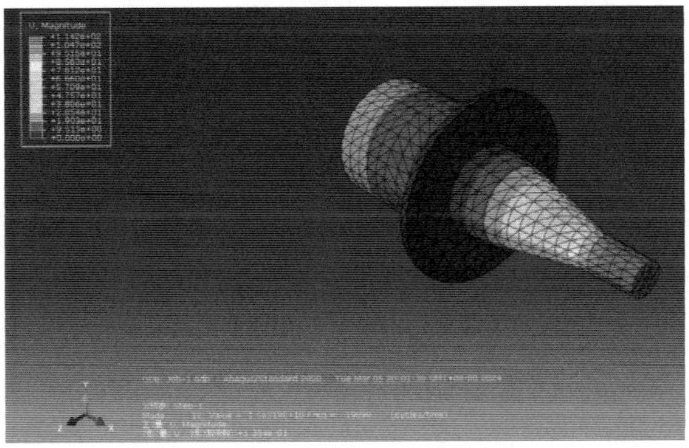

Figure 3 Simulation displacement contour map.

It can be seen that at the twelfth modal frequency, there is only axial displacement, with a frequency of 19, 899 Hz, which closely matches the design frequency of 20 kHz.

3. Vibration analysis

Two one-dimensional ultrasonic vibrations combine to generate a two-dimensional vibration on the tool head, which then transmits the two-dimensional vibration to the electrode wire. Therefore, it is necessary to analyze the generation of the two-dimensional vibration on the tool head and its transmission to the electrode wire.

3.1 Vibration analysis of the tool head

The sinusoidal motion of the tool head in the X-axis and Y-axis under the assistance of two-dimensional ultrasonic vibration in electrical discharge wire cutting is described by the following equations:

$$\begin{aligned} x &= A_1\cos(\omega t + \varphi_1) \\ y &= A_2\cos(\omega t + \varphi_2) \end{aligned} \tag{1}$$

Eliminating the variable t, the motion trajectory of the particle is obtained.

$$\frac{x^2}{A_1^2} + \frac{y^2}{A_2^2} - \frac{2xy}{A_1A_2}\cos\left(\varphi_2 - \varphi_1\right) = \sin^2\left(\varphi_2 - \varphi_1\right) \tag{2}$$

The Lissajous curve is closed when the following equation is satisfied.

$$\varphi_2 - \varphi_1 = \frac{(2k'+1)f_1 - (2k+1)f_2}{f_1}\pi \tag{3}$$

where k'=0, 1, 2,..., k=0, 1, 2,...

Some common Lissajous curves are shown in the figure 4. From the figure, it can be seen that under different conditions of phase difference and vibration frequency ratio between the X-axis and Y-axis, the uniformity of the motion trajectory of the two-dimensional ultrasonic vibration varies.

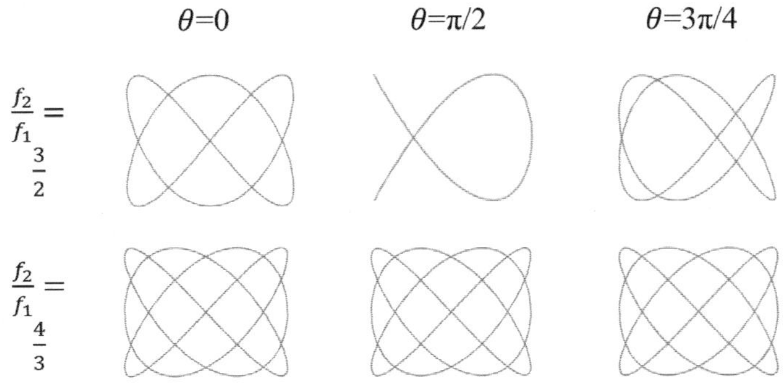

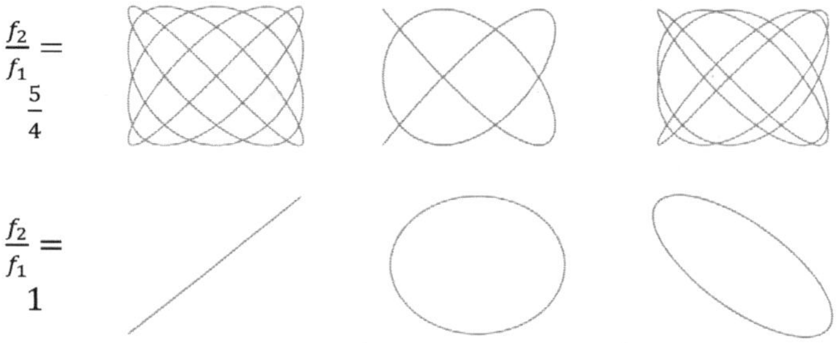

Figure 4 Lissajous curves under different frequency ratios and phase angle differences.

The complexity of the motion trajectory varies under different frequency ratios. Since the subsequent vibration needs to be transmitted along the electrode wire, complex trajectories are difficult to transmit on the electrode wire. Therefore, this paper mainly studies the vibration behavior under the condition where $\frac{f_2}{f_1} = 1, \varphi_2 - \varphi_1 \neq \mathrm{K}\pi$.

3.2 Electrode wire vibration analysis

In ultrasonic vibration-assisted WEDM-LS, the electrode wire moves between the upper and lower guide wheels. The electrode wire has uniformly distributed physical properties such as mass, elasticity, and damping. The electrode wire is considered a string fixed at both ends and subjected to a tensile force during the machining process.

As shown in the figure 5, a string of length L is fixed between two points. The axial direction of the string is the (x)-axis, and the linear mass density of the string is (x). The string is subjected to a constant tension force T. Under the influence of a distributed force f(x, t), it undergoes small vibrations along the y-axis. The effects of the string's weight and the damping force are neglected. The displacement at point x is denoted as y=y(x, t). We consider a small element dx of the string. Analyzing the motion of this element, we can apply Newton's second law of motion to obtain:

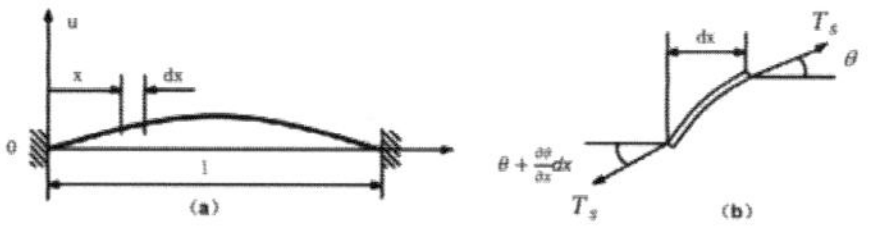

Figure 5 Stress diagram of electrode wire.

$$\rho_l \frac{\partial^2 U(x,t)}{\partial t^2} = T_s \frac{\partial^2 U(x,t)}{\partial x^2} (0 < x < l) \tag{4}$$

The excitation force caused by ultrasonic vibration is generally considered to be in the form of harmonic motion (F is the amplitude of the ultrasonic force and the frequency of the ultrasonic vibration). To analyze the motion pattern of the electrode wire caused by ultrasonic vibration, the following partial differential equation is derived from Newton's second law:

$$T_s \frac{\partial^2 U_u(x,t)}{\partial x^2} - \rho_l \frac{\partial^2 U_u(x,t)}{\partial t^2} - \mu \frac{\partial U_u(x,t)}{\partial t} + F(x,t) = 0 \tag{5}$$

Among them, T_s is the tension force of the electrode wire, ρ_l is the density of the electrode wire, and μ is the viscous damping coefficient. The excitation force caused by ultrasonic vibration is generally considered to be in the form $F = F_0 \cos wt$ of harmonic motion. To analyze the motion pattern of the electrode wire caused by ultrasonic vibration, it is assumed that the electrode wire only vibrates in the medium liquid under the action of ultrasonic vibration. Newton's second law has the following partial differential equation:

Journal of Physics: Conference Series **2992** (2025) 012015 doi:10.1088/1742-6596/2992/1/012015

By solving the above partial differential equation with boundary conditions $U_u(0,t) = 0, U_u(L,t) = 0$ and initial conditions $\frac{\partial U_u(x,t)}{\partial t}|_{t=0} = 0, U_u(x,t) = 0$, the vibration response of any point on the electrode wire can be obtained.

$$U_u(x,t) = \frac{2}{L}\sum_{i=1}^{\infty} \frac{\sin\frac{\omega_i}{a}x}{\rho_l}\int_0^t \sin\left(\frac{\omega_i}{a}\xi\right)\int \frac{e^{n(t-\tau)}}{\sqrt{\omega_i^2-n^2}}F(\xi,\eta)\sin\sqrt{\omega^2-n^2}(t-\tau)d\xi d\tau \tag{6}$$

$$U_u(x,t) = \frac{F_0}{l\rho_l}\sum_{i=1}^{\infty}\frac{\sin\left(\frac{\omega_i}{a}x\right)\sin\left(\frac{\omega t}{a}\eta\right)}{\sqrt{\omega_i^2-n^2}}\left\{A_n + C_n + (B_n + D_n)\left[\cos at - e^{nt}\cos\left(\sqrt{\omega_i^2 - n^2}\right)t\right]\right\} \tag{7}$$

where

$$A_n = \frac{n\left(e^{nt}\sin\left(\sqrt{\omega_i^2-n^2}\right)t\right)-\sin\omega t}{\omega^2-2\omega\sqrt{\omega_i^2-n^2}+\omega_i^2} \qquad B_n = \frac{\sqrt{\omega_i^2-n^2}-\omega}{\omega^2-2\omega\sqrt{\omega_i^2-n^2}+\omega^2} \tag{8}$$

$$C_n = \frac{n\left(e^{nt}\sin\left(\sqrt{\omega_i^2-n^2}\right)t+\sin\omega t\right)}{\omega^2+2\omega\sqrt{\omega_i^2-n^2}+\omega_i^2} \qquad D_n = \frac{\sqrt{\omega_i^2-n^2}+\omega}{\omega^2-2\omega\sqrt{\omega_i^2-n^2}+\omega^2} \tag{9}$$

4. Simulation analysis

The simulation was conducted using Abaqus software with an implicit dynamic solver. The overall vibration of the electrode wire within 0.001 seconds was analyzed, where the ultrasonic vibration frequency was 20 kHz, corresponding to a vibration period of 0.00005 s. A total of twenty cycles were analyzed, with each cycle divided into five steps, and each increment step was 0.00001 s. The parameters of the electrode wire are shown in the table1.

Table 1 Parameters of electrode wire.

Electrode Wire Parameters	Values
Diameter	0.2mm
Length	100mm
Material	Brass
Density	1.23E-08
Young's Modulus	106000
Poisson's Ratio	0.324

In Abaqus, ultrasonic force is presented in the form of displacement and loaded at the excitation point in the form of displacement after derivation. The amplitude chosen in this paper is 10. The velocity equation is as follows:

$$\begin{cases} V_X = -1256\sin(125600t) \\ V_Y = 1256\cos(125600t) \end{cases} \tag{10}$$

The excitation point is located 20 mm from the top end of the electrode wire. The vibration in the X and Y directions at the following positions—10 mm, 30 mm, 45 mm, 60 mm, 75 mm, and 90 mm—are analyzed as follows.

The vibration in one dimension is shown in the following figure 6:

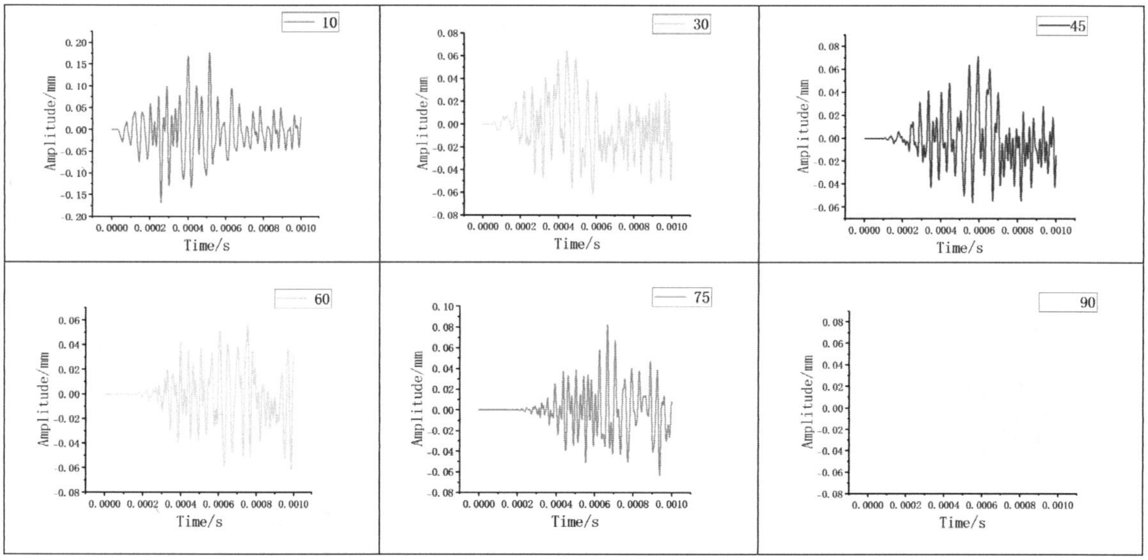

Figure 6 One-dimensional amplitude diagram.

As shown in the figure above, the electrode wire near the excitation point responds more quickly, but its amplitude decreases over time. However, its stability is better compared to positions farther from the excitation point. At positions farther from the excitation point (75 mm and 90 mm), the overall amplitude does not change significantly, but there are more dramatic variations in certain intervals.

The vibration in two dimensions is shown in the following figure 7 and figure 8.

In the two-dimensional X-direction vibration, compared to the one-dimensional vibration, the overall amplitude changes at positions farther from the excitation point.

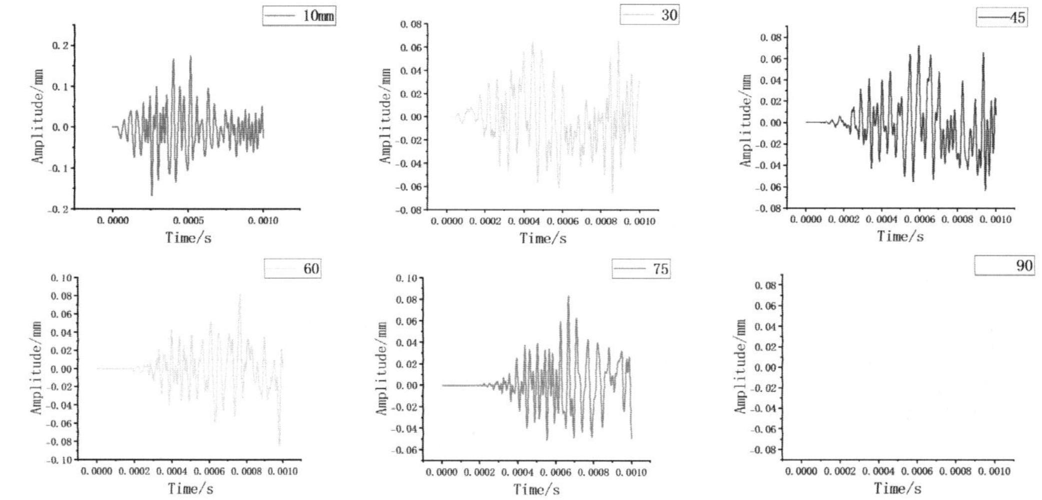

Figure 7 Two-dimensional X-direction vibration.

(75 mm, 90 mm) is not significant, and the stability is better.

Journal of Physics: Conference Series **2992** (2025) 012015 doi:10.1088/1742-6596/2992/1/012015

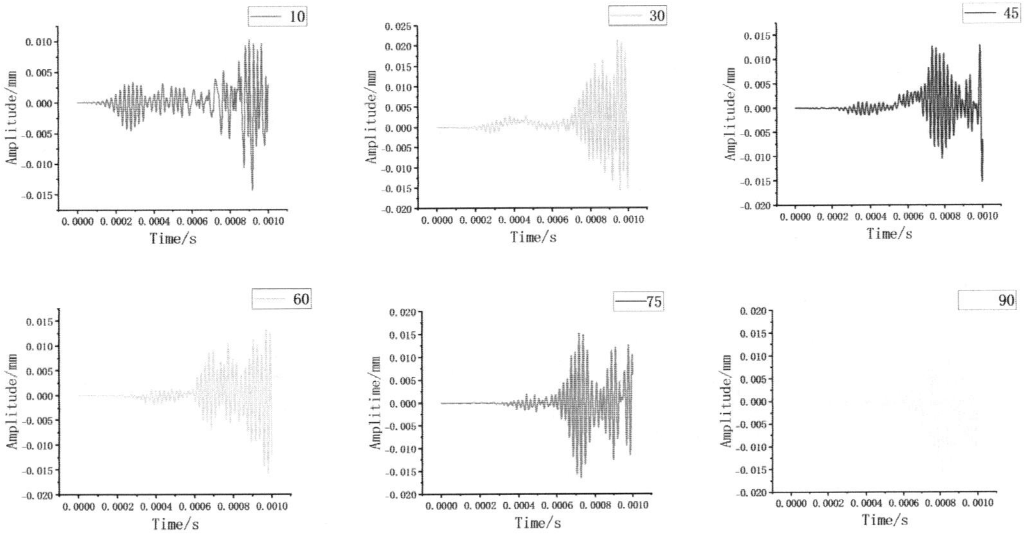

Figure 8 Two-dimensional Y-direction vibration.

In the two-dimensional Y-direction, the amplitude variation at positions farther from the excitation point (75 mm, 90 mm) gradually stabilizes. In summary, it can be concluded that the vibration effect in the two-dimensional system is better at positions between 75 mm and 90 mm.

The conditions of one-dimensional and two-dimensional vibrations at the moment of 0.0008 are selected, as shown in figs. 9, 10 and 11.

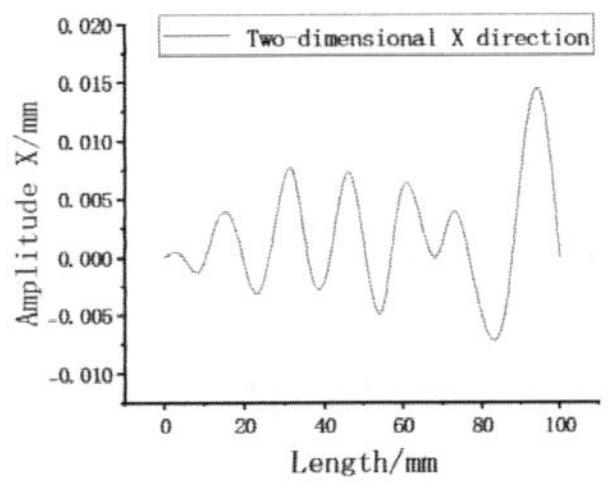

Figure 9 Vibration state of two-dimensional X-ray electrode wire.

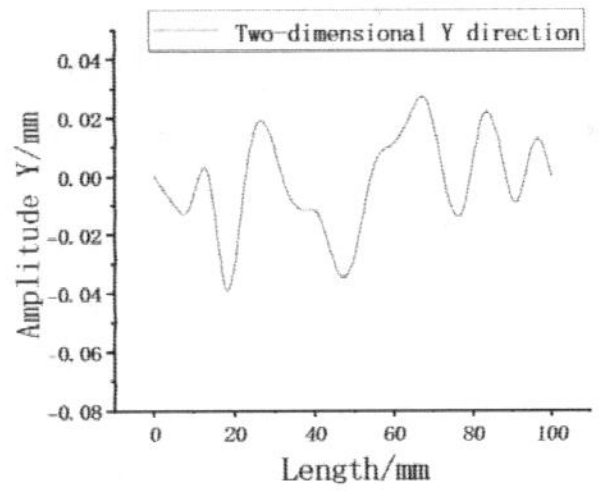

Figure 10 Vibration state of two-dimensional Y-ray electrode wire.

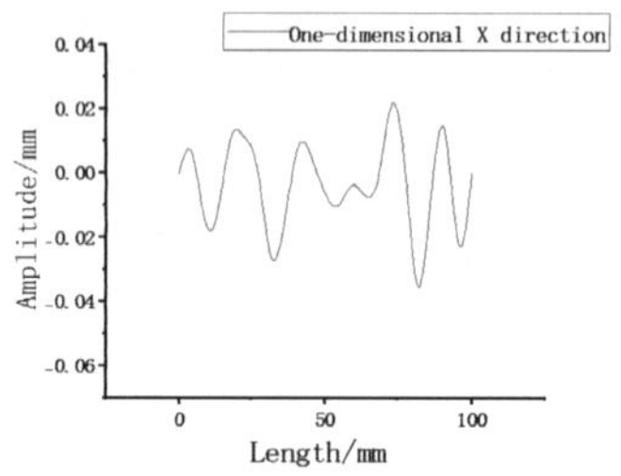

Figure 11 Vibration state of one-dimensional electrode wire.

In two-dimensional vibrations, there are vibrations in two directions, so when cutting along two perpendicular planes, the cutting quality of both planes can be ensured. However, in one-dimensional vibration, there is only vibration in one direction. When the cutting direction is aligned with the vibration direction or when the cutting direction is perpendicular to the vibration direction, there will be differences in the surface quality of the processed material.

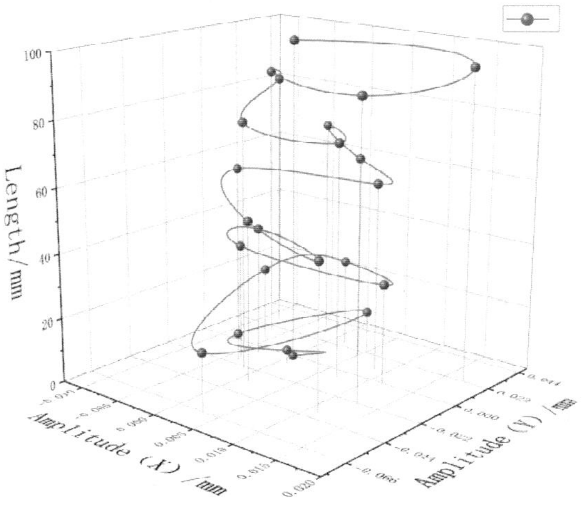

Figure 12 Two-dimensional compound vibration.

As shown in figure 12, when vibrations in the X and Y directions are combined, the resulting motion is not the circular or elliptical pattern we previously analyzed, but rather an irregular spiral ascending shape, with particularly large amplitude changes at both ends. This indicates the presence of coupling effects during the vibration transmission process. Although a 10 N tension force was set during the simulation, it was still insufficient to dissipate the vibration at both ends, allowing it to continue vibrating.

5. Conclusion

1. Through comparative analysis, a two-dimensional ultrasonic vibration scheme was designed. The amplitude rod was also designed.

2. The mechanism of two-dimensional vibration transmission was analyzed. The tool head's motion trajectory as well as the vibration transmission on the electrode wire were studied.

3. A simulation analysis was performed on the vibration of the electrode wire. By comparing the two-dimensional and one-dimensional vibration of the electrode wire, the results show that in one-dimensional ultrasonic vibration, the amplitude decays over time when close to the excitation point, but its stability is better compared to positions farther from the excitation point. At positions farther from the excitation point (75 mm, 90 mm), the overall amplitude variation is small, but significant

Journal of Physics: Conference Series **2992** (2025) 012015 doi:10.1088/1742-6596/2992/1/012015

changes occur in certain intervals. In two-dimensional ultrasonic vibration, there is no obvious amplitude decay throughout the entire process. The overall amplitude variation at positions farther from the excitation point (75 mm, 90 mm) remains small, and it exhibits better stability. Coupling effects exist during vibration transmission. Although a 10 N tension force was applied during the simulation, it was insufficient to dissipate the vibration at both ends, allowing the vibration to continue.

Reference

[1] Zhang S. Study on ultrasonic wire-cut machining [D]. null, 2005.
[2] Xu M. Study on Ultrasonic vibration-Gas dielectric EDM Composite Machining technology and mechanism.2007, Shandong University.
[3] Dong C. Study on Surface Strengthening Technology and Mechanism of Electrospark Discharge assisted by Ultrasonic Vibration, Shandong University, 2011.
[4] Chang W. Study on Ultrasonic vibration assisted EDM Machining Technology and Mechanism.2012, Shandong University.
[5] Zhang Z. Development and Experimental Research of Micro-EDM Machine aided by Rotary Ultrasound [D]. null, 2016.
[6] Wang J, Zhu X, Zou C. Study on Effect of Ultrasonic Vibration of Workpiece on Surface Quality of TC4 titanium alloy wireline-cutting with fast wireline-EDM [J]. Hot working technology., 2017, 46(06): 160-162. DOI: 10.14158/j.cnki.1001-3814.2017.06.044.
[7] Gao, C., Z. Liu. A study of ultrasonically aided micro-electrical-discharge machining by the application of workpiece vibration. Journal of Materials Processing Technology, 2003. 139(1-3): p. 226-228.
[8] Hoang, K.T., S.H. Yang. A study on the effect of different vibration-assisted methods in micro-WEDM. Journal of Materials Processing Technology, 2013. 213(9): p. 1616-1622.
[9] Shabgard, M.R., H. Alenabi. Ultrasonic Assisted Electrical Discharge Machining of Ti–6Al–4V Alloy. Materials and Manufacturing Processes, 2015. 30(8): p.991-1000.
[10] Wang Y, Chu L, Chai H, et al. Research on the machining characteristics of vertical ultrasonic vibration assisted WEDM-LS [J]. Materials and Manufacturing Processes, 2021(13/16): 36.

Experimental setup adapted to the ultrasonic vibration-assisted preparation process for thermoplastic FMLs and its validation

1st Hang Yin [a*], 2nd Yuqin Guo [b*], 3rd Dongjie Yang and 4th Fuzhu Li

School of Mechanical Engineering, Jiangsu University, Jingkou, Zhenjiang, China

* Corresponding author: [a]e-mail: 1149832641@qq.com
* Corresponding author: [b]e-mail: guoyuqin@ujs.edu.com

Abstract—The common hot-pressing process for fabricating thermoplastic fiber metal laminates (FMLs) often leads to problems such as poor interface properties and a high risk of delamination. This is mainly because the resin melts with high viscosity and struggles to flow and fill into the micro-pits on the metal surface or the micro-gaps between the reinforcing continuous fiber filaments and bundles. To overcome these challenges, this paper presents an integrated process for the preparation and interlayer interfacial modification of thermoplastic FMLs with the assistance of ultrasonic vibration. Subsequently, the corresponding experimental setup has been independently designed, developed, and validated. By integrating the obtained temperature and pressure curves with the scanning electron microscope (SEM) images of the cross-sections of the produced FMLs samples, it can be concluded that the devised experimental setup was able to achieve the desired processing results. Specifically, it achieved a metal/composite interface with excellent mechanical interlocking and a composite core where the fiber filaments and bundles were thoroughly infiltrated. Additionally, the recorded pressure change curve over time can effectively represent the typical physical phenomena and their evolution during the process test. These results are of great value as they provide crucial information for realizing process monitoring and optimization in future work, especially from the perspectives of pressure and temperature.

1. Introduction

Titanium-based thermoplastic fiber metal laminates (Ti-based FMLs), which are regarded as one of the advanced heterogeneous hybrid composites, are fabricated through the alternative stacking of thermoplastic continuous fiber-reinforced composites and titanium metal sheets, followed by hot-pressing and solidification procedures. Due to the high specific strength as well as the excellent resistance to corrosion, fatigue, and impact, Ti-based FMLs have been widely used in multiple industries, such as aerospace, marine, and rail transportation [1~3].

However, during hot-pressing, the thermoplastic resin melts with high viscosity and is hard to flow and fill into the gaps between filaments/bundles in the continuous fiber reinforcement, which results in unsatisfied interfacial bonding properties in Ti-based FMLs. Furthermore, the significant differences in thermal and physical properties of various constituent materials in Ti-based FMLs readily caused a large internal stress within interlayers. This will increase greatly the risk of delamination failure during the subsequent component forming and service stages, which damages the mechanical integrity of components and threatens the safety and reliability of equipment [4~6]. Therefore, in our research, the ultrasonic vibration-assisted preparation and interfacial modification process (UVPIM) was conceived for thermoplastic FMLs. This process endeavors to swiftly heat and liquefy the intermediate thermoplastic resin matrix by leveraging the energy supplied by ultrasonic vibration. Subsequently, the ultrasonic vibration exerts a localized and transient impact, facilitating the flow and infiltration of the resin melt into the voids within the carbon fiber bundles and the micro-pits on the metal sheet surface (micropores). Upon cooling and solidification, enhanced interfacial bonding characteristics can be attained.

Unfortunately, the determination of appropriate process conditions during the UVPIM process poses a significant challenge. This is because the thermoplastic resin matrix exhibits high viscosity, and it is unable to flow and fill into the micropores adequately if the ultrasonic vibration does not supply sufficient heat. Conversely, if ultrasonic vibration generates excessive heat or remains for too long a time, the risk of thermal degradation of the resin matrix will rise tremendously, which weakens the mechanical properties of FML components [7~9]. Therefore, the present work mainly concentrates on developing an experimental setup adapted to the UVPIM process of thermoplastic FMLs, which can monitor the change of temperature and pressure in real time during the UVPIM process. The corresponding process experiments are conducted to verify both the feasibility and effectiveness of the technique and setup.

2. UVPIM process Principle and experimental setup

As illustrated in Figure.1, the principle of the proposed UVPIM process for thermoplastic FMLs is: ① building the thermoplastic FMLs stack which consists of upper titanium sheet, thermoplastic resin film, continuous carbon fiber fabric, and lower titanium sheet; ② exerting High-frequency ultrasonic vibration by ultrasonic tool head so as to heat and melt the intermediate resin film layer; ③ executing the localized transient impact to promote the infiltration and filling of thermoplastic resin melt with high-viscosity into the microporous within fiber-reinforcement and metal sheet surface.

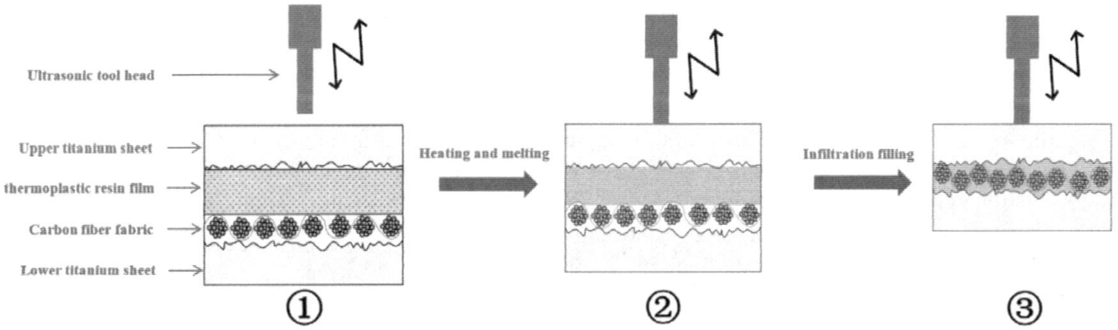

Figure.1. Principle of the thermoplastic FMLs preparation and interlayer interfacial modification integrated process assisted by ultrasonic vibration.

The experimental setup developed for the UVPIM process of thermoplastic FMLs is shown in Figure. 2. It consists of four main components, such as three-coordinate table, ultrasonic processing unit, temperature and pressure detection unit, and specialized fixture.

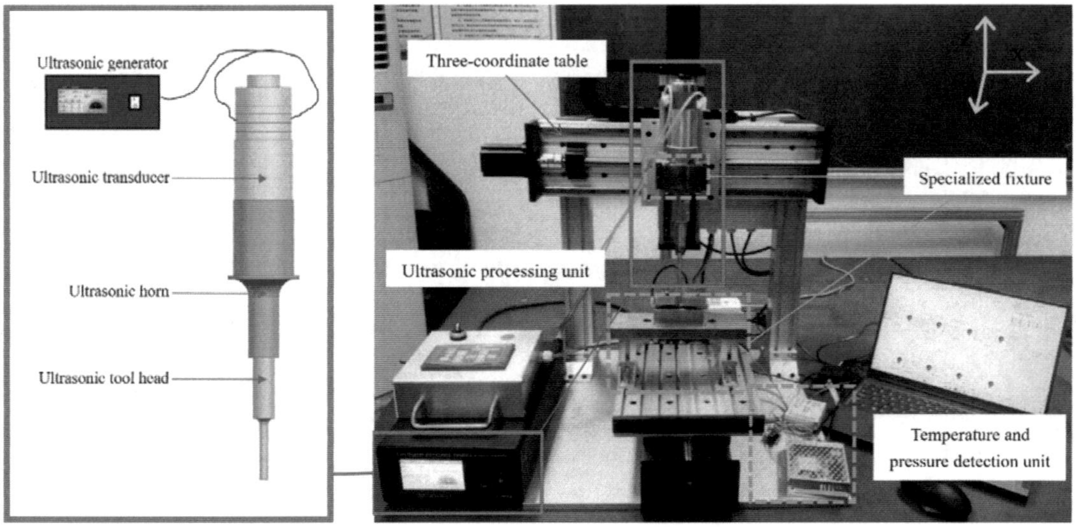

Figure.2. Experimental Setup of the thermoplastic FMLs preparation and interlayer interfacial modification integrated process assisted by ultrasonic vibration.

Here, the ultrasonic processing unit depicted on the left side of Figure. 2 comprises an ultrasonic generator, a transducer, an ultrasonic horn, and an ultrasonic tool head. The ultrasonic generator utilized is the S3000TC model, which features automatic frequency tracking. It has a power output of 3000W and a frequency tracking range spanning from 15kHz to 40kHz. The transducer is a sandwich-type piezoelectric transducer with an operating frequency of 20kHz [10]. Considering the proposed UVPIM process for thermoplastic FMLs requires the horn to have low acoustic resistance, high alternating load capacity, and strong fatigue resistance, thus the primary ultrasonic horn connecting with the transducer is made of 45# steel, while the secondary ultrasonic tool head directedly contacting with FMLs is made of TC4 titanium alloy [11]. The structural dimensions and physical picture of the final ultrasonic impact tool are given in Figure. 5.

As illustrated in Figure. 3, the pressure detection module employs a BCM-H3 planar force transducer, which exhibits a measurement range of 0 - 500N and an accuracy of 0.2% F-S. This transducer generates a pressure signal through a BSQ-3 analog transmitter. The temperature detection module consists of a WRNK-191 armored thermocouple with a probe diameter of 1mm, a measurement range of 0 - 800°C, and an accuracy of ±0.75°C. Both the pressure and temperature detection modules are connected to the DAQ122 data acquisition card, enabling 8-channel synchronous acquisition at a sampling rate of 200 kbps. The data is then displayed in real-time by the upper computer software, facilitating the observation of the temperature and pressure change curves.

The specialized fixture includes the tooling part for fastening the ultrasonic vibration system and the fixture part for fixing the FMLs specimen. The former as shown in Figure. 4 (a) is assembled by front, rear, and lower clamping blocks as well as a PTFE flange washer to fix the ultrasonic vibration system on the Z-axis moving rail of the three-coordinate table via the flange on the ultrasonic impact tool. Furthermore, to alleviate the possible negative influence of slight ultrasonic vibrations at the flange on the operational performance of the experimental setup, a polytetrafluoroethylene (PTFE) flange washer is placed between the flange on the ultrasonic impact tool and the lower clamping block for vibration damping purposes. Regarding the latter, as demonstrated in Figure. 4(b), a circular central slot is fabricated on the mounting block to accommodate the planar force sensor and a thermocouple layout plate is positioned above it. On the thermocouple layout plate, rectangular through-slots, each measuring 3 mm in width and 3 mm in depth, are machined at both the lateral and longitudinal centers. Along these slots, 21 through-holes with a diameter of Ø1.2mm are symmetrically arranged, with the spacing varying from 1.5mm at the center to 4mm at the edge. These holes are designed to install K-type thermocouples, which are used to monitor the temperature changes at different locations on the relevant region of the UVPIM thermoplastic FMLs.

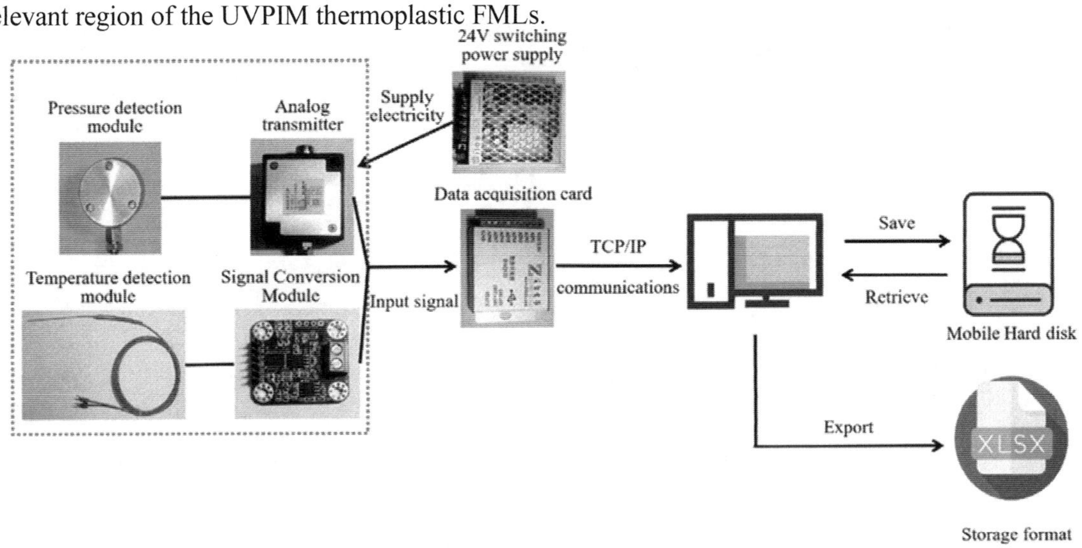

Figure.3 Detection Unit of Temperature and Pressure.

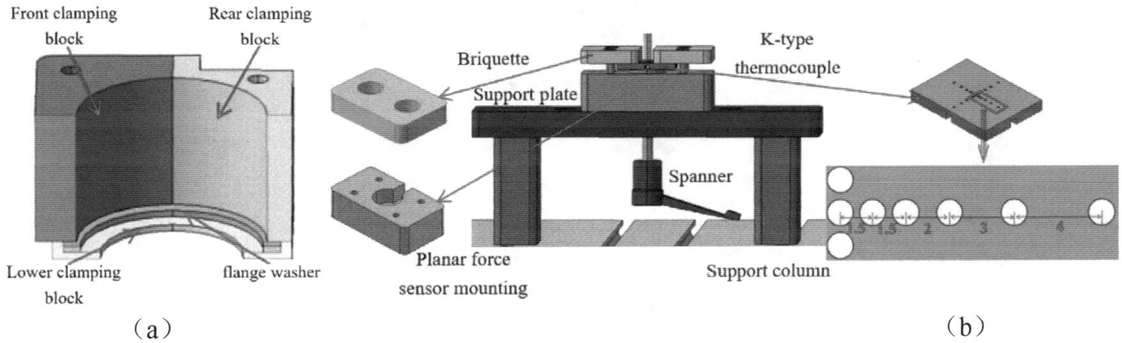

Figure.4 Illustration of Jig and Fixture Tooling for Ultrasonic Vibration System (a) and FMLs specimen clamping (b).

3. Design of ultrasonic impact tool

Based on the material selection of the ultrasonic tool head described in Section 2 and the fluctuation equations of longitude vibration in slender rods with variable cross-sections shown in Equi. (1)-(4), the half-wavelength stepped horn theory is applied to design the primary ultrasonic horn and the secondary ultrasonic tool head. Here, according to the operating frequency (20 kHz) and size of the output end face of the selected transducer, the related dimensions indicated in the ultrasonic impact tool are determined as follows: $D_1 = \phi 54mm$, $D_2 = \phi 31mm$, $l_1 = l_2 = 64.5mm$, $D_3 = \phi 20mm$, $D_4 = \phi 8mm$, $l_3 = l_4 = 62.5mm$.

Furthermore, to enhance the energy output of the ultrasonic impact tool, a cylindrical end structure $\phi 6mm \times 1mm$ is designed at its output end. Arc transitions of $R_1 = 14mm$ and $R_2 = 5mm$ are applied at the sudden change in cross-sectional area of the primary ultrasonic horn and the secondary ultrasonic tool head respectively, in order to minimize stress concentration and reduce the risk of fatigue failure. The final manufactured ultrasonic impact tool is shown in Figure. 5(a).

$$\frac{\partial^2 \xi}{\partial x^2} + \frac{1}{S}\frac{\partial S}{\partial x}\frac{\partial \xi}{\partial x} + k^2 \xi = 0 \tag{1}$$

$$l_1 = l_2 = \frac{\lambda_{45\#}}{2} \tag{2}$$

$$l_3 = l_4 = \frac{\lambda_{Ti}}{2} \tag{3}$$

$$\lambda_{45\#} = \frac{c_{45\#}}{f} \quad \lambda_{Ti} = \frac{c_{Ti}}{f} \tag{4}$$

Where, S represents the cross-sectional area of the ultrasonic horn or ultrasonic tool head, $k = \omega/c$ is the number of circular waves, $\omega = 2\pi f$ is the circular frequency, and f is the ultrasonic vibration frequency. $c = \sqrt{E/\rho}$ is the longitudinal wave propagation speed in an ultrasonic horn or ultrasonic tool head. $c_{45\#}$ and c_{Ti} denote the longitudinal wave velocities in 45# steel and TC4 titanium alloy respectively. $\lambda_{45\#}$ and λ_{Ti} indicate the corresponding longitudinal wavelength in the above two materials.

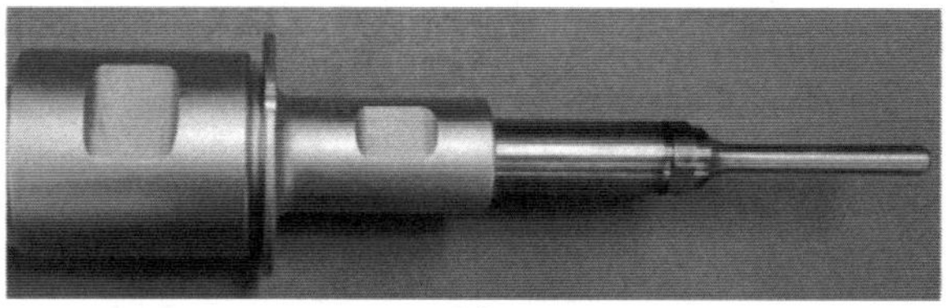

(a)

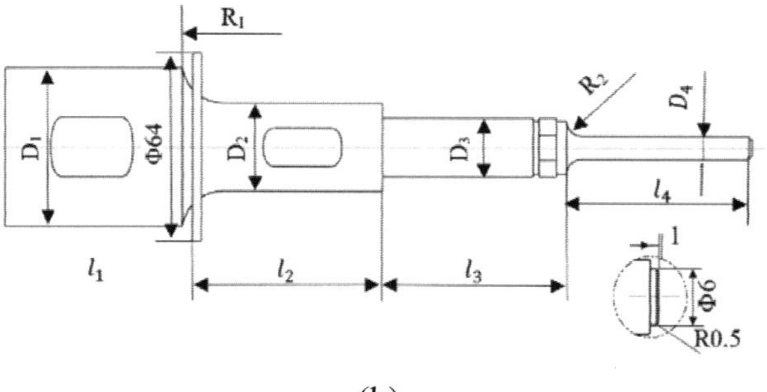

(b)

Figure.5. Ultrasonic Horn: (a) Diagram of Structural Dimensions; (b) Image of the real object.

For the designed ultrasonic impact tool, modal and harmonic response analyses were carried out within Ansys Workbench software. The results indicate that the natural frequency in the longitudinal vibration mode is 20,009 Hz. This meets the resonance requirement of the ultrasonic vibration system, as the deviation from the operating frequency of 20 kHz is less than 0.1%. Moreover, the maximum equivalent stress, which amounts to 753.6 MPa, is located at the stepped cross-section transition of the secondary ultrasonic tool head. This value is significantly lower than the yield strength of TC4 (1092 MPa). The maximum output amplitude of the finally assembled ultrasonic impact tool, as depicted in Figure. 5(b), was measured to be 15 using the JY-J200 digital ultrasonic amplitude measuring instrument. These findings confirm that the designed ultrasonic impact tool fulfills the requirements for the ultrasonic vibration pressure infiltration molding (UVPIM) process of thermoplastic fiber metal laminates (FMLs) in terms of frequency, strength, and ultrasonic energy input.

4. Process and setup Verification

To verify the feasibility of UVPIM thermoplastic FMLs, FML stacking specimens were prepared as the following step: (1) TC4 sheet with 0.2 mm thick was sanded with 300 mesh sandpaper, cut into rectangles of 40 mm × 30 mm , cleaned and dried; (2) 300 mesh PEEK powder was evenly spread in a mold and heated up to 360 °C on a JF-956A microcomputer heating platform, then the molding press, cooling and demolding were executed in sequence to achieve the PEEK thermoplastic resin film with 40 mm × 30 mm × 0.2 mm thick; (3) FMLs stacking specimen was fabricated by stacking the processed TC4 sheet, PEEK films and bidirectional braided carbon fiber fabrics with identical sizes in the order as illustrated in Figure. 1. Then, UVPIM process was carried for the FMLs stacking specimen at the ultrasonic amplitude of 12 μm and the table movement speed of 2 mm/s by using the experimental setup shown in Figure. 2, and the change curves of temperature and pressure versus time were recorded as shown in Figure. 6. After cooling, encapsulating with transparent epoxy resin, and water cutting, SEM observation was performed for FMLs specimens to estimate its cross-sectional morphology by comparing with the counterpart obtained via hot-pressing, as shown in Figure. 7.

By integrating Figure. 6(a) and (b), it can be observed that during the ultrasonic vibration pressure infiltration molding (UVPIM) process of thermoplastic fiber metal laminates (FMLs), the lower titanium plate experiences a nonlinear temperature increase, rising from room temperature to 67.1°C commencing at 1.3 seconds. Simultaneously, the pressure, after a sharp ascent from its initial value to 219 N, remains nearly constant. This phase is designated as Stage I, signifying that the stacked FMLs specimen has been compacted, while the intermediate polyether ether ketone (PEEK) films within it have not yet reached the temperature required for softening. Subsequently, from 3.4 seconds, the pressure begins a slow and nonlinear increase until it attains a peak value of 222.4 N at 7.7 seconds. This stage is referred to as Stage II and involves the combined effect of work hardening resulting from the local plastic deformation of the upper titanium sheet and the softening and melting of the intermediate PEEK films due to ultrasonic heating. After 7.7 seconds, the gradual decline of pressure to 217.6 N indicates that the intermediate PEEK films have started to melt and infiltrate into the micropores of the carbon fiber fabrics and the surface of the TC4 sheet. This is known as Stage III, which is then followed by a rapid pressure drop at 14.8 seconds, suggesting that the upper titanium sheet has been pierced, marking the end of the UVPIM process. These results confirm that the developed experimental setup can not only reflect properly the typical physical phenomena and their evolution during UVPIM thermoplastic FMLs but also track in real time the change of temperature and pressure.

Compared with the metal/composite interface of FMLs sample produced by hot pressing shown in Figure. 7(a), it is evident that, a desirable mechanical interlocking effect without cracking on metal-composite interface can be obtained successfully by using the UVPIM process as shown in Figure. 7(c). Furthermore, from Figure. 7(d), it is found that under the action of ultrasonic vibration exerted by the UVPIM process, a good wrapping of resin matrix around the continuous fiber filaments/bundles is formed in a composite core with no visible pores, unlike the reference sample in Figure. 7(b). Moreover, the sufficient melting of the intermediate PEEK films during UVPIM thermoplastic FMLs can be confirmed further by the presence of resin melt droplets, fiber indentations, and fiber pull-out holes in the cross-section.

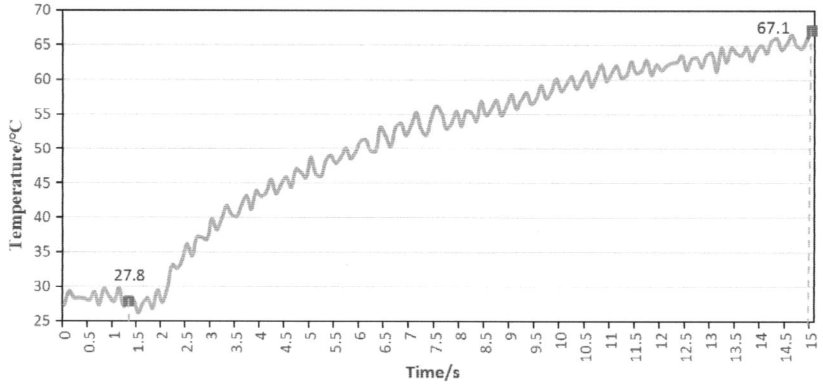

(a)

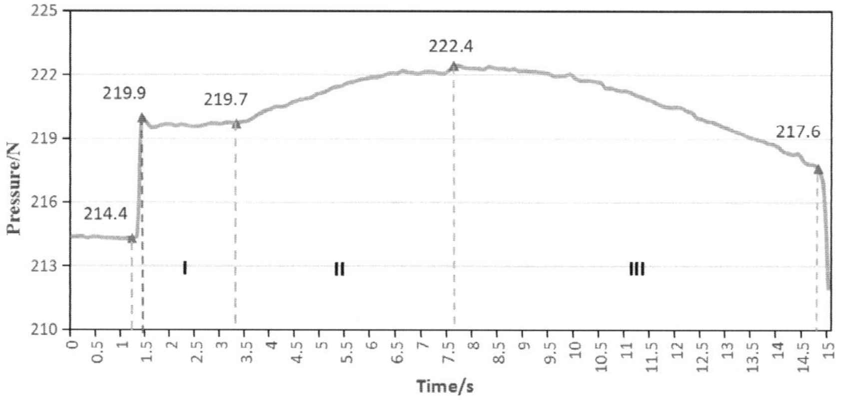

(b)

Figure.6 Change Curve of Temperature (a) and pressure （b） versus time.

The above results from UVPIM process experiments and sample observations demonstrate that the developed experimental setup is enable to achieve the desired process effects, and also verify the feasibility and effectivity of the proposed UVPIM thermoplastic FMLs process. And, the recorded change curve of pressure versus time can reflect the typical physical phenomena and their key evolution (i.e., Stage I, II, III.). Such data is of great significance as it provides a dependable information reservoir for subsequent process monitoring and optimization endeavors, thereby facilitating the refinement and enhancement of the overall UVPIM process. However, it should be notable that the peak temperature recorded in Figure. 6(a) was only 67.1°C much lower than the melting point of the PEEK resin matrix, which disagrees with the observed resin melting in Figure. 7(d). This can be attributed to the heating produced by ultrasonic vibration failure to transfer from composite core to outer TC4 sheet surface at an enough rapid speed within an extremely short duration of UVPIM thermoplastic FMLs because thermocouples are arranged to contact directly with the lower titanium sheet rather than composite core. This indicates that a significant temperature gradient in the thickness direction of the stacking FMLs from top to bottom needs to be considered carefully to promote temperature measurement accuracy and sensitivity in future work so that the monitoring and optimization based on both pressure and temperature can be utilized for improving the quality and mechanical properties of the FMLs materials produced.

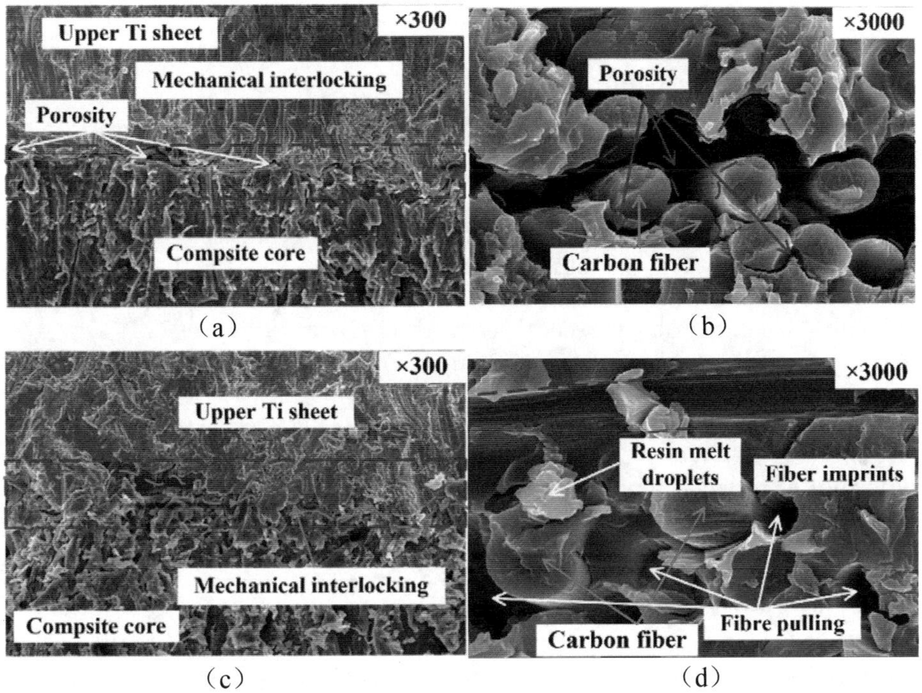

Figure.7 SEM Images of Cross Sections of FML Specimens: (a) Metal/Composite Interface for Reference Group, (b) Composite Core for Reference Group, (c) Metal/Composite Interface for Experimental Group, (d) Composite Core for Experimental Group.

5. Conclusions

For the proposed UVPIM thermoplastic FMLs process, an experimental setup was independently developed to record the real-time change in temperature and pressure. By process experiments, the following conclusions were obtained:

(1) The devised experimental setup effectively attained the anticipated process outcomes. It established a robust mechanical interlocking mechanism at the interface between the metal and composite and ensured a comprehensive infiltration of the resin around the fiber filaments and bundles within the composite core. This verifies both the feasibility and the dependability of the UVPIM process.

(2) The documented pressure-time curves effectively captured the essential phases of the process, namely the initial compaction stage, the stage of localized plastic deformation and softening, and subsequently, the final stage of resin penetration. This offers significant insights for comprehending the physical progression of the process.

(3) The unit's dual-parameter monitoring framework, which utilizes both temperature and pressure, lays a solid foundation for future process control. It enables real-time optimization to enhance the quality and mechanical properties of FML materials.

Acknowledgment

The authors extend their sincere gratitude for the financial support received from the Weaponry Pre-Research Fund (80923010201). Additionally, the authors wish to convey their appreciation to the editors and anonymous reviewers for their valuable suggestions that have contributed to the enhancement of the quality of this article.

References

[1] Zhang G ,Teng C . (2024). Current status and development trend of advanced structural materials technology in aerospace field. Journal of Aeronautical Materials, 44(02):1-12. https://doi.org/10.11868/j.issn.1005-5053.2024.000050.
[2] Du X ,Man X . (2023). Study on the application of composites in marine ships. Fiber Composites, 40(02):107-110. https://lib.cqvip.com/Qikan/Article/Detail?id=7110021912.
[3] Qin M ,Zhang W ,Ji X ,Wu M ,Qin N ,Su C. (2024). Application research of composite materials in aerospace and rail transit in China. Hot Working Technology, 53(06):6-9. https://doi.org/10.14158/j.cnki.1001-3814.20230966.
[4] Wang Y ,Cao J ,Yang C ,Wei Q ,Zhao L ,Hu L. (2023). Research progress in forming methods of super hybrid fiber metal laminates. Journal of Netshape Forming Engineering, 15(03):19-35. https://doi.org/10.3969/j.issn.1674-6457.2023.03.003.
[5] Xie M ,Zhan L ,Ma B ,Hui S. (2024). Classification of fiber metal laminates (FMLs), adhesion theories and methods for improving interfacial adhesion: A review. Thin-Walled Structures, 198:111744-. https://doi.org/10.1016/J.TWS.2024.111744.
[6] Guo Y, Xu X, Chen Z, Yin H, Li F. (2024). Deformation mechanisms and Surface/Interface characteristics of titanium-based thermoplastic FMLs under ultrasonic impact loading. Ultrasonics, 138: 107217. https://doi.org/10.1016/J.ULTRAS.2023.107217.
[7] Zhou Z, Chen C . (2024).Research advances in the ultrasonic-assisting adhesive bonding. Journal of Manufacturing Processes, 110134-160. https://doi.org/10.1016/j.jmapro.2023.12.050.
[8] Dong H ,Sun X ,Li W ,Jia R ,Zhao X. (2024). Research on 3D printing of continuous carbon fiber composite materials based on ultrasonic control of infiltration. Manufacturing Technology & Machine Tool, (01):13-18+25. https://doi.org/10.19287/j.mtmt.1005-2402.2024.01.001.
[9] Tian S ,Liu X ,Wang W ,Cui Q ,Hao J. (2023). Research status of ultrasonic vibration system and its application progress in plastic forming field. Journal of Mechanical Engineering,59(20):198-214. https://doi.org/10.3901/JME.2023.20.198.
[10] Chen Z. (2022). Ultrasonic impact simulation and experimental study of Ti-based thermoplastic FMLs materials . Zhenjiang: Jiangsu University. https://doi.org/10.27170/d.cnki.gjsuu.2022.000994.
[11] Liu Z ,Zhang T ,Zhang Y ,Dai Q. (2021) The Design of ultrasonic rolling device and the study of experiment. Machinery Design & Manufacture, (01):84-87. https://doi.org/10.19356/j.cnki.1001-3997.2021.01.020.

Optimization design of stiffened plates for crossbeam of a certain type of gantry machining center

Huaming Mai [1], Bo Li [2], Weiwei Zhang [*3] and Ling Yin [4]

[1] School of Mechanical Engineering, Dongguan University of Technology, Dongguan, China
[2] School of Mechanical Engineering, Dongguan University of Technology, Dongguan, China
[3] School of Mechanical Engineering, Dongguan University of Technology, Dongguan, China
[4] DGUT-CNAM Institute, Dongguan University of Technology, Dongguan, China

* Corresponding author: zwwps@126.com

Abstract - A certain type of gantry machining center was selected as the research object, and three stiffened plate structures of crossbeam were designed. By comparing the static and dynamic performance of different crossbeams, the "X"-shaped stiffened plate was determined to be the optimal crossbeam structure. On this basis, a sensitivity analysis of feature dimensions was conducted, identifying nine key dimensions of the crossbeam. The optimization objective was to obtain the minimum deformation of the crossbeam, and the multi-objective optimization was used for dimension design. The results indicate that, while maintaining the same mass of the crossbeam, the total deformation of the optimized crossbeam decreased by 14.2%, while the first-order natural frequency increased by 4.1%. This provides a reference for the optimization design of the crossbeam of the gantry machining center.

1. Introduction

As the "industrial mother machine", CNC machine tools play an important strategic role in the manufacturing industry. With the development of Industry 4.0 and the implementation of the 14th Five-Year Plan, higher requirements have been placed on the structural performance of CNC machine tools [1-2]. The crossbeam is an indispensable key component in machine tools, playing a role in supporting key parts such as the sliding seat and spindle boxes. Its performance directly affects the machining accuracy of the product. Therefore, improving the static and dynamic performance of the crossbeam has become one of the key issues in the design of CNC machine tools [3].

Many scholars have researched the structural design and optimization of crossbeams. For example, Hou et al. [4] obtained the static and dynamic performance of the transverse crossbeam of the gantry milling machine through the finite element, laying the foundation for further optimization design. Zhao et al. [5] redesigned the crossbeam stiffened plate based on the biomimetic principle of leaf veins, reducing the maximum static deformation of the crossbeam by 16%. Liu et al. [6] identified weak links in the crossbeam based on the principle of energy balance, improved the crossbeam structure, and increased the first four natural frequencies of the crossbeam by 8%. Jiang et al. [7] used topology optimization to optimize the crossbeam of a double-column vertical lathe, and the optimized crossbeam reduced static deformation by 12% and mass by 5%.

Although these studies have promoted the development of crossbeam optimization design, they were limited to select schemes and have not further optimized the dimensions related to crossbeam design. This paper takes a certain type of gantry machining center crossbeam as the research object, models the crossbeam in three dimensions, designs the crossbeam stiffened plate structure based on this, and then compares the performance of different stiffened plate structures through finite element analysis to determine the optimal crossbeam scheme. Finally, a sensitivity analysis is conducted on the key design dimensions, and multi-objective optimization is used to determine the final design dimensions, providing a reference for the manufacturing and production of machining center beams.

2. Establish the beam model

The assembly diagram of the crossbeam of a certain type of gantry machining center is shown in Figure 1. Its external dimensions are 6670 mm × 1250 mm × 1133 mm. The upper part of the crossbeam is equipped with a guide rail (X-direction guide rail), and there is also a guide rail at the lower end on the same side. These two guide rails jointly constrain the sliding seat, driving the spindle to move along the X direction. The sliding pillow and sliding seat cooperate to drive the spindle to move in the Y direction. In actual working conditions, the external forces acting on the crossbeam are the gravity of the slider and slide seat, as well as the cutting force generated during the operation of the machining center. The cutting force has three components in the X, Y, and Z directions: tangential component F_t, radial component F_r, and axial component F_a. The axial component force is generally small and can be ignored [8].

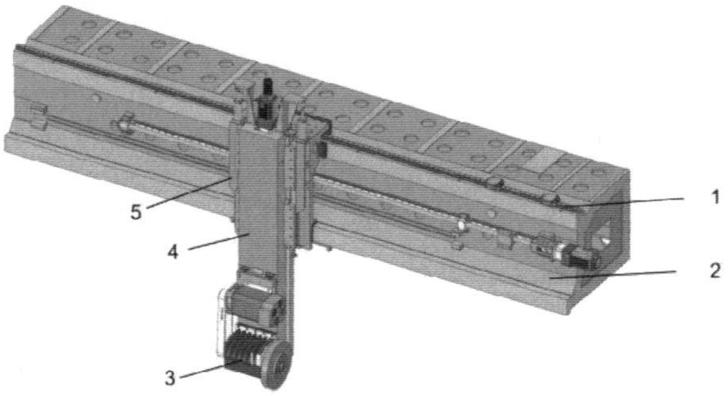

Figure 1 Assembly diagram of crossbeam
1-Guide rail; 2- Crossbeam; 3- Spindle; 4- Sliding pillow; 5- Sliding seat.

The cutting force can be estimated by an empirical Formula[8]:

$$F_t = C_F\left(V_W \cdot f_r \cdot \frac{B}{V_c}\right) + \mu \cdot F_r$$
$$F_r = C_F \cdot \left(\frac{\pi}{2}\right) \cdot \left(V_W \cdot f_r \cdot \frac{B}{V_c}\right) \cdot \tan\alpha \tag{1}$$

where V_W is the workpiece speed; V_c is the grinding wheel speed, f_r is the radial feed rate of the grinding wheel; B is the grinding width; C_F is the force required to remove per unit area; α is the cone half angle; μ is the friction coefficient between workpiece and grinding wheel. The parameters of the machining center during operation are grinding wheel linear velocity V_c =30 m/s, radial feed f_r =0.3 mm/s, workpiece velocity V_W=0.3 m/s, and grinding width B=25 mm. The friction coefficient between the workpiece and the grinding wheel is μ=0.35 and the half angle of the abrasive particles top is α=40°. The most used cast iron material has a C_F=4900 N/mm^2. After calculation, the radial grinding force F_r=484.40 N and the tangential grinding force F_t=537.04 N. The machining accuracy of a machining center largely depends on the stiffness of the crossbeam. To meet the stiffness conditions, optimize the stiffened plate of the crossbeam.

3. Finite element analysis and optimization results of crossbeam

3.1 Selection of crossbeam stiffened plate

This paper focuses on the static and dynamic characteristics of the crossbeam under the "X"-shaped, "V"-shaped, and "W"-shaped stiffened plates. Figure 2 shows the crossbeam with the "X"-shaped, "V"-shaped, and "W"-shaped stiffened plates, all of which have a thickness of 28 mm.

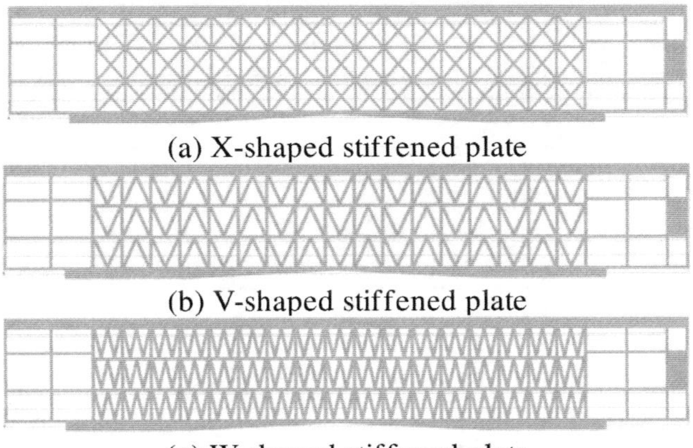

(a) X-shaped stiffened plate

(b) V-shaped stiffened plate

(c) W-shaped stiffened plate

Figure 2 Crossbeam stiffened Plate Structure.

Perform finite element modeling of the second section crossbeam model. The material used for the crossbeam is HT300, and its specific property parameters are set as follows: elastic modulus of 130 GPa, Poisson's ratio of 0.25, and density of 7300 kg/m³. The analysis considers the working condition with the sliding seat in the middle of the crossbeam. Figure 3 shows the loads and constraints on the crossbeam. In the figure, *A* represents the cutting force exerted on the crossbeam, *B* represents the gravitational force of the crossbeam itself, *C* represents the gravitational force exerted on the crossbeam by the sliding seat and sliding pillow, *D* represents the cutting torque applied to the crossbeam, and *E* represents that the contact surface between the crossbeam and the column is a fixed constraint.

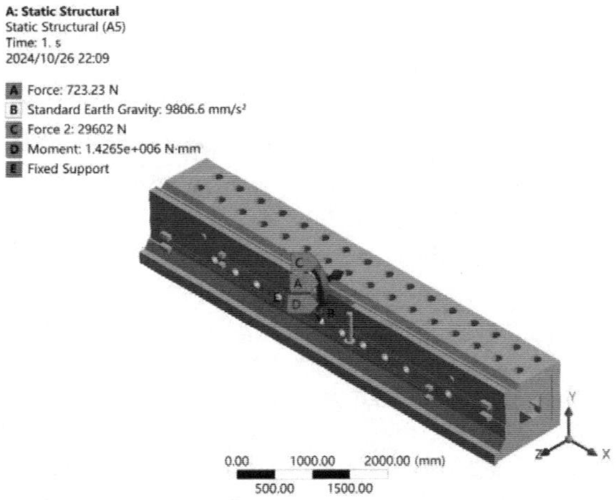

Figure 3 Load and constraints on the crossbeam.

Journal of Physics: Conference Series **2992** (2025) 012017 doi:10.1088/1742-6596/2992/1/012017

The dynamic and static performance of the crossbeams with three different stiffened plates shown in Figure 2 is analyzed and compared with the performance of the original crossbeam. The results are presented in Table 1. According to the analysis in Table 1, under the same loading conditions, the "X"-shaped stiffened plate performs the best in static analysis. Although the "W"-shaped stiffened plate exhibits the best dynamic performance, its corresponding mass is also the largest. All things considered, the "X" - shaped stiffened plate is used as the stiffened plate for the new crossbeam. Since the factors considered are qualitative, the dimensions of the crossbeam do not reach the best. Therefore, in the subsequent optimization plan for the crossbeam, the key design dimensions are optimized to determine the specific values of the crossbeam dimensions.

Table 1 Static and dynamic analysis results of crossbeam.

Stiffened plate type	Total deformation/mm	Mass/kg	First-order natural frequency/Hz
Original crossbeam	0.023132	13166	100.59
X-shaped stiffened plate	0.020473	13297	101.82
V-shaped stiffened plate	0.022557	13263	101.57
W-shaped stiffened plate	0.021208	13756	102.26

3.2 Sensitivity analysis

Fourteen feature dimensions were preliminarily identified as the key dimensions of the crossbeam, as shown in Figure 4. Among them, D_1, D_7, D_{10}, D_{14} are the wall thickness dimensions of the crossbeam, D_2, D_3, D_4, D_5, D_6, D_8, D_9, D_{11}, D_{12} are the thickness dimensions of the crossbeam stiffened plate, and D_{13} is the opening diameter.

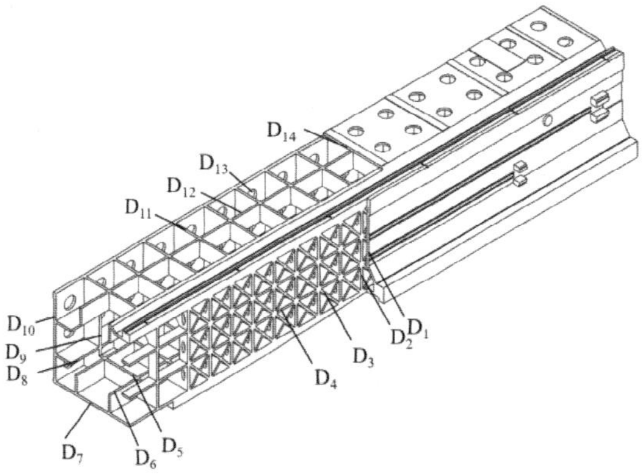

Figure 4 Dimension of crossbeam.

Sensitivity analysis was conducted on the selected dimensions of the crossbeam, and the results are shown in Figure 5. In the sensitivity analysis chart, dimensions with a sensitivity value exceeding 0.2 are defined as key dimensions that have a significant impact on the performance of the crossbeam. Positive numbers in the figure indicate that the response results increase with the increase of design parameters, while negative numbers indicate that the response results decrease with the decrease of design parameters. The larger the value, the more significant the impact of the parameter on the performance of the crossbeam. From Figure 5 (a), D3 and D_2 have a significant impact on the maximum deformation. Figure 5 (b) shows that D_{10}, D_7, D_1, D_{14}, D_3 , D_{13} and D_5 have a significant impact on mass. In Figure 5 (c), D_{10}, D_5, D_2, D_9, D_{13}, and D_{14} have a significant impact on the first-order natural frequency. Therefore, the above nine key dimensions are selected as the design variables for optimizing the key dimensions of the crossbeam structure, and their initial values and variation ranges are shown in Table 2.

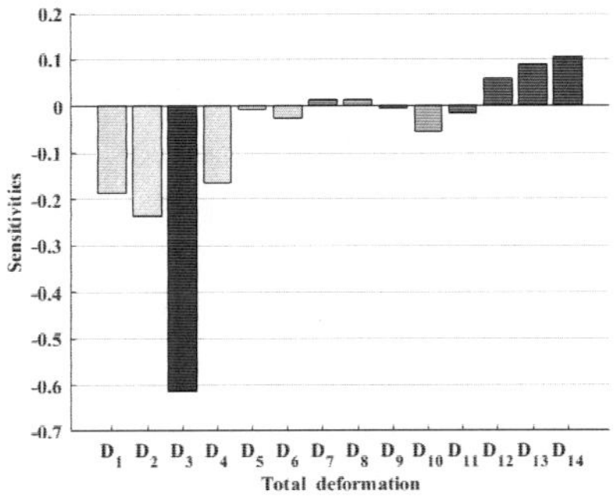

(a) Total deformation

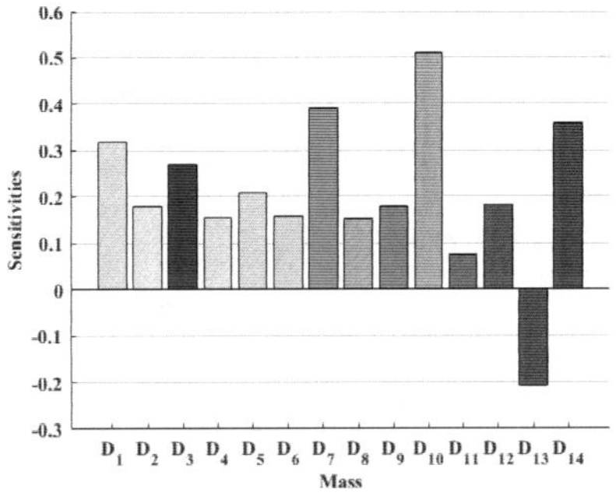

(b) Mass

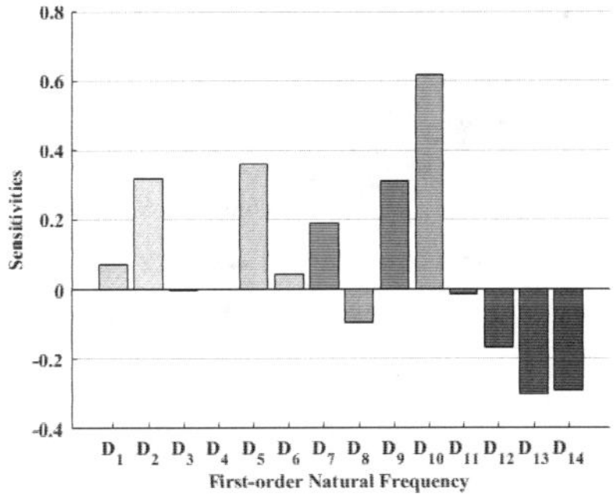

(c) First-order natural frequency

Figure 5 Sensitivity analysis results of crossbeam.

Table 2 Key dimensions and variation range of crossbeam.

parameter	Initial/mm	Change range/mm	parameter	Initial/mm	Change range/mm
D_1	28	22.4-33.6	D_9	28	22.4-33.6
D_2	28	22.4-33.6	D_{10}	28	22.4-33.6
D_3	28	22.4-33.6	D_{13}	150	120-180
D_5	25	20-30	D_{14}	28	22.4-33.6
D_7	28	22.4-33.6			

3.3 Optimization result analysis
The optimal space-filling (OSF) was used to design the nine key dimensions above. This method can evenly distribute the design points within the design interval and obtain more complete experimental data. After calculation, a total of 357 sets of design points were generated. The Kriging was used to interpolate and fit the experimental data, establish a response surface model, and predict the response

results of the interpolated points. This paper obtained the minimum deformation of the crossbeam as the objective function and selected the key dimensions of the crossbeam as the optimization variables. At the same time, to ensure that the dynamic performance of the crossbeam meets the design standards, the first-order natural frequency should be higher than that of the original crossbeam and the mass should not be greater than that of the original crossbeam. The multi-objective genetic algorithm (MOGA) was used to screen the data, and a total of three optimal solutions were generated. The optimal dimensions are input into the finite element model for verification, and the optimization results are shown in Table 3.

Table 3 Finite element analysis results of the optimized crossbeam.

schemes	Total deformation/mm	Mass/kg	First-order natural frequency/Hz
1	0.019842	13145	104.69
2	0.020063	13167	104.60
3	0.019902	13148	104.27

According to the optimization results in Table 3, Scheme 1 outperforms the other two schemes in terms of total deformation, mass, and first-order natural frequency. Therefore, Scheme 1 is chosen as the optimization result for the crossbeam. Figure 6 (a) shows the total deformation of the optimized crossbeam, and Figure 6 (b) shows the first-order mode shape of the crossbeam.

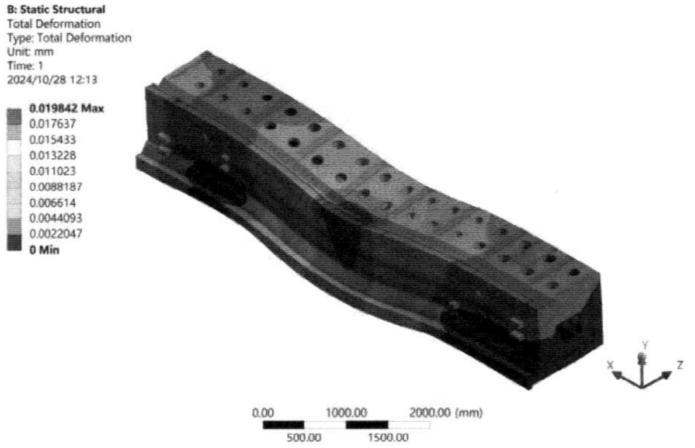

(a) Total deformation

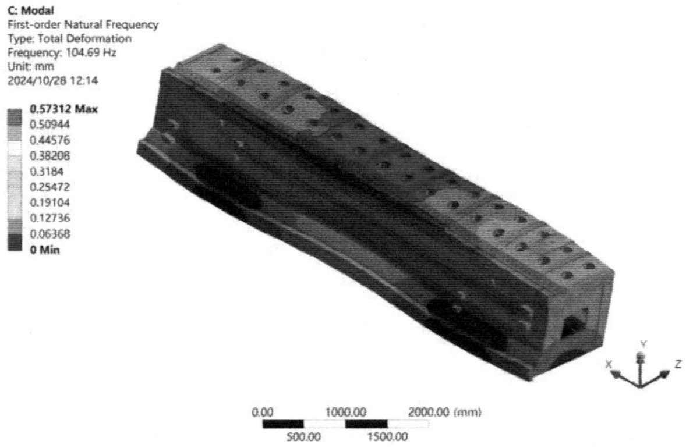

(b) First-order mode shape

Figure 6 After optimization total deformation and first-order mode shape of the crossbeam.

Compared with the original crossbeam, the total deformation of the optimized crossbeam is 0.019842 mm, while the total deformation of the original crossbeam is 0.023132 mm, a decrease of 14.2%. The first-order natural frequency of the optimized crossbeam is 104.69Hz, while the first-order natural frequency of the original crossbeam is 100.59Hz, an increase of 4.1%.

4. Conclusion

This paper optimized the design of the crossbeam stiffened plate types for the gantry machining center. Through the analysis of the dynamic and static mechanical properties of the crossbeam, it is found that the "W" - shaped stiffened plate has better vibration resistance performance but the highest mass, while the "X"-shaped stiffened plate has the best static performance and smaller mass. Therefore, the X-shaped stiffened plate was selected as the stiffened plate structure for the crossbeam. On this basis, sensitivity analysis was used to determine the key dimensions of the crossbeam, such as wall thickness and rib thickness. Finally, to obtain the minimum static deformation of the crossbeam, multi-objective optimization was used to design its key dimensions. The results indicate that the stiffness and vibration resistance of the optimized crossbeam had been improved, providing a reference for the optimization design of the crossbeam in gantry machining centers.

Acknowledgments

This work was supported by the National Key R&D Program of China (Grant No. 2023YFC2206504).

References

[1] Wakai, N., Takasugi, K., & Asakawa, N. (2022). Condition for machining feasibility for a five-axis machining center. Precision Engineering, 74, 414-425.
[2] Genbao, Z. H. A. N. G., & Tianhong, L. U. O. (2023). Necessity of CNC machine tools transforming to high-end. Computer Integrated Manufacturing System, 29(8), 2862.
[3] Besharati, S. R., Dabbagh, V., Amini, H., Sarhan, A. A., Akbari, J., Hamdi, M., & Ong, Z. C. (2016). Multi-objective selection and structural optimization of the gantry in a gantry machine tool for improving static, dynamic, and weight and cost performance. Concurrent Engineering, 24(1), 83-93.
[4] Hou, H. L., Qiu, Z. H., & Zhao, Y. Q. (2006). Static & dynamic analysis for a beam of high-speed cutting machine tool. Mechanical Design and Manufacturing, 44(5), 38-39.
[5] Zhao, L., Ma, J., Chen, W., & Guo, H. (2011). Lightweight design and verification of gantry machining center crossbeam based on structural bionics. Journal of Bionic Engineering, 8(2), 201-206.
[6] Liu, J., Wang Z. T., & Ruan Y. K. (2021). Dynamic optimization of machine tool structure based on energy balance principle. Manufacturing Technology& Machine Tool. (03),31-34.
[7] Jiang S., Jia R., Wang J., & Gao F. (2022). Research on lightweight design of cross beam structure of double-column vertical lathe (04):143-148
[8] Chen G. (2018). Research on Dynamic Characteristics and Structural Optimization Design of High-speed Grinding Machin.

Structural performance study on aluminum foil coating steel roller shaft based on finite element method

Bo Shen[1], Xin'an Zhou[2*], Zhenyu Guo[3] and Jiazhong Xu[4]

[1,2,4] Engineering Research Center of Aerospace High-end Manufacturing-Universities of Shaanxi Province, Shaanxi Institute of Technology, Xi'an, China
[3] Jilin Dongguang Aowei Automotive Brake System Co., Ltd. Changchun, China

Corresponding author: ZhouXinan_Cn@163.com

Abstract—This study establishes a 3D steel roller shaft model to investigate the failure mechanism of the steel roller shaft and analyzes the maximum stress point of coating the steel roller shaft by applying loads in ABAQUS software based on real working conditions. The results show that the maximum stress point of the coating steel roller shaft under simulated working conditions is at the junction between the large end and small end of the steel roller shaft, and the maximum stress value is 366 MPa, exceeding the yield strength of 45# steel. The position is the same as the damage position of the failed coating steel roller shaft in the factory, indicating the correctness of the simulation analysis. At the same time, the structural optimization analysis is performed, and the service life of the optimized structure is twice or even longer than that of the original one, indicating that the optimization can significantly improve the service life of the steel roller shaft.

1. Introduction

A hydrophilic aluminum foil coating line is a special equipment that applies chemical coating on the surface of aluminum foil to enhance the hydrophilic property of aluminum foil. The production process includes: cleaning - drying - coating - drying - cooling - coating - drying - cooling, and finally, rolling up [1,2]. Coating is a core process of the equipment. During the coating process, a coating steel roller (hereinafter referred to as "steel roller") is often used to evenly apply the coating on the surface of the aluminum foil [3]. The steel roller rotates around the center under the tension of the aluminum foil to complete the coating process on the aluminum foil surface, which determines the surface quality of the coated aluminum foil [4]. However, structural failures occurred to the aluminum foil coating steel roller shaft during actual production, causing suspension of the coating line and certain economic losses. In order to find out the reasons for the failure of aluminum foil coating steel roller shaft, a 3D steel roller shaft model was established, and finite element meshing and load application in ABAQUS software were conducted according to the real working conditions to analyze the maximum stress point of the coating steel roller shaft. Then the results were compared with the damage point of the failed aluminum foil coating steel roller shaft to further identify the structural safety of the aluminum foil coating steel roller shaft and failure factors, thus providing a reference basis for the structural improvement of the design and optimization of the aluminum foil coating steel roller shaft.

2. Structure of aluminum foil coating steel roller

The structure of the aluminum foil coating steel roller mainly contains the coating roller shaft and the roller sleeve, as shown in Figure 1. The roller shaft and roller sleeve are welded together, and the roller sleeve has a hollow structure [5]. During the working process, the roller sleeve contacts with the aluminum foil and rotates. The aluminum foil has a tension of 400 kg to ensure the flatness of the aluminum foil coating.

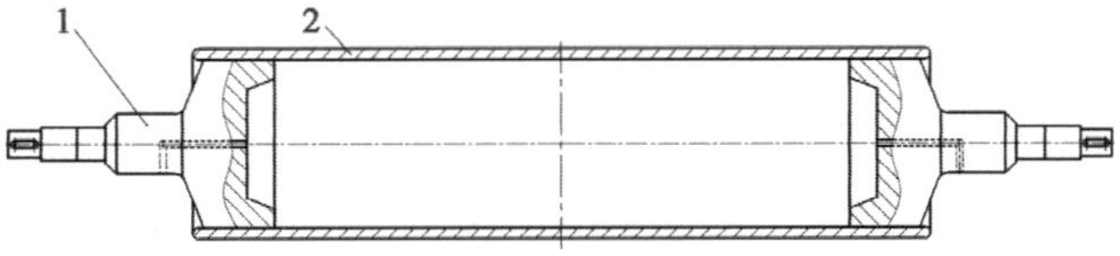

1 - Coating roller shaft; 2 - Roller sleeve.

Figure 1. Structure of aluminum foil coating steel roller.

3. Finite element analysis of the structure of aluminum foil coating steel roller

Table 1. Material property parameter.

Parameter	Value
Steel roller material	45 steel
Elastic modulus /GPa	210
Poisson ratio	0.31
Maximum tensile strength / MPa	600
Maximum yield strength / MPa	355

Finite element analysis (FEA) is to use mathematical approximation to simulate the loading condition of mechanical structures, i.e. to infinitely approximate the real situation of an unknown quantity with a number of regular small elements. This method has been widely used in engineering fields, and has achieved significant results [6-9]. ABAQUS is a powerful finite element analysis software, which is widely used in engineering and scientific fields, especially in structural analysis, thermal analysis and fluid analysis. It has flexible modeling capabilities and rich material models, which can simulate complex engineering phenomena. In the structural analysis of aluminum foil coated steel rolls, ABAQUS can effectively deal with the contact and interaction of multi-layer materials and provide accurate stress, strain and deformation analysis. Its advantage is that it can consider various factors such as adhesion, heat conduction and fatigue performance between aluminum foil and steel roller, and provide an important basis for design optimization and performance evaluation. In addition, the user interface of ABAQUS is friendly and suitable for engineers at different levels, thus improving the efficiency and accuracy of analysis.

Table 2. The main parameters of grid division.

Parameter	Value
Mesh parameter	Mesh based on mixed curvature
Mesh-density	High
Maximum tetrahedral mesh side length /mm	20
Minimum tetrahedral mesh side length /mm	1
Mesh transition mode	Smooth transition

Parameter	Value
Number of mesh nodes	3 107 991
Number of meshes	2 328 491
Jacobi	4
Average quality of mesh	0.72

In actual production, structural failure may occur to the aluminum foil coating roller shaft. Figure 2(a) shows a structurally good steel roller shaft, and Figure 2(b) shows a structurally failed steel roller shaft. The steel roller shaft was fractured at the transition position where the bearing mounting surface meets the large end of the steel roller shaft. In order to study the failure mechanism and provide a theoretical basis for structural optimization, this paper applied the finite element method for the performance analysis.

3.1 Meshing and material definition

(a) Structurally good steel roller shaft (b) Structurally failed steel roller shaft

Figure 2. steel roller shaft.

A 3D model of the aluminum foil coating steel roller was established in the 3D software UG NX 12.0 (Figure 3), saved as the general 3D format *.step. Pretreatments such as meshing on the steel roller model were performed using the professional meshing software Hypermrsh. The *.step format 3D steel roller model was imported into Hypermrsh, and tetrahedral unstructured meshing with strong structural adaptability was used for the meshing of the steel roller [10]. The results are shown in Figure 4. The material properties of the steel roller were defined. The material properties are shown in Table 1, and the main parameters of meshing are shown in Table 2.

Figure 3. 3D model of aluminum foil coating steel roller.

After meshing, the model was imported into ABAQUS finite element analysis software. Fixed constraints were applied at both ends of the steel roller based on its actual working conditions, and a linear uniform force of 4000 N was applied to the roller sleeve to form the mechanical constraints and load application scheme of the steel roller model.

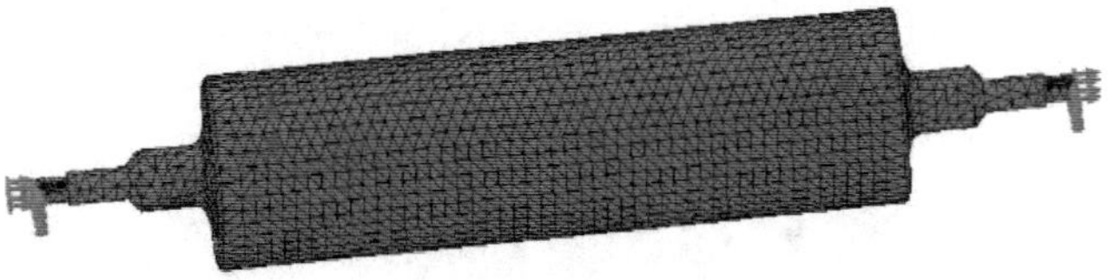

Figure 4. Meshing results of aluminum foil coating steel roller.

3.2 Analysis results

The stress simulation results of the steel roller are shown in Figure 5. Under the action of linear uniform load, the steel roller was bent and deformed, and the maximum axial stress of the steel roller occurred at the end of the shaft, i.e., the bearing mounted position. The maximum stress is 366 MPa, larger than the maximum yield strength of the material, which is the main factor leading to failure and deformation [11]. This position is consistent with the actual failure location of the steel roller in the factory, indicating that the simulation analysis is of reference significance.

Figure 5. Stress of aluminum foil coating steel roller.

4. Finite element optimization analysis of aluminum foil coating steel roller structure

According to the finite element analysis results of the coating steel roller, the existing coating steel roller was optimized and designed. The coating roller shaft was changed from the original 45# steel to a forge piece and tempered to improve the stiffness and toughness [12]. Finite element analysis was performed using the same method as in Section 2, and the results are shown in Figure 6. After optimization, the maximum stress is reduced to 192 MPa, which is smaller than the yield strength, and the overall stress distribution is more uniform and can meet the engineering requirements.

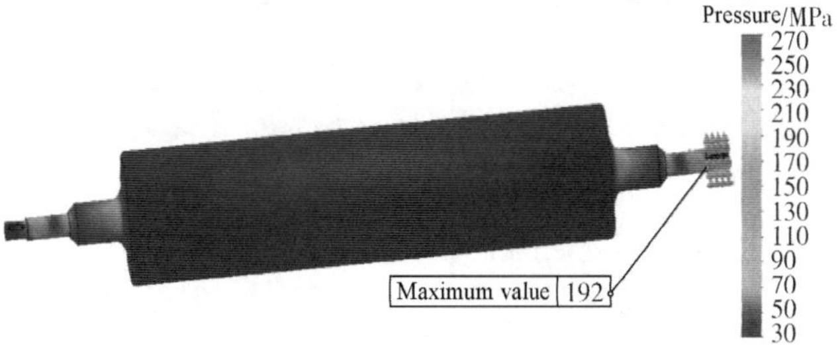

Figure 6. Stress of aluminum foil coating steel roller after optimization.

5. Factory experiment

In order to verify the optimization effect, the optimized steel roller was installed on the aluminum foil coating line of a company in Anhui Province (Figure 7). Then the experiment on the life time was carried out according to the normal production cycle and the results were compared with the failure cycle of the previous experiment, as shown in Figure 8.

As shown in Figure 8, before optimization, the deformation of the steel roller reaches 1.2 mm after 3 months of use in the factory, which begins to affect the normal use of the equipment. After optimization, the steel roller can still work normally after 6 months of use and can meet operational requirements. The service life of the optimized steel roller is more than 2 times of the original one, which is of great significance in improving the stability of the equipment and the production efficiency of the enterprise.

Figure 7. experimental site.

6. Discussions

Based on the technical method of finite element analysis, the stress, strain and deformation of steel roller shaft in the process of aluminum foil coating can be effectively studied. Through the simulation of different material properties and geometric parameters, the mechanical behavior of steel roller shafts under working conditions can be revealed, and then its design can be optimized. The research shows that the structural strength of a steel roller shaft is closely related to its geometry, material properties and coating pressure. By using finite element software for numerical simulation, the stress distribution and deformation of steel roller under different working conditions can be obtained, which is very important to improve the quality and efficiency of aluminum foil coating. In addition, through the analysis of the simulation results, potential weak links can be identified, and improvement measures can be proposed to enhance its structural performance. In summary, the research on the structural performance of aluminum foil coated steel roller based on finite elements not only provides an important basis for theoretical research, but also provides effective technical support for practical engineering applications.

Through systematic experimental research on aluminum foil coated steel rollers, we can deeply understand their mechanical properties, thermal properties and corrosion resistance under different working conditions. The experiment mainly includes static and dynamic loading tests to evaluate its bearing capacity and fatigue life. At the same time, combined with the finite element analysis method, the stress distribution and deformation characteristics of the roller are simulated and verified. In addition, wear resistance and adhesion tests were carried out for different coating materials and thicknesses to ensure the uniformity and durability of the coating. Through these studies, it can provide a theoretical basis for the design optimization of aluminum foil coated steel roller shafts, so as to promote the development of coating technology and improve production efficiency and product quality.

From the material point of view, the alloy composition and heat treatment process used by the steel roller may not be enough to resist the high temperature and chemical corrosion generated during the coating process, resulting in a decline in material performance. Secondly, from the design point of view, if the structure of the steel roller fails to distribute the stress reasonably or does not take into account the thermal expansion effect, it may lead to local stress concentration, thus causing cracks. Finally, at the technical level, improper setting of operating parameters such as coating speed and pressure may also aggravate the wear and fatigue of steel rollers. Therefore, in order to prevent the occurrence of similar failures, it is recommended to consider alloys with higher corrosion resistance and thermal stability in material selection, optimize the design of steel rolls to ensure uniform stress distribution, and strengthen the training of operators to ensure that the technical parameters in the production process are set reasonably to improve the service life and reliability of steel rolls.

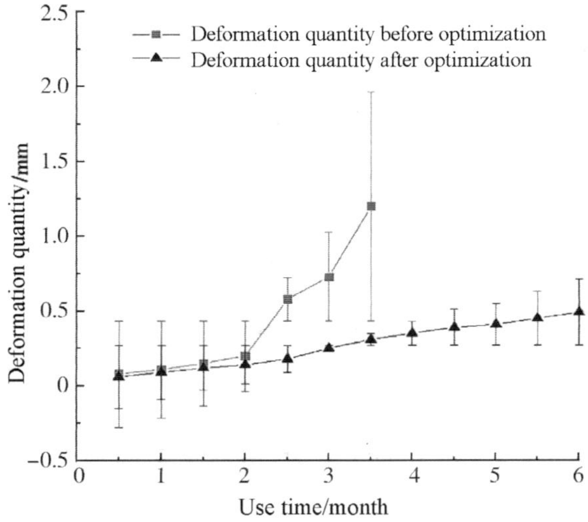

Figure 8. Use time of steel roller.

7. Conclusion

It is found that the aluminum foil coating steel roller has the problems of fast wear speed in the working process, which seriously affects the service life. In this study, the coating roller shaft was optimized from the original 45# steel to forgings, and quenched and tempered. The simplified three-dimensional model of the steel roller shaft before and after optimization was established, the mechanical finite element analysis was carried out, and the optimized steel roller shaft was tested. The results show that the mechanical model results of the steel roller shaft are basically consistent with the failure parts of the steel roller shaft on site, which verifies the credibility of the mechanical model and simulation results of the steel roller shaft. After optimization, the life of the steel rod is more than twice that of the original, which significantly improves the stability of the equipment and has a positive significance for improving the competitiveness of the product.

In this study, the optimization of aluminum foil coating steel roller is only optimized from the material, and the structural optimization is small. The follow-up research can start from the structural optimization to further promote the stability research of aluminum foil steel roller. It is suggested that the research on coating materials should be strengthened in practical applications, and the development of new high-performance coatings should be explored to meet the needs under different process conditions. In addition, the follow-up research can focus on the optimization of the coating process to improve the adhesion and uniformity of the coating, thereby enhancing the service life and stability of the roll.

Acknowledgment

This work was financially supported by the General Special Project of Shaanxi Provincial Department of Education of China (24JK0322, 23JK0318) and the 2024 Scientific Research Project of Shaanxi Institute of Technology of China (Pt24-05, Gfy24-20).

References

[1] M. Sundararajan, M. Devarajan, M. Jaafar, “A novel sealing and high scratch resistant nanorod Ni-P coating on anodic aluminum oxide,” Materials Letters, Vol. 289, pp. 129425, April 2021.

[2] H. M. Sun, X. T. Li, Z. Y. Hu, et al, "Hydrophilic-hydrophobic silver nanowire-paper based SERS substrate for in-situ detection of furazolidone under various environments," Applied Surface Science, Vol. 556, pp. 149748, August 2021.
[3] K. B. Liu, Z. Y. Xu, G. T. Meng, et al, "Heat transfer enhancement of three-dimensional oscillating heat pipe based on evaporation surface hydrophilicity regulation for thermal management," International Journal of Heat and Mass Transfer, Vol. 231, pp. 125911, October 2024.
[4] Q. F. Xu and J. N. Wang, "A superhydrophobic coating on aluminium foil with an anti-corrosive property," New Journal of Chemistry, Vol. 33, pp. 734-738, December 2018.
[5] X. Zhou, E. P. Wan and P. Zhang, "Numerical Simulation and Design of Flow Field of Water-Cooled Roller Based on FLUENT," Applied Mechanics and Materials, Vol. 233, pp. 11-16, November 2012.
[6] G. S. Lewis, D. Mischler, H. Wee, et al, "Finite Element Analysis of Fracture Fixation," Curr Osteoporos Rep, Vol. 19, pp. 403-416, August 2021.
[7] N. Kardani, A. Zhou, M. Nazem, et al, "Improved prediction of slope stability using a hybrid stacking ensemble method based on finite element analysis and field data," Journal of Rock Mechanics and Geotechnical Engineering, Vol. 13, pp. 188-201, February 2021.
[8] T. Strouboulis, K. Copps, I. Babuška, "The generalized finite element method," Computer Methods in Applied Mechanics and Engineering, Vol. 190, pp. 4081-4193, May 2001.
[9] J. Ryatt, M. Ramulu, "Prediction of tensile failure of stochastic tow-based discontinuous composites via mesoscale finite element analysis," Composite Structures, Vol. 279, pp.114769, January 2022
[10] L. G. Guo, M. Fahs, B. Koohbor, et al, "Coupling mixed hybrid and extended finite element methods for the simulation of hydro-mechanical processes in fractured porous media," Computers and Geotechnics, Vol. 161, pp. 105575, September 2023.
[11] Li. Yan, L. Y. Guan, D. Wang, et al, "Application and Prospect of Wear Simulation Based on ABAQUS: A Review," Lubricants, Vol. 12, pp. 57, February 2024.
[12] Lei Shi, Chun Dong Zhu, Xin Liu, and Yi Zhang, "Optimum design of the double roll rotary forging machine frame," Mechanical Sciences, Vol. 11, pp. 101–114, November 2020.

Research and analysis on the reliability of large vertical CNC lathe based on improved FMECA

Yiran Dong, Dechao Wang, Chengdao Piao and Shengkui Zheng *

Engineering College of Yanbian University, Yanji. China

*Corresponding author e-mail: ZSK-0618@163.com

Abstract—A large vertical CNC lathe is a highly automated precision machine that consists of several subsystems. The performance and status of these subsystems directly affect the reliability of the machine tool. In this paper, the improved FMECA method is used to calculate the hazard degree of each subsystem and statistically identify its main failure parts. Then the reliability evaluation indexes of the two subsystems that are most prone to failure are calculated, and the reliability evaluation of this large vertical CNC lathe is carried out according to the reliability indexes obtained. The results show that the reliability evaluation grade of the spindle system and automatic tool changer system of this large vertical CNC lathe is "very high", and the reliability of this large vertical CNC lathe is also very high.

1. Introduction

Vertical CNC lathe is a kind of precision equipment mainly used in the field of mechanical engineering, and its cutting function is the focus, which is a kind of high-efficiency and high-precision CNC machine tool. With the continuous development of industrial automation and digitalization, large vertical CNC lathes can be used in a variety of manufacturing fields, such as automotive, aerospace, mold, electronics, medical, instrumentation, etc. Vertical CNC lathes can ensure efficient production work with high reliability. Among the important indicators of CNC lathe performance is reliability[1-4]. The reliability analysis method can determine the overall reliability level of the CNC machine tool, find out the existing problems, and propose improvement methods[5]. Therefore, it is necessary to conduct reliability research and analysis in the field of CNC machine tools.

Famous foreign scholars and experts have studied the reliability of CNC machine tools, such as Vasilev pointed out that CNC machine tools should be manufactured and designed with clear specifications, which reflects the importance of machine tool reliability in the life cycle of the machine tool[6]. Keller in the UK established a machine tool failure database and analyzed the machine tool data in the database to determine the failure-prone parts of the machine tool[7], and Bowles John B proposed to use the cost of wear and tear as the basis for judging the order of the RPN[8].

To analyze each subsystem of the CNC lathe more accurately and comprehensively, this paper proposes to improve FMECA to analyze the reliability of a certain model of foreign vertical CNC lathe, and to rank and compare the hazards of each subsystem, aiming to find out the subsystems that are prone to failures of CNC lathes.

Journal of Physics: Conference Series **2992** (2025) 012019 doi:10.1088/1742-6596/2992/1/012019

2. Vertical CNC lathe fault classification

2.1 Determining who to evaluate

In this paper, a foreign model of a vertical CNC lathe as a research object, as shown in Figure 1. This model of vertical CNC lathe uses the FANUC18i system, its spindle motor power is 22 KW, the maximum spindle speed can reach 1125 r/min, the number of tool storage is 12, the maximum turning diameter is 950 mm, the maximum cutting length is 850 mm. In this paper, the failure data of 25 large vertical CNC lathes of this model were collected and their reliability was studied by using the timed truncation method.

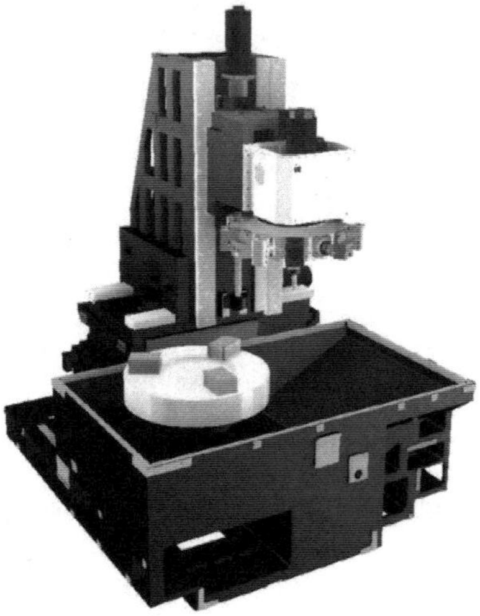

Figure 1. Vertical Cutting Center.

2.2 Vertical CNC lathe subsystem classification

In studying the reliability of vertical CNC lathes, we consider it as a whole and categorize it into 9 subsystems according to the different functions of each part. The categorization of the vertical CNC lathe and the failure data of each subsystem are shown in Table 1.

Table 1. Number and frequency of subsystem failures.

Serial number	System part	Abbreviation code	Ordinal number	Proportions (%)
1	Spindle system	S	103	41.2
2	Automatic tool changer system	T	45	18.0
3	Feed system	J	28	11.2
4	Guarding system	Q	22	8.8
5	Numerical Control system	NC	16	6.4
6	Electrical system	V	13	5.2
7	Chip removal system	K	8	3.2
8	Lubrication system	R	8	3.2
9	Hydraulic system	Y	7	2.8

From the results of the statistical data, the vertical CNC lathe in two years, had a total of 250 failures, of which the automatic tool changer system and the spindle system failures are more frequent, accounting for about half of the total number of failures.

3. Vertical CNC lathe subsystem FMECA analysis

There are a variety of methods used to analyze the reliability of CNC lathes, one of which is the Failure Mode Effect and Criticality Analysis (FMECA) method[9-14]. The advantages of FMECA are its comprehensiveness, preventive nature, operationalization, and flexibility. It not only considers the probability of occurrence and propagation path of the fault but also evaluates the impact of the fault to help identify potential failure modes and take preventive measures. It comprehensively analyzes all kinds of failures that may occur from top to bottom of the product, as well as their possible consequences and hazards, and identifies the subsystems that are prone to failures, and solves and improves them by summarizing and summarizing the methodology.

3.1 FMEA Analysis of Vertical CNC Lathes

Machine tool failure will be manifested in a certain phenomenon, the phenomenon can be summarized and categorized to analyze the causes of failure. Summarize the failure mode according to the cause of failure, different kinds of failure causes may cause the same failure mode, now according to the cause of failure will be divided into the following failure mode, as shown in Table 2.

Table 2. Typical fault modes and classification of fault causes.

Failure mode	Cause of failure
Functional type	No movement of moving parts
	Indexing not in place
	Moving parts jammed
Process type	Working accuracy exceeds the standard
	Geometric accuracy exceeds the standard.
Seepage type	Liquid, gas, oil leakage
	Liquid, gas, oil clogging
State type	Loss of device function
	Unstable performance
Out-of-regulation type	Spindle misalignment
Damaged type	Cable break
	Cable short circuit
Loose type	Poor continuity
Other type	Failure to operate properly
	False alarm
	Poor lubrication

3.2 Improved FMECA methodology

In the traditional FMECA methodology, faults are usually categorized into different classes based on their frequency and their severity. However, to avoid the influence of differences in classifications on the results of the study, this paper proposes an improved FMECA methodology to quantify the factors affecting maintenance costs during the maintenance process. This approach helps to more accurately assess the impact of maintenance costs on system reliability and provides more reliable data for decision-making and improvement.

3.3 Vertical CNC Lathe Hazard Degree

The harm caused by each failure during the repair process is usually measured against the cost of repair. The cost of maintenance in the maintenance process is mainly divided into three aspects: First is the downtime, the downtime caused by the failure of the production schedule will have a different size of impact; second is the frequency of failures, the higher the frequency indicates that this type of failure may have a design flaw, and this type of failure needs to be focused on; third is the degree of difficulty of repair. These three areas encompass the full cost of the repair process. By quantifying these three aspects it is possible to illustrate more precisely the impact of each fault in the production process.

The maintenance cost hazard degree is:

$$c_{ij} = q_{ij} f_{ij} o_{ij} \#(1)$$

In the formula:

$$q_{ij} = 100 \times \left(t_{ij} / \sum t_{ij} \right) \#(2)$$

$$f_{ij} = 100 \times \left(n_{ij} / \sum n_{ij} \right) \#(3)$$

$$c_i = \sum_{j=1}^{n} c_{ij} \#(4)$$

In the formula: $\boldsymbol{i}$ is the subsystem serial number, *and* $\boldsymbol{j}$ is the fault cause serial number. $\boldsymbol{c_{ij}}$ is the hazard degree for the $\boldsymbol{j}$th cause of failure of the $\boldsymbol{i}$th subsystem; $\boldsymbol{c_i}$ is the hazard degree of the $\boldsymbol{i}$th subsystem to the whole machine tool; $\boldsymbol{q_{ij}}$ is the severity of the $\boldsymbol{j}$th cause of failure of the $\boldsymbol{i}$th subsystem; $\boldsymbol{f_{ij}}$ is the occurrence degree of the $\boldsymbol{j}$th cause of failure of the $\boldsymbol{i}$th subsystem; $\boldsymbol{o_{ij}}$ is the difficulty of repairing the $\boldsymbol{j}$th cause of failure of the $\boldsymbol{i}$th subsystem; $\boldsymbol{t_{ij}}$ is the number of failures due to the $\boldsymbol{j}$th cause of failure of the $\boldsymbol{i}$th subsystem; $\sum \boldsymbol{t_{ij}}$ is the total downtime; $\boldsymbol{n_{ij}}$ is the number of failures caused by the $\boldsymbol{j}$th cause of failure for the $\boldsymbol{i}$th subsystem; $\sum \boldsymbol{n_{ij}}$ is the total number of failures. $\boldsymbol{o_{ij}}$ is based on expert experience and opinion, and the specific grades are shown in Table 3.

Table 3. Maintenance difficulty level.

Level	Value	Evaluation Criteria
Low	1	No need to replace parts, machining suspended
Medium	2	No need to replace parts, machining is interrupted
High	3	Need to replace parts
Very High	4	Requires manufacturer's visit for repair

3.4 Hazard analysis by subsystem

Calculated using the above improved FMECA method, the FMECA values for each subsystem and their hazard ranking were obtained, as shown in Table 4.

Table 4. Parameters and hazard levels of each subsystem.

Serial number	Abbreviation code	Failure mode	Cause of failure	n_{ij}	t_{ij}	f_{ij}	q_{ij}	O_{ij}	c_{ij}	c_i
1	S	Process type	Working accuracy exceeds the standard	12	162	4.8	4.5	2	43.2	483.5
			Geometric accuracy exceeds the standard	10	123	4.0	3.4	1	13.6	
		Functional type	No movement of moving parts	44	704	17.6	19.7	1	346.7	
		Out-of-regulation type	Spindle misalignment	13	143	5.2	4.0	2	41.6	
		Other type	False alarm	24	144	9.6	4.0	1	38.4	
2	T	Functional type	Moving parts jammed	18	378	7.2	10.6	2	152.6	245.8
		Loose type	Poor continuity	6	42	2.4	1.2	1	2.9	

		State type	Loss of device function	7	175	2.8	4.9	4	54.9	
		Other type	Failure to operate properly	14	210	5.6	5.9	1	35.4	
3	J	Functional type	No movement of moving parts	13	221	5.2	6.2	1	32.2	70
			Indexing not in place	7	56	2.8	1.6	1	4.5	
			Moving parts jammed	8	184	3.2	5.2	2	33.3	
4	Q	State type	Loss of device function	6	168	2.4	4.7	3	33.9	49.3
		Loose type	Poor continuity	4	20	1.6	0.6	1	1.0	
		Seepage type	Liquid, gas, oil leakage	10	130	4.0	3.6	1	14.4	
5	NC	State type	Loss of device function	5	110	2.0	3.1	3	18.6	29.4
		Damaged type	Cable break	2	38	0.8	1.1	3	2.6	
			Cable short circuit	2	34	0.8	1.0	4	3.2	
		Other type	Failure to operate properly	7	63	2.8	1.8	1	5.0	
6	V	Damaged type	Cable break	7	91	2.8	2.5	3	7.5	19.1
		Loose type	Poor continuity	6	102	2.4	2.9	4	11.6	
7	K	State type	Loss of device function	2	50	0.8	1.4	4	5.6	7.8
			Unstable performance	5	40	2.0	1.1	1	2.2	
8	R	Other type	Poor lubrication	8	48	3.2	1.3	2	8.3	8.3
9	Y	Seepage type	Liquid, gas, oil leakage	4	72	1.6	2.0	1	3.2	7.5
			Liquid, gas, oil clogging	3	63	1.2	1.8	2	7.5	

Through the calculation results in Table 4, it can be seen that: the spindle system is the most hazardous (483.5), and the automatic tool changer system is the second most hazardous (245.8), and these two sub-systems are the important parts of the large-scale vertical CNC lathe, so they should be paid attention to and improve them.

4. Calculation and evaluation of reliability indicators

The statistics of the first failure time of the spindle system of these 25 CNC lathes are shown in Table 5.

Table 5. First Fault Time of Spindle System.

246	1844	572	618	3078	54	1866
644	2712	504	5716	926	2406	804
5438	2156	1578	96	2608	134	4906
90	3716	366	694	374	96	6348
574	3698	3984	72	488	3054	454
96	260	922	3754	844	19504	2756
1774	3404	5836	3010	7746	144	3426
3012	1480	794	3258	1560	1238	378
594	72	2734	108	204	458	2634
386	806	368	3312	428	1634	144

4.1 Lathes Determining the type of distribution

The first failure time of the spindle system in Table 5 can be imported into Matlab to obtain four frequency distributions, as shown in Figure 2. Then analysis and comparison, the first failure time of the spindle system of this large vertical CNC lathe is in line with Weibull distribution.

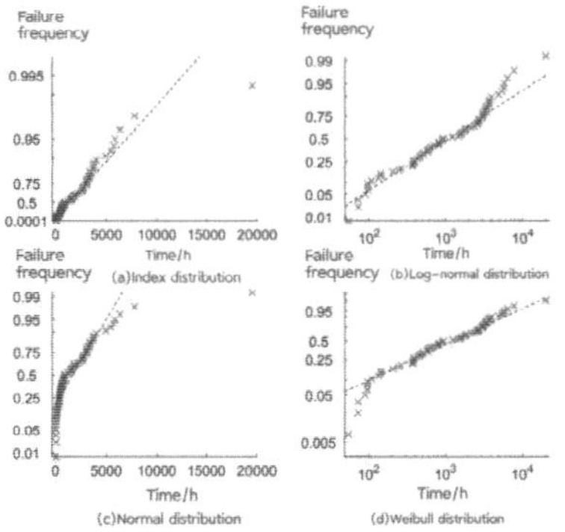

Figure 2. Distribution of first failure time and failure probability.

4.2 Parameter estimation and experimentation

The Weibull distribution better characterizes the failure data of mechanical products. When selecting the Weibull distribution model for the first failure time of the spindle system of the CNC lathe, the most used distribution models are the two-parameter and three-parameter Weibull distribution models, because the data difference in the first failure time of the spindle system of the CNC lathe is too large, and to more accurately show the reliability of this CNC lathe, the two-parameter Weibull distribution is selected. Parameter estimation of the size parameter n and shape parameter m using the least squares method yields m = 0.8723 and n = 2635.4.

The two parameters are calibrated using the D-detection method with a critical value of:

$$D_{n,0.1} = \frac{1.22}{\sqrt{n}} = 0.126 \geq max\left|F(t_i) - \bar{F}(t_i)\right| = 0.049 \#(5)$$

Thus, it is known that the first failure time of this CNC lathe spindle system obeys a two-parameter Weibull distribution as a function of:

$$F(t) = 1 - exp\left[\left(-\frac{t}{2635.4}\right)^{0.8723}\right] \#(6)$$

In the formula, t is the time between failures.

4.3 Analysis and estimation of MTTFF

The mean time to first failure (MTTFF) can most intuitively reflect the quality of CNC machine tools, so it is necessary to analyze and estimate the MTTFF of the spindle system of this model of a large vertical CNC lathe. Since the MTTFF of the spindle system of this model CNC machine tool obeys the two-parameter Weibull distribution, the application of the point estimation method can obtain its value:

$$\text{MTTFF} = n * \Gamma(1 + 1/m) = 2570.7455\text{h}$$

According to the experience of experts as well as related research, the reliability evaluation index level is divided, and the specific level is shown in Table 6.

Table 6. Liability Evaluation Indicators (MTTFF) and Level Classification.

Grade	**Norm MTTFF/h**
Very high	1500
High	1000
Medium	700
Low	500
Very low	300

From the above, the MTTFF value of the spindle system of this large vertical CNC lathe is 2570.7455, which belongs to the "very high" grade. Then the above algorithm to calculate the MTTFF value of the automatic tool changer system is 1985.3621, the same belongs to the "very high" level, far more than the highest level of domestic CNC lathes.

4.4 Analysis and estimation of MTBF

Mean Time Between Failure (MTBF) is a measure of a product's reliability. It reflects the time quality of a product and is an indication of the ability of a product to maintain functionality for a specified period of time. Similarly, the collected failure data are organized, and invalid data are eliminated to get the data of the spindle system failure interval time as shown in Table 7.

Table 7. Main spindle system failure interval time.

336	744	2064	4152	1512	48	168
1512	216	48	1676	3144	504	96
840	24	4800	6256	956	1058	512
10264	255	360	794	266	14236	496
1060	459	12364	8946	954	688	140
6952	964	436	7458	2462	694	452
1964	2678	356	4102	3628	342	1066
7952	3654	982	2108	652	8022	2014
544	1658	428	2698	2126	872	1754
636	866	1924	306	970	1668	406

Then it is imported into Matlab, and the type of distribution of the failure interval time of this model of CNC machine tool is obtained to conform to the lognormal distribution, as shown in Figure 3.

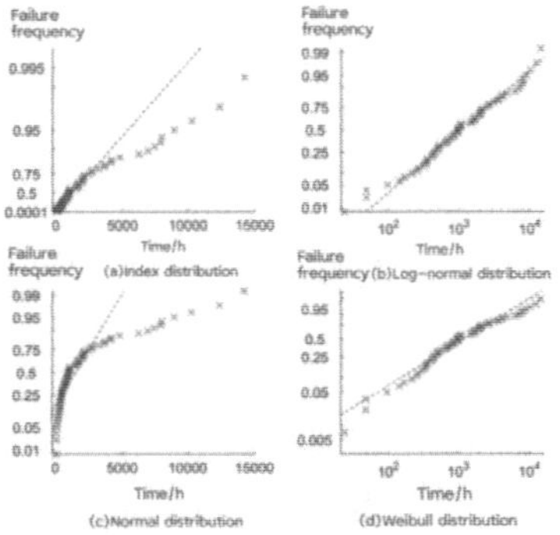

Figure 3. Fault Interval Time Fault Probability Distribution Diagram.

Like MTTFF, the estimation of positional and shape parameters by the least squares method gives $\mu = 6.9837$, $\sigma^2 = 2.3416$. Hence its fault interval distribution function:

$$F\left(t,\mu,\sigma^2\right)=\int_0^t \frac{1}{6.983x\times\sqrt{2\pi}} e^{\left(-\frac{(lnt-6.983)^2}{2\times 2.342}\right)} dx$$
$$=\phi\frac{(lnt-6.983)}{1.53} \#(7)$$

The mean time between failures is calculated as:

$$MTBF = E(t) = e^{\left(\mu+\sigma^2\right)} \#(8)$$

MTBF = 3082.1421h is obtained from Equation (8).

According to the experience of experts as well as relevant research, the reliability evaluation index level is divided, and the specific level is shown in Table 8.

Table 8. Reliability Evaluation Indicators (MTBF) and Level Classification.

Grade	**Norm MTTFF/h**
Very high	1500
High	1000
Medium	700
Low	500
Very low	300

From the above, the MTBF value of the spindle system of this large vertical CNC lathe is 3082.1421, which belongs to the "very high" level, and then the MTBF value of the automatic tool changer system calculated by the above algorithm is 2452.3984, which also belongs to the "very high" level. The MTBF value of the automatic tool changer system is 2452.3984 calculated by the above algorithm.

4.5 Comprehensive Reliability Evaluation

From the analysis and estimation of the above two reliability indexes, the reliability of the spindle system and the automatic tool changer system in this large vertical CNC lathe belongs to the "very high" grade, and it has far exceeded the highest grade of the domestic CNC lathe. This reliability evaluation index level is applied to the whole CNC lathe, but the spindle system and the automatic tool changer system are very important in the CNC lathe, so the reliability evaluation indexes of the spindle system and the automatic tool changer system of the CNC machine can be used instead of the reliability evaluation indexes of the whole lathe. From the above results, the reliability of this model of a large vertical CNC lathe is high.

5. Conclude

In this paper, the improved FMECA method is utilized to calculate the hazards of each subsystem and to find out the subsystems that are prone to failure, which reduces the subjective factors and makes the evaluation results more comprehensive and objective. Through the example, it is found that the spindle system and automatic tool changing system are more prone to failure. The improved FMECA is easy to calculate and simple to operate, the calculation results are in line with the objective reality, and it is more applicable to the comprehensive evaluation in different fields, which has an important reference value in the selection of evaluation methods in reliability research and analysis.

This paper takes a large vertical CNC lathe abroad as an example to calculate the parameter estimation and test for the first failure time of the spindle system, finds that the first failure time of the spindle system obeys the Weibull distribution, and calculates the MTTFF value of it as 2570.7455, and uses the same method to calculate the MTTFF value of the automatic feeding system with the second-ranking in terms of the degree of jeopardy as 1985.3621; and the MTBF value of it is 3082.1421. The MTBF value of the spindle system is calculated as 3082.1421, and the MTBF value of the automatic feeding system ranked second in terms of hazard degree is calculated as 2452.3984 by the same method. The above index values show that the average first failure time of the spindle system and automatic feeding system of this large vertical CNC lathe is higher than that of the domestic CNC machine tools, and its average failure interval time is lower than that of the domestic CNC machine tools, so it shows that its reliability is higher.

Acknowledgments

Supported by Yanbian University Science and Technology Program (602022130; 602020025); Jilin Province Higher Education Higher Education Research Project (JGJX2020D50; JGJ24D0058).

References

[1] YANG Z J， CHEN C H， CHEN F, Progress in the Research of Reliability Technology of Machine Tools［J］. Journal of Mechanical Engineering，2013，49(20):130－139.
[2] ZHANG K， LIU C S， LI l. Measures to Elevate Reliability of Horizontal Machining Center［J］. Machine Tool &Hydraulics，2012，40(3):157－161.
[3] Dalian University of Technology; Researchers Submit Patent Application, "Reliability Evaluation Method for Cnc Machine Tools Based on Bayes and Fault Tree", for Approval (USPTO 20200232885)[J]. Journal of Engineering,2020.
[4] Zongyi Mu, Genbao Zhang, Yan Ran, Shengyong Zhang, Jian Li 0039. A Reliability Statistical Evaluation Method of CNC Machine Tools Considering the Mission and Load Profile.[J]. IEEE Access,2019,7.
[5] LI Y, ZHANG X, RAN Y. Early failure modeling and analysis of CNC machine tools [J]. The international journal of advanced manufacturing technology, 2021, 112: 2731-54.
[6] Vasilev V S. Cost efficiency approach to the selection of NC machine tool reliability standards[J]. Soviet Engineering Research，1985，5(9): 49 -50.
[7] Keller A Z， Kamath A R R， Perera U D. Reliability analysis of CNC machine tools[J]. Reliability Engineering，1982，3 (6): 449-473.
[8] Bowles J B. An assessment of RPN prioritization in a failure modes effects and criticality analysis[C].Proceedings of Annual Reliability and Maintainability Symposium，2003: 380-386.
[9] SHEN G X， LI H Y， ZHANG Y Z. Failure Analysis of Turret System of NC Lathe［J］. Machine Tool & Hydraulics，2011，39(19):126－129.
[10] WANG D M， YU L M. Criticality Analysis of Machining Centers Based on Fuzzy Mathematic［J］. Machinery Design &Manufacture，2009(11):215－216.
[11] LIU Z R， WANG H J， LI Y. Evaluation method of system hazard based on numerical control machine tool FME- CA and improved RPN[J].Manufacturing Technology & Machine Tool，2020(11):40-45.
[12] YANG Xiaojian. Reliability analysis and maintenance of valves and actuators in pressure gas station based on FMECA[J]. Petroleum Engineering Construction, 2024,50(02):28-32.
[13] Lahri V, Shaw K, Ishizaka A. Sustainable Supply Chain Network Design Problem: Using the Integrated Bwm, Topsis, Possibilistic Programming, and ε-constrained Methods[J]. Expert Systems with Applications, 2021, 168: 114373.
[14] Yang Z, Guo J, Tian H L. Weakness Ranking Method for Subsystems of Heavy-Duty Machine Tools Based on FMECA Information[J]. Chinese Journal of Mechanical Engineering, 2021, 34(1): 167-178.

Tool wear detection of milling cutter based on heterogeneous data decision-level fusion

Jingshu Wang*, Weihan Zhou, Jinghua Ma and Qingfeng Wang

College of Mechanical Engineering, Chongqing University of Technology, Chongqing, China

*Corresponding author's e-mail: donot@cqut.edu.cn

Abstract—The traditional single-sensor approach for detecting tool wear in flat-end milling cutters presents limitations, often resulting in misdiagnoses and missed detections. To address this issue, this paper proposes a novel method that extracts time-frequency features from heterogeneous data and applies enhanced BP and ResNet neural networks for preliminary diagnosis. The improved D-S evidence theory is then employed for decision-level data fusion, achieving an average diagnostic accuracy of 95.60%, significantly surpassing the performance of single-method detection approaches.

1. Introduction

As a critical component in machining, the milling cutter undergoes rapid wear due to friction and impact, which affects both product quality and production schedules [1]. Therefore, monitoring tool wear is essential. Scholars such as Liu [2], Li [3], and Wang [4] have explored advanced fault detection methods, including singular value decomposition, BPNN, and kernel PCA with t-SNE. Additionally, machine vision techniques applied by Liu [5-8] offer new perspectives for wear detection.

This paper proposes a flat-end mill wear detection model based on decision-level fusion of heterogeneous data. The proposed method, which combines vibration and image signals for wear detection in flat-end mills, demonstrates higher accuracy and robustness compared to single-method detection models. The approach includes key processes such as feature extraction from vibration signals, training of sub-diagnostic models, and decision-level fusion of heterogeneous data outputs. The diagnostic algorithm presented in this study is capable of identifying various wear categories in milling cutters, enhancing the reliability of wear detection.

2. Data Processing

2.1 Vibration Signal

In the milling process, the statistical feature parameters of acceleration vibration signals from the fixture and spindle vary with different wear states, indicating the wear degree of the flat-end mill. For the individual raw vibration signal data that are vast and vary in length, the sensitive fault indicators are selected for the wear diagnostic model. To obtain a greater number of training samples of identical length and enhance the generalization capability of the vibration signal sub-diagnostic model, a sliding sampling method is employed to divide the vibration signals into samples, as shown in Figure. 1.

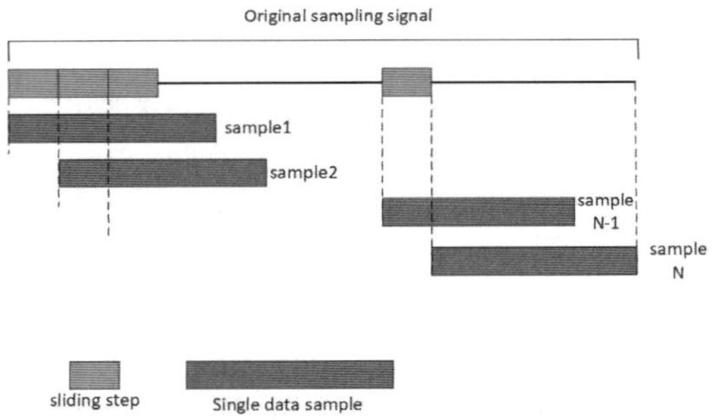

Figure. 1. Siding sampling method.

The sliding step is set to be 8000 data points, with each sample length being 10000 data points. By sliding to intercept the data signal segment, 3000 data samples are obtained for each wear state. A total of 15000 data samples are obtained, covering five categories, new, slight, moderate, severe, and broken. Feature parameters are extracted according to the vibration signal characteristics. After extracting the feature parameters, each feature parameter is normalized to eliminate the amplitude differences between data features. The feature parameter values are set as $X_i=\{X_i(j), j=1, 2, \ldots, k\}$, $i=1, 2, \ldots, m$, where m represents the feature number, and k indicates the wear state number of the milling cutter. The normalization formula is as

$$X = \frac{X_i - X_{\min}}{X_{\max} - X_{\min}} \tag{1}$$

Random sampling is employed to construct the experimental sample set. For each state sample, the training-to-test sample ratio is set at 8:2, with both training and test samples obtained through equal probability random sampling. These samples are then used as inputs for the subsequent experimental vibration signal sub-diagnostic model. The probability vector from the output layer of the neural network serves as the input for the final decision fusion model.

2.2 GA-BP neural network

The genetic algorithm (GA) is applied to optimize the parameters of the backpropagation neural network (BP) resulting in a GA-BP neural network with an adaptive structure and optimized parameters. This study uses GA to optimize BP's parameters, calculating hidden layer neuron j's function signal.

$$y_i(n) = f\left(\sum_{i=0}^{m} \omega_{ij}(n)x_i - \theta_j + s_j\right) \tag{2}$$

In the formula, x_i is the input signal, y_i is the output signal, s_j is the external input signal, and ω_{ij} represents the connection weight from the *i-th* to the *j-th*. The initial threshold of the *j-th* neuron is represented as θ_j.

$$\min(\omega_1,\omega_2,b_1,b_2) = \frac{1}{2}\sum_{i=1}^{n}\sum_{j=1}^{N_i}[y_i(j) - \hat{y}_i(j)]^2 \tag{3}$$

In formula (3), $\omega_1 \in R^{m\times p}$, $\omega_2 \in R^{p\times n}$, $b_1 \in R^p$, $b_2 \in R^n$. ω_1, b_1 are the weight matrix and threshold matrix of the input layer; ω_2, b_2 are the weight matrix and threshold matrix of the output layer. *m, n, p* are the number of layers in the input layer, output layer, and hidden layer, and N_i represents the number of samples in the *i-th* hidden layer.

2.3 Image signal

Capturing images of flat-end mills using industrial cameras can result in distortions such as color degradation and noise, caused by the complex surface structure of the cutters, camera instability, and high reflectivity of the tool. To address these issues, enhanced image processing techniques are employed, including the application of a 5×5 adaptive filter window to eliminate Gaussian noise and salt-and-pepper noise from the captured images [9], as shown in Figure. 2.

However, conventional denoising algorithms tend to blur image edges and reduce the clarity of hierarchical details. Therefore, this paper develops an adaptive contrast enhancement method to improve the distinction of wear areas. This method offers localized adjustments and preserves image details more effectively than traditional approaches.

$$M(i,j)=\frac{\sum_{s=i-n}^{i+n}\sum_{k=j-m}^{j+m} f(s,k)}{(2n+1)(2m+1)} \tag{4}$$

$$\sigma^2(i,j)=\frac{\sum_{s=i-n}^{i+n}\sum_{k=j-m}^{j+m} \left(f(s,k)-M(i,j)\right)^2}{(2n+1)(2m+1)} \tag{5}$$

In equations (4) and (5), $f(s, k)$ represents the pixel value at the coordinate point (s, k), $M(i, j)$ denotes the local mean within a window of size $[(2n+1)(2m+1)]$ centered at (i, j), $\sigma^2(i, j)$ is the local variance, and $\sigma(i, j)$ is the standard deviation.

The image is enhanced by the following calculation method:

$$I(i,j)=M(i,j)+G(f(i,j)-M(i,j)) \tag{6}$$

$$G=\frac{\alpha M}{\sigma^2(i,j)} \tag{7}$$

In Equations (6) and (7), $I(i, j)$ is the enhanced pixel value, M is the global mean, and α is the coefficient parameter set to α=0.4.

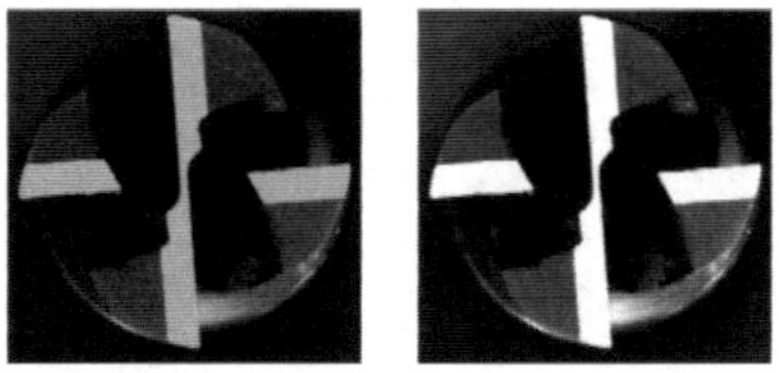

(a) Image of bottom blade

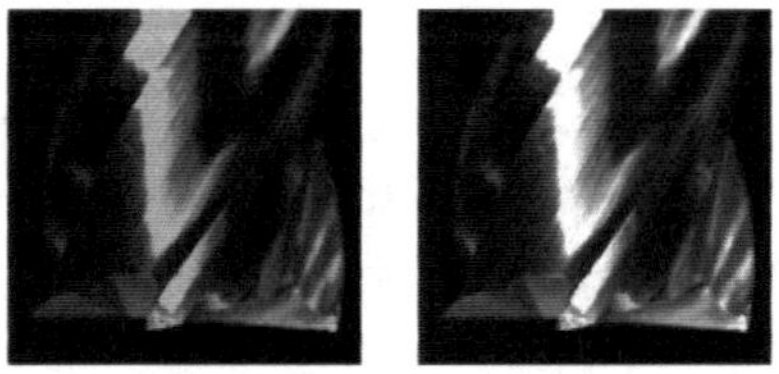

(b) Image of side blade

Figure. 2. Comparison of image processing.

2.4 Residual neural network classifier

The processed images (cropped, denoised, and contrast-enhanced) are divided into an 8:2 ratio for training and testing. ResNet-34 is employed to develop the milling cutter wear diagnostic model, as illustrated in Figure. 3.

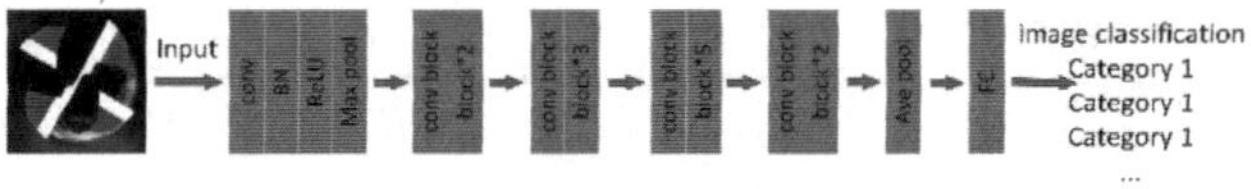

Figure. 3. ResNet-34 network architecture diagram.

3. Decision level Fusion Diagnostic Model

This paper introduces a weighted fusion algorithm based on D-S evidence theory to enhance the accuracy and robustness of flat-end mill wear diagnosis through heterogeneous data fusion. The proposed method combines the diagnostic results from ResNet-34 and BP neural network using a weighted K-coefficient approach to address evidence conflicts. By dynamically adjusting the weights, the algorithm achieves intelligent data fusion, outperforming traditional D-S theory in terms of reliability, particularly under complex conditions.

3.1 The basic principles of D-S evidence theory

The decision fusion algorithm presented in this study improves upon traditional D-S evidence theory by incorporating a weighted K-coefficient method. Traditional D-S theory combines multiple pieces of evidence to reach likely conclusions but struggles when faced with conflicting evidence. The improved algorithm uses the average support degree as a reference weight, redistributing conflicting evidence probabilities to corresponding conclusions. This approach mitigates the impact of conflicts and ensures more reliable fusion outcomes [10].

3.2 Probability allocation strategy

The BPA involves assigning probabilities to subsets A within the framework Θ using the function $m(x)$, while adhering to specific conditions.

$$\begin{cases} m(\varnothing) = 0 \\ \sum_{A\subseteq\Theta} m(A) = 1 \end{cases} \tag{8}$$

In both ResNet and the BPNN, the Softmax layer maps neuron outputs to a probability distribution for classification. This is achieved through exponentiation and normalization, ensuring that the output values fall within the range [0,1], with their sum equal to 1, thereby reflecting a valid probability distribution. The probability of sample x belonging to category i is represented by the output values of the Softmax layer.

3.3 Decision Fusion Method

Within the framework Θ, where there are n evidence bodies, BPA functions m_1 and m_2 are defined. The combined total probability for subset A is represented as:

$$m(A) = (1-K)\frac{\sum_{A_i \cap B_j} m_1(A_i)m_2(B_j)}{1-K} + Kf(A) \tag{9}$$

In formula (9), A_i and B_j, represent the focal elements corresponding to evidence bodies A and B respectively. $f(A) = \sum_{i=1}^{n} \frac{m_i(A)}{n}$, while $K = \sum_{B\cap C=\emptyset} m_1(B)\cdot m_2(C)$, and K represents the contradiction factor, quantifying conflict in evidence. Larger K indicates more severe conflict.

The decision-level fusion model classifies samples by selecting the state with the highest final probability, thereby identifying the milling cutter category with the maximum probability as the wear diagnosis result.

$$m(X_s) = \max(m(X_i)), s \in [1, m] \tag{10}$$

In formula (10), the final output wear state of the *i-th* flat-end milling cutter is denoted as the *s-th* state.

4. Experimentation

4.1 Milling experiment

The efficacy of the proposed heterogeneous data fusion diagnostic model is validated on a 5-axis CNC machining center. The cutter, shown in Figure. 4, is a 4-edge tungsten steel HSS tool with the following dimensions: a 30 mm blade length, 75 mm total length, 10 mm blade diameter, and 10 mm shank diameter. The wear states are categorized into five levels: new, mild, moderate, severe, and broken.

During the milling experiment, the spindle speed is set to 1600 rpm, the feed rate is 200 mm/min, and the milling depth is 0.5 mm. The vibration signal is sampled at a frequency of 25.6 kHz.

An LED light source is selected for the visual system due to its high brightness, ease of installation, flexible structure, and suitability for small spaces within machine tools. This study utilizes the AFT-RL5428-29W light source, with parameters detailed in Table 1.

Table 1. Light source performance parameters.

Camera parameters	Parameter			
	Size	***Thickness***	***Voltage***	***Power***
Parameter values	53.5mm/28.5mm	24mm	9v	2.88w

(a)New (b) Mild (c)Moderate (d) Severe (e) Broken

Figure. 4. Five types of flat end milling cutters with different degrees of wear.

The heterogeneous data fusion diagnostic model is tested on 45 steel blocks during milling operations on the 5-axis CNC machining center. The original vibration signal is captured during the milling process, as illustrated in Figure. 5a, while post-milling images of the cutter's edges and sides are shown in Figure. 5b.

(a) Layout of vibration sensors

(b) Layout of Image shooting

Figure. 5. Machine tool experimental platform.

4.2 Experimental results and analysis

The test set is divided into five groups, each containing 120 vibration and image data samples per wear state, resulting in a total of 600 samples. Figure. 6 presents a confusion matrix example for the first test set, showing outputs from both sub-models and the fusion model. Tags 1 to 5 represent the wear states: new, mild, moderate, severe, and broken.

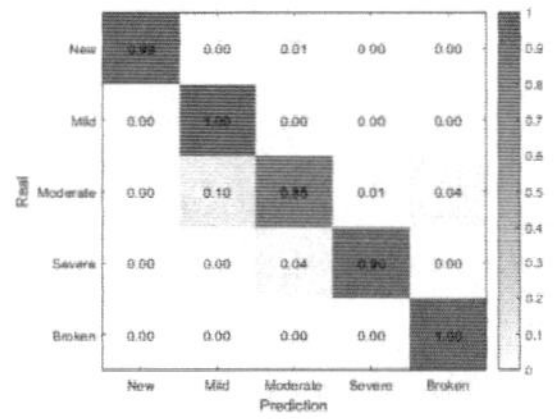

(a) Vibration signal sub-models.

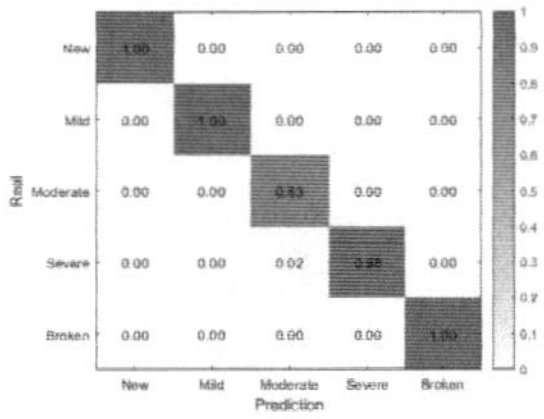

(b) Machine vision sub-models

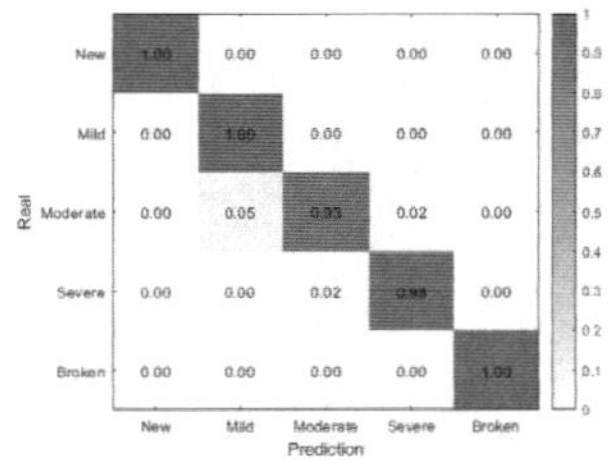

(c) Decision fusion model

Figure. 6. The confusion matrix of the first experimental data.

The confusion matrix for the first test set highlights that single sub-models face challenges in achieving comprehensive and accurate diagnoses of milling cutter wear. While the models perform well in identifying new, severe, and broken states, they struggle with distinguishing mild and moderate wear, leading to frequent misclassifications.

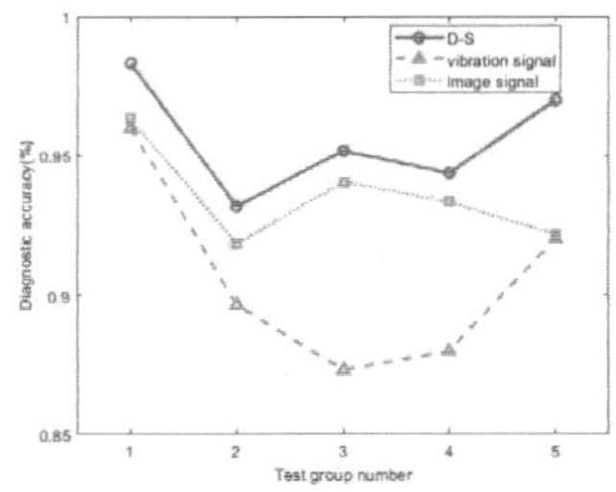

Figure. 7. Comparison of diagnostic accuracy among five test sets.

Figure. 7 demonstrates that the fused diagnostic model outperforms individual sub-models in accuracy across all five test sets. The vibration sub-model exhibits notably lower performance in sets 3 and 4, with accuracy rates of 87.33% and 88.00%, respectively. In contrast, the machine vision sub-model shows minimal variation in accuracy across sets 1, 5, and others. This discrepancy underscores the limitations of single sub-models in detecting all types of cutter wear. The fusion model, however, enhances overall performance, particularly when individual sub-model outputs are accurate. The analysis reveals the superior accuracy of the fusion model, achieving 90.60% for the vibration sub-model, 93.54% for the vision sub-model, and 95.60% for the fusion model. The fusion model integrates sub-model outputs through an internal correction mechanism, providing enhanced robustness and reliability. Notably, it maintains high accuracy rates of 95.17% and 94.36%, even when the vibration sub-model exhibits lower performance in sets 3 and 4.

5. Conclusions

This study addresses the limitations of single fault diagnosis classifiers in accurately identifying flat-end milling cutter wear states. By integrating vibration signals and machine vision data, a decision-level fusion model is developed and validated. The following key conclusions are drawn: (1) Sub-model-specific probability allocation strategies enhance the objectivity and rationality of probability assignments, improving diagnostic accuracy. (2) The fusion model leverages D-S evidence theory for effective wear state differentiation, achieving an average diagnostic rate of 95.60%, outperforming individual sub-models. The model offers a new approach for CNC tool wear identification.

Acknowledgment

Thank for Natural Science Foundation of Chongqing, (CSTB2022NSCQ-MSX1000), National Natural Foundation of China (52205144), and the Chongqing University of Technology Graduate Education High Quality Development Project (gzlcx20242012).

References

[1] Malekian M, Park S S, Jun M B G . Tool wear monitoring of micro-milling operation[J]. Journal of Materials Processing Technology, 2009, 209(10): 4903-4914.
[2] LIU Xiangnan, ZHAO Xuezhi, SHANGGUAN Wenbin. The impact features extraction of rolling bearing under strong background noise[J]. Journal of Vibration Engineering,2021,34(1):202~210.
[3] LI J, YAO X, WANG X, et al. Multiscale local features learning based on BP neural network for rolling bearing intelligent fault diagnosis[J]. Measurement, 2020, 153:107419.
[4] WANG Wangwang, DENG Linfeng, ZHAO Rongzhen, WU Yaochun. Fault feature extraction of rolling bearing integrating KPCA and t-SNE[J]. Journal of Vibration Engineering,2021,34(2):421~430.
[5] LIU Yirong, WANG Fuan, LV Jiechao, et al . A novel method for tool identification and wear condition assessment based on multisensordata[J] . Applied Sciences, 2020, 10(8): 2746 .
[6] SUN P, YANG Z, JIANG Y, et al. A Fault diagnosis method of modular analog circuit based on SVDD and D–S evidence theory[J]. Sensors, 2021, 21(20): 6889.
[7] QIU Y, FENG Y, INFIELD D. Fault diagnosis of wind turbine with SCADA alarms based multidimensional information processing method[J]. Renewable Energy, 2020, 145: 1923-1931.
[8] Mustafa K., Emin S., Munish K., et al. A state-of-the-art review on sensors and signal processing systems in mechanical machining processes[J]. International Journal of Advanced Manufacturing Technology, 2021, 116:2711-2735.
[9] P. M. Narendra and R. C. Fitch, Real-Time Adaptive Contrast Enhancement. Transactions on Pattern Analysis and Machine Intelligence, vol. PAMI-3, no. 6, pp. 655-661, Nov. 1981.
[10] WU T F, LIN C J, WENG R C. Probability estimates for multiclass classification by pairwise coupling[J]. Journal of Machine Learning Research, 2004, 5: 975-1005.

Journal of Physics: Conference Series **2992** (2025) 012021 doi:10.1088/1742-6596/2992/1/012021

Acoustic emission monitoring of the hot temperature tensile behavior of 20CrMo steel

Qingshan Li [1, 2], Xumeng Xie [1, 2*], Wenjie Pan [1], Huiquan Cao [1], Hao Wu and Xinhua Xia [1, 2]

[1] Zhejiang Academy of Special Equipment Science, Hangzhou 310009, China
[2] Key Laboratory of Special Equipment Safety Testing Technology of Zhejiang Province, Hangzhou 310009, China

* Corresponding author's e-mail: xiexm@zjtj.org

Abstract: Acoustic emission characteristics of 20CrMo steel during hot temperature tensile tests at 650, 700, and 750°C were investigated in this study. During the elastic deformation and fracture stage, a large number of high-amplitude AE signals were recorded for all samples. Upon entering the work hardening stage, few high-amplitude AE signals were detected for samples at 650°C and 700°C while a large number of AE signals were recorded for the sample at 750°C. The AE counts began to decrease after entering the necking stage for the samples at 650°C and 700°C. However, the counts decreased but still remained at a certain level for the sample at 750°C after entering the necking stage. Cumulative counts and cumulative counts rate were higher for the sample at 750°C. It was attributed to the increased ease of dislocation movement and improvement of ductility.

1. Introduction

20CrMo steel is a Chinese GB/T 3077 steel with good heat resistance, hardenability, and weldability. It is used in the manufacture of non-corrosive media in chemical equipment and high-pressure pipes at moderate temperatures. 20CrMo steel is commonly utilized for carburizing components, such as gears and shafts, which maintains its tribological performance effectively across a wide temperature range without experiencing any significant degradation [1-3].

Acoustic Emission (AE) is a phenomenon where materials emit transient elastic waves due to the rapid release of strain energy from microstructural changes. AE technique is a highly effective non-destructive testing method used for the inspection and structural health monitoring (SHM) of various types of equipment, including pressure vessels, pressure pipelines, and recreational facilities. AE is also employed to study the tensile behavior of materials, providing valuable insights into the material's response to stress and deformation without causing any damage to the test specimen. Therefore, AE is a crucial tool for monitoring material behavior during tensile tests. Numerous researchers have used AE parameters to give a deeper comprehension of the tensile deformation of metals [4-6], rocks [7], composite [8], and concrete [9]. Yu et al. [4] used AE to evaluate AE parameters in the tensile test of AZ31B magnesium alloy. They concluded that the yield signal exhibited a longer duration compared to the fracture signal, while the energy emitted by the yield signal was found to be less than that of the fracture signal. Mohammad et al. [5] analyzed the AE characteristics of ss304 steel at high temperatures with a waveguide, and higher amplitudes. A greater fast Fourier transform amplitude were observed in the fracture region. However, little research has been conducted on AE

Journal of Physics: Conference Series **2992** (2025) 012021 doi:10.1088/1742-6596/2992/1/012021

characteristics at high temperatures due to the limitations of high-temperature conditions and inconvenience.

This study delves into the effects of elevated temperatures on the mechanical properties of 20CrMo employing AE techniques during tensile testing. The relationship of AE characteristics with temperature was established. The influence of temperature on AE characteristics was also discussed.

2. Experimentation

2.1. Materials and welding process

The chemical composition of 20CrMo steel is presented in Table 1. A cylindrical sample with a gauge length of 50 mm and a diameter of 10 mm was fabricated from the 20CrMo ingot for a hot temperature tensile test.

Table 1. Chemical composition of 20CrMo (mass%).

C	Si	Mn	Cr	Mo	P	S
0.20	0.24	0.58	0.91	0.17	0.017	0.002

2.2. AE test during tensile test

The tensile test of 20CrMo steel was investigated with an ETM305D tensile tester at 650, 700, and 750°C at a strain rate of 5 mm/min. The AE signals were captured using a system from Physical Acoustic Corporation (PAC). For the monitoring process, a WD differential wideband sensor known for its exceptional sensitivity and broad bandwidth was chosen for this test.

In general, waveguide rods are considered for coupling in high-temperature conditions. However, the space inside the high-temperature tensile test furnace is limited, making it difficult to install waveguide rods. Specific equipment is required to weld them. Therefore, this study designed a dedicated device for acoustic emission monitoring of cylindrical specimens in high-temperature tensile tests, as shown in Figure 1. The lower rod is machined with two symmetrical planar structures, one for the coupling plane of the sensor and the other for the installation of the sensor clamping fixture. The cylindrical specimen is connected to the upper and lower pull rods within the furnace through thread. Once the preset temperature is reached, the temperature of the lower pull rod plane is measured to ensure it is below the normal operating temperature of the AE sensor.

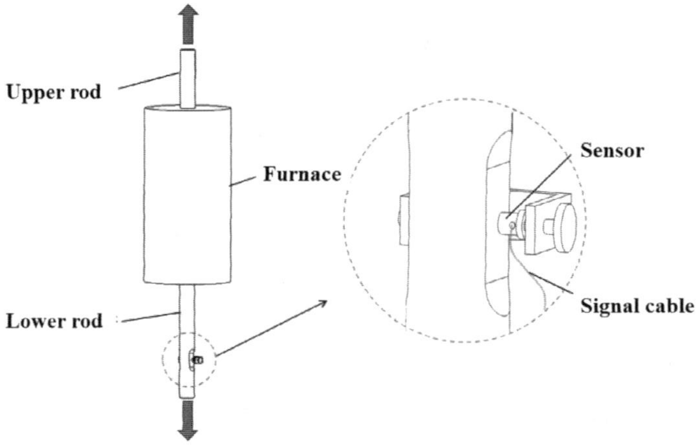

Figure 1. Experimental setup for tensile test and AE monitoring.

An amplitude threshold of 40 dB was established to effectively filter out ambient noise. To verify the proper coupling of the sensor with the specimen and to ensure that the AE system was functioning correctly, a pencil lead break test was conducted on the specimen's surface.

3. Result and discussion

3.1. AE signals

AE signals are detected by sensors that are attached to the surface of the lower rod. Figure 2 illustrates two typical AE waveforms from this study, which depicts the change in voltage over time. These two waveforms are burst signal and continuous signal, respectively. Burst signals are characterized by a distinct start and finish, which sets them apart from the underlying background noise. Therefore, burst signals can be used for detecting and analyzing events such as material fractures, crack growth, and other discontinuities in structures. Continuous signals vary smoothly over time without any abrupt changes or interruptions, and the macroscopic yielding process of metals is an important source of continuous acoustic emission [10]. In actual research processes, burst and continuous AE signals often do not exist independently but appear mixed. By examining the characteristic parameters, such as amplitude, count, and duration, AE signals can be effectively identified and interpreted with changes in microstructure of materials.

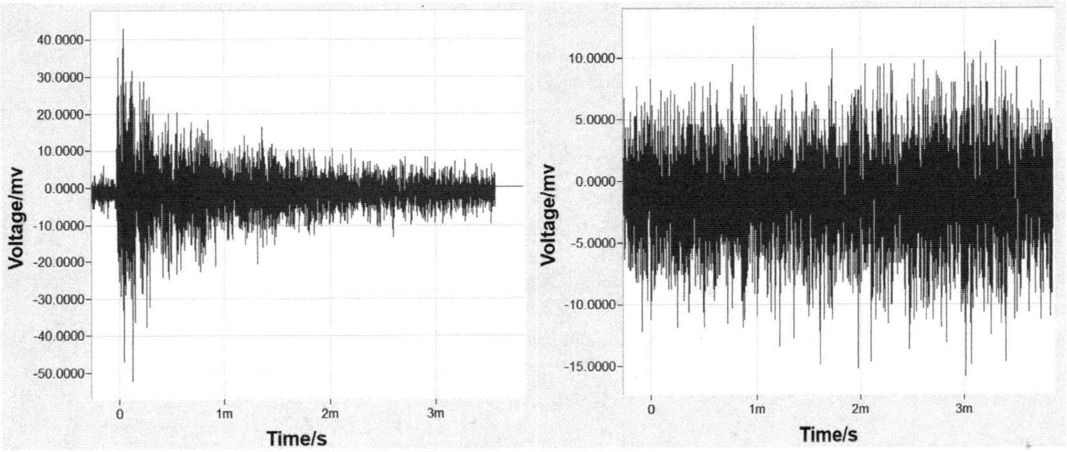

(a) Burst signals (b) Continuous signals

Figure 2. AE waveform.

3.2. Tensile properties and AE characteristics

The graph depicting the variation of AE amplitude and stress over time for 20CrMo is presented in Figure 3. It is noted that the tensile strength of 20CrMo gradually diminished with the increase in temperature, whereas the duration of the tensile process progressively increased. Different from what is typically observed in plain carbon steels at room temperature, this study did not record a yield plateau. The amplitude of AE signals, which is a crucial parameter for characterizing the intensity of these signals, is not solely dependent on the threshold.

Each tensile process can be divided into four stages: the elastic deformation stage (I), the work hardening stage (II), the necking stage (III), and the fracture stage (IV). During the elastic deformation stage, samples under each test condition generated a large number of high-amplitude AE signals.

Upon entering the work hardening stage, the AE signals of the samples at 650°C and 700°C rapidly decrease. Few high-amplitude AE signals were detected for samples at 650°C and 700°C during the necking stage while a large number of AE signals were recorded for the sample at 750°C during the necking stage. For the sample at 750 °C, the boundary between the elastic deformation stage and the work hardening stage is not easily identifiable. Moreover, many high-amplitude AE signals (>60 db) were observed at 250-350 s. For low-carbon steel like 20CrMo, the microstructure is primarily composed of pearlite and ferrite below 700 °C. When the temperature exceeds 727 °C and enters the two-phase region, the microstructure consists of ferrite and austenite. At high temperatures, the deformation of ferrite and austenite may generate more elastic waves, hence more acoustic emission signals are detected. Apart from that, during the fracture stage, a large number of high-amplitude AE signals are detected, which is consistent with the conclusions of another researcher [11].

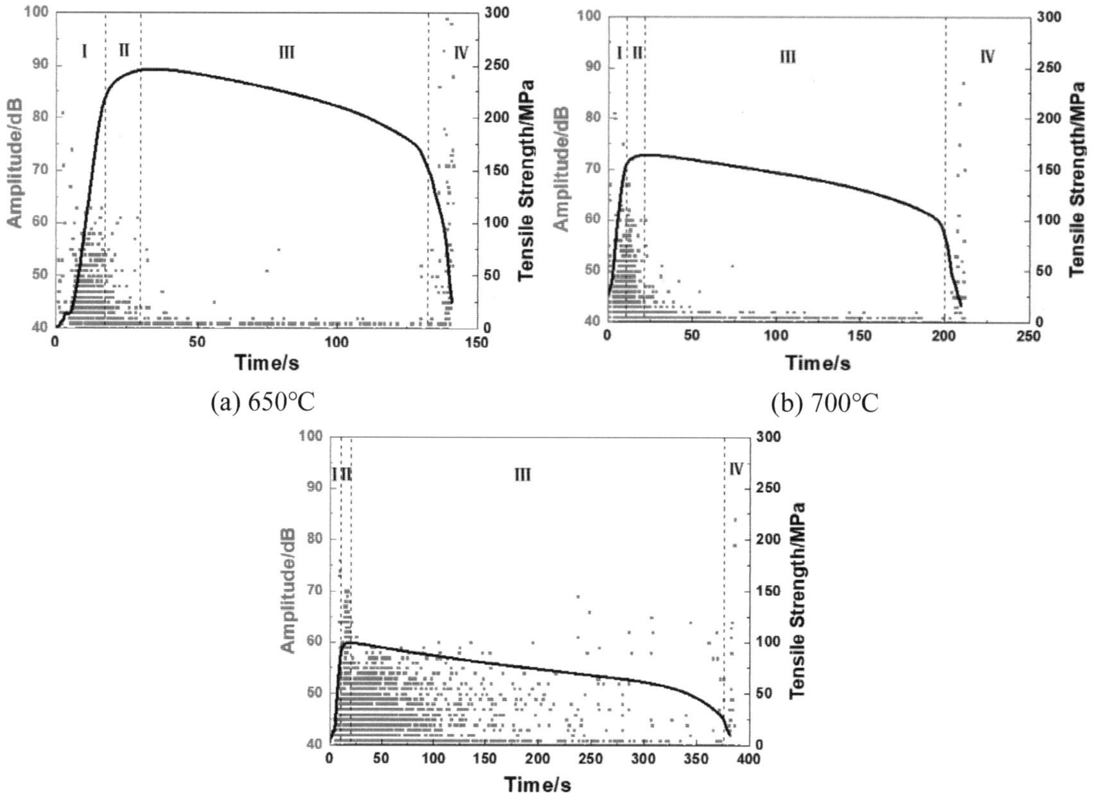

(a) 650°C (b) 700°C

(c) 750°C

Figure 3. AE amplitude and stress versus time of 20CrMo.

The AE counts refer to the number of pulses produced by the measurement circuit whenever the amplitude of an AE signal surpasses a predetermined threshold. This parameter indicates the frequency of AE events, specifically the counts of instances where the signal amplitude crosses the threshold during a defined time interval. Early studies have shown AE counts are closely correlated with the damage progression [12]. The AE counts captured in this research are presented in Figure 4. AE counts exhibit significant peaks at the initiation and the end of the tensile test. The peak of the AE counts occurred around 25 s for the samples at 650°C and 700°C. For the sample at 750°C, it occurred at around 10 s. Similar to the trend of amplitude changes, the AE counts begin to decrease after entering the necking stage for the samples at 650°C and 700°C. However, although the count decreases, it still remains at a certain level after entering the necking stage for the sample at 750°C. Cumulative counts are displayed in Figure 3(d). It was noted that total cumulative counts and cumulative counts rate were higher for the sample at 750°C. The cumulative counts and cumulative counts rate of the sample at 650 °C were a little larger than those of the sample at 700°C. The generation of AE signals is related to the movement of dislocations and the movement of dislocations becomes easier as the temperature is raised to 750°C. Moreover, plasticity improves as the temperature increases, which also leads to the detection of more AE signals.

The analysis mentioned above all belongs to time-domain analysis, so it inevitably has limitations. Therefore, in future work, we will combine frequency-domain analysis and even machine learning methods to conduct a more systematic analysis of AE waveforms.

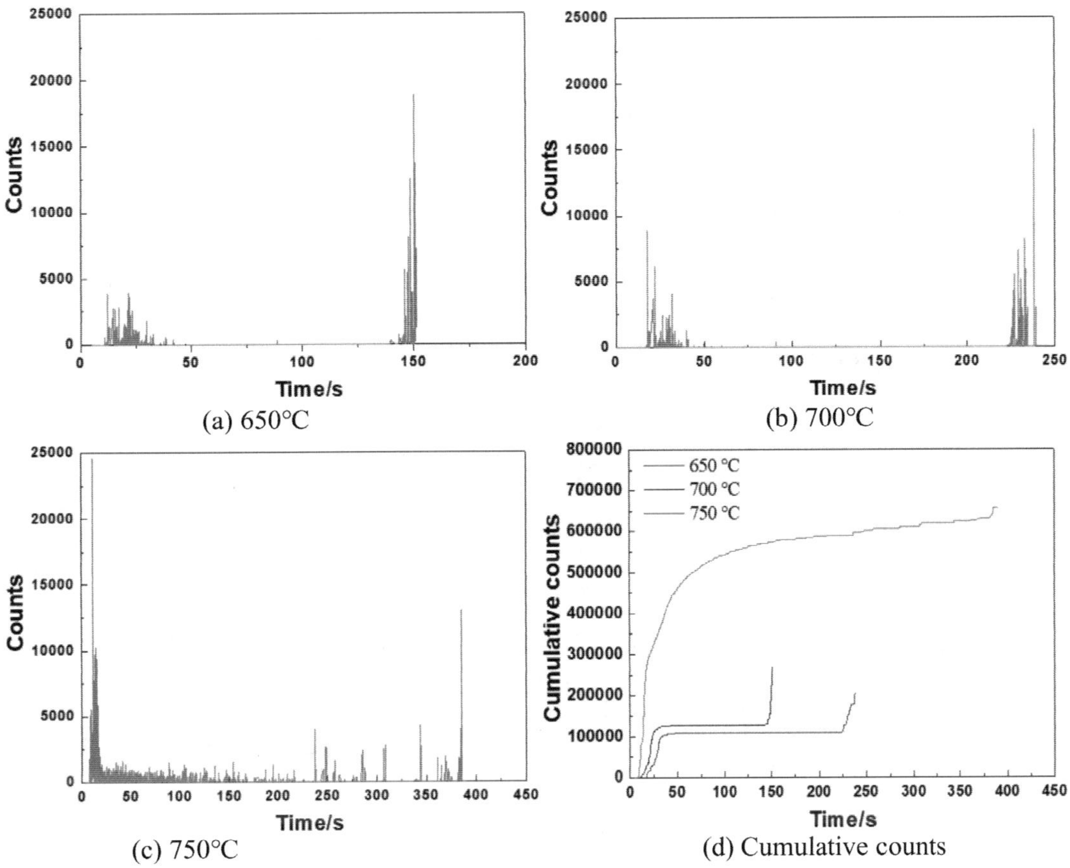

Figure 4. AE counts versus time.

4. Conclusion

1. During the elastic deformation and fracture stage, all 20CrMo samples generated a large number of high-amplitude AE signals.

2. Upon entering the work hardening stage, a few high-amplitude AE signals were detected for samples at 650°C and 700°C while a large number of AE signals were recorded for the sample at 750°C.

Similar to the trend of amplitude changes, the AE counts began to decrease after entering the necking stage for the samples at 650°C and 700°C. However, although the counts decreased, they still remained at a certain level after entering the necking stage for the sample at 750°C.

For the sample at 750°C, total cumulative counts and cumulative counts rate were higher. It was attributed to the increased ease of dislocation movement and improvement of ductility.

References

[1] Guan, Jian, et al. "The friction behavior and wear mechanism of RV reducer gear steel (20CrMo) subjected to three different heat treatment processes from − 20°C to 100°C." Journal of Materials Research and Technology 32 (2024): 2609-2623.

[2] Gao, Zhixin, et al. "Effects of Heat Treatment Processes on the Microstructure and Properties of 20CrMo Alloy Steel." International Conference on Mechanical Engineering and Applied Composite Materials. Singapore: Springer Nature Singapore, 2023.

[3] Gao, Zhixin, et al. "Effect of double quenching process on Microstructure and Properties of 20CrMoH Alloy Steel." Journal of Physics: Conference Series. Vol. 2845. No. 1. IOP Publishing, 2024.
[4] Yu, Jae-Hyeong, et al. "Analysis and Clustering of Acoustic Emission Signals in the Tensile Deformation of AZ31B." Metals and Materials International (2024): 1-16.
[5] Mohammad, Mazian, Abdul Rahim Othman, and Mohd Faisal Ismail. "Influence of high temperature during tensile test for stainless steel using acoustic emission." Process Safety Progress 43 (2024): S108-S115.
[6] Aftimi, Sabah, et al. "Acoustic Emission for Real-Time Monitoring of Interfacial Erosion-Corrosion of Austenitic Stainless Steel in Polluted Phosphoric Acid Environment." Wear (2024): 205666.
[7] Du, Kun, et al. "Understanding the evolution mechanism and classification criteria of tensile-shear cracks in rock failure process from acoustic emission (AE) characteristics." Engineering Fracture Mechanics 296 (2024): 109864.
[8] Morscher, Gregory N., et al. "Acoustic emission accuracy from a tensile test of a ceramic matrix composite." Journal of the American Ceramic Society 107.12 (2024): 8556-8571.
[9] Van Steen, Charlotte, Lotfollah Pahlavan, and Els Verstrynge. "Smart aggregates for acoustic emission monitoring of concrete cracking and reinforcement corrosion." Construction and Building Materials 443 (2024): 137644.
[10] Han, Zhiyuan, Hongyun Luo, and Hongwei Wang. "Effects of strain rate and notch on acoustic emission during the tensile deformation of a discontinuous yielding material." Materials Science and Engineering: A 528.13-14 (2011): 4372-4380.
[11] Chen, Jiangbo, et al. "Analysis of acoustic emission signal in the tensile process of weld-bonding joint and its damage model." Nondestructive Testing and Evaluation (2024): 1-19.
[12] Zhao, Peng, et al. "Correlation between acoustic emission detection and microstructural characterization for damage evolution." Engineering Fracture Mechanics 230 (2020): 106967.

Simulation study on the power performance of LPG-NH_3 dual-fuel engine

Zan Huang [1], Xiuhua Ma [1], Hua Xiao [*1] and Guoqing Zhang [2]

[1]Guangzhou Maritime University, School of Marine Equipment Engineering, Guangzhou, Guangdong, China
[2]Guangdong University of Technology, College of Materials and Energy, Guangzhou, Guangdong, China

* Corresponding author: 1091501434@qq.com

Abstract. In this paper, the effect of ammonia blending ratio from 0% to 100% on the engine power performance was investigated using numerical simulation at different speeds (1000, 2000 and 3000) rpm for a single cylinder ignition type internal combustion engine LPG-NH_3 dual fuel. It was found that the best engine power performance was achieved at 70% ammonia blending ratio with an increase in brake power and brake torque of around 18% compared to pure LPG.

1. Introduction

Internal Combustion Engines (ICEs), which are widely used in transportation worldwide, are a major contributor to fuel consumption and environmental pollution. For this reason, countries have successively implemented stricter regulations on emissions and fuel consumption [1]. So it is urgent to reduce the fuel consumption and emission of ICEs. Utilizing alternative fuels is a crucial approach to improving the fuel economy of ICEs and reducing their emissions [2].

Liquefied petroleum gas (LPG) has gained widespread recognition and acceptance as an alternative fuel for internal combustion engines due to its abundant resources, low cost, and reduced greenhouse gas emissions. LPG has become a stable alternative fuel for passenger vehicles, with combustion characteristics similar to gasoline but with a higher octane rating. Additionally, its volatility allows it to mix well with air, making it suitable for use in spark-ignition (SI) engines. Compared to traditional gasoline and diesel, LPG has a lower carbon content, which reduces emissions of carbon dioxide and hydrocarbons [1]. Kim et al. [3] innovatively used an experimental platform based on a standard 2000 cc four-cylinder gasoline port fuel injection internal combustion engine, modified to operate as a direct injection LPG internal combustion engine. Their research revealed that compared to previous gasoline-based studies, the use of LPG not only reduced HC and NOx emissions, but also significantly lowered particulate matter emissions, compared to the original gasoline-based study, demonstrating the environmental advantages of LPG.

Ammonia (NH_3) [4], as a carbon-free alternative fuel for ICEs, has shown promising potential for use in automotive ICEs [5], boilers [6], and fuel cells [7] due to its efficient and clean combustion characteristics. Additionally, the easy storage and transport characteristics of ammonia and the advantage of no carbon-based emissions have led to its rapid rise as a green fuel for decarbonisation and reduction of greenhouse gas emissions. Wang et al. [8] explored the combustion process of ammonia in

ICEs using a multiple-injection jet-ignition method with H_2. The results showed that by adjusting the H_2 injection strategy and optimising the spark timing, stable combustion and high thermal efficiency of NH_3-H_2 ICEs were successfully achieved, with a minimum hydrogen-to-energy ratio of even less than 3 %.

After an in-depth study, the dual-fuel model has been proven to be currently the most suitable and effective configuration for ammonia ICEs. The ignition fuels used in this combustion process are typically more reactive substances such as diesel, gasoline, hydrogen, acetylene, butane, liquefied natural gas, liquefied petroleum gas, and dimethyl ether (DME) [9-10]. This dual-fuel internal combustion engine [11-12] not only significantly improves operational efficiency and air utilization efficiency, but also has the distinct advantages of a comfortable working environment and good noise control. Additionally, it effectively reduces carbon emissions and environmental pollution, thus contributing to optimizing the energy architecture, promoting sustainable economic development, and reducing exhaust emissions [13].

To more precisely explore the potential for reducing pollutant emissions, we constructed an SI ICE simulation model, focusing on the impact of LPG-NH_3 dual-fuel mixtures on engine performance characteristics. We analyzed six different blending ratios in detail, examining the trends in engine performance characteristics. Additionally, an in-depth analysis of the optimal operating conditions for LPG-NH_3 dual-fuel combustion in ICEs has been carried out. This study aims to enrich the simulation research on LPG-NH_3 dual-fuel and provide valuable reference data for future experimental and simulation studies related to SI ICEs.

2. Model Establishment

In this paper, an intuitive zero-dimensional combustion model has been chosen and simulated using the GT-Power software which has the advantage of integrating this model. GT-Power is part of the GT-SUITE software and can be used for detailed numerical simulations of all types of ICEs, including heat transfer models, combustion models, and the operating mechanisms of the different subsystems. The main parameters of the selected engine are shown in Table 1. Based on these parameters and the actual structure, including the intake system, exhaust system, cylinder, and engine components, a gt-power based single-cylinder SI LPG-NH_3 engine model (Figure 1) was developed, which mainly consists of the air intake, oil supply, cylinder, exhaust, and engine components.

Table 1 Main Parameters of the LPG-NH_3 Engine.

Parameter	Value
Bore (mm)	82
Stroke (mm)	64
Valve system(mm)	8
Compression Ratio	8.5
Swept volume (cm^3)	338
Maximum ratio	3800
Intake Method	Naturally Aspirated
Fuel	Dual fuel

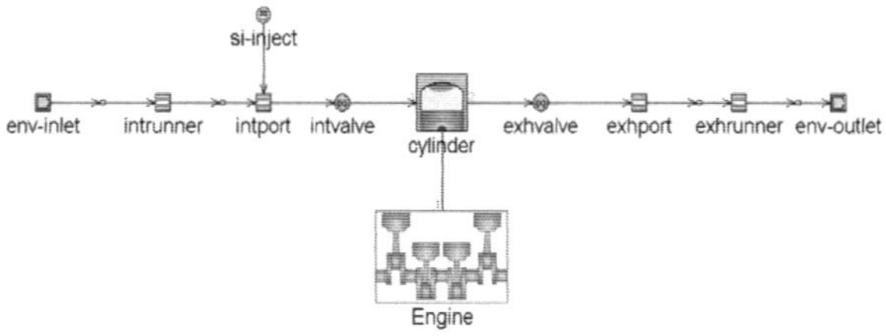

Figure 1. Gt-power single-cylinder SI LPG-NH_3 engine model.

2.1 *Heat Transfer Model – WoschniGT*

Woschni provides a standard for calculating the transient heat transfer coefficient. Currently, most studies on transient heat transfer coefficients, both domestically and internationally, utilize empirical formulas or similarity-based principles [14-15]. Woschni is one of the typical similarity criteria, and its criterion equation is as follows:

$$Nu=0.035Re^{0.8} \quad (1)$$

Its heat transfer coefficient equation is:

$$\alpha=820p^{0.8}\cdot T^{-0.53}\cdot D^{-0.2}\cdot\left[C_1\cdot C_p+C_2\frac{T_aV_h}{p_aV_a(p-p_0)}\right]^{0.8} \quad (2)$$

where p and T are the instantaneous pressure and temperature inside the cylinder; D is the cylinder bore; C_1 is the airflow velocity coefficient (which differs for intake/exhaust and compression/expansion phases); C_2 is the coefficient of the shape of the combustion chamber; C_p is the piston movement velocity; P_a, V_a, and T_a are the gas state parameters at the start of compression; p_0 is the pressure during engine motoring (understood as the pressure when the engine maintains speed without combustion); and V_h is the working volume of the cylinder.

2.2 *Combustion Model – Wiebe*

The Wiebe model, a semi-empirical formula, is used to calculate the combustion rate:

$$x=1-e^{-6.908y^{m+1}} \quad (3)$$

$$\frac{d_x}{d_\emptyset}=6.908\frac{m+1}{\emptyset_z}y^m e^{-6.908y^{m+1}} \quad (4)$$

where x is the percentage of fuel burned; y is the dimensionless time function; m is the combustion quality index.

2.3 *Model Validation*

After the test run, the engine system model showed good stability under different operating conditions. LPG fuel with a composition of 90% propane and 10% butane was used. The torque and power of the internal combustion engine at different speeds were simulated at an internal combustion engine speed of 1700-3200 rpm, and then the obtained results were compared with the data of the study by Çinar [16]. Based on the in-depth analysis of the power and torque trend plots (Figures 2 and 3).

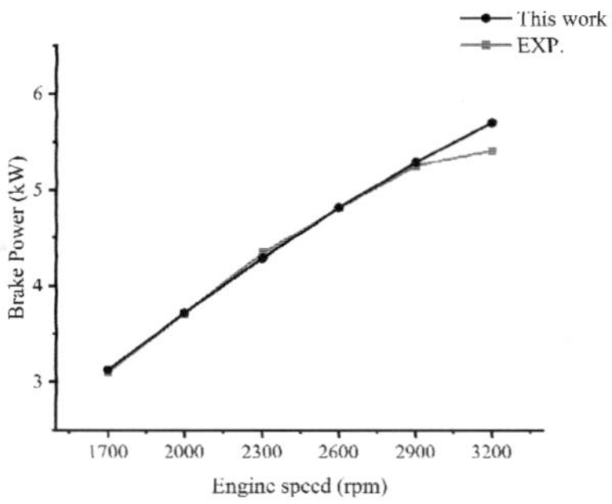

Figure 2. Brake power verification.

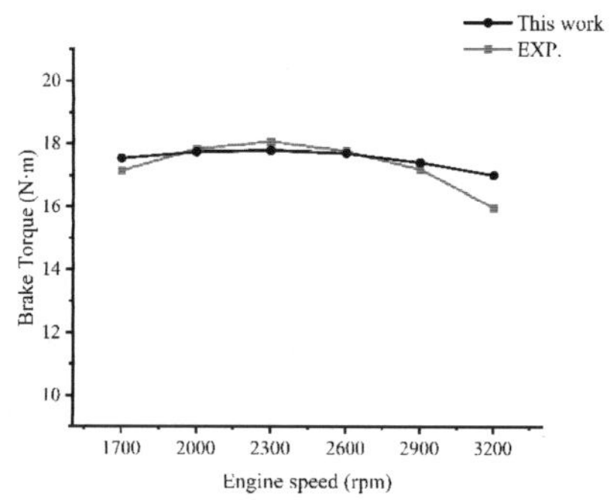

Figure 3. Brake torque verification.

By observing Figures 2 and 3, it can be seen that the overall brake torque and power values are consistent with Çinar's results, and the error is within a reasonable range. Therefore, it is considered that this model meets the requirements for subsequent research.

3. Simulation and analysis

3.1 Simulation

As a single-cylinder engine model, the external environment of the intake and exhaust system of this model is set at room temperature and standard atmospheric pressure, while the initial state inside the cylinder and piping is also set at room temperature and standard atmospheric pressure. The performance of the LPG-NH_3-fuelled engine and its emission impacts are investigated under a range of engine speeds from 1000 to 3000 rpm and ammonia dosing ratios from 0 to 100%. The simulation of different ammonia blending ratios in the mixture is achieved by varying the fuel input ratio by adjusting the fluid composition in the objector module, while keeping the rest of the parameters fixed.

3.2 Simulation results analysis

3.2.1. Impact of NH3 blending on engine power performance

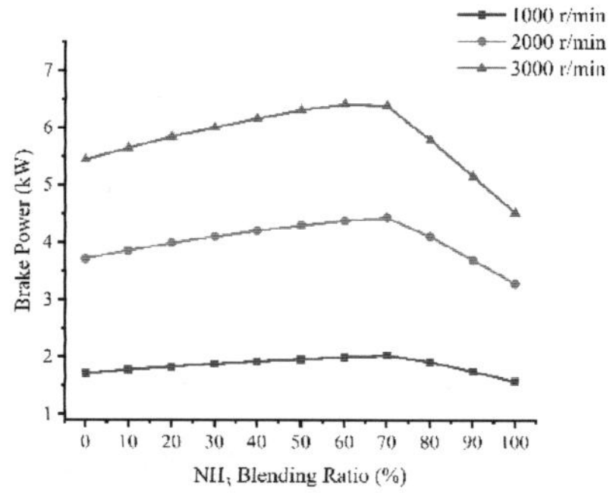

Figure 4. Simulation of brake Power with different NH_3 blending ratio.

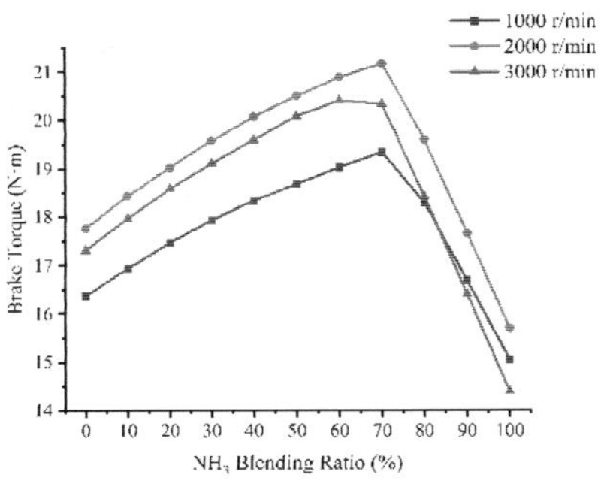

Figure 5. Simulation of brake torque with different NH_3 blending ratio.

Figures 4-5 show the impact curves of the NH_3 blending ratio on engine brake power (BP) and brake torque (BT) under three different speed conditions. With the increase in NH_3 blending ratio, the BP, and BT exhibit a trend of first rising and then falling across all conditions [16-17]. At low rpm, the change in power is relatively small. The power reaches a maximum value of 2.1 kW at a 70% NH_3 blending ratio and decreases to a minimum value of 1.6 kW at a 100% NH_3 blending ratio, which is an increase of 28.5%. The power increased by 18.7% compared to pure LPG. The maximum value of power at high rpm also occurs at 70% NH_3 blending ratio with 6.4 kW, while it decreases to 4.5 kW at 100% NH_3 blending ratio with an increase of 42%. Compared to pure LPG, there is an 18.5% increase in power at high rpm. Similarly, the torque reaches a maximum value of 19.3 N·m at 70% NH_3 blending ratio and decreases to 15.0 N·m at 100% NH_3 blending ratio with an increase of 28.7%. Compared to pure LPG, the torque increased by 17.7% at low rpm. The maximum value of torque was 20.3 N·m and the minimum value was 14.4 N·m at high rpm with an increase of 41%, compared to pure LPG with an increase of 17.3%.

The reasons for the curve changes shown in the figures after NH_3 blending are mainly as follows:

3.2.1.1 Change in Combustion Characteristics: The combustion characteristics of NH_3 are different from those of LPG. After blending NH_3, the fuel's combustion speed, combustion temperature, and flame propagation characteristics will change. Initially, increasing the NH_3 blending ratio may improve combustion efficiency, thereby enhancing engine performance. However, when the NH_3 ratio exceeds a certain level, the differences in combustion characteristics between NH_3 and LPG become apparent. A too-high blending ratio may lead to unstable combustion, reducing combustion efficiency and thus affecting engine performance.

3.2.1.2 Low Calorific Value of NH_3: Since the calorific value of ammonia (approximately 18.6 MJ/kg) is lower than that of LPG (approximately 46 MJ/kg), ammonia in blended fuels can nevertheless optimise the combustion process under certain operating conditions. At high temperatures, ammonia produces additional hydrogen, which contributes to the overall combustion efficiency of the fuel. Ammonia has a higher auto-ignition temperature and burns more slowly, resulting in a more uniform and smoother combustion process. Under proper combustion conditions, the presence of ammonia suppresses detonation and allows the engine to operate at higher compression ratios, resulting in higher BP and BT. However, the overall calorific value of NH_3-blended fuels also decreases when the blend ratio is increased to a certain point. A lower calorific value means that less energy is available per volume of fuel, which usually leads to a reduction in engine performance. The low calorific value of NH_3 significantly affects the overall energy density of the fuel, which in turn reduces engine performance.

3.2.1.3 Cooling Effect of NH_3: Since NH_3 has a certain cooling effect in the gas state, NH_3 blending may lead to lower gas temperatures, affecting the combustion process and the brake thermal efficiency of the engine. When the blending ratio reaches a certain level, this cooling effect may negatively impact the combustion process, further leading to a decline in engine performance.

3.2.1.4 Engine Design and Calibration: Engines are usually optimized for specific fuels during design and calibration. NH3 blending may require adjustments to the engine's fuel injection system, ignition system, etc., to adapt to the new fuel characteristics. If the adjustments are not made properly, the engine power performance output may be affected.

4. Conclusion

This paper primarily discusses the effects of different NH_3 blending ratios in LPG-NH_3 dual-fuel on engine power performance. With the increase of the NH_3 blending ratio, the BP and BT of the engine will show a tendency to increase and then decrease due to the change in combustion characteristics, the decrease of fuel energy density, the engine tuning and matching problems, etc. Simulation results indicate that compared to pure LPG, the engine performs best at a 70% NH_3 blending ratio. In practical applications, it is crucial to find the optimal NH_3 blending ratio to enhance engine performance.

Acknowledgment

Scientific Research Capacity Improvement Project of Key Construction Discipline of Guangdong Province (No. 2021ZDJS062)

References

[1] Kim J, Kim K, Oh S. An assessment of the ultra-lean combustion direct-injection LPG (liquefied petroleum gas) engine for passenger-car applications under the FTP-75 mode [J]. Fuel processing technology, 2016, 154:219-226.
[2] Liu H, Yu S, Wang T, et al. A systematic review on sustainability assessment of internal combustion engines [J]. Journal of cleaner production, 2024, 451:141996.
[3] Kim K, Kim J, Oh S, et al. Lower particulate matter emissions with a stoichiometric LPG direct injection engine [J]. Fuel, 2017, 187:197-210.
[4] Cardoso J S, Silva V, Rocha R C, et al. Ammonia as an energy vector: Current and future prospects for low-carbon fuel applications in internal combustion engines [J]. Journal of Cleaner Production, 2021, 296:126562.
[5] Chiong M, Chong C T, Ng J, et al. Advancements of combustion technologies in the ammonia-fuelled engines [J]. Energy Conversion and Management, 2021, 244:114460.
[6] Zhihua W, Zuochao Y U, Chenlin C, et al. Research progress on combustion characteristics of new zero carbon ammonia fuel [J]. Journal of Huazhong University of Science and Technology(Nature Science Edition), 2022, 50(7):24-40, 78.
[7] Wan Z, Tao Y, Shao J, et al. Ammonia as an effective hydrogen carrier and a clean fuel for solid oxide fuel cells [J]. Energy Conversion and Management, 2021, 228:113729.
[8] Wang Z, Qi Y, Sun Q, et al. Ammonia combustion using hydrogen jet ignition (AHJI) in internal combustion engines [J]. Energy (Oxford), 2024, 291:130407.
[9] Dimitriou P, Javaid R. A review of ammonia as a compression ignition engine fuel [J]. International Journal of Hydrogen Energy, 2020, 45(11):7098-7118.
[10] Xiao H, Li H. Experimental and kinetic modeling study of the laminar burning velocity of NH3/DME/air premixed flames [J]. Combustion and Flame, 2022, 245:112372.
[11] Wang G, Chen C, Beshiwork B A, et al. Developing a low-carbon hybrid of ammonia fuel cell and internal combustion engine for carbon neutrality [J]. Applications in Energy and Combustion Science, 2023, 16:100214.
[12] Oh S, Park C, Oh J, et al. Combustion, emissions, and performance of natural gas–ammonia dual-fuel spark-ignited engine at full-load condition [J]. Energy, 2022, 258:124837.
[13] Boretti A. Numerical study of the substitutional diesel fuel energy in a dual fuel diesel-LPG engine with two direct injectors per cylinder [J]. Fuel Processing Technology, 2017, 161:41-51.

[14] Talib Hashem G, Al-Dawody M F, Sarris I E. The characteristics of gasoline engines with the use of LPG: An experimental and numerical study [J]. International Journal of Thermofluids, 2023, 18:100316.
[15] Wang X, Cai Y, Li Z, et al. GT-POWER Internal Combustion Engine Performance Simulation [M]. Xi'an University of Electronic Science and Technology Press, 2022.
[16] Çinar C, Şahin F, Can Ö, et al. A comparison of performance and exhaust emissions with different valve lift profiles between gasoline and LPG fuels in a SI engine [J]. Applied Thermal Engineering, 2016, 107:1261-1268.
[17] Zhong R. Optimization Study of CNG Motorcycle Engine Performance Based on GT-Power [J]. 2010.

The buffer protection mechanism of heterogeneous metal laminated structures during severe overloading

Yu'Nan Chen, Peng Zhang, Hao'Nan Guo, Yunbo Shi *, Rui Zhao, Xiangyu Li, Yu Wang and Tao Guo

Science and Technology on Electronic Test & Measurement Laboratory, North University of China, Taiyuan, 030051, China

*Corresponding author's e-mail: shiyunbo@nuc.edu.cn

Abstract: As missile-borne electronics face growing severe high overload environments, the matching protective structure also meets higher challenges. Thus, a laminated configuration with varying wave impedance is fabricated using appropriate metallic materials to facilitate energy absorption buffering. Through an examination of stress wave propagation within the laminated structure, an energy absorption model is formulated. This model, in conjunction with an assessment of the filtering properties of the laminated structure, guides the preparation of Al-Fe heterogeneous metallic laminated materials for dynamic impact testing. Subsequently, an optimization process is conducted to design a metal-laminated protective structure that maximizes the engineering application capabilities of metal-laminated materials in protective structures.

1. Introduction

The ability of electronic devices to endure high overload conditions in various fields like space, maritime, electronics, biology, medicine, and especially missile applications is increasingly being challenged [1]. High overload stress harms electronic equipment by damaging circuits and components, leading to system failure. To boost instrument and equipment reliability, it is key to protecting sensitive parts effectively. Currently, prevalent methods for safeguarding these components consist of encapsulation protection and structural reinforcement[2]. Improving buffering performance through structural reinforcement and studying laminated materials under impact could lead to innovative solutions [3].

Zhao et al. [4] conducted a study on the mechanical characteristics of three types of metal laminated materials featuring varying wave impedance arrangements using the Hopkinson bar method, which exhibited the highest energy absorption efficiency, reaching up to 73%. Zhang et al. [5] investigated the mechanical properties of carbon fiber-reinforced laminates through high-speed impact testing. Zhu et al. [6] explored the dynamic response of ceramic/ultra-high molecular weight polyethylene laminated composites when subjected to rifle bullet impacts, noting a maximum energy absorption efficiency of 8.73%.

On the other hand, the reduction in stress wave intensity and energy attenuation in multilayer composite materials due to the reflection and transmission at the interface between soft and hard layers has been examined and elucidated by various researchers in the context of filtering. Liu et al. [7] discovered that interleaved composite materials exhibit notable filtering properties. Nemat-Nasser et al. [8] improved the filtering bandwidth of composite materials through structural optimization. Li et al. [9] discovered that non-linear filling joints exhibit a notable high-frequency filtering impact on stress waves. The utilization of laminated materials can attenuate the propagation of stress waves through the creation of interlayer gaps. This mechanism facilitates the filtration of stress waves.

Current research lacks an in-depth analysis of laminated materials' mechanical processes and their energy absorption. This study examines stress wave propagation in layered structures, focusing on layers and interfaces. It develops models for stress wave attenuation and energy absorption efficiency, investigates filtering properties in electronic packaging, and tests Al-Fe metal layers for impact. It also seeks ways to boost layered structures' protective capabilities using energy absorption and filtering principles.

2. Analysis of stress wave attenuation and energy absorption in laminated structures

2.1 Analysis of stress wave attenuation in laminated structures

Through the stress propagation analysis in the laminated structure, the stress attenuation model is obtained. Assuming the attenuation coefficient of any homogeneous material is α, the intensity of internal stress waves σ decays with the propagation distance x as [10]:

$$\sigma = \sigma^* \beta = \sigma^* e^{-\alpha x} \tag{1}$$

where σ^* represents the initial strength of the stress wave, and β denotes the attenuation rate of the stress wave within the material. When stress waves propagate from one material to another, there is [13]:

$$\begin{cases} \sigma_R = F\sigma_I \\ \sigma_T = T\sigma_I \end{cases} \tag{2}$$

where σ_I, σ_R, and σ_T respectively represent the incident wave, reflected wave, and transmitted wave. T is the transmission coefficient. $F=T-1$ is the reflection coefficient.

From layer i to layer $i+1$, the transmission coefficient is given by $T_i=2/(1+R_i)$, where the wave impedance ratio is $R_i=(\rho C)_i/(\rho C)_{i+1}$, with $C=\sqrt{E/\rho}$ representing the material wave velocity, and E and ρ representing Young's modulus and density, respectively. The transmission coefficient reflects the attenuation of stress waves at the interface between two different materials.

For an N-layered laminated structure, Y_1, Y_2, ..., Y_{N-1}, and Y_N respectively represent the interfacial layers between the first and second layers, the second, ..., and so on up to the (N-1)-th and N-th layers. When the left end of the laminated structure is subjected to a constant sudden load σ^*, after multiple layers of attenuation, the peak value of the stress wave output at the other end can be expressed as:

$$\begin{aligned} \sigma_{\text{out}} &= T_{Y_1} T_{Y_2} \cdots T_{Y_{N-1}} \beta_1 \beta_2 \cdots \beta_N \sigma^* \\ &= \frac{2}{1+R_1} \frac{2}{1+R_2} \cdots \frac{2}{1+R_{N-1}} e^{-(\alpha_1 h_1 + \alpha_2 h_2 + \cdots + \alpha_N h_N)} \sigma^* \end{aligned} \tag{3}$$

2.2 Analysis of energy absorption in laminated materials

Based on the attenuation law of stress wave, the energy absorption of laminated materials is analyzed using dynamic impact tests with a Split Hopkinson Pressure Bar setup, as shown in Figure 1.

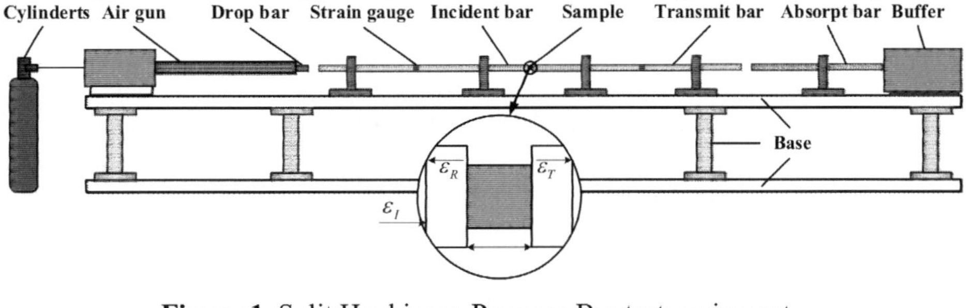

Figure 1. Split Hopkinson Pressure Bar test equipment.

By using strain gauges on the incident and transmitted rods, the strain information ε_I and ε_T of the incident and transmitted waves can be obtained. According to the Hopkinson bar testing principle [2], the deformation caused by the impact during the test is in its elastic response stage, so the stress and strain comply with $\sigma=E\varepsilon$.

The strain energy can be calculated as:

$$W=\int_0^{\varepsilon}\sigma d\varepsilon=\sum_i(\sigma_i\bullet d\varepsilon_i)=\frac{E}{2}\sum_i\varepsilon_i^2 \tag{4}$$

The energy absorption efficiency can be defined as:

$$\eta=1-\frac{W_T}{W_I} \tag{5}$$

Combining Equation (3), for a given energy absorption efficiency η^*, the thickness of each material in the laminate structure should be satisfied.

$$\alpha_1h_1+\alpha_2h_2+\cdots+\alpha_Nh_N=\frac{1}{2}\ln\left[\frac{4R_1}{(1+R_1)^2}\frac{4R_2}{(1+R_2)^2}\cdots\frac{4R_{N-1}}{(1+R_{N-1})^2}\Big/\left(1-\eta^*\right)\right] \tag{6}$$

3. Analysis of the filter characteristics of laminated structure

3.1 Filter band analysis

Through the analysis of the filtering performance of the laminated material, it helps the design of the laminated structure. For the emission test of electronic packaging equipment, the length of its shell is denoted as L_s. The elastic modulus and density of the packaging shell material for electronic devices are represented as E_s and ρ_s, respectively. The stress wave pulse width generated by the packaging structure after impact is [11]:

$$\tau_s=\frac{2L_s}{C_s} \tag{7}$$

where C_s is the stress wave velocity, and the frequency of the stress wave generated is $f_s=1/\tau_s$. Based on the size and material parameters of the electronic device packaging structure (as shown in Table 1 [12]), the generated stress wave pulse width is approximately 0.078 ms. The frequency is approximately 12.8 kHz.

Table 1. Parameters of electronic device packaging structure.

Material	L_s (mm)	E_s (GPa)	ρ_s (kg/m^3)
35CrMnSiA	200	194	7850

Based on empirical findings [1], the high overload shock produced post-projectile impact typically exhibits a pulse width of approximately several milliseconds and a frequency in the range of hundreds of hertz. This differs significantly from the pulse width and frequency of the stress wave generated. Consequently, through frequency band analysis, a tailored laminated structure is devised to attenuate high-frequency stress waves, thereby enabling the mitigation and safeguarding of electronic equipment.

3.2 Analysis of laminated structures

Through the above analysis, the cutoff frequency for filtering out the high-frequency signal is set to $f_{\text{cut}}=f_s$. The corresponding period is $T_{\text{cut}}=1/f_{\text{cut}}$. Combined with the wave velocity formula [11], the critical wavelength for filtering stress waves is:

$$\lambda_{\text{filter}}=C_{\text{lam}}\cdot\frac{1}{f_{\text{cut}}}=\sqrt{\frac{E_{\text{lam}}}{\rho_{\text{lam}}}}\cdot\frac{1}{f_{\text{cut}}} \tag{8}$$

where C_{lam} represents the wave impedance of the laminated structure, and E_{lam} and ρ_{lam} are Young's modulus and density. In order to ensure that the laminated structure can filter high-frequency waves, the wavelength should not exceed the total thickness of the laminated structure L_{lam} [9]. Combining Equations (7) and (8), we can obtain:

$$\frac{C_{\mathrm{lam}}}{C_s} \leq \frac{L_{\mathrm{lam}}}{2L_s} \tag{9}$$

Solving the equation indicates the laminated structure can effectively reduce stress waves from impacts. This finding helps choose the right materials and layer thicknesses for the laminated structure.

4. Structure design of metal laminated protective

4.1 Preparation of Al-Fe metal laminates

Equations (3) and (6) explain how material impedance differences in laminated structures affect stress wave attenuation and energy absorption. To boost electronic device protection, composites of common materials like Al and Fe are tested for their protective performance. This study focuses on using Al and 430-SS in a semi-solid reaction to determine the optimal reaction temperature for the preparation of iron-based intermetallic compound layered composite materials [14], as shown in Figure 2(a).

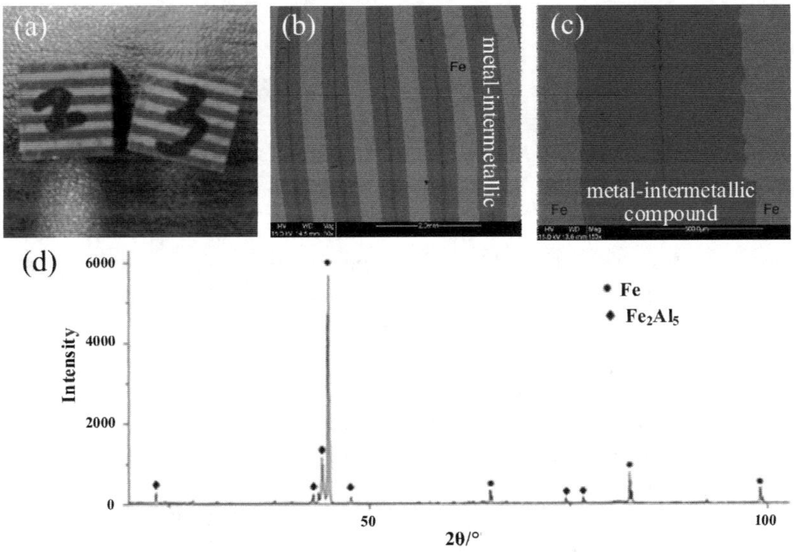

Figure 2. (a) Al-Fe metal laminate sample (a) Low magnification (b) High magnification (c) XRD analysis results.

Intermetallic compound particles started to emerge at the Al/430 stainless steel interface during the diffusion bonding phase in the early stages of the reaction. Then the nucleation sites of these intermetallic compound particles at the interface gradually multiplied, eventually merging to create a continuous intermetallic compound layer within the reaction medium. Subsequently, as the intermetallic compound layer continued to grow, the growth interface steadily progressed towards the interior of the Al layer, ultimately involving the entire Al layer in the formation of the intermetallic compound layer. Microscopic examinations were carried out at both low and high magnifications on the metal laminated composite materials that were prepared, as shown in Figure 2(a) and Figure 2(b). The images illustrate the alternating stacking of the intermetallic compound layer and metal layer, which are tightly bonded to form a compact structure. Additionally, the XRD analysis results of the laminated composite materials are presented in Figure 2(c). Notably, no peak corresponding to the presence of aluminum element was detected, indicating the absence of residual elemental Al in the iron-based laminated composite materials. The intermetallic compound layer predominantly consists of Fe_2Al_5.

4.2 Dynamic mechanics testing and protective performance analysis

Impact tests on three Al-Fe composite material sets were performed using the Split Hopkinson Pressure Bar, depicted in Figure 1. Within this experimental configuration, a correlation between voltage and strain is established, as both incident and transmitted bars experience elastic deformation. This linear correlation facilitates the use of voltage amplitude to indicate both strain and stress intensity.

The voltage-time curves obtained from the impact test of specimen No. 2 are shown in Figure 3. The red and blue curves represent the voltage-time signals collected from the strain gauges on the incident rod and the transmission rod, respectively.

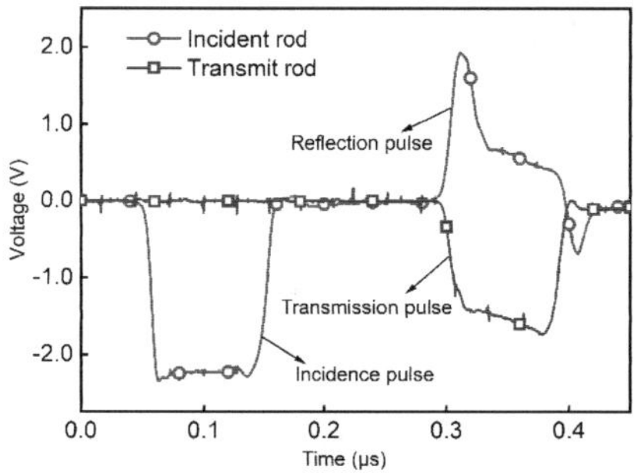

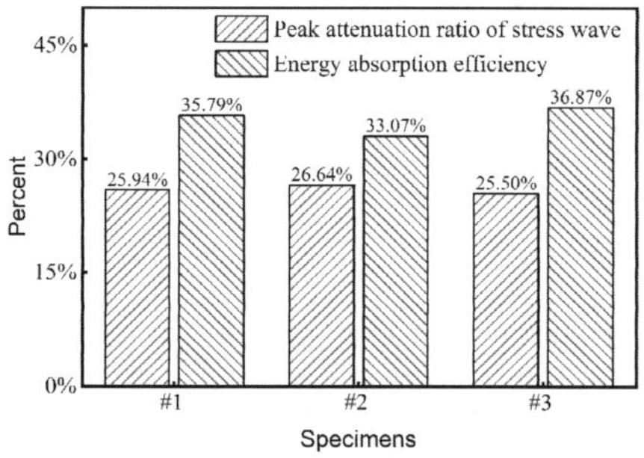

Figure 3. Voltage-time curve. **Figure 4.** Peak attenuation ratio and energy absorption efficiency.

From Figure 3, it is evident that the maximum transmitted voltage is notably lower than the maximum incident voltage. Given that the Hopkinson bar is operating within the elastic response phase, the attenuation ratio of the peak stress wave value can be determined by comparing the peak voltage values. Additionally, the ratio of strain energy can be derived by comparing the voltage curve with the area under the envelope along the horizontal axis. The outcomes of these calculations are depicted in Figure 4.

Figure 4 shows that the peak stress attenuation ratio and energy absorption efficiency of the three specimens are relatively close, with mean values of 26.02% and 35.24%, respectively.

The Al-Fe metal laminated structure consists primarily of Fe and Fe_2Al_5, with detailed material parameters outlined in Table 2. The material attenuation coefficient is established through the fitting of simulation test outcomes from the Hopkinson bar. The laminated material is applied to the Hopkinson bar, with a stainless-steel cylindrical bullet chosen to impact the incident end. Five stress peak values are evenly chosen along the bar. The attenuation coefficient of the metal material is derived by fitting the peak-displacement curve.

Table 2. Al-Fe metal laminate material parameters.

Laminated materials	Density (kg/m^3)	Young's modulus (GPa)	Attenuation coefficient	Thickness (mm)
Fe	7880	190	0.37	1
Fe_2Al_5	4150	125	0.19	1

According to Equations (3) and (6), combined with Table 2, the peak stress attenuation ratio and energy absorption efficiency at each interface of the laminated material can be calculated in Figure 5.

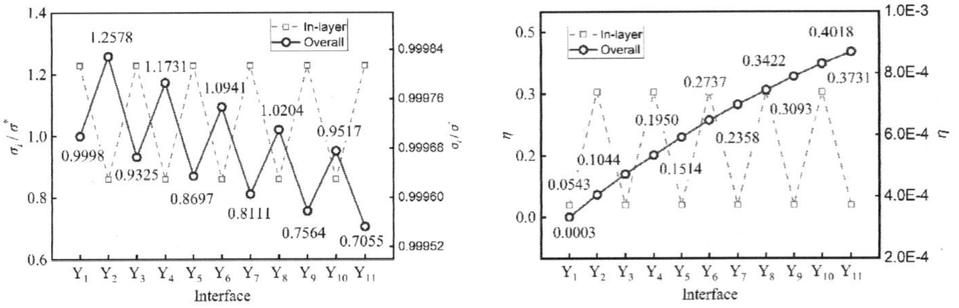

(a) (b)

Figure 5. (a) Stress wave peak attenuation curve (b) Energy absorption efficiency curve.

From the graphical representation in Figure 5(a), the attenuation curve of the peak stress wave demonstrates a distinct step-like pattern at the boundary between materials with high and low impedance, gradually diminishing until it stabilizes at 0.7055. The associated peak attenuation rate is calculated at 29.45%, with the attenuation predominantly concentrated around 0.99 within the layer, indicating that the attenuation of the peak stress wave primarily occurs at the interface. In Figure 5(b), the energy absorption efficiency displays a nearly linear correlation with the interface, peaking at 40.18%. Similarly, the energy absorption efficiency within the layer exhibits minimal overall influence. The theoretical analysis solely accounts for the primary laminated materials. In practical applications, the Al-Fe intermetallic compound may encompass components beyond Fe_2Al_5, resulting in an increased peak attenuation rate and energy absorption efficiency. These findings align with the deductions drawn from empirical observations.

The 11-layer Al-Fe metal laminate structure depicted in Figure 2, in conjunction with Table 2, exhibits an overall theoretical Young's modulus and density of 155 GPa and 5850 kg/m^3, respectively. Analysis based on Equation (9) and Table 1 reveals that for the Al-Fe metal laminate structure to effectively filter high-frequency stress waves, the thickness of the laminate structure must meet the condition $L_{Lam}\geq 414$mm. However, the actual prepared sample has a thickness of only 11 mm, significantly below the critical value, resulting in suboptimal stress wave filtering. On the other hand, with a thickness of 11 mm, Equation (9) dictates that the critical wave velocity of the laminate structure should adhere to $c_{Lam}\leq 137$m/s. Typically, metal materials exhibit wave velocities in the order of thousands, making those with velocities in the hundreds range scarce. In such cases, porous materials, rubber, or fiber materials may be considered as alternatives. To enhance the filtering efficacy of the laminate protection structure, further optimization of the structure's thickness or the properties of the laminate material may be warranted.

5 Conclusion

Stress wave analysis in metal laminates shows greater intensity reduction at interfaces than in layers. In the 11-layer Al-Fe laminate, stress wave damping reached 26.02% and energy absorption 35.24%, slightly under theoretical values. The filtering properties of the laminated structure showed modest filtering effects, suggesting that performance could be improved by optimizing layer thicknesses or selecting metals with higher wave impedance to boost the structure's overall filtering capabilities. This work provides new methods and ideas for the protective structure of electronic equipment and promotes the engineering application of protection technology.

Acknowledgments

Financial support from the National Natural Science Foundation of China (No. 52375553).

References

[1] Chen Y, Shi Y, Zhao R, et al. (2024) Study on the equivalent impact method of long/short pulsewidth loading based on cumulative energy, IEEE Trans. Instrum. Meas. 73: 1–10. 10.1109/TIM.2024.3350118.
[2] Shi Y, Wen X, Zhao Y, et al. (2020) Investigation and experiment of high shock packaging technology for High-G MEMS accelerometer, IEEE Sensors. J. 1–1. 10.1109/JSEN.2020.2987971.
[3] Etri H E, Polat M Z, Özyilmaz E B. (2024) Classification, applications, and impact behavior of composite materials: A review, Mech. Adv. Mater. Struc. 1–44. 10.1080/15376494.2024.2326665.
[4] Zhao W, Liu B, Wang Y, et al. (2023) Design of shock wave attenuation effects on multi-impedance-matched laminated composites, J. Mater. Res. Technol. 23: 5846–5860. 10.1016/j.jmrt.2023.02.143.
[5] Zhang N, Qian X, Zhang Q, et al. (2024) On strain rate effect and high-velocity impact behavior of carbon fiber reinforced laminated composites, Thin Wall Struct. 194: 111328. 10.1016/j.tws.2023.111328.
[6] Zhu Y, Ma J, Fan Z, et al. (2024) Ballistic impact performance and blunt injury assessment of ceramic/UHMWPE laminate composite body armor, J. Mater. Res. Technol. 29: 1703–1728. 10.1016/j.jmrt.2024.01.140.
[7] Liu J, Zhang H, Gao Y, et al. (2024) Reinforcement hybridization in staggered composites enhances wave attenuation performance, J. Mech. Behav. Biomed. 152: 106435. 10.1016/j.jmbbm.2024.106435.
[8] Nemat-Nasser S, Sadeghi H, Amirkhizi A V, et al. (2015) Phononic layered composites for stress-wave attenuation, Mech. Res. Commun. 68L: 65–69. 10.1016/j.mechrescom.2015.05.001.
[9] Li X F, Li H B, Li J C, et al. (2018) Research on transient wave propagation across nonlinear joints filled with granular materials, Rock Mech. Rock Eng. 51: 2373–2393. 10.1007/s00603-018-1471-8.
[10] Pu H X, Zhou J Z, Jiang G Q, et al. (2024) Stress wave propagation and forming mechanism of the stiffened plate by laser peening, J. Manuf. Process 112: 329–338. 10.1016/j.jmapro.2024.01.045.
[11] Gao J, Guo Y, Li Y, et al. (2024) The split-Hopkinson (Kolsky) bar technique, Elsevier 183–217.
[12] Ren J, Xu Y, Zhao X, et al. (2018) Dynamic mechanical behaviors and failure thresholds of ultra-high strength low-alloy steel under strain rate 0.001/s to 106/s, Mater. Sci. Eng. A 719: 178–191. 10.1016/j.msea.2018.02.019.
[13] Chen Y, Shi Y, Zhao R, et al. (2024) Research on the equivalent impact test method for different pulse widths based on micro-electro-mechanical system devices, Sensors Actuat. A-Phys. 372: 115362. 10.1016/j.sna.2024.115362.
[14] Wang Y, Liu X, Wang H, et al. (2020) The effect of oxides on Fe/Al interfacial reaction in Metal-Intermetallic Laminate (MIL) composites, J. Alloy Compd. 845: 156268. 10.1016/j.jallcom.2020.156268.

Journal of Physics: Conference Series **2992** (2025) 012024 doi:10.1088/1742-6596/2992/1/012024

Performance study of 430 stainless steel/6061 aluminum alloy laser welded joints

Haiting Peng [1], Zhanlei Wang [2] and Qiaobo Feng [3,*]

[1,2,3]College of Energy and Mechanical Engineering, Shanghai University of Electric Power, Shanghai, China

[*]Corresponding author:ce-mail:fengqb@163.com

Abstract: In this paper, 430 stainless steel and 6061 aluminum alloy sheets were used as the research objects, and the lap welding methods of "steel on top and aluminum on bottom" were welded by laser pulse welding equipment, and the impact of peak power and defocus amount on the microstructure, macroscopic morphology and mechanical properties of steel/aluminum dissimilar welded joints were investigated. The studies show that the weld joint after welding is in the shape of fish scales, and with the increase of peak welding power, the degree of oxidation of the weld becomes more serious and the spatter phenomenon occurs at higher power. The cross-section of the weld presents an inverted triangle with a width at the top and a narrow bottom, and there is an obvious distinction between nucleation regions. The tensile resistance of the welded joint increases with the increase of the peak welding power, and 647.2 N is the maximum. The change in the amount of defocus will also affect the maximum tensile force, and with the increase of the defocus amount, the maximum tensile force will increase first and then decrease, and the maximum is 530.4 N.

1. Introduction

With the proposal of the "double carbon" strategic goal and the rapid development and popularization of new energy vehicles in recent years, the national industrial manufacturing level has reached a new level, but environmental protection and energy issues are still unavoidable topics. Automotive lightweight technology has become the key for major automobile companies to enhance product competitiveness, and lightweight can solve key problems such as automobile energy consumption, cruising range, and environmental impact; At present, the main ways of automobile lightweight are new materials (using high-strength steel, aluminum alloy and composite materials, etc.), new processes (using laser welding and hydroforming, etc.) and new designs (structural optimization, etc.)[1-4]. In this paper, the laser welding process between stainless steel and aluminum alloy is used, and the materials and processes are used to explore a new way of lightweight. Austenitic stainless steels rich in nickel are now widely used internationally, but due to the scarcity of nickel resources and the increase in manufacturing costs, it is becoming more and more important to seek ferritic stainless steels with no or fewer nickel elements[5]. Ferritic stainless steel has low nickel content, low cost, small coefficient of thermal expansion, and good corrosion resistance, and is widely used in the automotive industry and home appliance industry[6]; 430 stainless steel is one of the earliest ferritic stainless steel types, which are widely used in home appliances, kitchen appliances, and automobile parts[7] and 6061 aluminum alloy is a Mg/Si alloy, which has the characteristics of medium strength, good plasticity, and excellent corrosion resistance[8]. In addition, 6061 aluminum alloy has excellent weldability and cold workability and is widely used in rail transit, the automotive industry, aerospace, and other manufacturing fields[9, 10].

Steel/aluminum dissimilar materials can give full play to the advantages of stainless steel plate and aluminum alloy at the same time through welding and are widely used in the field of automotive lightweight. However, steel/aluminum dissimilar metal welded joints are prone to brittle and hard intermediate metal compounds (IMCs), which greatly affect the mechanical properties of the joints, thus inhibiting their practical industrial applications [11-13]. Yuan[14] et al. studied the effects of different powers on steel/aluminum dissimilar metal joints, and found that the power of steel/aluminum dissimilar metal joints is too low or too high will affect the welding quality. When the power is too low, the effective welding between the plates is affected, and when the power is too high, the welded joint becomes tighter, but it also introduces more weld spatter, which greatly reduces the quality of the connection. Huang[15]et al, found that laser power, welding speed, focal length, and gas flow rate are the most important factors affecting weld deformation. As the welding speed, focal length, and gas flow rate increase, the laser power increases and decreases, and the distortion also increases. Within the range of experimental parameters, the optimal welding parameters of the minimum weld bending deformation and the mechanical properties of the joint are obtained. Shi et al. found that Cu and Ni preferentially replaced the Fe atoms in the Fe/Al interface, which improved the stability of the alloy system[16]. From the tensile test, the addition of Cu and Ni can improve the average tensile strength and elongation of welded joints to varying degrees. Yan et al. introduced the application of a magnetic field into the laser welding system and studied the effects of adding a permanent magnetic field on weld morphology, internal porosity defects, fracture morphology, intermetallic compound composition segregation, and mechanical properties of weld joints[17]. It is found that the tensile strength of the weld is significantly improved after the addition of a permanent magnetic field. The Lorentz force generated by the magnetic field stirs the molten pool, thereby improving the morphology of the weld, the number of pores inside the weld is also significantly reduced, the phase of the needle-like $FeAl_3$ compound is refined, and the growth of the Fe/Al metal compound is also inhibited, which effectively improves the mechanical properties of the joint.

In this paper, the effects of different peak welding power and defocusing amounts on the microstructure and tensile and shear properties of 430 stainless steel with a thickness of 0.5 mm and 6061 aluminum alloy sheets with a thickness of 1 mm were discussed by laser welding, to provide technical support and reference for the welding of steel/aluminum dissimilar metals.

2. Materials and methods

In this paper, steel/aluminum dissimilar metal materials were used: 430 stainless steel was used on the steel side, and 6061 aluminum alloy was used on the aluminum alloy side, the former thickness was 0.5 mm, and the latter was 1 mm, and the main chemical composition and microstructure of the two plates were shown in Tables 1 and 2 and Figure 1. It can be seen from the microstructure and morphology of the base metal in Figure 1(a) that the microstructure of 430 stainless steel is mainly slat-like ferrite, and the black dots distributed between ferrite grains are carbides. In Figure 1(b), the structure of 6061 aluminum alloy is mainly a-Al grains, and the black part in the middle is a Si/Mg compound.

Table 1. Chemical composition of 430 stainless steel.

Element	C	Si	Mn	P	S	Ni	Cr	Fe
Mass fraction/%	0.12	0.29	0.34	0.02	0.02	0.15	16.2	Bal

Table 2. Chemical composition of 6061 aluminum alloy.

Element	Si	Fe	Cu	Mn	Mg	Cr	Zn	Ti	Al
Mass fraction/%	0.72	0.4	0.21	0.09	1.1	0.15	0.02	0.02	Bal

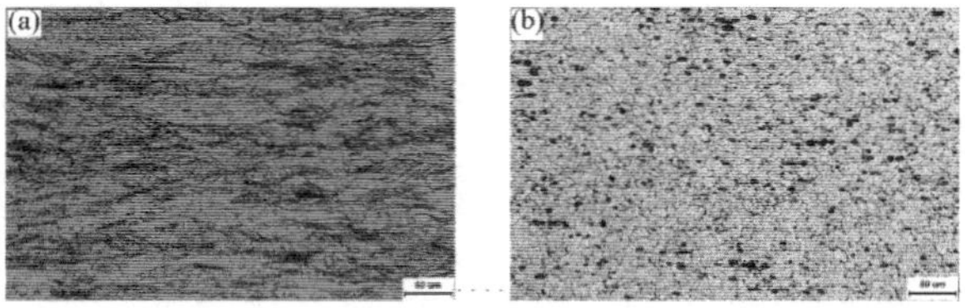

Figure 1 Tissue morphology of the base metal.

The 430 stainless steel and 6061 aluminum alloy sheets are processed into 100 mm × 30 mm sizes, and the base metal needs to be cleaned, degreased, and dried before welding, and then a suitable fixture is adopted when the base metal specimen is clamped, to minimize the welding error caused by the thermal deformation caused by the process of lap welding of steel/aluminum dissimilar metal sheets [18]; the welding parameters are controlled by low-power pulse laser welding equipment, and the welding equipment adopts JHM-4GX-300 pulse laser welding machine and IRB1410 robot. As shown in Figure 2(a); the laser welding machine has a maximum power of 300 W, a rated frequency of 50 Hz, and a laser wavelength of 1064 nm. The specimen adopts the lap joint method of "steel on aluminum and below", and the specific welding schematic diagram is shown in Figure 2 (b); after the preliminary sample test welding analysis, The welding frequency is 50 Hz, the input ratio is 80%, the welding speed is 10 mm/s, the defocusing amount is -1.3 mm ~ -1.5 mm, the argon flow rate is 5 L/min, and the peak welding power is in the range of 3.45 kW ~ 3.6 kW.

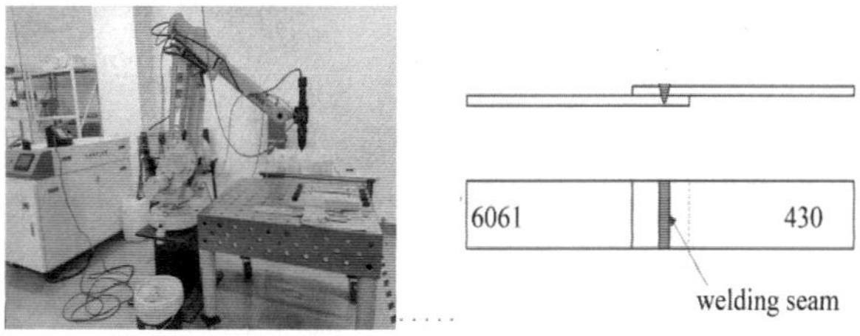

Figure 2 Welding equipment and welding schematic diagram.

The welded sample is extracted by wire cutting, the sample is extracted, cleaned and dried, and then placed into the metallographic sample inlay machine for the inlay to obtain the metallographic sample, and the metallographic sample is coarsely ground with sandpaper, finely ground, and polished, and then the residual impurities are cleaned up by acetone and absolute ethanol, and finally dried with a dryer and placed on an optical metallographic microscope for observation; by adjusting the aperture magnification, the macroscopic morphology and microstructure of the specimen weld can be observed. The tensile and shear performance test is carried out on the mechanical property testing machine, the tensile specimen size is determined according to the latest national standard GB/T 228.1-2021, the size of the wire-cut tensile specimen after welding is shown in Figure 3, the tensile rate is 2 mm/min, and the three groups of specimens are tested under the same parameters, and their average values are taken as the final research results to ensure the accuracy of the results.

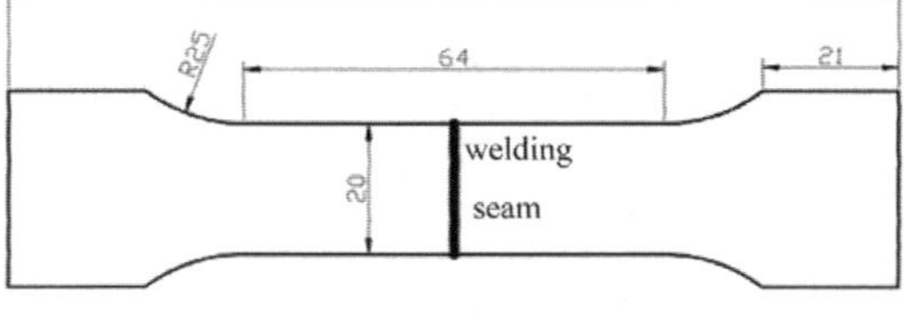

Figure 3 Size of tensile specimen.

3. Results & discussion

3.1 Macroscopic topography of the joint

The frontal macromorphology of the steel/aluminum lap welded joint is shown in Figure 4, where (a), (b), (c) and (d) are the surface morphology of the weld under the same other process parameters (welding speed 10 mm/s, defocus amount -1.5 mm, welding frequency 50 Hz) and four different peak welding powers (3.45 kW, 3.50 kW, 3.55 kW, 3.60 kW). When the peak welding power increases from 3.45 kW to 3.55 kW, it can be seen that the surface weld melt width increases, which is due to the increase of welding power will increase the input heat of the weld, thereby increasing the area of the weld pool. Not only that, with the difference in welding power, the weld has different degrees of oxidation and the texture is slightly disordered. When the peak welding power is 3.60 kW, there is a slight spatter in the weld, mainly due to the excessive peak welding power and the internal spatter of the welded joint[19], the texture is also clearly messy. In Figure 5, (a), (b), and (c) is the surface morphology of the weld with three different defocusing amounts (-1.3 mm, -1.4 mm, -1.5 mm) when the peak welding power is 3.45 kW, the welding speed is 10 mm/s, and the welding frequency is 50 Hz. When the defocusing amount is -1.3 mm and -1.4 mm, the weld can no longer see the regular pattern, there are different degrees of spatter, and the surface quality of the weld is poor. When the defocusing amount is -1.5 mm, the grain is clear, there is no spatter, and the weld quality is good. Due to the thickness of the aluminum side, and the high reflectivity of the aluminum alloy, the low power of the welding equipment leads to the lack of burn-through of the back of the welded joint, and there is no difference on the back of the welded joint under various process parameters, so it is not presented here.

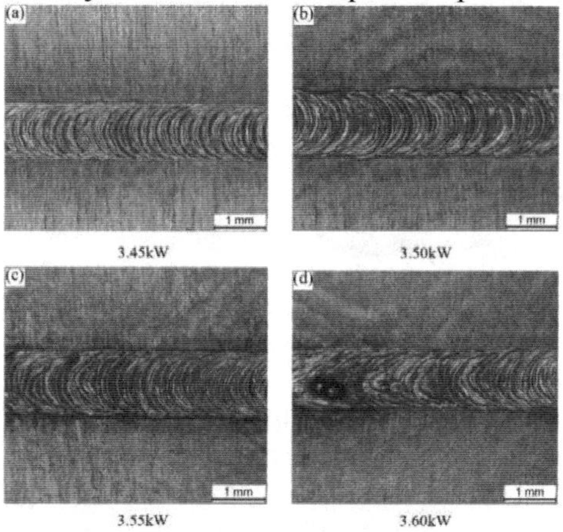

Figure 4 Frontal macroscopic topography of the welded joint at different peak welding powers.

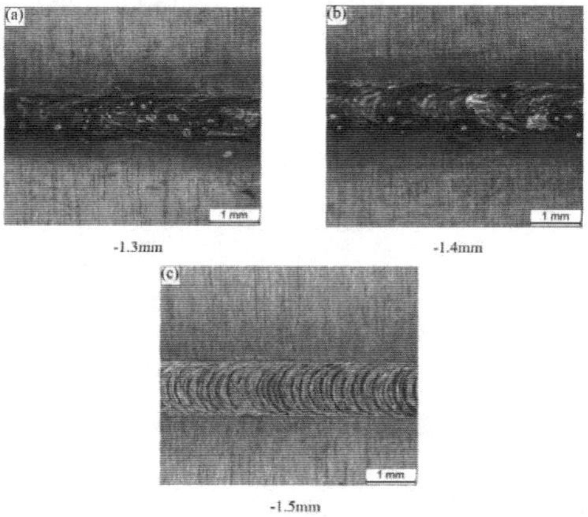

Figure 5 Macroscopic morphology of the front side of the welded joint under different defocus amounts.

3.2 Microstructure of joints

The cross-sectional microstructure of the steel/aluminum lap welded joint is shown in Figure 6(a), due to the input of laser energy, the stainless steel in the liquid state penetrates downward into the aluminum alloy side, thereby realizing the effective connection of the two plates, and then due to the thick aluminum alloy, and the strong reflectivity of the aluminum, and the weak light absorption capacity [20], the lower plate is not completely penetrated and welded, and the cross-section of the joint presents an inverted triangle with wide upper and narrow lower narrow. During the welding process, no visible cracks or other defects were found, which illustrates the excellent performance of the entire process. In the core of the weld, three different structures can be seen, namely the fusion zone (FZ), the heat-affected zone (HAZ), and the base material zone (BM). The fusion zone is composed of columnar ferrite, and it can be seen from Figure 6(b) that the direction of the columnar crystals is perpendicular to the fusion line, which is due to the large temperature gradient, fast cooling rate, and well-developed columnar crystals[20]. The heat-affected zone (HAZ) is mainly composed of equiaxed crystals. Figure 6(c) shows the morphology of the aluminum side melt pool at the bottom, and there are obvious defects such as porosity and cracks at the bottom, because the solubility of steel and aluminum varies greatly, and it can form brittle and hard intermetallic compounds, so there are cracks; At the same time, due to the incomplete evaporation of the gel layer, pores and inclusions are formed at the bottom of the molten pool.

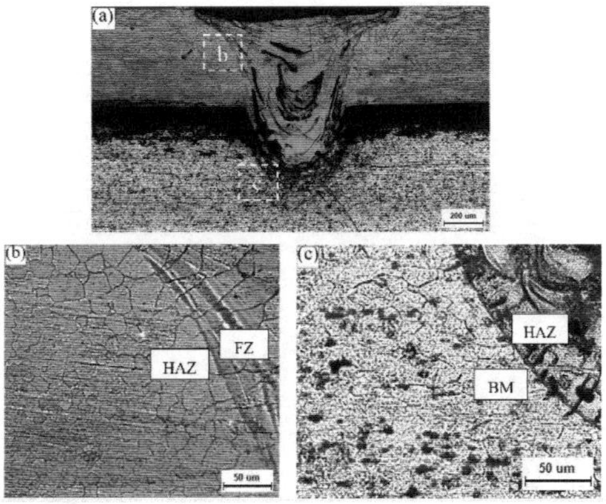

Figure 6 Microstructure and morphology of the cross-sectional weld joint.

3.3 Mechanical properties of joints

Figure 7 (a) is the hardness dotting position of the joint cross-section, line A is the hardness dotting line of 430 stainless steel side, and the interval between points is 0.1 mm, line B is the hardness dotting line of 6061 aluminum alloy side, and the interval between points is 0.1 mm, and line C is the vertical hardness dotting line in the center of the weld fusion zone, and the interval between points is 0.1 mm. As can be seen from Figure 7(b), the hardness distribution on the steel side presents a "W" shape, and the hardness on both sides is higher, which is the hardness of 430 stainless steel base metal, and the hardness value decreases first, then increases and then decreases in the middle, because of the difference between the heat-affected zone (HAZ) and the fusion zone (FZ) in the region. The hardness softening phenomenon occurs in the heat-affected zone (HAZ), because the heat-affected zone is mainly composed of equiaxed crystals, and the fusion zone is mainly composed of columnar crystals, which have more crystal interfaces than equiaxed crystals, and have less segregation and dense structure, so its strength is higher and the hardness is greater. The hardness distribution of the aluminum alloy side presents a "several" type, with large middle hardness and low hardness on both sides; The hardness of the 6061-aluminum alloy base metal on both sides is significantly increased in the middle, because there is an intermediate metal compound layer in the middle area, and the intermediate Fe/Al compound has the characteristics of brittle and hard. As can be seen from Figure 7(c), the hardness from the first half of the weld surface is higher, with an average of about 250 HV, while the hardness of the second half is lower, with an average of about 100 HV, with obvious abrupt changes in the middle. This is mainly because the mutation point is located in the steel/aluminum interface line, the front hardness is large because the main component of the fusion zone is stainless steel, which has high hardness, and the rear dot position is offset from the fusion zone, which is located in the base metal area on the aluminum side, and the main component is aluminum with low hardness.

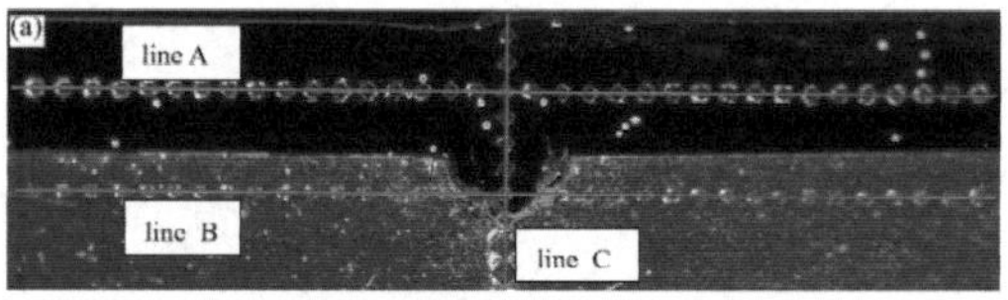

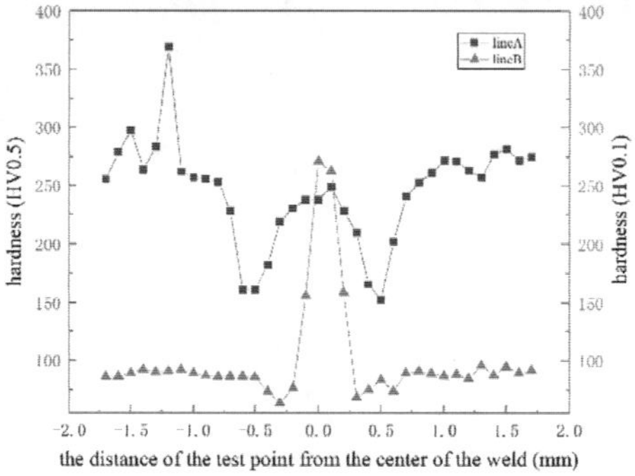

(b)

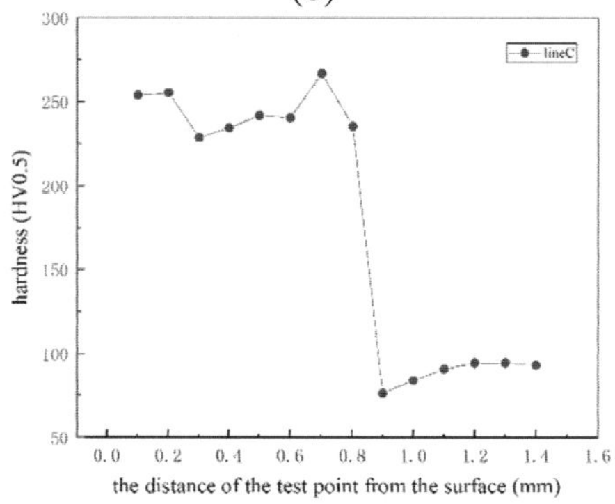

(c)

Figure 7 Distribution of joint hardness.

Figure 8 shows the influence of different peak welding power and different defocusing amounts on the tensile and shear force of the welded joint. As shown in Figure 8(a), when other process parameters remain unchanged and the peak welding power gradually increases, the tensile force of the welded joint also increases. The minimum tensile force is 405.6 N at a peak power of 3.45 kW. The maximum tensile force is 647.2 N at a peak power of 3.60 kW. This is because the peak power increases and the heat input of the joint increases, which increases the effective fusion width of the steel/aluminum interface and thus the tensile strength as the aluminum side is not completely burned through. As shown in Figure 8(b), when other process parameters remain unchanged, When the defocusing amount increases from -1.50 mm to -1.30 mm, the tensile force of the welded joint increases first and then decreases. The maximum tensile strength is 530.4 N at -1.40 mm. The minimum tensile strength is 342.9 N at -1.30 mm. This is because the difference in the amount of defocusing affects the heat input of the melt pool, which in turn affects the effective fusion width of the steel/aluminum interface, and ultimately the mechanical properties of the joint.

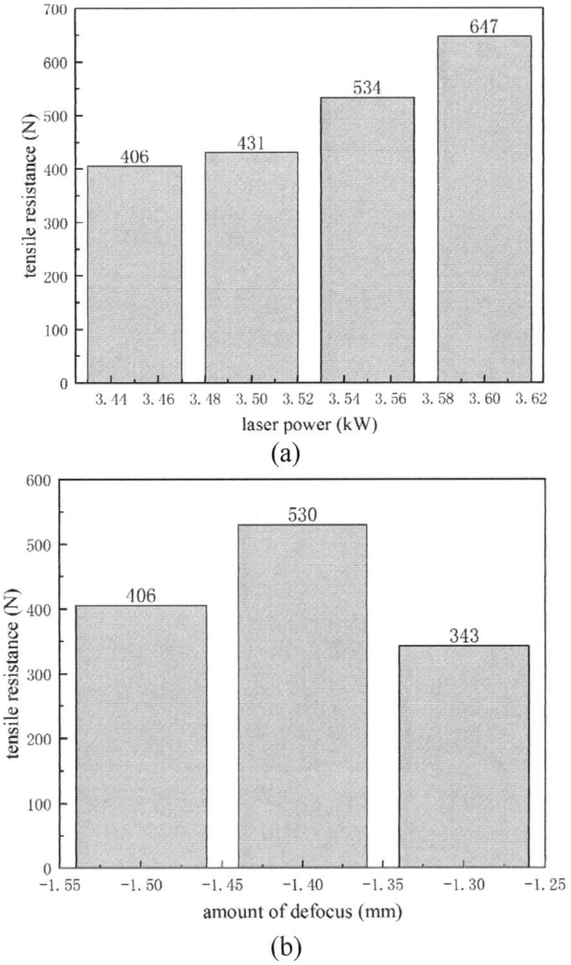

Figure 8 The effect of peak welding power and defocus on the tensile shear force of welded joints.

4. Conclusions

The lap welding method of 430 stainless steel and 6061 aluminum alloy was carried out by using low-power pulsed laser welding equipment, and the effects of different peak welding power and defocusing amount on the macroscopic morphology and mechanical properties of steel/aluminum dissimilar metal welded joints were compared. The microstructure morphology was observed. (1) The change of peak welding power and defocusing amount affects the heat energy input, change the size of the molten pool, and the texture of the weld surface gradually becomes disordered and irregular, and the spatter phenomenon occurs from a regular "fish scale pattern". (2) With the input of laser energy, the surface of the joint weld sags; the cross-section of the joint appears as an inverted triangle with a width at the top and a narrow bottom, and there are three obvious areas at the joint, which are the fusion zone, the heat-affected zone, and the base metal zone. The fusion zone is mainly columnar crystals, while the heat-affected zone is equiaxed, and the bottom of the fusion zone is an intermediate metal compound layer. (3) The hardness distribution of stainless steel is a "W" shape, in which the hardness mutation point is located near the fusion line, and the hardness will also change due to the difference in regional tissue structure. When the peak welding power increases, the tensile strength of the joint will also increase accordingly; However, when the amount of defocus increases, the tensile force of the joint rises first and then decreases.

Acknowledgments
This article was supported by the following funds: National Natural Science Foundation of China-Nucleation mechanism and interface failure control method of resistance point welding joints of ultra-high-strength austenitic stainless steel heterogeneous materials (Approval number:52175343).

References

[1] WANG T, DONG R, ZHANG S, et al.(2020)Research on Lightweight Design of Automobile Collision Safety Structure Based on Multiple Materials. Journal of Physics: Conference Series,1670(1): 012004.
[2] LIU D F, TAO J. (2011)Application of Automobile Lightweight Alloys and the Development of its Die-Casting Technology.Advanced Materials Research,308-310: 785-789.
[3] LI H, LI X.(2012) The Present Situation and the Development Trend of New Materials Used in Automobile Lightweight. Applied Mechanics and Materials, 189: 58-62.
[4] LYU R, JIANG X, MINORU O, et al. (2018)Lightweight design of automobile frame based on magnesium alloy. IOP Conference Series: Materials Science and Engineering, 372(1): 012047.
[5] YANG Z, LONG L H, LIANG W L, et al. (2022)An overview of the novel heat-resistant ferritic stainless steels.Tungsten,5(4): 467-480.
[6] GUO D H, MA D W, ZHANG Z Y, et al. (2012)Application Key Technologies of Ferritic Stainless Steel for Traffic Sign Substrate Plate. Advanced Materials Research, 510: 227-232.
[7] UYSAL A. (2016)Investigation of surface roughness in MQL milling of 430 ferritic stainless steel by using nano MoS2 particle reinforced cutting fluid. DÜMF Mühendislik Dergisi,(1): 151-158.
[8] RANJITH KUMAR S, DINESH KRISHNAA S, DINESH KRISHNA M, et al. (2021)Investigation on corrosion behavior of aluminum 6061-T6 alloy in acidic, alkaline and salt medium. Materials Today: Proceedings, 45: 1878-1881.
[9] CHRISTOPHER W, ERICH N, MICHAEL K, et al. (2023)Investigations of Plasma Metal Deposition (PMD) of 6061 and 7075 Aluminum Alloys for Aerospace and Automotive Applications.BHM Berg- und Hüttenmännische Monatshefte ,168(5): 209-218.
[10] YANG T S, LUO K C. (2017)Friction Factor of 6061 Aluminum Alloy and Application to the Finite Element Analysis of Wheel Forging. Key Engineering Materials, 739: 231-234.
[11] KAI X, QI-RUI C, GUO-QIANG L, et al.(2019) Research progress in friction stir welding of steel and aluminum. IOP Conference Series: Earth and Environmental Science, 295(4): 042085.
[12] ZHOU D, XU S, ZHANG L, et al. (2016)Microstructure, mechanical properties, and electronic simulations of steel/aluminum alloy joint during deep penetration laser welding. The International Journal of Advanced Manufacturing Technology, 89(1-4): 377-387.
[13] EVDOKIMOV A, OBROSOV A, OSSENBRINK R, et al. (2018)Mechanical properties of dissimilar steel-aluminum welds. Materials Science and Engineering: 722: 242-254.
[14] YUAN R, DENG S, CUI H, et al. (2019)Interface characterization and mechanical properties of dual-beam laser welding-brazing Al/steel dissimilar metals. Journal of Manufacturing Processes, 40: 37-45.
[15] HUANG Z, LUO Z, AO S, et al. (2018)Effect of Laser Welding Parameters on Weld Bowing Distortion of Thin Plates. High-Temperature Materials and Processes,37(4): 299-311.
[16] SHI Y, LI Z, YIN C, et al. (2022)Effect of alloying elements Cu and Ni on mechanical properties of steel/aluminum laser welded joints. Optik, 255:168707.
[17] YAN F, WANG X, CHAI F, et al. (2019)Improvement of microstructure and performance for steel/Al welds produced by magnetic field assisted laser welding. Optics & Laser Technology,113: 164-170.
[18] MATSUOKA S, OKAMOTO Y, OKADA A. (2013)Influence of Weld Bead Geometry on Thermal Deformation in Laser Micro-Welding. Procedia CIRP,6: 492-497.

[19] LIU L, SHI J, HOU Z, et al. (2018)Effect of distance between the heat sources on the molten pool stability and burn-through during the pulse laser-GTA hybrid welding process. Journal of Manufacturing Processes, 34: 697-705.
[20] XU K, ZHANG S Q.(2018) Microstructure Analysis of Laser Bonding Welding on Steel-Aluminum. Materials Science Forum,4554(911-911): 28-33.

Journal of Physics: Conference Series **2992** (2025) 012025 doi:10.1088/1742-6596/2992/1/012025

Damping performance analysis of torsional Shear thickening fluid damper based on FLUENT

Weijie Zhu [1], Kaihui Chen [2], Chuanze Liang [1] and Yafeng Qiu [1]*

[1] School of Mechanical Engineering, Nanjing University of Science and Technology, Nanjing, China
[2] Interdisciplinary Center for Fundamental and Frontier Sciences, Nanjing University of Science and Technology, Nanjing, China
WeiJie Zhu and KaiHui Chen are co-first authors

* Corresponding author's email: nilghcn@sina.com

Abstract: Shear Thickening Fluid (STF) dampers, as a kind of passive dampers, have been applied in various vibration and shock absorption fields due to the good self-adaptation of their medium. In this paper, a newly designed torsional STF damper is taken as the research object, and the fluid domain model of the damper is established by using the pre-processing software WOKBENCH/MESHING and FLUENT. For the unique rheological characteristics of STF, a segmented function model combining the Logistic function and power function is proposed to fit the viscosity shear rate profile of STF. Comparative experimental results show that the proposed segmented function fits the viscosity-shear rate curves of shear-thickening liquids well. The dynamic viscosity of STF is defined using UDF. The damping performance of this non-Newtonian flow is simulated using the dynamic mesh method. The damping characteristics of the STF damper are analyzed by comparing the damping moment output and hysteresis curve characteristics of the damper under different loading conditions. The simulation results show that under static loading, the damping torque output of the damper can adaptively output a damping torque that matches the rotor speed. Under dynamic loading, the increase of frequency and angle amplitude increases the damping torque output of the damper. Compared with the elliptical hysteresis curve of the damper at low frequency, the shape of the hysteresis curve of the damper at high frequency is in the form of a spindle. The energy dissipation capacity of the damper is obviously improved.

1. Introduction

Shear Thickening Fluids (STF), as a non-Newtonian fluid, is a new type of functional material with complex rheological properties. At the low shear rate, the viscosity of shear thickening fluids is small; when the shear rate exceeds its critical shear rate, the thickening phenomenon occurs, and its viscosity will increase rapidly; when the shear rate is large, the viscosity reaches dozens of times the low shear rate to present a class of solid state [1]. When the shear rate is large, the viscosity reaches tens of times the low shear rate to present a solid-like state. Several studies have shown that in the study of shear-thickened fluids (STFs), factors such as the size of nanoparticles, their distribution in the medium, the ambient temperature, and the magnitude of the applied shear stress have a significant effect on their thickening behavior [2-4]. Researchers have used rheology combined with optical methods, neutron diffraction techniques, and numerical simulations in order to gain insight into the state transition and thickening mechanisms of STF when subjected to shear. Among these studies, the "particle cluster" theory has been recognized by many scholars [5]. According to the particle cluster theory, when the STF is subjected to shear, the colloidal particles inside the STF will gather together to form particle clusters when the shear rate increases, and the formation of particle clusters will lead to an increase in the friction between the particles, which is manifested as a rise in the viscosity of the fluid when the number and size of the particle clusters are large. In view of the unique mechanical adaptability and energy-absorbing ability of STF to impact loads, it has been more and more widely used as a vibration-damping and impact-resistant material in the fields of civil engineering, aerospace, military equipment, and human body protection.

STF damper adopts shear thickening liquid as base fluid, which can adaptively adjust the damping force, and is a new type of passive damper. Compared to traditional dampers that use common liquids such as silicone oil as the base fluid, STF dampers can provide more controllable damping force and a wider range of damping force [6]. The STF damper can provide more controllable damping force and a wider range of damping force. The output damping force of conventional viscous dampers is limited by the dimensions of their main structure, and thus they are only suitable for a specific range of vibration shocks. Research on STF dampers has mostly focused on the application of STF materials to traditional piston dampers. Scholars have fully investigated the effects of piston structural features such as gaps and orifices, as well as the effects of various loading conditions on the energy dissipation performance of piston STF dampers [7-9]. In this paper, a newly designed torsional STF damper is taken as the research object. FLUENT simulation software is used to simulate the damping effect of the STF damper by combining it with the control program written using UDF (User-Defined Function).

2. Principle of operation of STF damper

The torsional STF damper mainly consists of a stator and a rotor, which are filled with shear-thickened damping fluid. When the torsion-type damper works, its rotor rotates along the axis. In the process of rotation, the rotor is subjected to the damping force from the damping fluid to generate damping torque. As shown in Figure 1, the torsional STF damper mainly consists of ball screws, shape memory alloy springs, connectors, rotor shafts, gear disks, sealing compartments, and seals. The damper connects the ball screw nut to the controlled structure through a connector. When the controlled structure vibrates, the connector drives the ball screw nut to rotate the rotor shaft. The rotor shaft is connected to the gear rotor inside the chamber through a key. When the rotor shaft rotates, it will drive the gear disk to rotate at the same speed, and then drive the STF liquid inside the chamber to flow in the circumferential direction, which generates a frictional resistance to dissipate the energy in the process of flowing, to achieve the purpose of structural vibration damping.

3. STF mechanical modeling study

Regarding the resistance analysis and simulation calculations of torsional STF dampers, the most fundamental calculations are still based on the material properties such as viscosity and critical strain rate of the shear thickening fluid. Stable shear thickening fluid was prepared with silica nanopowder as well as polyethylene glycol PEG200 at room temperature $25\,^{\circ}C$, which was loaded and whose data collected by a rheometer. PEG200 has a long molecular chain, is easy to combine with the dispersed phase particles, and has a certain viscosity and suspension ability. It is widely used in STF dispersions because of its long molecular chain, easy to combine with particles in the dispersed phase, and has a certain viscosity and suspending ability. The variation of STF material viscosity with the shear rate when the shear rate was changed was tested as follows in Figure 2. From the figure, it can be seen that STF usually exhibits three characteristic rheological regions under different steady-state shear effects [10]: the shear thinning region at a low shear rate, the shear thickening region at a medium shear rate (between the critical shear rate and peak shear rate) and the second shear thinning region at higher shear rate. When the shear rate reaches the critical shear rate, the STF shear thickening effect occurs, and its viscosity increases slightly with the increase of shear rate; when the shear rate exceeds 60.26 s^{-1}, the STF system undergoes an obvious thickening effect, and its viscosity increases rapidly; with the shear rate increasing further to 160.2 s^{-1}, the viscosity of the STF system reaches the peak value of 28.3 $Pa{\cdot}s$; thereafter, with the increase of shear rate, the viscosity of the STF system reaches the peak value of 28.3; the viscosity of the STF system increases with the increase of shear rate, and the viscosity of the STF system increases with the increase of shear rate. After that, with the increase in shear rate, the viscosity of the STF system began to decrease slowly.

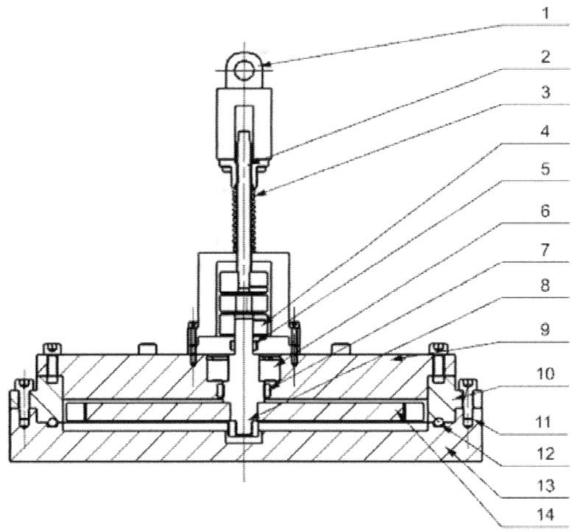

1. Connector; 2. Ball screw sub; 3. Damping spring; 4. Coupling; 5. Shaft end retaining ring; 6. Rotor bearing; 7. Sealing felt; 8. Rotor shaft; 9. Upper-end cap; 10. Sealing chamber; 11. Screw; 12. Sealing O-ring; 13. Lower end cap; 14. Rotor

Figure 1 Sketch of the damper.

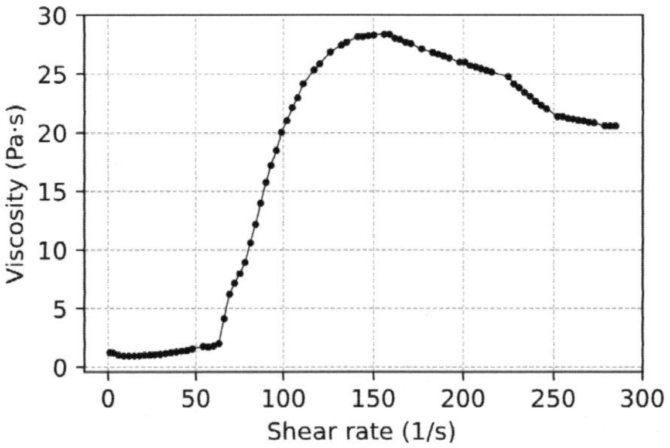

Figure 2 Viscosity-shear rate variation curve of STF system [11].

The most commonly used intrinsic viscosity model for shear-thickening materials is the power law model. This model generalizes the nonlinear growth trend of fluid viscosity as a power exponential growth. According to this model, the viscosity function can be expressed as Equation (1).

$$\eta(\dot{\gamma}) = K \cdot \dot{\gamma}^{(n-1)} \tag{1}$$

where η is the viscosity ($Pa{\cdot}s$); K is the material correlation constant; $\dot{\gamma}$ is the shear rate (s^{-1}); n is the power law exponent. From Equation (1), it can be seen that this function can show the shear thickening performance of STF, i.e., the viscosity of the system increases rapidly with the increase of the shear rate in an exponential trend, but it cannot be fitted to the shear thinning phenomenon presented by the STF after exceeding the peak shear rate. Other common viscoplastic fluid models include: Carreau model [12], Bingham model [13], Herschel-Bulkley model [14], nonlinear bi-viscosity model, etc. In the shear-thickening region, the shear stress increases with the increase of shear rate, showing an obvious "S-shaped" curve, and the viscosity of STF in the two shear-thinning regions shows a decreasing trend approximating to the power function and exponential function with the change of shear rate. Therefore, for the STF system, its rheological constitutive model can be described by the viscosity segmentation function combining the logistic function and the power function as in Equation (2) below.

$$\eta(\dot{\gamma})=\begin{cases}\eta_{\mathrm{I}}(\dot{\gamma})=b_1\dot{\gamma}^{a_1} & \dot{\gamma}\le\dot{\gamma}_c\\ \eta_{\mathrm{II}}(\dot{\gamma})=\dfrac{c}{1+\exp\left[a_2(\dot{\gamma}+b_2)\right]}+d & \dot{\gamma}_c\le\dot{\gamma}\le\dot{\gamma}_{\max}\\ \eta_{\mathrm{III}}(\dot{\gamma})=b_3\dot{\gamma}^{a_3} & \dot{\gamma}_{\max}\le\dot{\gamma}\end{cases} \tag{2}$$

where $\dot{\gamma}_c$ is the critical shear rate of $60.26\,s^{-1}$; $\dot{\gamma}_{\max}$ is the peak viscosity corresponding to a shear rate of 160.2 s^{-1}; a_1, b_i, and d are the parameters that characterize the fitted curves. For the logistic function of the shear thickening stage, when the parameters c and d are kept constant, the change of parameter a_2 affects the degree of distribution of the curves in the lateral direction, while the change of parameter b_2 affects the relative position of the left and right of the S-shaped curves. The coordinates of the critical viscosity point and the peak viscosity point (the two end points of the shear thickening region) are introduced into the viscosity description model as constraints. The expression of the viscosity function phenomenon model is simplified as shown in Equation (3):

$$\eta(\dot{\gamma})=\begin{cases}\eta_{\mathrm{I}}(\dot{\gamma})=\eta_c\left(\dfrac{\dot{\gamma}}{\dot{\gamma}_c}\right)^{a_1} & \dot{\gamma}\le\dot{\gamma}_c\\ \eta_{II}(\dot{\gamma})=\eta_c+(\eta_{\max}-\eta_c)\bullet L & \dot{\gamma}_c\le\dot{\gamma}\le\dot{\gamma}_{\max}\\ \eta_{\mathrm{III}}(\dot{\gamma})=\eta_{\max}\left(\dfrac{\dot{\gamma}}{\dot{\gamma}_{\max}}\right)^{a_3} & \dot{\gamma}_{\max}\le\dot{\gamma}\end{cases} \tag{3}$$

where L is called the "moderator" and its expression is shown in Equation (4). L can be seen as a combination of the numerator and denominator of the logistic function, which are multiplied by $\eta_{\max}-\eta_c$ to form the overall form of the viscosity function. This "adjustment factor" plays a role in regulating and influencing the rate and shape of the viscosity function as the shear rate changes.

$$L=\frac{1+\mathrm{e}^{a_2(\dot{\gamma}_{\max}-\dot{\gamma}_c+b_2)}}{1+\mathrm{e}^{a_2(\dot{\gamma}-\dot{\gamma}_c+b_2)}}\cdot\frac{\mathrm{e}^{a_2(\dot{\gamma}-\dot{\gamma}_c+b_2)}-\mathrm{e}^{a_2b_2}}{\mathrm{e}^{a_2(\dot{\gamma}_{\max}-\dot{\gamma}_c+b_2)}-\mathrm{e}^{a_2b_2}} \tag{4}$$

The Levenberg-Marquardt algorithm was used to fit the model and optimize the parameters yielding the following coefficients for each interval of the segmentation function. a_1 : 0.020781308045412766. a_2 : 0.06247553709726486. b_2 : -81.20853356733792. a_3 : -0.53379262441021. Viscosity is in better agreement with the experimental values. In conclusion, the segmented function model can accurately fit the viscosity of STF materials in both the initial stage of the shear thickening liquid and the shear thickening stage. The main consideration in the use of STF materials is also the performance of the shear thickening section of STF materials, so the viscosity segmented function model can be applied to the intrinsic model of STF systems.

In order to facilitate the subsequent fluid simulation of the STF damper and monitoring of the damping force, we brought the coefficients of each interval of the fitted segmented function into the viscosity function formula and wrote the viscosity function of the shear-thickening fluid in C form in advance.

4. Simulation calculation of damping force for STF damper

4.1. Fluid simulation preprocessing and solution setup analysis

Firstly, the 3D modeling software UG is used to model and model simplification of the STF damper, and the X_T file is imported into WORKBENCH. The torsional STF damper is subjected to watershed extraction and the naming of each body and face in Spaceclaim. Then the mesh division of the flow field region is carried out by MESHING. Since the tetrahedral mesh can better adapt to the complex structure and is suitable for dynamic mesh simulation, the mesh is all unstructured with tetrahedral. At the same time, the expanded boundary layer meshing was performed at the walls of both basins. The divided tetrahedral mesh is shown in Figure 3 below.

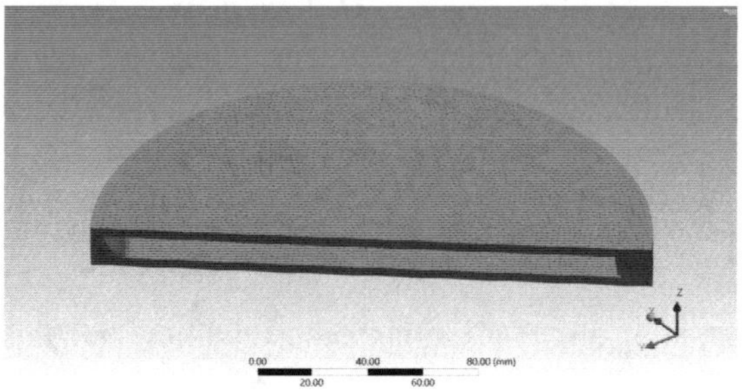

Figure 3 Mesh delineation model of the fluid domain of the damper.

Each wall section of the entire flow field region, the fluid domain, and the gear wall are named wall, fluid, and moving_wall, respectively. In the actual simulation, the viscosity of the shear-thickening fluid varies due to different shear rates because of the different positions from the center of rotation. Since the viscosity segmentation function of the shear thickening fluid has been fitted in the previous section, the viscosity change of the fluid is considered to be controlled using the UDF [15]. The first step is to obtain the current cell (c) and time (t); then we calculate the shear rate; according to the different cases of shear rate, we use different formulas in the segmentation function to calculate the dynamic viscosity; finally, we return the calculated viscosity value. c_STRAIN_RATE_MAG(c, t) is a macro function provided by FLUENT to calculate the shear rate of the current cell. c denotes the current cell, is a pointer to the current cell. t denotes the current time, usually in the calculation of dynamic parameters which needs to take into account the effect of time change.

UDF is a C-based development environment. The Dynamic Viscosity code includes the "udf.h" header file, which ensures that the prototypes and definitions of the FLUENT UDF library can be used in the code. The variables and data types of the dynamic viscosity program as well as the function definitions and return values are in standard C program format and can therefore be directly mounted into FLUENT for compilation. We select and use this customized dynamic viscosity function when setting the viscosity of a material in FLUENT.

During the working process of the torsional shear-thickening liquid damper, the structure of the flow field domain changes with time and torsion angle. At the same time, the viscosity distribution and flow velocity distribution of the shear thickening fluid will also have an effect on the structure of the flow field domain. The dynamic mesh method can allow the mesh to adaptively adjust with the change of flow structure to ensure that the simulated geometry is consistent with the actual situation. By dynamically adjusting the mesh to fit the flow structure and geometric deformation, the dynamic mesh method can improve the accuracy and stability of the simulation, which can reduce the numerical error and improve the reliability of the simulation results.

4.2. Damping force solution and calculation results

In order to simulate and calculate the damping torque of the output STF damper in the working process, the results of simulating and calculating the damping torque of the torsion type STF damper under different working conditions with the wall surface of the fluid rotating domain as the monitoring object are shown in Table 1 below.

Table 1 Rotor node speed and corresponding resisting moment.

Rotor speed (rad/s)	2	3	4	5	6	8	10	12	20
Output torque (N·m)	0.84	3.30	6.37	10.33	16.27	20.26	23.79	25.24	31.87

The continuous curve of the damper output resisting moment versus the input angular velocity is shown in Figure 4 above. It can be seen that there is an obvious nonlinear relationship between the STF output resisting moment and the input angular velocity under the rotational speed of 0~10 rad/s. Overall, the damper output resisting moment increases with the increase of input angular velocity. At 0~2rad/s rotational speed, the resisting moment increases slowly, the slope of the curve is small because the rotational speed is low, the shear thickening fluid in the chamber is working under a low shear rate and its viscosity is also low. Within 3rad/s~6rad/s, the resisting moment increases by 12.97 N•m, and the increase of resisting moment is 393.03%. In this interval, most of the shear thickening fluid inside the rotor outer circle is in the thickening state. The viscous-plastic force plays a major role in the output resisting moment. In the interval from 8 rad/s to 12 rad/s, the resisting moment only increases by about 5 N•m, and the increase of resisting moment decreases by 24.58%. Because the viscoplastic force of the fluid is larger than the inertial force at smaller velocities, the change in viscosity can greatly affect the change in damping force. When the angular velocity exceeds 10 rad/s, the linear velocity increases everywhere in the fluid domain, and the inertia force of the fluid increases. The shear thickening zone is formed only in the middle region of the rotor. The other regions cannot be thickened due to the shear rate being too large, so the magnitude of the dynamic viscosity has a limited effect on the damping force.

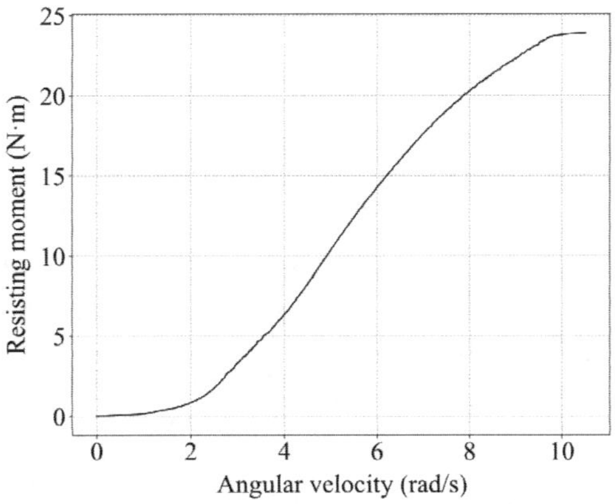

Figure 4 Plot of damper output resisting moment versus input angular velocity.

In the dynamic performance analysis, in order to study the mechanical properties of the torsional STF damper under different dynamic loads, the loading simulation analysis of the torsional STF damper is carried out by using a sine wave waveform, which is shown in Equation (5).

$$\theta = A \bullet \sin(2\pi f t) \tag{5}$$

where A is the corner amplitude (°) and f is the frequency (Hz). The torsional STF damper is loaded sinusoidally by changing the frequency and corner amplitude of sinusoidal loading. The angle amplitude of sinusoidal loading is taken as 15°, 45°, and 60°. The frequency of sinusoidal loading is taken as 0.1Hz, 0.5Hz and 1Hz. The results of the simulation calculations are shown in Figure 5 and Figure 6 below.

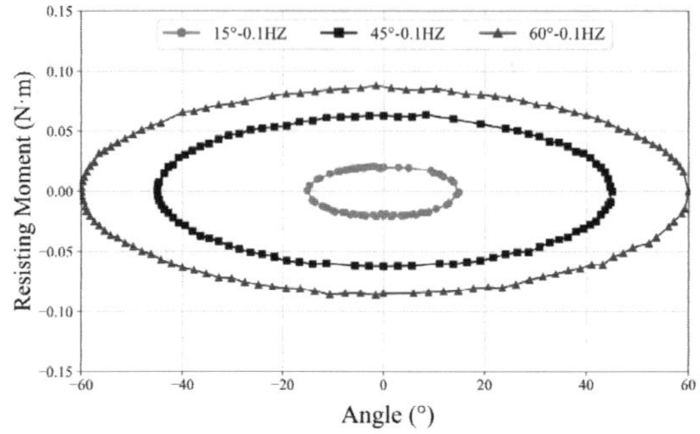

Figure 5 Torque-angle hysteresis curves for different angle amplitudes at 0.1 HZ.

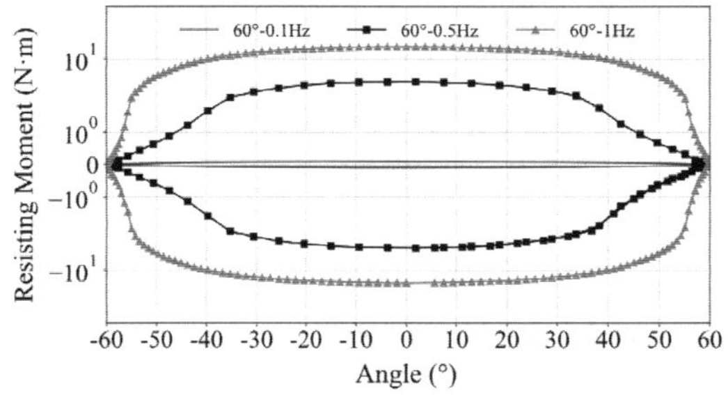

Figure 6 Torque-angle hysteresis curves at different loading frequencies.

As a whole, the damping torque and hysteresis curve area of the torsion-type STF damper increase with increasing corner amplitude and loading frequency, i.e., the energy dissipation capacity increases. Meanwhile, the damping torque of the torsion-type STF damper increases with the increase of the corner amplitude at low loading frequency; the hysteresis curve of the damper is presented as an ellipse when the frequency is 0.1 Hz, which is due to the fact that the shear rate of the fluid in the torsion-type STF damper is small at this time. The STF material does not undergo the phenomenon of shear thickening. The change of its viscosity is relatively small, which is approximately caused by the fixed value. With the gradual increase of the loading frequency, the fluid shear rate in the torsional STF damper increases, the STF material begins to occur shear thickening phenomenon, the force of the torsional STF damper shows a nonlinear growth, the shape of the hysteresis curve is presented as a spindle, and the energy dissipation capacity of the torsional STF damper is significantly improved relative to the low frequency and low rotational angle amplitude.

5. Conclusion

The simulation results show that the proposed segmented function combining the logistic function and power function fits the viscosity-shear rate curve of the shear thickening liquid well for the self-developed design of the torsional STF damper and the shear thickening liquid medium. At the same time, the damper output can adaptively output a resisting torque that matches the rotor speed. In a certain range (0~10 rad/s relatively low rotational speed), most of the internal watershed of the damper is in the thickening state under the action of the corresponding shear stress, and the dynamic viscosity has a large influence on the damping force output from the damper. In the case of dynamic loading of the damper, both frequency and corner amplitude have an effect on the damping moment output from the damper. The effect of frequency on the damping moment is more obvious.

For the unique rheological properties of STF, the segmented function model combining the Logistic function and power function can more accurately describe the nonlinear viscosity change of STF under different shear rates, especially the behavior of thickening effect; torsional STF damper can adaptively output damping torque under low and high-speed conditions, which provides more choices for structural vibration damping. The UDF program for dynamic viscosity and the UDF program for rigid-body motion with boundary motion in a dynamic mesh is not only written specifically for the model used in the paper, but it is also applicable to other damper models. The simulations can be performed by modifying the parameters of the program for different shear-thickening fluids and damper structures.

Acknowledgments
Project funded by the National Natural Science Foundation of China (Project Grant No. 12202199).

References
[1] WANG P, YU K, QIAN K, et al. 2021 Application of shear thickening fluid in damping and vibration reduction technology. Materials Guide, 35(17): 17218-24.
[2] FREYA R, J CHARLY R, SHANMUGAM A, et al. 2023 Investigation of Rheological Properties in Shear Thickening Fluid for Seismic Dampers. Materials Today: Proceedings, 80: 1506-13.
[3] NAGY-GYöRGY P, HŐS C. 2021 Predicting the damping characteristics of vibration dampers employing generalized shear thickening fluids. Journal of Sound and Vibration, 506: 116116.
[4] JUNG I Y, RYU J, OH J, et al. 2021 Experimental investigation on displacement amplification mechanism of steel wire rope-pulley damping systems with viscous damper. Engineering Structures, 248: 113206.
[5] ZHOU H, GUO C Y, ZONG L H, et al. 2013 Shear thickening fluid and damper performance study. Vibration and Shock, 32(18): 15-20.
[6] LIU Q, FAN J Q. 2023 Progress of shear thickening fluid and its application in cushioning and vibration damping. Journal of Functional Materials and Devices, 29(03): 178-86.
[7] LI X, ZHAO Y L, ZHU C H, et al. 2022 Comparative analysis of STF damper performance tests with different structural characteristics. China Shipbuilding, 63(5): 233-44.
[8] ZHAO Q, HE Y, YAO H, et al. 2018 Dynamic performance and mechanical model analysis of a shear thickening fluid damper. Smart Materials and Structures, 27(7).
[9] YEH F-Y, CHANG K-C, CHEN T-W, et al. 2014 The dynamic performance of a shear thickening fluid viscous damper. Journal of the Chinese Institute of Engineers, 37(8): 983-94 37(8): 983-94.
[10] YANG S F, HUANG X M, MING Yang, et al. 2023 Mechanical modeling and experimental study of a novel magnetorheological-shear thickening damper. Research. Journal of Mechanical Engineering, 59(16): 418-26.
[11] Huang, C C. 2021 Design and performance study of torsional damper based on shear thickening fluid. Harbin Institute of Technology University of Harbin.
[12] KHAN M, SALAHUDDIN T, CHU Y-M. 2021 Analysis of the Carreau fluid model presenting physical properties along different molecular axes near an anisotropic rough surface. International Communications in Heat and Mass Transfer, 123: 105233.
[13] BINGHAM E C. 1922 Fluidity and Plasticity. Fluidity and plasticity.
[14] HERSCHEL W H, BULKLEY R. 1926 K von Gummi-Benzollösungen. K-Z, 39(4): 291-300.
[15] WU J Q, HOU B L. 2017 Damping Characterization of Magnetorheological Dampers Based on FLUENT. Arms Equipment Journal of Engineering, 38(11): 142-5.

Workspace analysis of an R (2-UPR/2-UPS) hybrid machine tool

Zuwang Yin, Dong Wang, Xiaoguang Zhou*, Jiahua Wu

School of Electronics and Automation, Century College Beijing University of Posts and Telecommunications, Beijing, China

*Corresponding author's e-mail: 572113157@qq.com

Abstract: This paper proposes a hybrid machine tool with an R (2-UPR/2-UPS) asymmetric layout. Its workspace is analyzed, that is, the set of accessible positions of the spindle tool head point under the condition that the motion range of each hinge is met. The workspace of the mechanism is an important component for analyzing the movement of the mechanism and an important indicator for evaluating the processing performance of the machine tool. An intuitive understanding of the workspace and analysis of the tool's posture and position performance in the workspace are conducive to fully utilizing the processing capacity of the machine tool and improving the processing efficiency of the machine tool.

1. Introduction

The hybrid machine tool is a combination of parallel mechanisms, traditional machine tools, and robot technology. It takes into account the characteristics of traditional machine tools such as large working space and flexible transmission. At the same time, it also integrates the characteristics of parallel mechanisms such as a large stiffness-to-mass ratio, multi-degree-of-freedom coupling output, and good dynamic performance. It effectively avoids the shortcomings of the relatively small stiffness of traditional machine tools and the narrow motion space of parallel mechanisms. For example, some parallel machine tools in Germany, Sweden, and other countries have achieved industrial promotion[1]. The Tricept series parallel machine tool products of Neos Robotics of Sweden account for about 70% of the current parallel machine tool market[2]. Volvo, Volkswagen, General Motors, OPEL, and other companies have used them in the processing and manufacturing of their products[3]. The working space of machine tools, as an important indicator of machine tools, limits the application field of machine tools[4].

The hybrid machine tool with an R (2-UPR/2-UPS) asymmetric layout is shown in Figure. 1, and a schematic diagram of its mechanism is shown in Figure. 2.

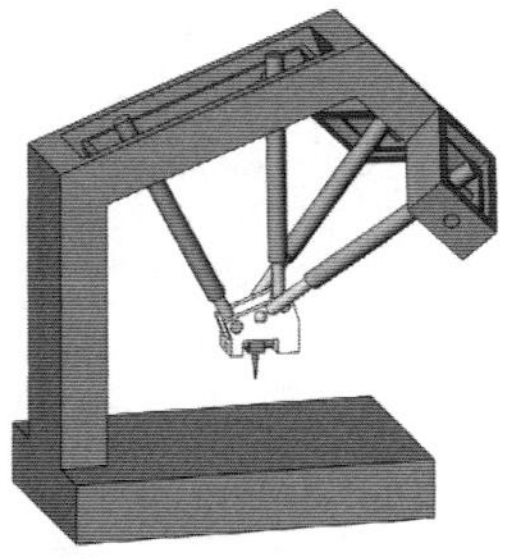

Figure. 1 Schematic diagram of the hybrid machine tool structure.

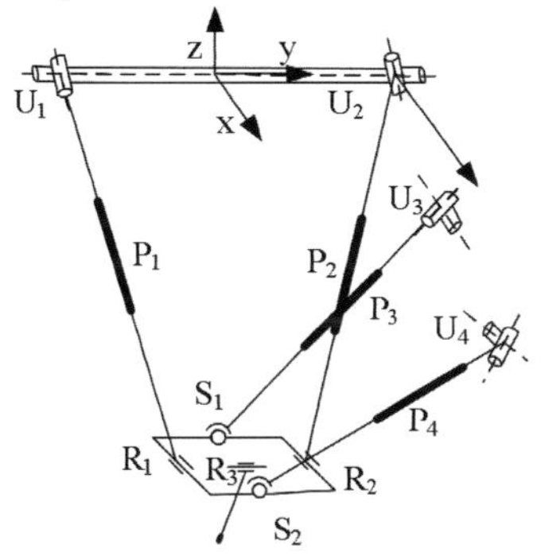

Figure. 2 Schematic diagram of the hybrid mechanism structure.

2. Modeling of the machine tool workspace

2.1 Definition of tool posture.

The working space of a machine tool is related to the position and posture of the tool, the range of motion of each kinematic pair, and the position and size of each hinge point of the machine tool. To analyze the working space of machine tools, we must first determine the parameters of these machine tools, and then determine the set of tooltip points that meet the conditions based on the analytical relationship of the mechanism[5].

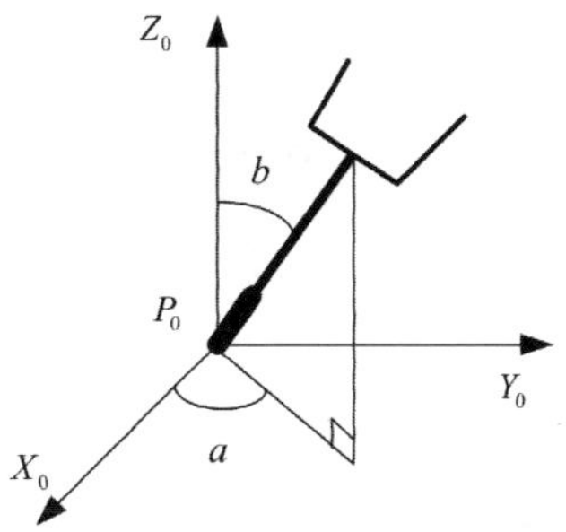

Figure. 3 Tool coordinate system.

The Euler angle method is used to express the spindle tool posture. For ease of understanding, as shown in Figure. 3, the position coordinates and posture of the spindle tool are expressed by $\boldsymbol{P}_0$ the coordinates of the point and the precession angle a and nutation angle b when analyzing the workspace.

Journal of Physics: Conference Series **2992** (2025) 012026 doi:10.1088/1742-6596/2992/1/012026

2.2 Constraints of each kinematic pair of machine tools

The kinematic pairs of machine tools are divided into four categories: rotating pairs, translation pairs, universal joints, and ball pairs. The working space is mainly limited by the range of motion of these four types of kinematic pairs.

1) The rod length of each linear drive will change during the motion $\boldsymbol{L}_i(i=1,2,3,4)$. When the stroke is the smallest $\boldsymbol{L}_{\min}$, the rod length is the smallest. When the stroke is the largest, the rod length should satisfy the relationship $\boldsymbol{L}_{\max}$.

$$\boldsymbol{L}_{\min} \le \boldsymbol{L}_i \le \boldsymbol{L}_{\max} \tag{1}$$

2) The two UPR branches of the rotation angle limiting mechanism of the rotating pair are connected to the moving platform through the rotating pair and connected to the fixed platform through the universal joint. The universal joint is a two-stage rotating pair. The parallel mechanism of the machine tool in the plane where the two branches are located can be regarded as a four-bar mechanism. The two rotating pairs and the two-stage rotating pairs of the two universal joints are the four rotating pairs of the four-bar mechanism.

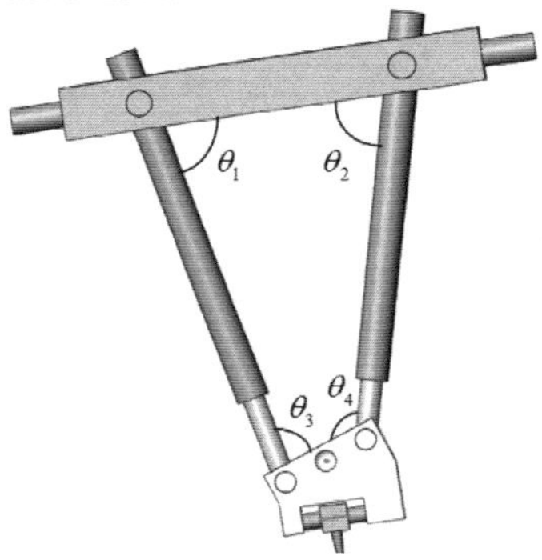

Figure. 4 Two upr branches.

In order to prevent the mechanism from being singular or close to being singular, as shown in Figure. 4 $\theta_i(i=1,2,3,4)$, it should move within a certain range.

$$\theta_{\min} \le \theta_i \le \theta_{\max} \tag{2}$$

This four-bar mechanism rotates around the two colinear primary revolute pairs of the universal joint, as shown in Figure. 5, and its swing angle is θ_0.

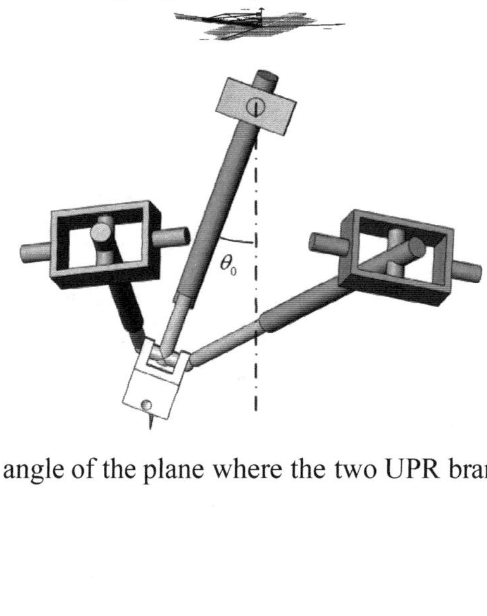

Figure. 5 Swing angle of the plane where the two UPR branches are located.

3) Limiting mechanism of two-stage rotation angle of the universal joint. The left and right UPS branches are connected to the fixed platform through the universal joint. The universal joint has two-stage rotation pairs. The coordinate system of the universal joint is established as shown in Figure. 6.

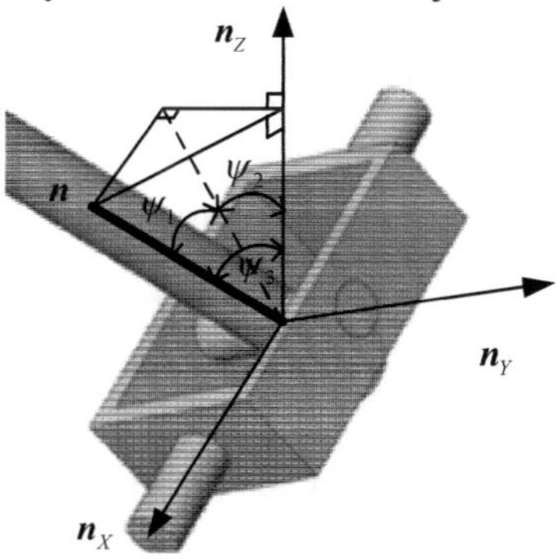

Figure. 6 U-type two-stage turning angle.

In the initial state, the direction of the primary rotation pair of the universal joint $\boldsymbol{n}_X$ coincides with the direction vector; $\boldsymbol{n}_Y$ and the direction of the secondary rotation pair coincides with the direction vector. The direction vector of the linear drive is $\boldsymbol{n}$, so the rotation angle of the primary and secondary rotation pairs of the universal joint is as follows[6]:

$$\psi_1 = \arccos\frac{n\cdot n_Z}{\sqrt{1-\left(n\cdot n_X\right)^2}} = \arctan_2\left(\frac{n\cdot n_Z}{n\cdot n_Y}\right) \tag{3}$$

$$\psi_2 = 90°\text{-}\arccos\left(n\cdot n_X\right) = \arcsin(n\cdot n_X) \tag{4}$$

Due to structural limitations, the primary and secondary rotation pairs should satisfy the following relationship:

$$\psi_{1\min} \le \psi_1 \le \psi_{1\max} \tag{5}$$

$$\psi_{2\min} \le \psi_2 \le \psi_{2\max} \tag{6}$$

4) The two UPS branches of the angle limiting mechanism of the ball pair are connected to the moving platform through the ball pair. The angle of the ball pair is shown in Figure. 7:

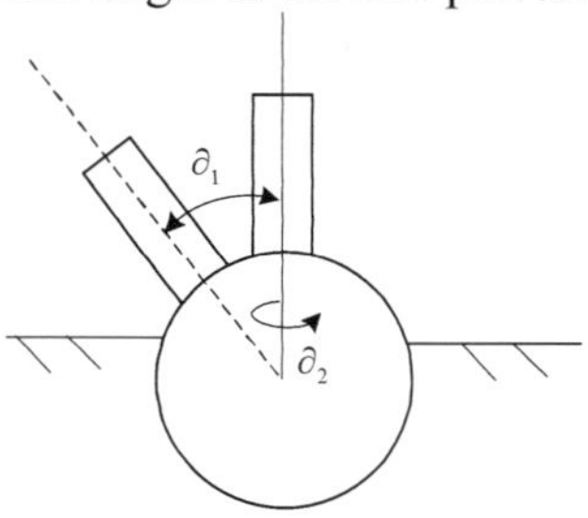

Figure. 7 Ball secondary rotation angle.

∂_1 is the deflection angle of the ball pair. ∂_2 is the rotation angle of the ball pair. The ball pair can rotate a full circle ∂_2=360°, so it satisfies:

$$\partial_{1\min} \le \partial_1 \le \partial_{1\max} \tag{7}$$

5) Limitation of the rotation angle of the tandem swing head. The rotation angle between the tandem swing head and the moving platform λ is shown in Figure. 8.

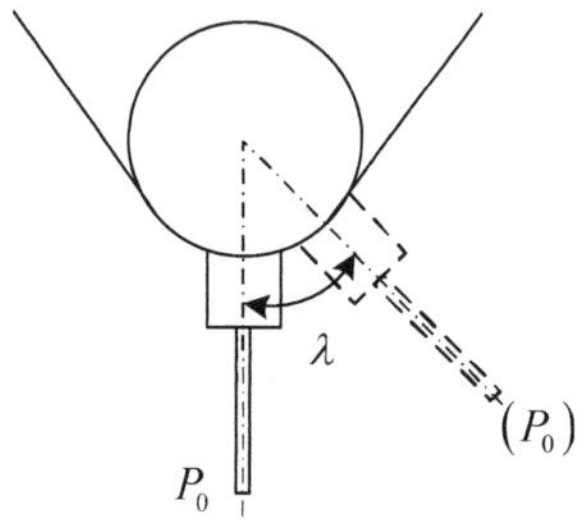

Figure. 8 Swing motor rotation angle.

Due to the structural limitations of the moving platform, the turning angle λ should meet the following requirements:

$$\lambda_{\min} \le \lambda \le \lambda_{\max} \tag{8}$$

2.3 Other Parameter definitions of machine tool workspace

The workspace is the set of tooltip points that satisfy the above constraints P_0 . Since the workspace is generally a non-convex set, this section adopts the method of successive search in a rectangular coordinate system. The structural parameters are shown in Fig. 9.

1) Structural parameters of the mechanism. The hinge points on the fixed platform are in the fixed system coordinates A_{01}, A_{02}, A_{03}, A_{04}. The hinge points on the moving platform are in the moving system coordinates A_1, A_2, A_3, A_4, and the length of the series swing head l.

2) The kinematic limits of the mechanism. The length limit $\boldsymbol{L}_{\min}$ of each driving rod is $\theta_{\min}$. The angular limit of the four-bar mechanism rotational pair is $\boldsymbol{L}_{\max}$, θ_{max}. The plane swing limit is θ_{0max}, θ_{0max}. The angular limit of the primary rotational pair of the universal joint is $\psi_{1\min}$, $\psi_{1\max}$. The angular limit of the secondary rotational pair is $\psi_{2\min}$, $\psi_{2\min}$. The deflection angle limit of the ball pair is $\partial_{1\min}$, $\partial_{1\max}$. The swing angle limit of the series swing head motor is $\lambda_{\min}$, $\lambda_{\max}$.

3) The spindle tool posture range is $a_{\max}$, $\mathrm{b}_{\max}$.

4) Define the search range. Based on previous research on machine tool kinematics. The coordinates of the tooltip $P_0(P_{0x}\ P_{0y}\ P_{0z})$, $P_{0x} \in (P_{0x\min} \sim P_{0x\max})$, $P_{0y} \in (P_{0y\min} \sim P_{0y\max})$, $P_{0z} \in (P_{0z\min} \sim P_{0z\max})$, are searched step by step, and the motion of the mechanism at each point is calculated to see whether it is within the limit range. If it is within the limit range, it is retained; if not, it is removed. Finally, an image of the set of retained points is drawn, and the search range is adjusted based on the image.

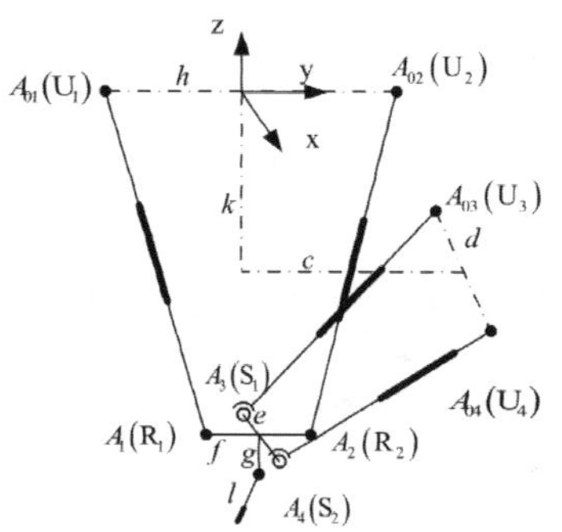

Figure. 9 Structural parameters of the hybrid mechanism.

Given example mechanism as Figure 9 are: h=250mm , k=323.41mm , d=409.2mm , e=69mm , f=80mm , g=80mm , l=125mm , L_{min}=400mm , L_{max}=1100mm , θ_{max}-θ_{min}=140° , θ_{0max}=35° , ψ_{1max}-ψ_{1min}=60°, ψ_{2max}-ψ_{2min}=60°, ∂_{1max}=30°, λ_{max}-λ_{min}=60°. The mechanism is rotated 20° clockwise around the axis as an analysis example. Matlab is used to calculate and draw the image of the workspace. In order to analyze the mechanism quickly, only the upper and lower surfaces of the figure. are retained during the drawing process. This method can be used to draw the figure quickly, and conveniently calculate the volume of the figure.

3. Generate the workspace

3.1 Accessible working space for machine tools

Based on previous research on machine tool kinematics, the reachable workspace is the set of points that the mathematical model searches for a when b_{max}=360° and b gives the values of a_{max}=360°. Analyzing the reachable space can help us understand the maximum motion space of the mechanism, which helps adjust the position and shape of the fixed parts of the mechanism to prevent interference and improve the working capacity of the mechanism[7]. The reachable space of the machine tool is the set of tooltip positions only when all joints meet the motion limit range. The search logic is shown in Figure. 10. Workspace of the machine tool can be obtained through MATLAB calculation as shown in Figure. 11.

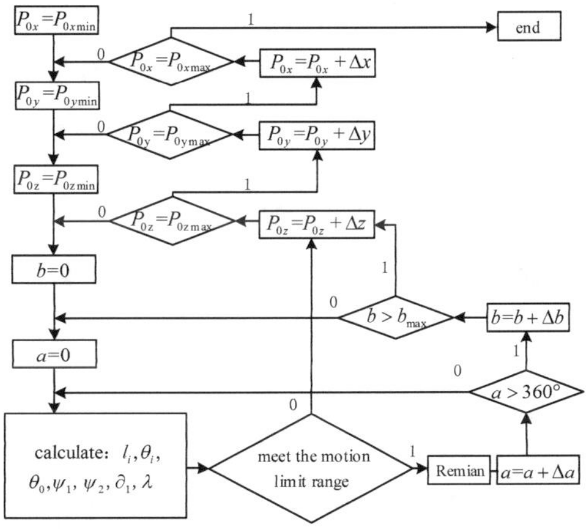

Figure. 10 Logic diagram of reachable space coordinate search.

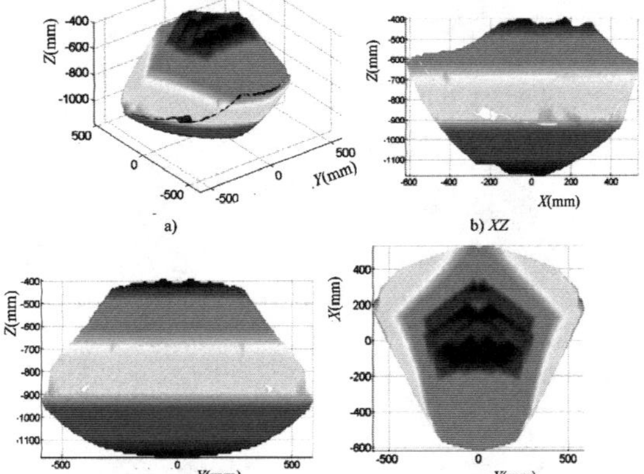

Figure. 11 Accessible space of machine tools.

Journal of Physics: Conference Series **2992** (2025) 012026 doi:10.1088/1742-6596/2992/1/012026

3.2 Flexible working space of machine tools

The flexible space of a machine tool is a collection of points that the tool head can reach, at which the tool can adjust its posture to a certain extent, that is, a workspace with posture change capability. In the actual use of machine tools, because the normal direction of the part surface is uncertain, it is often required that the machine tool execution end can not only reach the point but also reach it in different postures.

The flexible space of the machine tool is the position set of the tooltip when all joints meet the motion limit range and the tool can have a certain posture angle. The search logic is shown in Figure. 12[8].

Given the tool turnover angle ∂_1 = 360°, deflection angle ∂_2= 30°, and other parameters remain unchanged. The flexible space of the machine tool can be obtained through MATLAB calculation as shown in Figure. 13

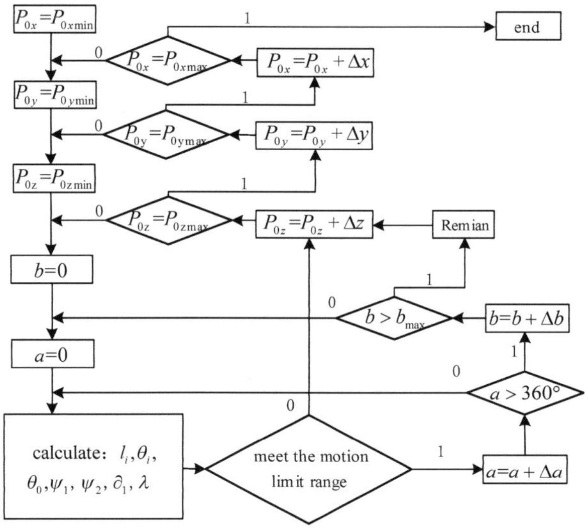

Figure. 12 Flexible spatial coordinate search logic diagram.

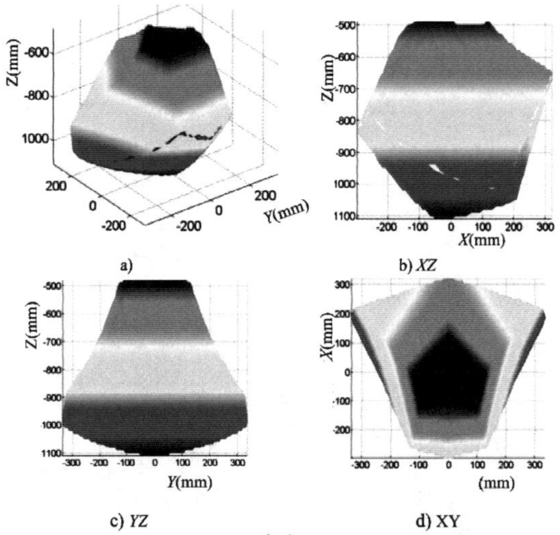

Figure. 13 Flexible space of machine tools.

4. Workspace analysis

4.1 Calculation and analysis of workspace volume

The volume of the workspace is an important indicator for evaluating the performance of a mechanism. The characteristics of the parallel mechanism determine that the workspace of the mechanism is small, and the ratio of the workspace to the machine tool volume is small, while the hybrid machine tool can significantly change this problem.

For the actual calculation of the volume of the mechanism working space, a numerical algorithm is used for analysis[9].

$$V=\sum_i\sum_j\left(h_{ij}\cdot\Delta\right) \tag{9}$$

h_{ij} is the distance in the vertical direction of the working space in the spatial rectangular coordinate system $\left(P_{0xi},P_{0yj}\right)$. Δ is unit area[10].

$$P_{0xi}=P_{0x\min}+i\cdot\Delta x \tag{10}$$

$$P_{0yj}=P_{0y\min}+j\cdot\Delta y \tag{11}$$

Δx, Δy the smaller the workspace, the higher the calculation accuracy, but the calculation time will be longer. In the flexible space calculation example, the flexible space of the machine tool can be obtained by calculation 1.406 m^3. The length, width and height of the machine tool in the calculation example are 1.5 m, 1.4 m, and 1.6 m respectively, and the machine tool volume is 3.36 m^3. Therefore, the ratio of the machine tool's workspace to the machine tool volume is 0.417. This is significantly improved compared to the 10%~20% volume ratio of the parallel machine tool.

4.2 Relationship between tool working posture and working space

Five-axis machine tools can be used to process parts with complex surface shapes, and the degree of surface undulation of the parts reflects the performance requirements of the machine tool. The processing of some surface features of parts requires a large tool inclination angle, which entails a large deflection angle of the machine tool. The workspace of a machine tool with greater flexibility is a subset of the workspace with less flexibility. The processing capabilities of the machine tool are different for different types of parts, so the relationship between the tool posture and the workspace is an important indicator of the machine tool's performance. Here, the tool deflection angle b is used as the change in tool posture.

The above example is analyzed by Matlab, and the relationship between the working posture of the machine tool and the working space volume is shown in Figure. 14.

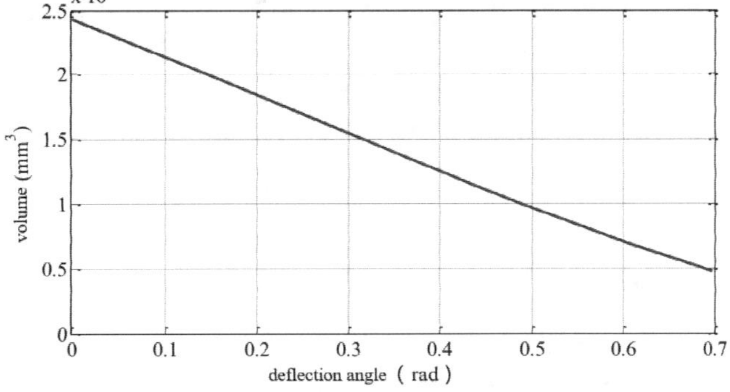

Figure. 14 Relationship between tool working posture and working space volume.

It can be seen that the volume of the flexible space of the mechanism has a great relationship with the required tool working posture size. The larger the deflection angle that the mechanism tool can achieve, the smaller the flexible working space of the mechanism. The relationship between the two is basically linear, and the appropriate parameters can be selected according to the requirements for the tool deflection angle and the requirements for the workspace volume of the workpiece to be processed.

4.3 Relationship between the range of motion of each hinge and the working space

In the actual engineering design of machine tools, the size of the working space of the mechanism is not only related to its performance requirements but also to the motion limits of each kinematic pair of the mechanism itself.

However, in engineering practice, the range of motion of each kinematic pair is a fixed value, and a larger range of motion often reduces the load-bearing capacity of the mechanism. In order to ensure mechanical performance, the structure of the mechanism often becomes more complicated, and the processing and production cost of parts increases. In practice, we optimize the design of the range of motion of each kinematic pair according to the needs so that it can meet the requirements.

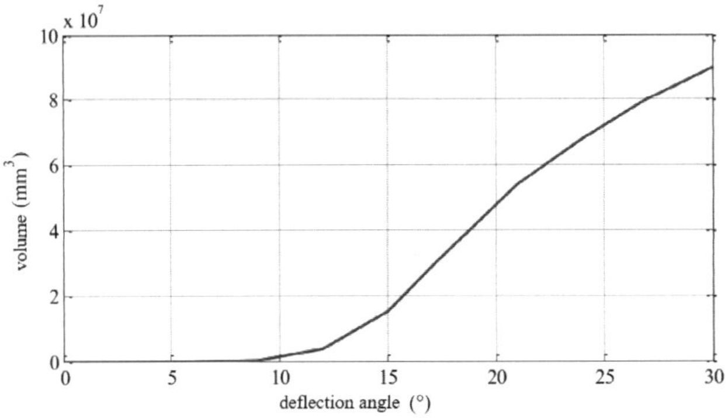

Figure. 15 Relationship between the ball pair rotation angle and the workspace volume.

The two ball pairs of the mechanism moving platform have a great influence on the attitude changeability of the moving platform. The rotation range of the spherical sliding pair is smaller than that of the three-rotation ball pair, so the arrangement and utilization of its limited rotation angle is particularly important. The relationship between the ball pair rotation angle and the workspace volume is shown in Figure. 15 by using Matlab to analyze the above example.

It can be seen that when the rotation angle of the ball pair of the mechanism is within a certain range, the working space will increase with the increase in the rotation angle of the ball pair, but the rotation range of the ball pair is generally not large.

The rotation angle of the ball pair has a great influence on the position of the moving platform, and if the spindle tool tends to obtain a larger posture deflection, the swing angle of the moving platform rotation pair and the moving platform series motor must have a larger range of motion. Generally, the larger range of motion of the rotation pair can meet the requirements, and the rotation pair of the moving platform series swing head motor is generally not too large due to the limitations of the motor, spindle, and cable. Using Matlab to analyze the above example, the relationship between the swing angle of the machine tool moving platform series motor and the workspace volume is shown in Figure. 16.

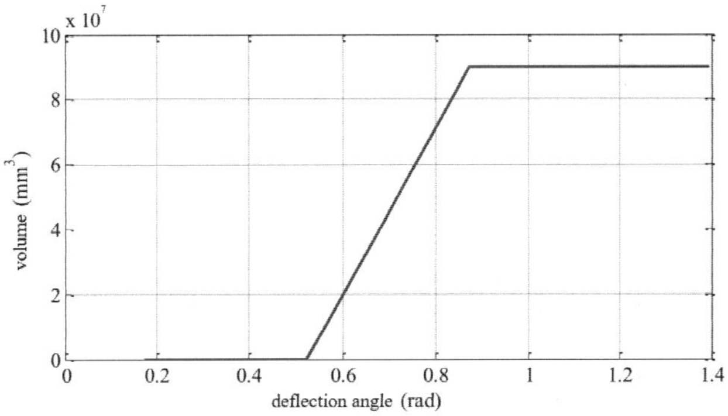

Figure. 16 Relationship between the swing head motor rotation angle and the workspace volume.

It can be seen that when the swing angle of the series motor is a small value, the volume of the working space of the mechanism is zero, indicating that the swing angle of the motor at each point of the mechanism movement cannot meet the requirements. When the mechanism reaches a certain value, it begins to increase linearly from zero, and when it increases to about 50°, the working space of the mechanism no longer changes, indicating that the limiting condition of the flexible space of the mechanism is no longer the swing angle of the motor.

5. Conclusions

This paper analyzes the workspace of this new hybrid machine tool. The structural parameters of each part of the machine tool are defined, and the range and type of motion of each kinematic pair are specified. The workspace graphics of the machine tool under given conditions and the workspace graphics of the machine tool under given tool working angles are obtained. The influence of the required tool angle and each kinematic pair on the workspace of the machine tool is analyzed, and the optimal joint motion scheme is selected from them. The research in this paper has done some basic work for the subsequent performance analysis of this mechanism.

References

[1] Xiangyu K, Liping W, Guang Y, et al. Research on Servo Matching of a Five-axis Hybrid Machine Tool [J]. *The International Journal of Advanced Manufacturing Technology*, 2023, 129(1-2):983-997.

[2] Krimpenis A A, Iordanidis M. D. Design and Analysis of a Desktop Multi-Axis Hybrid Milling-Filament Extrusion CNC Machine Tool for Non-Metallic Materials [J]. *Machines*, 2023, 11(6):

[3] Patra B, Nag A, Bandyopadhyay S. Analytical Determination of the Optimal Effective Regular Workspace of a 6-6 Stewart Platform Manipulator for a Specified Orientation Workspace [J]. *Mechanism and Machine Theory*, 2024, 203105791-105791.

[4] Mathematics; New Mathematics Findings from Ministry of Education Discussed (Workspace Analysis of a Hybrid Kinematic Machine Tool with High Rotational Applications) [J]. *Journal of Technology & Science*, 2018, 466-.

[5] Suryam L, Balakrishna B. Inverse Kinematic Analysis of 5-axis Hybrid Parallel Kinematic Machine Using CAD and Regression Analysis Approach [J]. *International Journal of Computer Aided Engineering and Technology*, 2020, 13(4):

[6] Huang Daoyang, Bo Ruifeng, Zhang Wangwang, et al. Position and Workspace Analysis of (2PSR + PUU)&RP Hybrid Machine Tool [J]. *Manufacturing Technology & Machine Tool*, 2022, (03): 92 - 97. DOI:10.19287/j.cnki.1005 - 2402. 2022.03.015.

[7] Huang Muhua. Design and Dynamic Analysis of a New Reconfigurable Mechanism - Based Parallel Machine Tool [D]. Guangxi University, 2019.

[8] Qin Caiyou, Huang Juan, Li Xiaoru. Performance Analysis of a Five - axis Linked Hybrid Machine Tool [J]. Journal of Mechanical Transmission, 2019, 43(02): 122 - 128. DOI:10.16578/j.issn.1004 - 2539.2019.02.023.
[9] Lu Yan. Configuration Design and Performance Analysis of a Five-axis Linked Hybrid Machine Tool with Large Swing Angle [D]. North University of China, 2017.
[10] Chen Zhizhou. Design and Analysis of Five-axis Hybrid Machine Tool Based on 2RPU/UPR Parallel Mechanism [D]. Yanshan University, 2017.

Journal of Physics: Conference Series **2992** (2025) 012027 doi:10.1088/1742-6596/2992/1/012027

Attitude wind stabilization control of inspection vehicle based on Stackelberg game theory

Chuanze Liang [1], Weijie Zhu [1] and Yafeng Qiu [1]*

[1] School of Mechanical Engineering, Nanjing University of Science and Technology, Nanjing, China

* Corresponding author's e-mail: nilghcn@sina.com

Abstract: Aiming at the problem that the electric power inspection vehicle is subject to wind field perturbation to produce attitude changes, resulting in the image quality of the UV detector carried by the vehicle being affected, a vehicle attitude anti-wind disturbance control method based on the Stackelberg game theory is proposed. According to the flight principle of the vehicle, the kinematics and dynamics model of attitude rotation and the wind field disturbance model are established, the control algorithm is designed, the optimal strategy in the game process is solved by using reinforcement learning and neural network (NN) to solve the problem of vehicle attitude affected by the disturbance, the robustness and adaptability of vehicle attitude control are enhanced, and the stability of the controller is proved based on Lyapunov theory. Finally, a simulation is carried out for the above problems, and the results verify the effectiveness of the designed control strategy.

1. Introduction

The vehicle is equipped with high-altitude reconnaissance capabilities and can carry high-resolution cameras and sensors. This allows it to quickly and comprehensively detect and monitor electric power facilities. It provides high-quality images and video data, is capable of rapid movement, and can easily fly over obstacles. As a result, it can cover large areas of electric power facilities efficiently, significantly improving inspection processes [1].

2. Attitude control modeling for quadrotors

2.1. Equations of rotational kinematics

The basic equations of kinematics are expressed as [2]:

$$\dot{\varepsilon} = \mu\omega \tag{1}$$

where $\varepsilon = [\varphi, \theta, \psi]$ denotes the attitude angle; $\omega = [p, q, r]$ denotes the angular velocity component of the vehicle in the body coordinate system; μ represents the rotation matrix [3].

The rotational kinematics equations are expressed as:

$$\begin{bmatrix} \dot{\varphi} \\ \dot{\theta} \\ \dot{\psi} \end{bmatrix} = \begin{bmatrix} 1 & \sin\varphi\tan\theta & \cos\varphi\tan\theta \\ 0 & \cos\varphi & -\sin\varphi \\ 0 & \sin\varphi/\cos\theta & \cos\phi/\cos\theta \end{bmatrix} \begin{bmatrix} p \\ q \\ r \end{bmatrix} \tag{2}$$

2.2. Equations of rotational dynamics

According to Eule's method, the basic equation of dynamics is expressed as:

$$M_{\Sigma} = I\dot{\omega} \tag{3}$$

where M_{Σ} denotes the sum of moments applied during the motion of the vehicle; I denotes the vector of inertia of the vehicle.

From the momentum conservation theorem, the combined external moment includes the aerodynamic moment M_b, gyroscopic moment M_g, and wind resistance moment M_W, which is expressed as:

$$\frac{dL}{dt} = M_b + M_g - M_W \tag{4}$$

where L denotes the total angular momentum of the entire vehicle rotation.

The aerodynamic drag moment due to wind interference can be expressed as:

$$M_W = 2^{-0.5} l F_W = 2^{-1.5} l \rho C_W S R_B^E R_W^B K_f V_A [1,0,0] \tag{5}$$

F_W is the wind field resistance in the inertial coordinate system, ρ is the air density, C_W denotes the wind field resistance coefficient, and V_A denotes the wind speed value in the airflow coordinate system.

Further analyzed by Coriolis' theorem, the transformed form of Equation (4) can be expressed as [4]:

$$I\dot{\omega} + \omega \times (I\omega) = M_b + M_g - M_W \tag{6}$$

The rotational dynamics equations of the vehicle in airframe coordinate system can be expressed as:

$$\begin{bmatrix} \dot{p} \\ \dot{q} \\ \dot{r} \end{bmatrix} = \begin{bmatrix} \left(\left(I_y - I_z\right) rq + U_2 - I_r q \Omega_r - M_{Wx}\right)\left(I_x\right)^{-1} \\ \left((I_z - I_x) pr + U_3 + I_r p \Omega_r - M_{Wy}\right)\left(I_y\right)^{-1} \\ \left(\left(I_x - I_y\right) pq + U_4 - M_{Wz}\right)\left(I_z\right)^{-1} \end{bmatrix} \tag{7}$$

where I_r denotes the moment of inertia of the propeller; I_x、I_y、I_z are the moment of inertia of the vehicle around the body axis; Ω_r is the sum of the relative rotational speeds of the four motors; U_1、U_2、U_3 are the virtual rotational torque.

2.3. Wind farm establishment

The Dryden turbulent wind field model with variable wind direction and the mean wind with variable wind speed are superimposed to establish the wind field model [5, 6]. After simulation, the wind speed of the wind field is obtained, as shown in Figure 1.

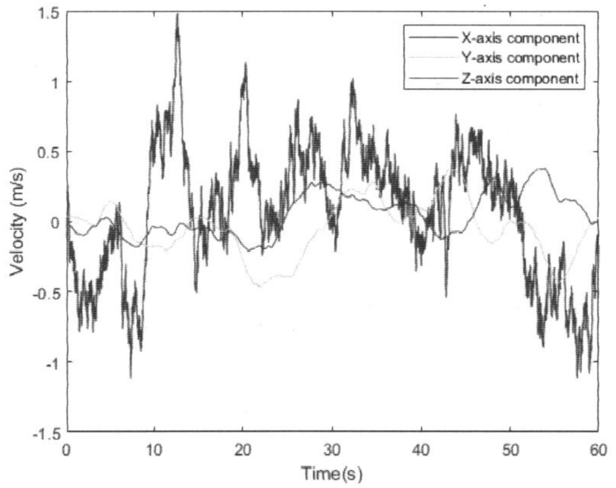

Figure 1. Velocity component of turbulent wind field.

3. Control algorithm design

3.1. Control model

Considering the unmodeled dynamics of the quadrotor itself and the presence of external noise, the mathematical form of its kinematic and dynamic model is expressed as:

$$\begin{cases} \dot{\xi} = \Upsilon(\xi)\omega \\ \dot{\omega} = f(\omega) + g\tau + \kappa(\xi,\omega,t) \end{cases} \tag{8}$$

where $\xi = [\varphi,\theta,\psi]$ represents the attitude component. $\omega = [p,q,r]$ represents the attitude angular velocity vector. $f(\omega)$ represents dynamics of the vehicle. g represents the control input gain. τ represents the control input. $\kappa(\xi,\omega,t)$ represents unmodeled dynamics for time-varying external disturbances and systems [7].

Further considering the external disturbances to the vehicle as well as the presence of system uncertainties, unmodeled dynamics, etc., the control input to the vehicle is described as:

$$\tau = \tau_c + \tau_d \tag{9}$$

where τ is the attitude angle controller to be designed and consists of two parts. τ_c is the control quantity used to stabilize the system; τ_d represents the control quantity to resist external disturbances and perturbations caused by unmodeled dynamics [8].

3.2. Control target

Attitude error vectors $e = \xi - \xi_t$ are introduced to evaluate the interaction between the vehicle command inputs and disturbances that affect its trajectory performance [9]. In order to ensure attitude error convergence, intermediate virtual control volumes are designed as follows [10]:

$$\omega_d = \Upsilon(\xi)^{-1}\left(-\Gamma_1 e + \dot{\xi}_d\right) \tag{10}$$

where Γ_1 is the positive definite gain matrix. The difference δ between actual velocity ω and desired control law ω_d is key to determine the convergence of e. The velocity error vector is denoted as:

$$\dot{\delta} = h(\delta) + g(\tau_c + \tau_d) \tag{11}$$

where $h(\delta)$ represents the unknown part of the dynamic system.

3.3. Controller design

An anti-jamming strategy is developed under the Stackelberg game. In the context of a non-cooperative game around the velocity error dynamics Equation (11), a sequential decision-making process is carried out between the jamming signal τ_d and the control input τ_c, described as follows [11]:

1) Decision making begins with the initial anti-interference controller input τ_c^0 to the vehicle.

2) Considering the control output of interference as a follower, it interacts with the anti-interference controller by maximizing the tracking errors e and δ, obtaining the optimal amount of control.

3) We consider the control inputs as the leader, which actively adjusts its strategy τ_c based on the control outputs τ_d^* of the followers, and chooses the optimal strategy τ_c^* to minimize error e and δ.

The leader's decision depends on the rational choices made by the followers. Both players work together to minimize their respective cost functions. The cost function of each player is defined as:

$$\begin{aligned} J_f(\delta_0,\tau_d,\tau_c) &= \int_0^\infty \left(\tau_d^T O \tau_d + \tau_c^T N \tau_c - \delta^T M \delta\right)\mathrm{d}s = \int_t^\infty r_f(\delta,\tau_d,\tau_c)\,\mathrm{d}s \\ J_l(\delta_0,\tau_d,\tau_c) &= \int_0^\infty \left(\delta^T O \delta + \tau_c^T N \tau_c + \Lambda(\tau_d)\right)\mathrm{d}s = \int_t^\infty r_l(\delta,\tau_d,\tau_c)\,\mathrm{d}s \end{aligned} \tag{12}$$

IOP Publishing
doi:10.1088/1742-6596/2992/1/012027

where G, Q, and R are positive definite matrices. $\Lambda(\tau_d)=\vartheta^T\dot{\lambda}$ is related to the follower and affects τ_c. The associated value function is expressed as follows:

$$
\begin{aligned}
V_f\left(\delta(t),\tau_d,\tau_c\right)&=\int_t^{\infty} r_f\left(\delta,\tau_d,\tau_c\right)\mathrm{d}s \\
V_l\left(\delta(t),\tau_d,\tau_c\right)&=\int_t^{\infty} r_l\left(\delta,\tau_d,\tau_c\right)\mathrm{d}s
\end{aligned} \tag{13}
$$

The follower's optimal value function is defined as follows:

$$
V_f^*\left(\delta_0\right)=\min_{\tau_d}\int_t^{\infty} r_f\left(\delta,\tau_d,\tau_c\right)\mathrm{d}s \tag{14}
$$

Based on it, the velocity error dynamics is calculated and the Hamiltonian function of the follower is:

$$
H_f(\delta,\nabla V_f,\tau_d,\tau_c)=r_f\left(\delta,\tau_d,\tau_c\right)+\nabla V_f^T(\eta(\delta)+g\tau_c+g\tau_d) \tag{15}
$$

where $\nabla V_f=\partial V_f/\partial\delta$ is the partial derivative of V_1 with respect to variable δ. Based on $\partial H_f/\partial\tau_d=0$, the optimal control output τ_d^* of the follower is:

$$
\tau_d^*=-\frac{1}{2}O^{-1}g^T\nabla V_f^* \tag{16}
$$

Introducing definition $\dot{\lambda}=\nabla\dot{V}_f=-\partial H_f/\partial\delta$ leads to the following covariance equation.

$$
\dot{\lambda}=2M\delta-\nabla\eta^T\nabla V_f \tag{17}
$$

where $\nabla\eta=\partial\eta(\delta)/\partial\delta$. The follower adopts the optimal strategy Equation (16) as its intended decision [12].

Subsequently, the leader's anti-interference controller formulates its policy by taking into account the followers' strategies Equation (16) and the co-state constraints Equation (17). Thus, the leader optimal control problem with constraints $\Lambda(\tau_d)=\vartheta^T\dot{\lambda}$ is formulated as:

$$
V_l^*\left(\delta_0\right)=\min_{\tau_c}\int_0^{\infty} r_l\left(\delta,\tau_d,\tau_c\right)\mathrm{d}s \tag{18}
$$

where ϑ is the designed Lagrange multiplier. The Hamiltonian function of the leader is as follows:

$$
H_l\left(\delta,\nabla V_l,\tau_d^*,\tau_c\right)=r_l\left(\delta,\tau_d^*,\tau_c\right)+\nabla V_l^T\left(S(\delta)+g\tau_c+g\tau_d^*\right) \tag{19}
$$

where $\nabla V_l=\partial V_l/\partial\delta$. Using the necessary condition of optimality, the optimal anti-influence control strategy and covariance equation of the leader can be obtained as follows:

$$
\tau_c^*=-\frac{1}{2}N^{-1}g^T\nabla V_l^* \tag{20}
$$

$$
\dot{\vartheta}=-\left[\frac{\partial H_l}{\partial\lambda}\right]^T=\frac{1}{2}gO^{-1}g\nabla V_l+\nabla S\vartheta \tag{21}
$$

For the minimum cost function V_f^* V_l^* under the optimal control input τ_c^* τ_d^*, after introducing the definition $\nabla V_f^*=\partial V_f^*/\partial\delta$, $\nabla V_l^*=\partial V_l^*/\partial\delta$, the Hamilton-Jacobi equation is obtained as:

$$
\begin{aligned}
0&=r_f\left(\delta,\tau_d^*,\tau_c\right)+\nabla V_f^{*T}\left(S(\delta)+g\tau_c+\tau_d^*\right) \\
0&=r_l\left(\delta,\tau_d^*,\tau_c^*\right)+\nabla V_l^{*T}\left(S(\delta)+g\tau_c^*+\tau_d^*\right)
\end{aligned} \tag{22}
$$

Substituting Equations (18) and (22) into Equation (24) yields:

$$\begin{aligned}0&=\frac{1}{4}\nabla V_f^{\ *}gO^{-1}g^T\nabla V_f^{\ *T}-\delta^T M\delta-\frac{1}{4}\nabla V_l^{*}gN^{-1}g^T\nabla V_l^{*T}+\nabla V_f^{\ *}S(\delta)-\frac{1}{2}gN^{-1}g^T\nabla V_l^{*T}-\frac{1}{2}gO^{-1}g^T\nabla V_f^{\ *T}\\0&=\delta^T M\delta+\frac{1}{4}\nabla V_l^{*}gN^{-1}g^T\nabla V_l^{*T}+2\vartheta^T M\delta-\vartheta^T\nabla S\nabla V_f+\nabla V_f^{\ *T}S(\delta)-\frac{1}{2}gN^{-1}g^T\nabla V_l^{*T}-\frac{1}{2}gO^{-1}g^T\nabla V_f^{\ *T}\end{aligned}\tag{23}$$

4. Algorithmic solution

This section employs the Stackelberg game approach in the solution process by means of an innovative NN-based IRL algorithm with a learning structure for action evaluation, which includes an additional NN to efficiently deal with the complexity arising from the unknown dynamics [13].

First, the optimal evaluation NN is developed to simulate the follower's optimized maximal function V_f^*. V_f^* and its gradient $\nabla V_f^{\ *}$ are represented as follows:

$$V_f^*=W_{fc}^{*T}\chi_{fc}+\upsilon_{fc},\nabla V_f^{\ *}=\nabla\chi_{fc}^T W_{fc}^*+\nabla\upsilon_{fc}\tag{24}$$

where $W_{fc}^*\in\Re^N$ is the optimal evaluation weights of the follower. $\chi_{fc}=\chi_{fc}(\delta)$ and $\upsilon_{fc}=\upsilon_{fc}(\delta)$ are the regression function and the approximation error.

The $\hat{V}_f$ to V_f^* estimate is the same as the $\nabla\hat{V}_f$ to ∇V_f^* estimate:

$$\hat{V}_f=\hat{W}_{fc}^T\chi_{fc},\nabla\hat{V}_f=\nabla\chi_{fc}^T\hat{W}_{fc}\tag{25}$$

where $\nabla\hat{V}_f=\partial\hat{V}_f/\partial\delta$. To ensure the stability of the attack strategy and the update of the value function, as opposed to employing the single update law, the follower's action NN is designed to be:

$$\hat{\tau}_d=-\frac{1}{2}O^{-1}g^T\nabla\chi_{fc}^T\hat{W}_{fa}\tag{26}$$

where $\hat{W}_{fa}$ is the actor weight used to estimate weight W_{fc}^*.

In order to avoid using unknown dynamics $\eta(\delta)$ throughout the learning process. the Bellman error equation for an arbitrary time integral1 is introduced as follows:

$$\sigma_{fc}^B=\int_{t-\eta}^{t}r_f(\delta,\hat{\tau}_d,\hat{\tau}_c)ds+\hat{W}_{fc}^T\Delta\chi_{fc}\tag{27}$$

where $r_f(\delta,\hat{\tau}_d,\hat{\tau}_c)=\hat{\tau}_d^T N\hat{\tau}_d-\delta^T M\delta-\hat{\tau}_c^T O\hat{\tau}_c$, and $\Delta\chi_{fc}=\chi_{fc}(t)-\chi_{fc}(t-\eta)$.

The objective function $\Sigma_{fc}=1/2\left(\sigma_{fc}^B\right)^2$ is set to minimize the error σ_{fc}^B by adjusting the critical weights W_{fc}^*. The ideal estimated update law for weights W_{ic}^* using the gradient descent method is:

$$\dot{\hat{W}}_{fc}=-\alpha_{fc}\frac{\Delta\varphi_{fc}}{(1+\Delta\varphi_{fc}^T\Delta\varphi_{fc})^2}\sigma_{fc}^B\tag{28}$$

where α_{fc} denotes the critic learning rate of the follower.

Second, since $\dot{\lambda}$ and ξ contain unknown information $\eta(\delta)$, a NN must be introduced to approximate, the unknown dynamics $\eta(\delta)$ and its gradient1 $\nabla\eta$ simulated by the NN are:

$$\eta(\delta)=\left(W_\eta^*\right)^T\chi_\eta+\upsilon_\eta,\nabla\eta=\nabla\chi_\eta^T W_\eta^*+\nabla\upsilon_\eta\tag{29}$$

where $W_\eta^{\ *}\in\Re^{N\times 3}$ is the bounded optimal weight of the dynamics $\eta(\delta)$, i.e., $\left\|W_\eta^*\right\|_F\le w_\eta$. $\chi_\eta=\chi_\eta(\delta)$. $\upsilon_\eta=\upsilon_\eta(\delta)$ denotes the bounded regression function and the approximation error respectively, i.e., $\left\|\chi_\eta\right\|\le b_{\chi\eta}$, $\left\|\upsilon_\eta\right\|\le b_{\upsilon\eta}$. $\nabla\chi_\eta=\partial\chi_\eta/\partial\delta$, $\nabla\upsilon_\eta=\partial\upsilon_{\eta h}/\partial\delta$.

Similarly, the estimation of the unknown dynamics and its gradient is denoted as:

$$\hat{\eta}(\delta)=\hat{W}_\eta^T\chi_\eta,\nabla\hat{\eta}=\nabla\chi_\eta^T\hat{W}_\eta\tag{30}$$

where $\nabla\hat{\eta} = \partial\hat{\eta}(\delta)/\partial\delta$. Item $\dot{\lambda}$ and ϑ can be redefined as follows:

$$\dot{\lambda} = 2M\delta - \hat{W}_\eta^T \nabla\chi_\eta \nabla\chi_{fc}^T \hat{W}_{fa}$$
$$\dot{\vartheta} = -\left(\frac{\partial H_2}{\partial\lambda}\right)^T = \frac{1}{2} gO^{-1} g \nabla\chi_{fc}^T \hat{W}_{fa} + \nabla\chi_\eta^T \hat{W}_\eta \vartheta \tag{31}$$

In order to fully utilize the state information and enhance the versatility of the vehicle, the optimal value function V_l^* is decomposed as follows:

$$V_l^* = \delta^T \Gamma_\delta \delta + 2(\sigma^T \Upsilon(\xi) + \hat{\eta}(\delta)^T)\Gamma_{\sigma\eta}\delta + \bar{V}_l^* \tag{32}$$

where $\bar{V}_l^* = V_l^* - \delta^T \Gamma_\delta \delta - 2(\sigma^T \Upsilon(\xi) + \hat{\eta}(\delta)^T)\Gamma_{\sigma\eta}\delta$, and $\Gamma_\delta = \Gamma_2 \Gamma_{\sigma\eta}$, $\Gamma_{\sigma\eta} = (g^T)^{-1} N^T g^{-1}$. By substituting the partial differentiation of Equation (32) with respect to δ into Equation (20), we get:

$$\tau_c^* = -g^{-1}(\Gamma_2\delta + \Upsilon(\xi)^T \sigma + \hat{W}_\eta^T \varphi_\eta) - \frac{1}{2} N^{-1} g^T \nabla\bar{V}_l^* \tag{33}$$

We design the optimal leader evaluation NN to model the value function and its gradient:

$$\bar{V}_l^* = W_{lc}^{*T} \chi_{lc} + \upsilon_{lc}, \nabla\bar{V}_l^* = \nabla\chi_{lc}^T W_{lc}^* + \nabla\upsilon_{lc} \tag{34}$$

where $W_{fc}^* \in \mathbb{R}^N$ is the leader's bounded optimal evaluation weight, i.e., $\|W_{lc}^*\| \le w_{lc}$. $\chi_{lc} = \chi_{lc}(\delta)$ and $\upsilon_{lc} = \upsilon_{lc}(\delta)$ denote the regression function and the approximation error. The approximations of $\bar{V}_l^*$ and $\nabla\bar{V}_l^*$ are:

$$\hat{\bar{V}}_l = \hat{W}_{lc}^T \chi_{lc}, \nabla\hat{\bar{V}}_l = \nabla\chi_{lc}^T \hat{W}_{lc} \tag{35}$$

The leader's action NN is denoted as:

$$\hat{\tau}_c = -g^{-1}(\Gamma_2\delta + \Upsilon(\xi)^T \sigma + \hat{W}_\eta^T \chi_\eta) - \frac{1}{2} N^{-1} g^T \nabla\chi_{lc}^T \hat{W}_{la} \tag{36}$$

where $\hat{W}_{la}$ is the action weight used to estimate weight W_{lc}^*.

The Bellman error equation for the leader is introduced as:

$$\sigma_{lc}^B = \int_{t-\eta}^{t} r_l(\delta, \hat{\tau}_d, \hat{\tau}_c) ds + \hat{W}_{lc}^T \Delta\chi_{lc} \tag{37}$$

where $r_l(\delta, \hat{\tau}_d, \hat{\tau}_c) = \delta^T M \delta + \tau_c^T N \tau_c + \Lambda(\tau_d)$ and $\Delta\chi_{lc} = \chi_{lc}(t) - \chi_{lc}(t-\eta)$.

$\Sigma_{lc} = 1/2(\sigma_{lc}^B)^2$ and $\Sigma_\eta = 1/2\delta^T\delta$ are set to facilitate the approximation of functions $\bar{V}_l^*$ and $\eta(\delta)$. The update law used for the ideal estimates of weights W_{lc}^* and W_η^* is derived as:

$$\dot{\hat{W}}_{lc} = -\alpha_{lc} \frac{\Delta\chi_{lc}}{(1 + \Delta\chi_{lc}^T \Delta\chi_{lc})^2} \sigma_{lc}^B \tag{38}$$

$$\dot{\hat{W}}_\eta = \alpha_\eta \chi_\eta \delta^T - \gamma_\eta \alpha_\eta \hat{W}_\eta \tag{39}$$

where α_{lc} and α_η denote the evaluation learning rate of the leader and $\eta(\delta)$. γ_η is a positive constant.

To ensure the stability of the policy update, the update law of the actor weights is formulated as:

$$\dot{\hat{W}}_{fa} = -\alpha_{fa}[\eta \nabla\chi_{fc} \nabla\chi_\eta^T \hat{W}_\eta \vartheta \Delta\chi_{lc}^T \hat{W}_{lc} m_2^{-2} - \eta Z_{fc} \nabla\chi_{fc}^T \hat{W}_{fa} \Delta\chi_{fc}^T \hat{W}_{fc} m_1^{-2} - 2^{-1} Z_{fc}\delta] - \gamma_{fa}\alpha_{fa}\hat{W}_{fa} \tag{40}$$

$$\dot{\hat{W}}_{la} = 2^{-1}\alpha_{2a}[2^{-}\eta Z_{lc} \nabla\chi_{lc}^T \hat{W}_{la} \Delta\chi_{lc}^T \hat{W}_{lc} m_2^{-2} + Z_{lc}\delta] - \gamma_{la}\alpha_{la}\hat{W}_{la} \tag{41}$$

where α_{fa} and α_{la} are participants action learning rates, $Z_{fc} = \nabla\chi_{fc} g O^{-1} g^T$, and $Z_{lc} = \nabla\chi_{lc} g N^{-1} g^T$. γ_{fa} is bounded. $\|Z_{fc}\| \le d_{fc}$, $\|Z_{lc}\| \le d_{lc}$, and γ_{la} is a positive parameter. $m_1 = 1 + \Delta\chi_{fc}^T \Delta\chi_{fc}$, and $m_2 = 1 + \Delta\chi_{lc}^T \Delta\chi_{lc}$.

5. Simulation verification

We add wind field to the system as a perturbation term. The controller is used to carry out simulation on the control of static and dynamic target value. The control effect is observed and analyzed.

The initial values are set: $\varphi(0)$ = -0.3, $\theta(0)$ = 0.4, and $\psi(0)$ = -0.6. The target is set: ϕ_t = 1, θ_t = 0, and ψ_t = -0.2. The simulation step size is T=0.05, and the time is 25 s. The simulation results are shown in Figure 2.

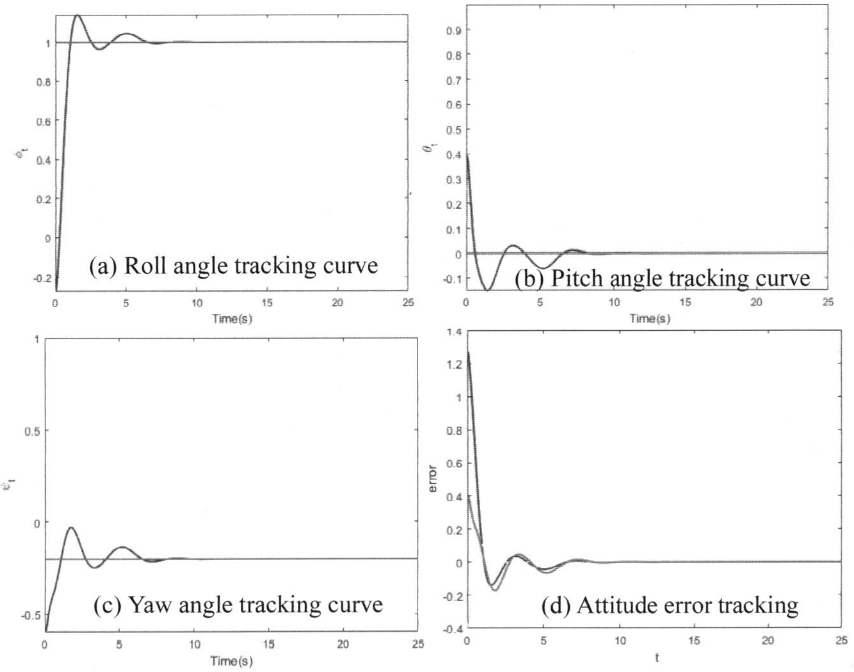

Figure 2. Simulation effect of static target value of attitude angle under wind field disturbance.

When tracking Roll, the rise time is 0.8s, the peak time is 1.1s, and the adjustment time is 1.9s. When tracking Pitch, the rise time is 0.7s, the peak time is 1.5s, and the adjustment time is 2.2s. When tracking Yaw, the rise time is 1s, the peak time is 1.9s, and the adjustment time is 2.5s.

The target is set to dynamically changing values: ϕ_t=0.2cost, θ_t=0.8sint, and ψ_t=-0.2sint. Other parameters remain unchanged. The simulation results are shown in Figure 3.

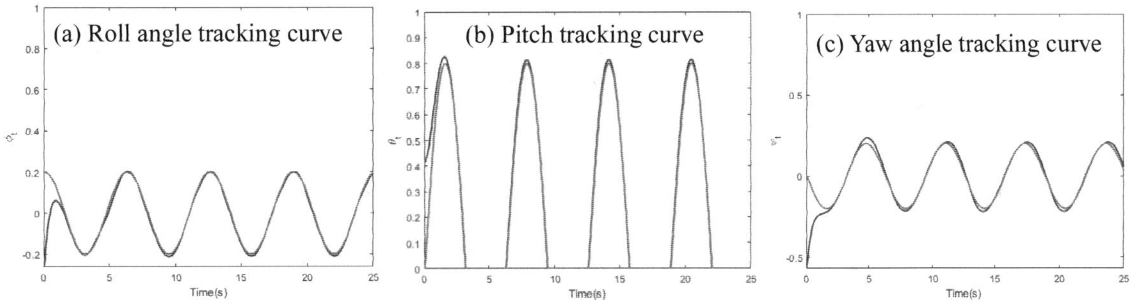

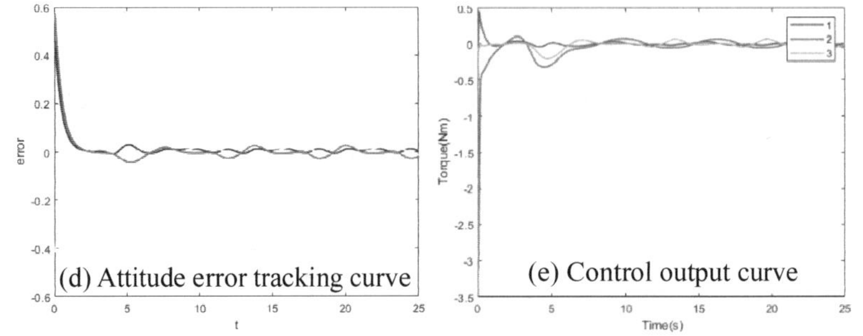

Figure 3. Simulation effect of the dynamic target value of attitude angle under wind field disturbance.

Under wind interference, the stabilization control curves of the attitude angles of each attitude angle do not undergo obvious vibration and rapidly converge to the target value. The error tends to be 0. All the signals are bounded. Accurate and stable tracking is achieved. The controller enables the attitude angle to effectively counteract the influence of external disturbances while keeping the tracking error within an acceptable range, which can well realize the purpose of stable control.

6. Conclusion

This paper focuses on the attitude control of a quadrotor UAV under wind. According to the analytically established UAV dynamics model with wind field perturbation, the attitude control algorithm of the quadrotor is designed based on the Stackelberg game theory. The optimal strategy in the game is solved using reinforcement learning and NN simulation. Finally, a simulation is carried out. The results verify the performance of the designed attitude controller.

References

[1] XU W, HUANG H. (2019) Mechanism analysis and solution of high voltage corona [J]. Radar and countermeasures, 39(01): 49-53+57. DOI: 10.19341/j.cnki.issn.1009-0401.2019.01.011.

[2] SUI Y, NING P F, NIU P J, et al. (2021) A review of research on mounted UAV power inspection technology for overhead transmission lines. Grid technology, 45(09): 3636-3648. DOI: 10.13335/j.1000-3673.pst.2020.1178.

[3] SHAO G W, LIU Z, FU J, et al. (2020) Research Progress of Overhead Transmission Line UAV Inspection Technology. High voltage technology, 46(01): 14-22. DOI: 10.13336/j.1003-6520.hve.20191227002.

[4] Nam D B, Thanh T N V, Duong M P. (2022) Lyapunov-Based Nonlinear Model Predictive Control for Attitude Trajectory Tracking of Unmanned Aerial Vehicles. International Journal of Aeronautical and Space Sciences, 24(2): 502-513.

[5] HOU T H, XING H Y, LIU Y. (2019) Orthogonal Wind Pressure Vector Decomposition Wind Measurement Based on Multi-rotor UAVs. Journal of Instrumentation, 40(10): 8. DOI: CNKI: SUN: YQXB.0.2019-10-023.

[6] ZHANG M. (2024) Application of machine vision technology for UAVs in inspection of ultra-ultra-high voltage transmission lines. Electronics, 53(09): 386-387.

[7] Biao X, Qingfeng M, Jianxin F, et al. (2023) Fault Tolerant Control of Drone Interceptors Using Command Filtered Backstepping and Fault Weighting Dynamic Control Allocation. Drones, 7(2): 106-106.

[8] SHEN J H. (2022) Wind disturbance impact analysis and wind-resistant control research on quadrotor UAV. Nanjing University of Aeronautics and Astronautics. DOI: 10.27239/d.cnki.gnhhu.2022.000614.

[9] Yang C, Cai X, Wu L, et al. (2024) Addressing Launch and Deployment Uncertainties in UAVs with ESO-Based Attitude Control. Drones, 8(8): 363-363.

[10] Najm A A, Ibraheem I K. (2020) Altitude and Attitude Stabilization of UAV Quadrotor System using Improved Active Disturbance Rejection Control. Arabian Journal for Science and Engineering, 45(3): 1985-1999. DOI: 10.1007/s13369-020-04355-3.
[11] Zhao Y, Ma Y, Hu S. (2021) USV Formation and Path-Following Control via Deep Reinforcement Learning with Random Braking. IEEE Transactions on Neural Networks and Learning Systems, PP (99): 1-11. DOI: 10.1109/TNNLS.2021.3068762.
[12] Shao S, Chen M, Hou J, et al. (2021) Event-Triggered-Based Discrete-Time Neural Control for a Quadrotor UAV Using Disturbance Observer. IEEE /ASME Transactions on Mechatronics, PP (99): 1-1. DOI: 10.1109/TMECH.2021.3051835.
[13] Yan Y, Cao H, Zhang B, et al. (2024) Large-Scale Solar-Powered UAV Attitude Control Using Deep Reinforcement Learning in Hardware-in-Loop Verification. Drones, 8(9): 428-428.

Research on nonlinear vibration characteristics of counter-rotating dual-rotor with multiple squeeze film dampers

Yan Li [1,2], Yan Ouyang [1,2], Huisong Wu [1,2], Xiping Kou [1,2] and Su Jia [1,2*]

[1] High Speed Aerodynamics Institute, China Aerodynamics Research and Development Center, Mianyang 621000, China

[2] National Key Laboratory of Aerospace Physics in Fluids, China Aerodynamics Research and Development Center, Mianyang 621000, China

[*] Corresponding author: Su Jia, Email: jiasutju@163.com

Abstract: In order to investigate the nonlinear vibration characteristics of a dual-rotor system with multiple squeeze film dampers, an aero-engine dual-rotor system with a small bypass ratio is taken as the research object. The dynamic model of dual-rotor with squeeze film dampers at 1#, 3#, and 5# Elastic supports is established. Furthermore, the nonlinear vibration and bistable characteristics of a dual-rotor system with multiple squeeze film dampers are analyzed and verified. The result shows that the degree of nonlinearity at different dampers mainly depends on the vibration mode shape and unbalance of the dual-rotor system. In order to reduce the nonlinear vibration, it is suggested to distribute the residual unbalance according to the mode orthogonality principle and choose the mode with the dominant vibration of the dampers in the aero-engine assembly process. What is more, an optimization method of residual unbalance is proposed and verified to be effective and reliable.

1. Introduction

The squeeze film damper (SFD) has been widely used in aero-engine because of its simple structure and good vibration reduction effect. However, the whirling film brings strong nonlinear problems to the rotor. At present, the dual-rotor has been the main structure of military or civil aero-engines in the world. In order to pursue better thrust-weight ratio performance, the structure of this aero-engine is becoming more and more flexible, with the vibration problem becoming more and more prominent. Minimizing the vibration of dual-rotor systems has become one of the most important things for engine designers. Therefore, it is necessary to install SFDs at multiple elastic supports, which undoubtedly further increases the nonlinear factor and leads to more complex vibration.

The bistable characteristic is a typical nonlinear vibration phenomenon caused by SFD [1], which is characterized by the existence of multiple steady-state solutions of rotor vibration in a specific interval, which leads to the vibration skip between these solutions during the operation of the rotor system, threatening the health of rotor seriously.

A series of works on rotor systems with SFDs have been developed by researchers [2-6]. Gunter et al. [3] investigated the design of SFD and established an SFD design method that can be used for the simple symmetric rotor. Hahn et al. [4] carried out simulation and experimental research on the stability of SFD and calculated the response of the rotor system with two dampers based on the harmonic balance method. Then a method based on Floquet theory to judge the damper stability is presented in this paper. The nonlinear vibration characteristics of rotor systems with SFD through experiments are studied and the influence of the static eccentricity of a damper on the unbalanced response is analyzed [5-8]. It can be seen that the nonlinear problem caused by SFDs has always been a hot spot for scholars, but on this basis, the nonlinear vibration of counter-rotating dual-rotors with multiple SFDs needs to be further studied.

In order to investigate the nonlinear vibration characteristics of a dual-rotor system with multiple SFDs, an aero-engine dual-rotor system with a small bypass ratio is taken as the research object, and the dynamic model of dual-rotor with squeeze film dampers at 1#, 3#, and 5# Elastic supports is established. In addition, the nonlinear vibration and bistable characteristics of dual-rotor systems with multiple squeeze film dampers are analyzed and verified. The design suggestion of residual unbalance according to the mode orthogonality principle has been developed, which is of great engineering practice.

2. Dynamic model of the dual-rotor with multiple SFDs

2.1. Dynamic model of squeeze film dampers

Figure 1 is an SFD dynamic model [1]. O_b stands for the initial position of the journal, O_j for the position of the journal after whirling, q_x and q_y for the displacement components in X and Y directions respectively, e for the relative eccentricity of SFD, Ω for the changing rate of the rotation angle, θ for the film simulation angle, and F_t and F_r for the relative tangential force and radial force of the film acting on shaft respectively.

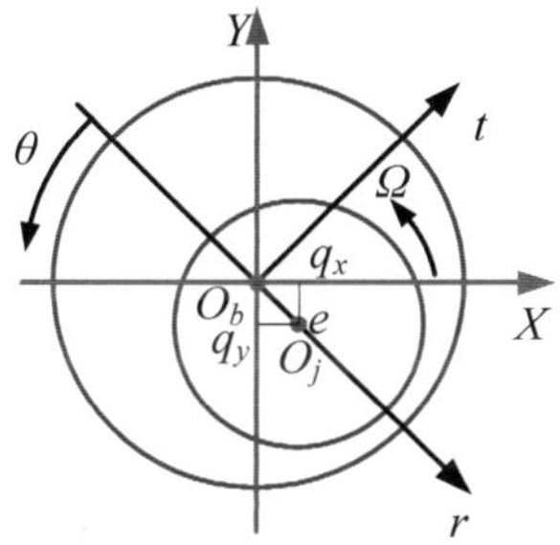

Figure 1. Dynamic model of SFD [1].

Based on the classical semi-fuel film model of short bearings [1], the expressions of relative tangential force and radial force of the film are as follows:

$$\begin{cases} F_r = \dfrac{\mu R L^3}{C^2}\left[\dfrac{\pi\dot{\varepsilon}(1+2\varepsilon^2)}{2(1-\varepsilon^2)^{2.5}} + \dfrac{2\Omega\varepsilon^2}{(1-\varepsilon^2)^2}\right] \\ F_t = \dfrac{\mu R L^3}{C^2}\left[\dfrac{\pi\dot{\varepsilon}\varepsilon}{(1-\varepsilon^2)^2} + \dfrac{\pi\Omega\varepsilon}{2(1-\varepsilon^2)^{1.5}}\right] \end{cases} \tag{1}$$

$$\varepsilon = \frac{\sqrt{q_x^2+q_y^2}}{C} \tag{2}$$

$$\dot{\varepsilon} = \frac{\dot{q}_x\cos\varphi + \dot{q}_y\sin\varphi}{C}, \quad \cos\varphi = \frac{q_x - x_s}{\sqrt{q_x^2+q_y^2}}, \quad \sin\varphi = \frac{q_y - y_s}{\sqrt{q_x^2+q_y^2}} \tag{3}$$

$$\Omega = \frac{-\dot{q}_x\sin\varphi + \dot{q}_y\cos\varphi}{\sqrt{q_x^2+q_y^2}} \tag{4}$$

where μ is the viscosity of lubricating oil; R is the semidiameter of the film; L is the length of the film, and C is the gap of the film. Therefore, the film force acting on the counter-rotating dual-rotor can be expressed as Equation (5).

$$\begin{cases} F_x^L = |F|\cos\varphi + F_f \sin\varphi \\ F_y^L = |F|\sin\varphi - F_f \cos\varphi \\ F_x^H = -|F|\cos\varphi - F_f \sin\varphi \\ F_y^H = -|F|\sin\varphi + F_f \cos\varphi \end{cases} \tag{5}$$

2.2. Dynamic model of counter-rotating double-rotor system with multiple SFDs

Taking an aero-engine with a small bypass ratio as a prototype, the finite element dynamic model of a similar test rig is established, as shown in Figure 2. The low-pressure rotor is 1-1-1 supporting structure, and 1-0-1 supporting structure for the high-pressure rotor. What is more, the 1#, 3#, and 5# supports are elastic with SFDs installed. The key parameters of elastic support elements and SFDs are shown in Tables 1 and 2. Other detailed information is shown in [9, 10].

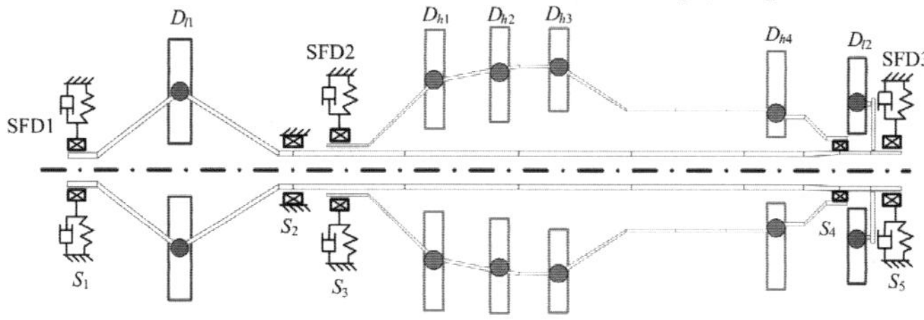

Figure 2. Finite element dynamic model of dual-rotor system.

Table 1. Parameters of elastic support elements.

Parameters	S_1	S_2	S_2	S_2	S_5
K/(N • m^{-1})	6.01×10^6	6.03×10^8	7.99×10^6	3.12×10^7	6.52×10^6
d/(Ns • m^{-1})	300	300	300	300	300

Table 2. Parameters of SFDs.

Parameters	R/mm	C/mm	L/mm	μ/(Pa·s)
SFD_1	62	0.24	9.8	0.022
SFD_2	63	0.235	10.4	0.022
SFD_3	78.6	0.21	9.1	0.022

According to Lagrange equation [8], the differential equation of motion of a dual-rotor system can be expressed as:

$$\boldsymbol{M}\ddot{\boldsymbol{q}} + \boldsymbol{C}\dot{\boldsymbol{q}} + \boldsymbol{K}\boldsymbol{q} = \boldsymbol{Q} \tag{6}$$

where $\boldsymbol{M}$ stands for the mass matrices of dual-rotor, $\boldsymbol{C}$ for damping, and $\boldsymbol{K}$ for stiffness. $\boldsymbol{q} = \left[\boldsymbol{q}^1 \ \boldsymbol{q}^2 \ \cdots \ \boldsymbol{q}^n\right]^{\mathrm{T}}$, $\boldsymbol{q}^n = \left[x \ y \ \theta_x \ \theta_y\right]^{\mathrm{T}}$, $\boldsymbol{Q} = \left[\boldsymbol{Q}^1 \ \boldsymbol{Q}^2 \ \cdots \ \boldsymbol{Q}^n\right]^{\mathrm{T}}$, and $\boldsymbol{Q}^n = \left[F_x \ F_y \ M_x \ M_y\right]^{\mathrm{T}}$. The upper corner number stands for the element node number. We submit the force of the SFDs in Equation (5) into the right end of Equation (6). Then the vibration response analysis of the dual-rotor system with multiple SFDs can be developed.

2.3. Dynamic characteristics of counter-rotating dual-rotor system

Based on the established dual-rotor dynamic model, the dynamic characteristics and vibration response are analyzed. The first three-order mode shape excited by a frequency equal to the low-pressure rotating speed are shown in Figure 3, where the solid lines stand for the mode shape of the dual-rotor, the dotted lines the moving trajectory, the triangle symbol represents the supports, and the circular symbol represents for the disk. Similarly, the first four-order mode shape excited by a frequency equal to the high-pressure rotating speed is shown in Figure 4.

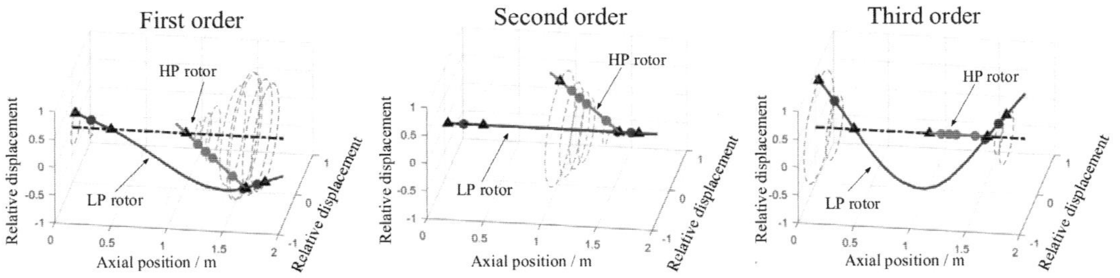

Figure 3. Mode shape of dual-rotor system with low-pressure rotor excitation.

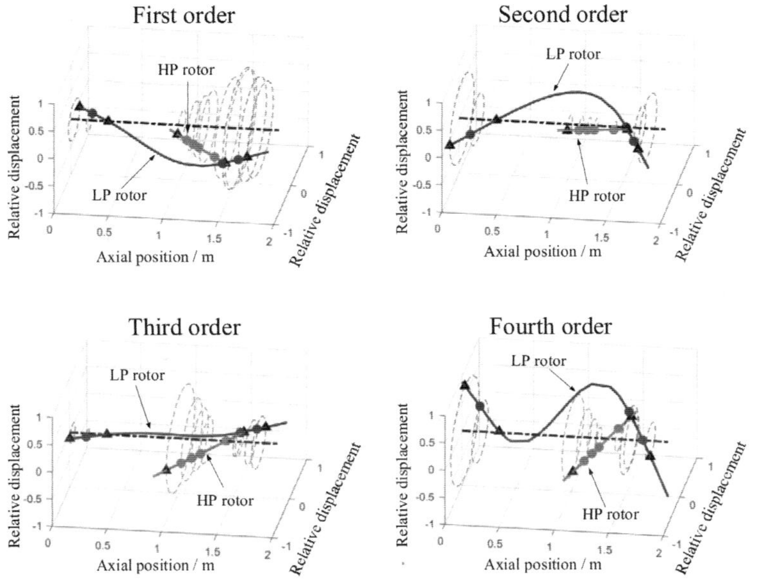

Figure 4. Mode shape of dual-rotor system with high-pressure rotor excitation.

3. Bistable Characteristics of dual-rotor with multiple SFDs

For the dual-rotor system with squeeze film dampers, if the unbalance exceeds the designed value, there will be strong nonlinearity in vibration characteristics because of the large eccentricity ratio of dampers, which will lead to the phenomenon of 'bistability'. Taking the vibration excitation by a low-pressure system as an example, Figure 5 shows the vibration response during acceleration and deceleration when the unbalance of the low-pressure turbine disk is 50g•cm, where Figure 5(a) is the simulation results. Figure 5(a) shows experimental results. The rate of rotating speed variation of the low-pressure rotor is 25 r/s^2. The vibration results are measured on a low-pressure turbine disk. It can be seen that the simulation results are in good agreement with the experimental results, the error of critical speed is less than 1%, and the error of vibration amplitude peak is 4.3%. In addition, the eccentricity ratio variation of the SFDs during acceleration and deceleration is shown in Figure 6. It is clear that the eccentricity ratio of SFD_3 is larger than 0.4 at the first-order critical speed. If the unbalance continues to increase, the phenomenon of 'bistability' may be found in the vibration results due to the excessive eccentricity ratio of SFD_3.

doi:10.1088/1742-6596/2992/1/012028

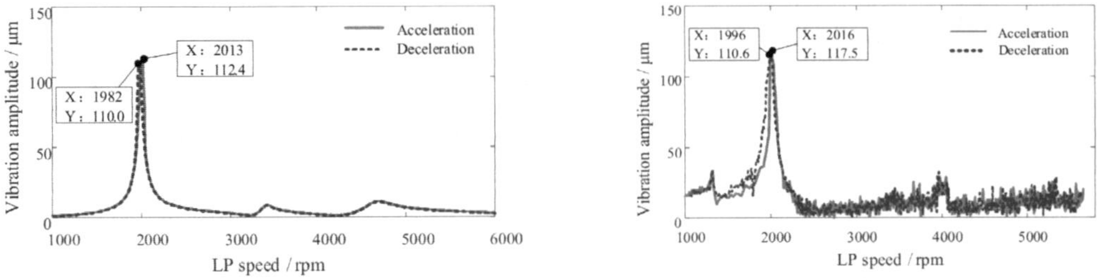

(a) (b)

Figure 5. Mode shape of dual-rotor system with low-pressure rotor excitation.

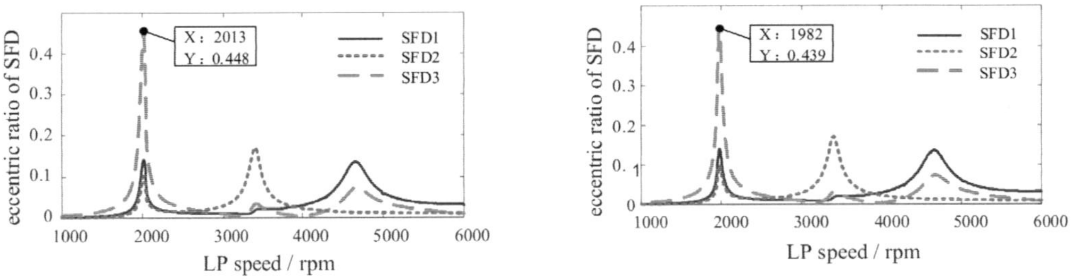

(a) (b)

Figure 6 Eccentricity ratio variation of the SFDs (U=50 g•cm).

Figure 7 shows the comparison between simulation and experimental results of vibration response during acceleration and deceleration when the unbalance of the turbine disk increases to 100 g•cm. Correspondingly, Figure 8 shows the eccentricity ratio variation of SFD_3. The results show that when the damper eccentricity ratio is larger than 0.7, the vibration response of the dual-rotor system shows a "bistable" phenomenon. Under the action of film stiffness, both the critical speeds during acceleration and deceleration have increased. In this case, there are multiple steady-state response solutions for the vibration of the dual-rotor system at the same speed. However, because the acceleration and deceleration of the rotor are continuously changing processes, the rotor response will first pass through the 'bistable' region along the continuous solution path, and then 'jump' to the next continuous steady-state solution.

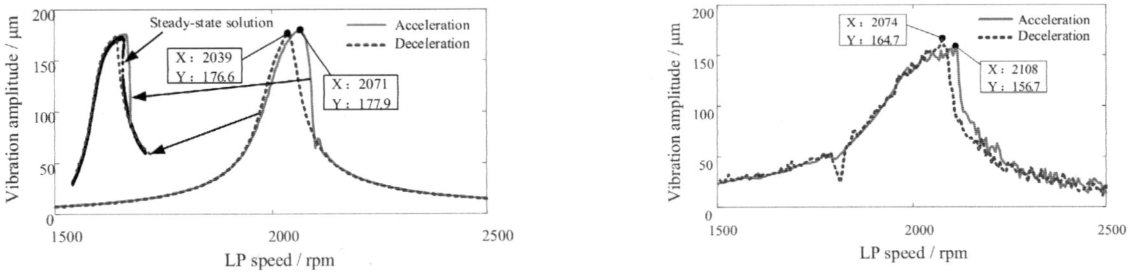

(a) (b)

Figure 7 Comparison of vibration response between simulation and experimental results (U = 100 g•cm).

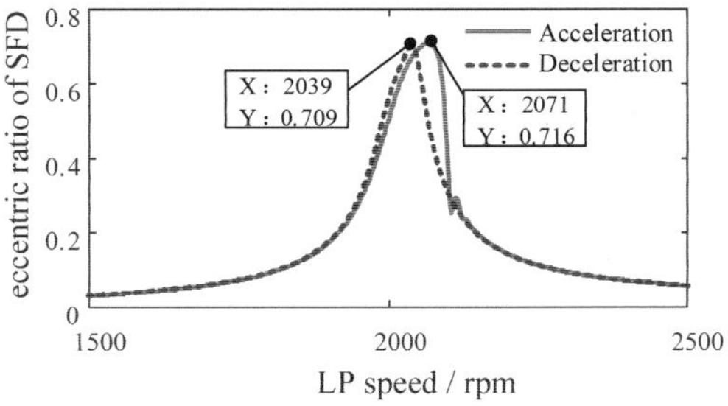

Figure 8. Eccentricity ratio variation of SFD_3 (U=100 g • cm).

When the unbalance of the LP turbine disk increases to 150 g•cm, the simulation and experimental results of vibration response are shown in Figure 9. In this situation, the critical speed of the dual-rotor system increases by 11.2% with the nonlinear action of SFDs, which is close to the design value of 15% for the safety margin in the design process of the aero-engine working range. A comprehensive analysis of Figure 7 to Figure 9 indicates that with the increase of unbalance and the eccentricity ratio of SFD, the 'bistable' skip speed of the dual-rotor system increases, and also the skip speed difference between acceleration and deceleration increases. More importantly, the vibration fluctuation after the skip becomes more violent. Even in this situation, the error of speed and vibration amplitude between simulation and experiment is less than 5%.

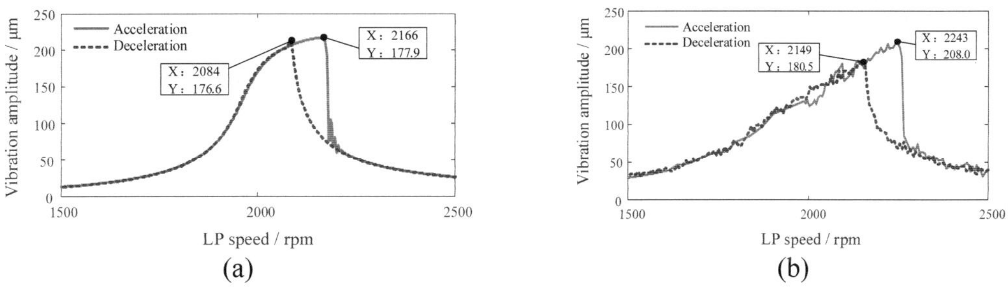

Figure 9 Comparison of vibration response between simulation and experimental results (U = 150 g•cm).

Similarly, SFD_2 and SFD_1 may also lead to 'bistability' at other critical speeds. Figure 10 shows the vibration response of the rotor system when "bistability" is observed in SFD_2 and SFD_1. The simulation parameters in Figure 10(a) show 60g•cm of unbalance on the first-stage disk of HP, with the eccentricity ratio variation of SFD_2 illustrated in Figure 10(b). Because the 3# suppor accounts for a large proportion in the third order mode shape with HP excitating, SFD_2 plays a major role in vibration reduction. Its eccentricity ratio has exceeded 0.7. Additionally, the simulation parameters in Figure 10(c) indicate an unbalance of 80 g•cm on the LP fan disk, with variations in the eccentricity ratio for SFD_1 and SFD_3, as shown in Figure 10(d). What is different is that the 1# suppor accounts for a large proportion in the third-order mode shape with LP exciting, so SFD_1 plays a major role in vibration reduction. As can be seen from Figure 10(d), when the eccentricity ratio of SFD_1 exceeds 0.7 and 'bistability' appears, although the eccentricity ratio of SFD_3 is still less than 0.4, 'bistability' still appears under the influence of SFD_1. It should be noted that the experimental conditions in Figure 10 are greatly higher than the damper design point, also the rotating speed is close to the design limit. Considering the experimental safety problem, no further experimental study has been carried out.

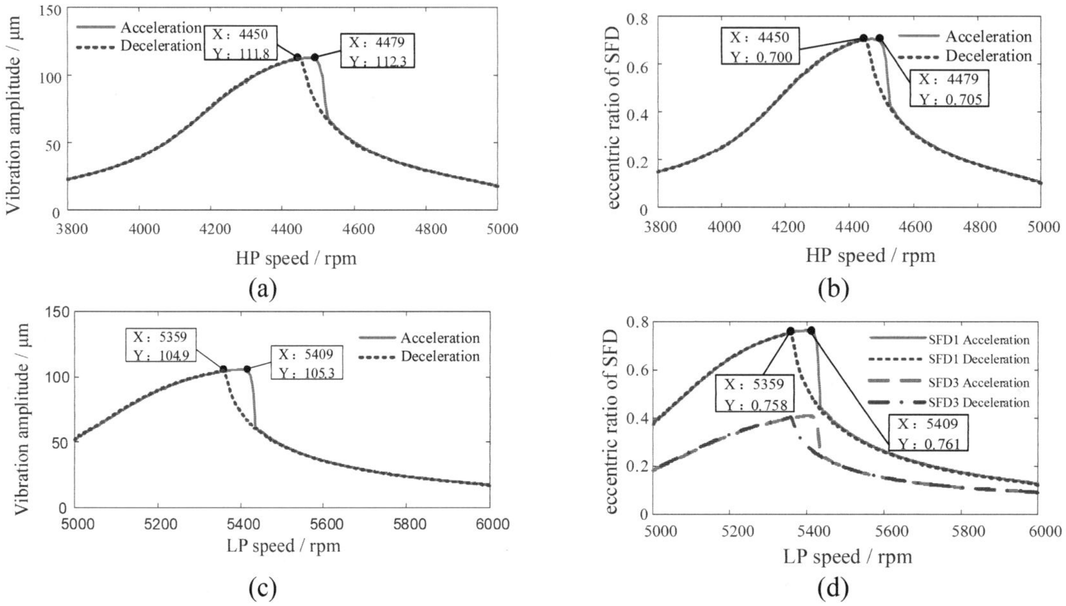

Figure 10 Vibration response of SFD_1 and SFD_2 system with bistable phenomenon.

Obviously, because of the multiple dampers in the dual-rotor system and many critical speeds in the working range, when the eccentricity ratio of any damper turns out to be too large, the vibration of the system will show the 'bistable' phenomenon. Therefore, the unbalanced distribution should be strictly controlled according to the damper position and the modal shape of the dual rotor to reduce the nonlinearity of the damper.

4. Residual unbalanced distribution based on modal orthogonality

The above research shows that the nonlinearity of the squeeze film damper is closely related to the unbalanced mass distribution on the rotor. For a newly produced engine, all levels of disks will generally carry out strict dynamic balance, but there will still be a certain amount of residual unbalance. When the residual unbalance on the multi-stage disks act together, violent vibration will be caused in the dual-rotor system.

Considering that the rotor vibration induced by unbalanced distribution is directly related to the mode shape, the optimization Equation (7) of residual unbalanced mass distribution of double rotors is constructed.

$$\begin{cases} f_{uL} = \min\left(\sum_{i=1}^{n} \zeta_i [\boldsymbol{\Phi}_i]^T U_L\right), & \zeta_i \in [0,1] \\ f_{uH} = \min\left(\sum_{i=1}^{n} \xi_i [\boldsymbol{\Phi}_i]^T U_H\right), & \xi_i \in [0,1] \end{cases} \tag{7}$$

In Equation (7), U is the unbalanced mass distribution, and $\boldsymbol{\Phi}_i$ is the i-th mode shape of the double rotor. The subscript L stands for low pressure and H for high pressure. ζ_i is optimization weight of the unbalance for the i-th LP excitation mode and ξ_i for the i-th HP excitation mode. When the unbalanced mass distribution of the HP/LP rotor system minimizes the value of Equation (7), the overall vibration of the dual-rotor system will reach the optimum.

Taking the HP system as an example, it is assumed that there are four unbalanced mass distribution sections on the high-pressure rotor, which are U_1, U_2, U_3, and U_4 respectively. According to Equation (7), the optimization objective of residual unbalanced mass distribution on the HP rotor can be expressed as Equation (8).

$$f_{uH} = \min\left(\sum_{i=1,3,4}^{n} \xi_i \begin{bmatrix} r_{i,1} & r_{i,2} & r_{i,3} & r_{i,4} \end{bmatrix}^T U_H \right) \begin{bmatrix} U_1 & U_2 & U_3 & U_4 \end{bmatrix} \tag{8}$$

According to the characteristics of each mode shape in Figure 5 and the influence of SFDs, ξ_i is taken as [0.2, 0, 1, 0.6]. $\begin{bmatrix} r_{i,1} & r_{i,2} & r_{i,3} & r_{i,4} \end{bmatrix}$ is the relative displacements of the four disks in the i-th mode shape. $\begin{bmatrix} U_1 & U_2 & U_3 & U_4 \end{bmatrix}$ is the residual unbalance vector of the four disks.

Table 3 shows the distribution of unbalanced mass on the high-pressure rotor. The phase of random distribution is randomly selected. The phase of optimal distribution is obtained by solving Equation (8). According to Table 3, the vibration response can be simulated, as shown in Figure 11. Figure 11(a) shows the vibration response with a random distribution of residual unbalance. Figure 11(b) is the optimized distribution. The result shows that the unbalanced response of the dual-rotor system has been greatly improved, with a maximum vibration reduction ratio of 69.2% in the first mode, 99% in the third mode, and 88.9% in the fourth mode.

Table 3. Residual unbalanced mass distribution of high-pressure rotors.

Disks	U_1	U_2	U_3	U_4
Residual unbalance /(kg m)	1.2×10^{-4}	1.3×10^{-4}	8×10^{-5}	1.6×10^{-4}
Randomly distributed phase/degree	0	20	0	100
Optimized distribution phase/degree	0	202	120	340

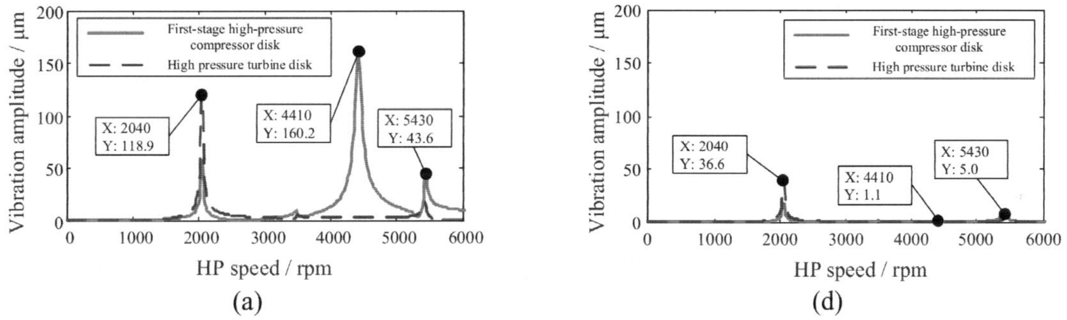

Figure 11 Vibration responses with different unbalanced mass distributions on the HP rotor.

According to the random distribution phase and the optimized distribution phase in Table 3, the vibration response of the double-rotor is tested. Figure 12 is the measured vibration response. The results show that, without changing the value of unbalance, the vibration amplitude is reduced by 43.3% after the distribution phase is optimized. To sum up, the optimization design method of residual unbalanced distribution established in this paper is accurate and effective.

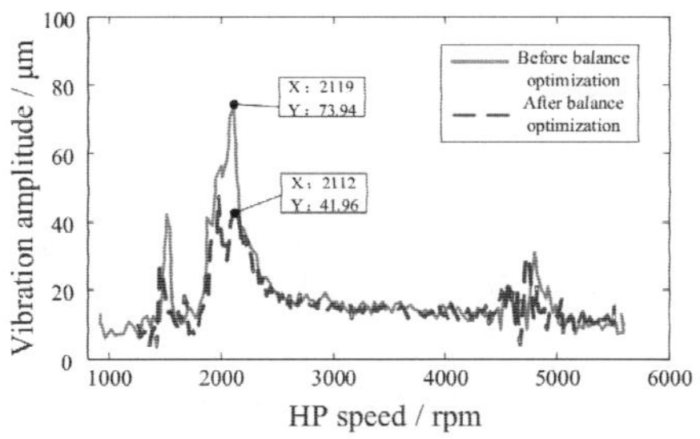

Figure 12. Vibration reduction effect after optimization design.

5. Conclusion

Based on the results and discussions presented above, the conclusions are obtained as below:

(1) For the dual-rotor system with multiple SFDs, the nonlinear characteristics of the dampers are closely related to the mode shape. When any damper shows a 'bistable' phenomenon, the nonlinear vibration 'jump' will be observed in the whole dual-rotor system.

(2) It is suggested that the imbalance should be strictly controlled to ensure that the eccentricity of any damper does not exceed 0.7 during the operation of the engine.

(3) When assembling a double-rotor system, the emphasis should be placed on the mode where the vibration of the damper position is dominant. The residual unbalance distribution optimization method based on modal orthogonality can effectively reduce the nonlinear vibration of the dual-rotor system.

(4) Multi-order critical speeds unavoidably exist in the working range of advanced aero-engines. In order to ensure that the dual-rotor system can still work reliably even when resonance occurs near the critical speeds, a comprehensive optimization method can be further studied with the vibration reduction performance of CFDs, unbalanced distribution, and rotor misalignment into consideration.

Acknowledgments

This work was financially supported by the Natural Science Foundation of Sichuan Province (2023NSFSC1295).

References

[1] R. Holmes. The control of engine vibration using squeeze film damper [J]. Journal of Engineering for Gas Turbines and Power, 1983, 105(3): 525-529. Doi: 10.1115/1.3227447.

[2] P. Bobello, P.M. Hai. A receptance harmonic balance technique for the computation of the vibration of a whole aero-engine model with nonlinear bearings [J]. Journal of Sound and Vibration, 2009, 324(1-2): 221-242. https://doi.org/10.1016/j.jsv.2009.01.039.

[3] L. Barrerr, E. J. Gunter, P. Allaire. Optimum bearing and support damping for unbalance response andstability of rotating machinery [J]. Journal of Engineering for Power, 1978, 100(1): 89-94. Doi: 10.1115/1.3446331.

[4] E. J. Hahn, P. Y. P. Chen. Harmonic Balance analysis of general squeeze film damped multidegree-of-freedom rotor bearing systems [J]. Journal of Tribology, 1994, 116(3): 499-507. Doi: 10.1115/1.2928872.

[5] Z. Liu, M. Liao, P. Cong, et al. Design method of squeeze film damper for aero-engine rotors [J]. Journal of Aerospace Power, 2015, 30(11): 2762-2770 (in Chinese). Doi: 10.13224/j.cnki.jasp.2015.11.026.

[6] Z. Liu, M. Liao, P. Cong, et al. Experimental investigation for effects of static eccentricity on vibration attenuating charateristics of squeeze film damper [J]. Journal of Propulsion Technology, 2016, 37(8): 1560-1568 (in Chinese). Doi: 10.13675/j.cnki.tjjs.2016.08.021.

[7] Z. Liu, M. Liao, W. Deng, et al. Dynamic similarity design of rotor system with squeeze film damper [J]. Journal of Aerospace Power, 2023, 38 (3): 546-557. Doi: 10.13224/j.cnki.jasp.20210462.

[8] Y. Li, M. Liao, Q. Zhao, et al. Effect of squeeze film damper concentricity and rubbing on rotor vibration characteristics [J]. Journal of Vibration and Shock, 2020, 39(1): 150-168 (in Chinese). Doi: 10.13465/j.cnki.jvs.2020.01.021.

[9] R. Wang, M. Liao, R. Cheng, et al. Workable mode optimization design method for dual-rotor system with inter-shaft bearing [J/OL]. Journal of Aerospace Power, 2022 (in Chinese). Doi: 10.13224/j.cnki.jasp.20220263.

[10] R. Wang, M. Liao, X. Lei, et al. Optimization design and experiment of workable mode for dual-rotor system with inter-shaft bearing [J]. Journal of Propulsion Technology, 2023, 44(5): 2208091 (in Chinese). Doi: 10.13675/j.cnki.tjjs.2208091.

Journal of Physics: Conference Series **2992** (2025) 012029 doi:10.1088/1742-6596/2992/1/012029

Study on slope parameter of fatigue SN curve for 2524-T3 aluminum alloy sheet

Tong Wang, Xianchao Li * and Fu Huang

Shanghai Aircraft Design and Research Institute, Commercial Aircraft Corporation of China, Ltd., Shanghai, 200135, P. R. China

*Corresponding author's e-mail: lixianchao@comac.cc

Abstract: The fatigue performance of 2524-T3 sheets with various geometric characteristics, stress ratios, and stress levels is studied in this paper by fatigue testing. An Ordinary Least Square (OLS) fitting for the fatigue SN curve was established. Furthermore, the slope parameter of the fatigue SN curve was derived as a statistical variable. The Bootstrap method was employed to investigate its statistical distribution. The results indicate that the statistics follow a normal distribution. The coefficient curve of the slope parameter is plotted with different stress levels and different numbers of test specimens. Combined with the fatigue results, the slope parameter for the 2524-T3 sheet with a 95 percent confidence level was determined for engineering calibration.

1. Introduction

The 2524 series aluminum alloy is the latest Al-Mg-Cu alloy developed by Alcoa in the 1990s. The copper and magnesium concentrations are reduced in the alloy to improve the purity of aluminum compared with the original 2000 series aluminum alloy, while further development of metallurgical processing technology reduces the content of iron, silicon, and other impurities, leading to better fatigue and damage tolerance properties. This aluminum alloy in T3 condition is now widely used in the new generation of civil transport aircraft such as the B777 and A380, replacing the original alloys such as 2024 and 2124 [1-2].

The fatigue properties of materials in the mid-cycle life region are necessary to be investigated to ensure the economy and safety of aircraft structures during their service life (usually 10^4 to 10^5 flight cycles). Bary *et al.* [3] have investigated the resistance to widespread fatigue damage (WFD) behavior of 2524-T3. Pang et al. [4] studied common defects, fatigue performance, influencing factors, and strengthening methods. Lu et al. [4] fitted the P-S-N curve of 7050 aluminum alloy with a three-parameter power function and plotted the S-N curve of 7150 aluminum alloy with a small sample size. Ye et al. [5] studied the fatigue performance of 2024-T62 aluminum alloy after different pre-corrosion times in air and saltwater environments, plotted the pre-corrosion influence coefficient curve, and analyzed the effect of corrosion damage on the material's fatigue performance. Ge et al. [6] compared the microstructure and fatigue crack propagation rate of 2524-T3 with domestic 2E12. Unlike traditional parametric approaches, the Bootstrap method offers a non-parametric solution that accounts for the variability in fatigue data more accurately. This technique presents a novel contribution to the study of aluminum alloy fatigue, enabling better estimation of fatigue life predictions by overcoming limitations in conventional statistical modeling.

2. Materials and experimental methods

2.1. Materials

The material for testing is a 2524-T3 aluminum alloy rolled sheet, with a gross thickness of δ= 6.0 mm and heat treated to T3 state. The chemical composition is shown in Table 1. The ultimate tensile strength is ~427MPa.

Table 1. Chemical Composition of 2524-T3 Alloy (mass fraction, %).

Alloy	Cu	Mg	Mn	Fe	Si	Ti	Cr	Al
2524	4.23	1.41	0.56	0.08	0.06	0.02	0.004	margin

2.2. Experimental methods

The geometries of the test specimens are demonstrated in Figure 1 with open-hole test pieces and lug test pieces with bushing protection. The tests were conducted with group method, i.e. each group of tests was tested at 3 stress levels, and 8 valid test data were guaranteed at each stress level. The standard fatigue test was conducted on Instron 8872 electro-hydraulic fatigue testing machine. Three stress ratio levels of 0.06, 0.5, and -0.5 were considered for the open-hole fatigue test, while the lug structure was tested only at a stress ratio of 0.06, as shown in Table 2.

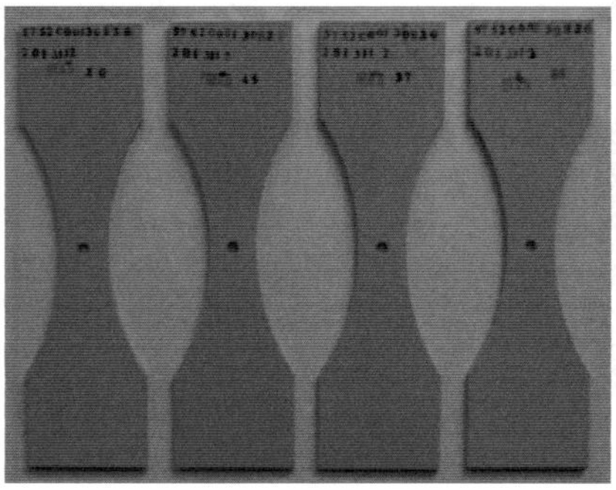

Figure 1. Schematic diagram of test specimens with (a) open-hole configuration and (b) lug configuration.

Table 2. Test specimen set-up.

Specimen configuration	Stress ratios	No. of stress level
	R=0.06	3
Open-hole test	R=-0.5	3
	R=0.5	3
Lug test	R=0.06	3

3. Results

The fatigue SN curve usually complies with the Basquin Equation (1). The general equation for engineering calibration is derived by logarithms on both sides of Equation (1). The value of k in Equation (2) is the value of the slope parameter of the fatigue SN curve described herein. The Ordinary Least Square (OLS) results of average test cycles in a double logarithmic coordinate system are illustrated in Figure 2.

$$\boldsymbol{\sigma^{\beta} \times N = constant} \quad (1)$$

where σ is the stress level, N is the fatigue life, and k and *constant* are the fitted parameters.

$$\mathbf{lg(\sigma) = -\frac{1}{\beta}lg(N) + \frac{1}{\beta}lg(c) \text{ or } lg(\sigma) = klg(N) + B} \quad \text{..................(2)}$$

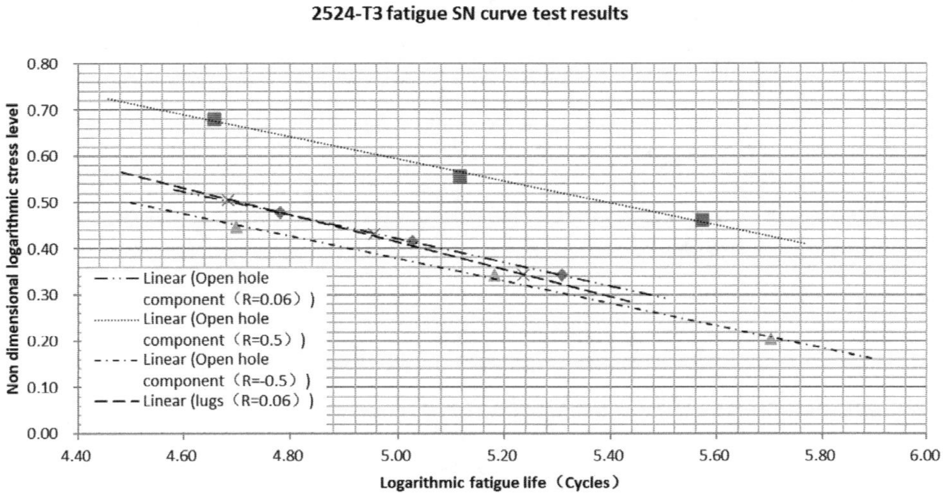

Figure 2. The SN curve fitting of 2524-T3 fatigue test.

According to the fitted results in Figure 2, a set of approximately parallel SN lines are obtained for different stress ratios and specimen types, which is because of the discrepancy of stress concentration factor with specimen configurations. According to Equation (2), the difference in stress concentration factor will result in a parallel shift of the fatigue SN curve without affecting the *k* value. As the stress ratio increases, the fatigue SN curve gradually flattens upward, which means the capacity of the structure bearing fatigue loads also rises. Therefore, the slope parameter from test results with different structural forms and stress ratios is investigated supposing from a same statistical matrix. In addition, the SN curve of the lug fatigue tests is very close to those of the open-hole fatigue tests, indicating that the fatigue qualities of the two types of structures are basically comparable. The coefficients and the fitting results of slope parameters from each test group are given in Table 3.

Table 3 The coefficients and fitted results of slope parameters from each test group specimen configurations.

	Stress ratios	Stress levels	Dispersion coefficients (CV)	Slope parameters of SN curves (k)
Open-hole specimen	0.06	I	0.018	-0.256
		II	0.013	
		III	0.020	
	0.5	I	0.022	-0.239
		II	0.008	
		III	0.008	
	-0.5	I	0.028	-0.243
		II	0.024	
		III	0.026	
Lug specimen	0.06	I	0.024	-0.295
		II	0.010	
		III	0.010	

4. Distribution of slope parameter estimates

4.1. The coefficients of 2524-T3

Bootstrap, proposed by Jarque and Bradley in 1981 [5], is a numerical approximation-like algorithm for estimating standard errors, confidence intervals, and biases in statistics after the development of computer technology. Then, based on the 12 sets of coefficients in Table 3, which constitute a Bootstrap world, the steps for solving the Bootstrap confidence intervals for the estimates of the mean values of the coefficients are as follows:

a) Bootstrap samples (B=5000) with a capacity of 12 are drawn independently and the mean value of each sample is calculated to obtain $\overline{x}_i$(i=1, 2, …, B).

b) We arrange $\overline{x}_i$ in descending order, $\overline{x}_{(1)} \leq \overline{x}_{(2)} \leq \cdots \leq \overline{x}_{(B)}$.

c) We suppose ξ_1=[B× $\frac{\alpha}{2}$] and ξ_1=[B× $\left(1-\frac{\alpha}{2}\right)$]. Then the 1-α confidence interval estimate for $\bar{x}$ is $\left(\overline{x}_{(\xi_1)}, \overline{x}_{(\xi_2)}\right)$.

The process was implemented through computer programming, and the resulting distribution histogram is shown in Figure 3. The one-sided tolerance limit with 95% confidence is 0.021.

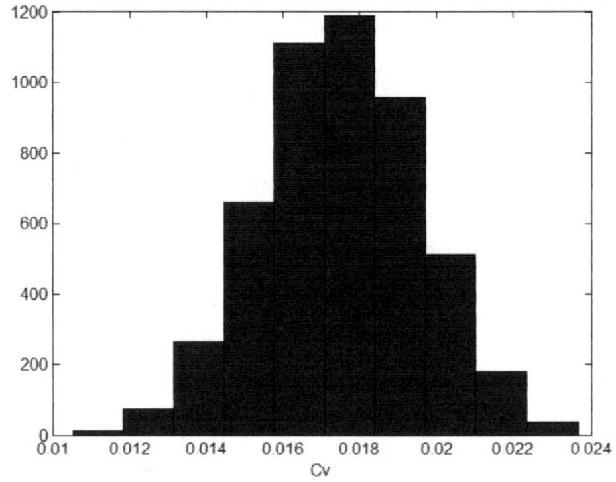

Figure 3. The bootstrap distribution histogram of coefficients.

4.2. The distribution of slope parameters

It is assumed that Equation (3) applies to the fatigue SN curve. Therefore, the coefficient is taken as 0.021 with a 95% confidence level obtained in *Section 4.1*. The fatigue life obeys a lognormal distribution, lg(n)~U(lg(N), and 0.021×lg(N)).

$$\mathbf{lg(\sigma) = -0.222\,lg(N) + 3.25} \quad (3)$$

The slope parameters are sampled from fatigue life matrices at 100 MPa, 150 MPa, and 200 MPa using the parametric Bootstraps method with 8 samples at each stress level. The real value of the slope parameter is obtained by OLS fitting of the mean values for a total of 5000 statistical simulations. If $\gamma=\frac{1}{k}$, the Jarque-Bera test was used to examine the samples and showed that γ follows a normal distribution with a coefficient of 0.039, as illustrated in Figure 4.

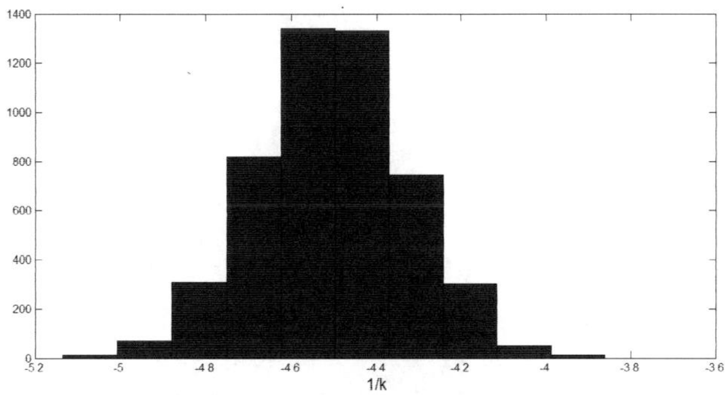

Figure 4. Bootstraps histogram of the SN curve slope parameters.

The influence of stress levels and sample size on the coefficient is also considered respectively, as shown in Figure 5. In this case, for the fatigue test to measure the slope parameter of the SN curve, the data obtained from 2 stress levels are more discrete than those of 3 stress levels. However, continuing increases in stress levels do not improve the coefficient significantly.

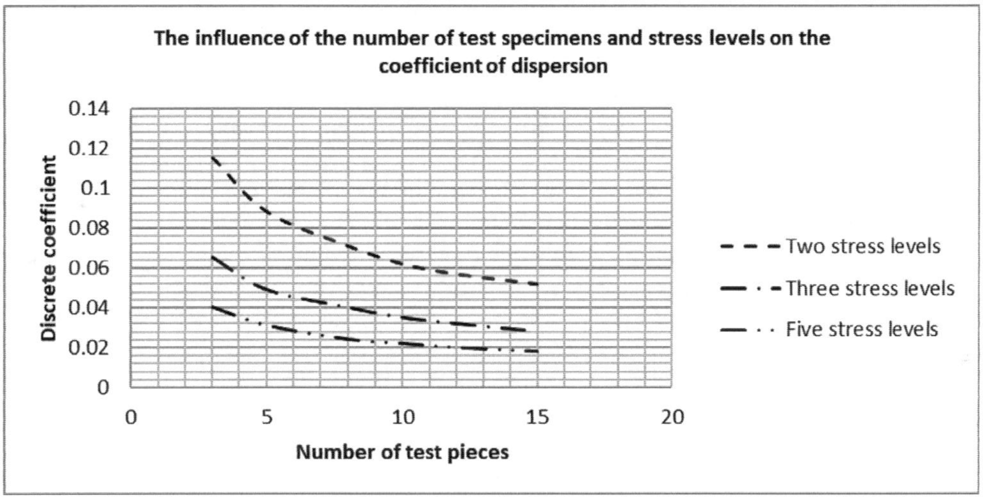

Figure 5. The influence of specimen number and stress level on the coefficient of γ estimates.

Based on the above analysis of Bootstraps simulation results, it is shown that under the current experimental condition, the coefficient of the γ estimate is 0.039. According to the relationship between the coefficient and the confidence interval [6], an estimate of the slope parameter with a 95% confidence level (5% error) can be obtained from the average estimation of 4 to 5 test groups. As for 2524-T3, the estimate of γ with 95% confidence is 3.90, *i.e.*, the slope parameter with 95% confidence is 0.256. In Figure 5, the dispersion of the coefficient across varying stress levels and sample quantities is presented. Notably, the coefficient dispersion increases under higher stress conditions, indicating heightened sensitivity to loading levels. In contrast, increasing the sample size stabilizes the coefficient dispersion, which leads to more reliable estimates for fatigue life prediction. These trends are crucial for understanding the impact of sample size and stress conditions on the analysis accuracy.

5. Summary

1. The fatigue SN curves of 2024-T3 with open-hole and lug configuration were obtained from the tests. The results show that although the structures and stress ratios are different, the SN curves are roughly parallel in the mid-cycle life region, which suggests that the material is consistent with current engineering experience.

Journal of Physics: Conference Series **2992** (2025) 012029 doi:10.1088/1742-6596/2992/1/012029

2. The dispersion in fatigue life leads to dispersion in the slope parameter of the fatigue SN curve during testing. The distribution characteristics of the slope parameter were investigated with the Bootstraps method. It was found that the inverse of the slope parameter (γ) conformed to a normal distribution. The coefficients were given for different stress levels and numbers of test pieces.

3. Based on the given coefficients combined with the results of the tests in this paper, an estimate of the 2524-T3 with 95% confidence is given, i.e., the slope parameter of the fatigue SN curve for the 2524-T3 is 0.256.

4. The results of this study have important implications for the practical use of 2524-T3 aluminum alloy in aerospace structures. By accurately determining the fatigue SN curve slope parameter, engineers can better predict the material's fatigue life under various operational conditions. This allows for more reliable design of components subjected to cyclic loading, such as aircraft fuselage and wing structures. Furthermore, the enhanced understanding of fatigue performance provided by the Bootstrap method could inform the development of fatigue life prediction models, optimizing the use of aluminum alloys in safety-critical applications.

References

[1] Starke Jr E A, Staley J T. (1996). Application of modern aluminum alloys to aircraft. Progress in aerospace sciences, 32(2-3), 131-172.

[2] Warner T. (2006). Recently-Developed Aluminium Solutions for Aerospace Applications. Materials Science Forum, 529-521(2): 1271-1278.

[3] Bray G H, Bucci R J, Kulak M, et al. (1998). Benefits of Improved Fuselage Skin Sheet Alloy 2524-T3 in Multisided Damage Scenarios. Light Metal Age, 12: 20-28

[4] Pang Jiayao, Zhang Cui, Cheng Wei, Research Progress in Fatigue Behavior of Aluminum Alloy, 2022-06-20

[5] LU Xiangfeng, GU Peipei, PENG Fei, ZHENG Xu, HE Kezhun, Study on fatigue strength of 7050-T7451 and 7150-T7751 aluminum alloys, 2023-03-20

[6] YE Yuanheng, LIU Xing, MU Xianlian, ZHANG Hongfei, HE Weiping, Effect of Corrosion Environment on Fatigue Properties of 2024-T62 Aluminum Alloy, 2024-05-25

Journal of Physics: Conference Series **2992** (2025) 012030 doi:10.1088/1742-6596/2992/1/012030

Rolling bearing fault diagnosis based on fusing composite multiscale entropy and random forests

*** Hongyuan Chu and Hua Liu**

Shanghai Institute of Special Equipment Inspection and Technical Research, Shanghai, China

*Corresponding author's e-mail: Hongyuan Chu: chuhy@ssei.cn

Abstract: This paper proposes a fault diagnosis approach for rolling bearings that integrates composite multi-scale entropy (CMSE) with a random forest (RF) model. Initially, composite multi-scale analysis is applied to address information loss issues in the coarse-grained process. Additionally, fuzzy entropy, approximate entropy, and sample entropy are combined to create a feature extraction method based on these measures. The resulting feature matrix undergoes feature selection. A random forest model is subsequently constructed to analyze the multi-scale fault features for diagnosing bearing faults. To assess the effectiveness of the proposed approach, a comparison with traditional methods is conducted, using the rolling bearing datasets from Case Western Reserve University (CWRU). The results indicate a fault identification accuracy of 100%, demonstrating that the proposed method significantly outperforms traditional approaches in terms of both recognition accuracy and diagnostic efficiency.

1. Introduction

In modern industrial systems, rolling bearings play a crucial role as part of the rotating components. Due to the demanding operating conditions and harsh environments, these bearings are prone to various faults, making their diagnosis vital for reliable operation [1]. Under real working conditions, rolling bearing vibration signals are often affected by substantial noise, resulting in non-linear and irregular characteristics. Conventional methods, such as time-frequency domain feature extraction [2], often fall short of effectively handling these complexities. To address these challenges, recent studies have introduced analytical techniques such as sample entropy (SE) [3], approximate entropy (ApEn) [4], and permutation entropy (PE) [5], which have demonstrated promising results in enhancing the accuracy of rolling bearing fault diagnosis.

To address the limitations of single-scale entropy analysis, multi-scale analysis techniques have been introduced [6]. Zheng et al. [7] presented the Composite Multi-scale Permutation Entropy (CMPE), enhancing the objectivity and precision of feature extraction for rolling bearings. Additionally, Zheng et al. [8] employed a composite multi-scale weighted replacement entropy approach for feature extraction, which better preserves useful information compared to earlier versions of multi-scale entropy.

In rolling bearing fault diagnosis, pattern recognition using Random Forest (RF) [9], a well-established integrated learning method, offers high efficiency and strong performance, while effectively mitigating overfitting issues and providing interpretable parameters. For these reasons, this paper employs Random Forest as the classifier to diagnose rolling bearing faults.

In summary, this paper proposes a rolling bearing diagnostic approach that utilizes composite multiscale entropy (CMSE) for feature extraction, followed by feature selection and pattern recognition using a Random Forest (RF) multi-classifier. The proposed method aims to accurately determine the fault status of rolling bearings. Experimental results validate that the proposed diagnostic method achieves a high accuracy rate, demonstrating its effectiveness in fault diagnosis.

2. Composite multiscale entropy

2.1 Multi-scale fuzzy entropy

Fuzzy entropy measures the size of the probability that a new pattern will be generated; the larger the value of the measure is, the greater the probability that a new pattern will be generated, i.e., the greater the sequence complexity.

For a one-dimensional time series $X = \{x_1, x_2, \dots, x_N\}$ containing N variables, we define the phase space dimension m and the similarity tolerance limit r. We reconstruct the phase space as follows:

$$X(i) = [x(i), x(i+1), \dots, x(i+m+1)] - x_0(i), i = 1, 2, \dots, N-m+1 \quad (1)$$

where $x_0(i) = \frac{1}{m}\sum_{j=0}^{m-1} x(i+j)$.

The fuzzy entropy value of a single scale can be obtained by fuzzy entropy calculation. After coarse-graining, the dynamic characteristics of the signal can be obtained from the multi-dimensional time scale. The principle of calculation is shown in **Figure 1**, with the following formula:

$$y_{i,j}^{(\tau)} = \frac{1}{\tau}\sum_{i=(j-1)\tau+1}^{j\tau+i-1} x_j \quad (2)$$

$$j = 1, 2, \dots, N/\tau, \quad i = 1, 2, \dots, \tau .$$

In Equation (2): N is the length of the time series, $\tau = 1, 2, \dots,$ i is the scale factor, and N/τ is rounding.

After that, the entropy values of all new sequences can be calculated sequentially to obtain the multi-scale fuzzy entropy:

$$\mathrm{MPE}(m, n, r, \tau) = \mathrm{FuzEn}(y_j^{(\tau)}, m, n, r, \tau) \quad (3)$$

Figure 1. $\tau = 2$ Schematic diagram of the coarse granulation process.

In Equation (3): m is the window size, n is the length of the original time series data, and r is the dimension.

2.2 Composite multiscale entropy

The composite multiscale entropy (CME) algorithm is introduced to address the shortcomings of the traditional multiscale entropy (MSE) approach. Taking fuzzy entropy as an example, the calculation starts with obtaining the τ fuzzy entropy value based on Equation (3), followed by computing the average of k entropy values. The algorithm's principle is illustrated in **Figure 2**, with the calculation formula provided below:

$$\mathrm{CMPE}(m, n, r, \tau) = \frac{1}{\tau}\sum_{k=1}^{\tau} \mathrm{FuzEn}(y_j^{(\tau)}, m, n, r, \tau) \quad (4)$$

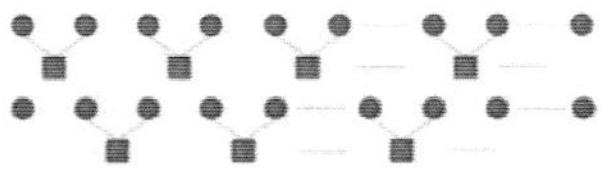

Figure 2. $\tau = 2$ Schematic diagram of the composite multiscale process.

3. Feature filtering method

When dealing with multi-dimensional data, feature selection helps to minimize the size of the training dataset, enhance model training speed, and reduce complexity, thereby mitigating overfitting. In this study, the Pearson Correlation Coefficient (PCC) [10] is employed as the feature selection technique within the filtering method, with the correlation coefficient r calculated as follows:

$$r = \frac{\sum_{i=1}^{n}(X_i - \bar{X})(Y_i - \bar{Y})}{\sqrt{\sum_{i=1}^{n}(X_i - \bar{X})^2}\sqrt{\sum_{i=1}^{n}(Y_i - \bar{Y})^2}} \quad (5)$$

where X and Y are the values of the two variables. $\bar{X}$ and $\bar{Y}$ are the average values of the two variables, indicating that the two are not correlated. The Pearson correlation coefficient is determined by calculating the difference between a variable and its mean, followed by multiplying this difference by the sum of the product and its square. In this study, features with correlation coefficients exceeding 0.85 are selected to construct the feature matrix.

4. Random forest

Random Forest is a well-known machine learning algorithm that utilizes integrated learning by combining several classifiers to achieve improved results. For classification tasks, it employs a voting mechanism, where the category chosen by the majority of sub-models is considered the final classification. The schematic representation of the random forest is shown in Figure. 3.

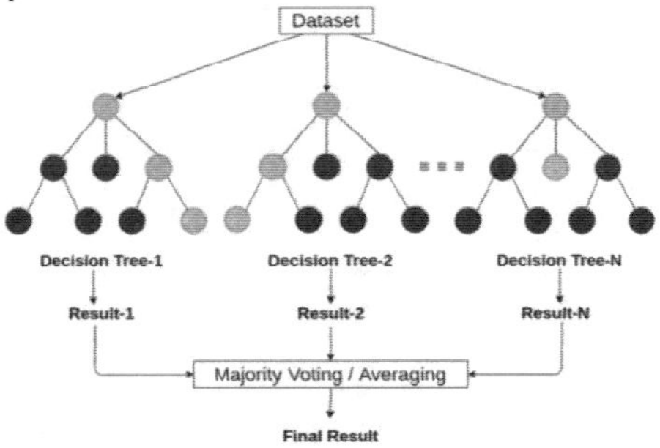

Figure 3. Schematic diagram of the random forest algorithm.

5. Rolling bearing failure based on fusing composite multiscale entropy and random forests

Building on the preceding analysis, a diagnostic model utilizing Composite Multiscale Entropy (CMSE) and Random Forest (RF) is proposed for fault detection in rolling bearings. The approach involves the following key steps:

(1) We classify the states of rolling bearings into I classes. Each state has N samples.

(2) We divide the N samples of each state into equal intervals, generate n subintervals for each sample, calculate the CMSE entropy value of each state sample, and use the composite multi-scale arrangement entropy as the eigenvectors to form the feature matrix, to generate the feature matrix $E = [E1, E2, \ldots, Ei]$;

(3) We analyze the feature matrix by Pearson's coefficient method $X = [X1, X2, \ldots, X3]$;

(4) The M training samples and N test samples, along with their respective class I labels, are fed into the Random Forest (RF) classifier for recognition analysis, followed by the calculation of the rolling bearing fault identification accuracy.

6. Analysis of experimental data

To validate the effectiveness of the proposed method, experiments were conducted using the bearing dataset from Case Western Reserve University (CWRU). Figure 4 provides an illustration of the bearing test platform utilized for this experimental validation.

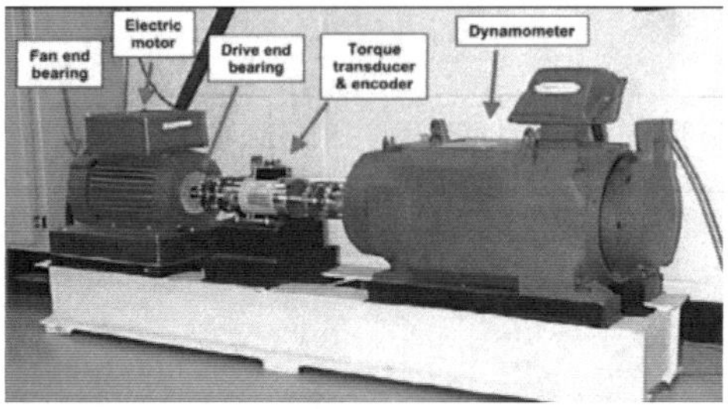

Figure 4. Bearing test platform.

The rolling bearings tested were 6205-2RS deep groove bearings with single point failures of 0.177 mm and 0.5334mm in diameter and 0.2794 mm in depth set by EDM. The data were obtained from one of the four states: normal rolling bearing (Norm), ball element failure (BE), outer ring failure (OR), and inner ring failure (IR) at a rotational speed of 1730 r/min and a sampling frequency of 12 kHz. A detailed description of the class labels is given in Table 1.

Table 1 Description of the rolling bearing class label.

Statuses	Fault diameter	Class label
Norm	0	0
Ir1	0.1778	1
Ir2	0.5334	2
Or1	0.1778	3
Or2	0.5334	4
Be	0.1778	5

The time domain waveform of a rolling bearing is shown in **Figure 5**:

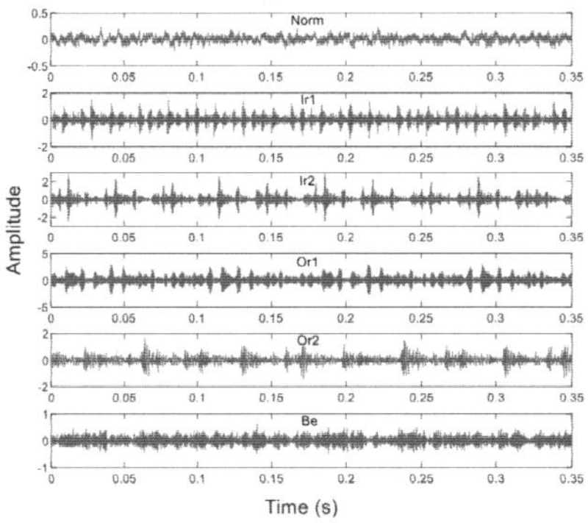

Figure 5. Time domain waveforms of rolling bearings.

Distinguishing fault conditions by simply observing the time-domain waveforms of the six states is challenging. Therefore, the characteristic parameters for different bearing states are extracted following the CMPE calculation steps. Each signal is divided into six distinct states, and, for instance, further split into 50 sub-intervals with a length of 2048 for each sub-interval. Initially, multi-scale decomposition is performed. Figure 6 illustrates the time-domain plots after decomposition when the scale parameters are set to 1 and 2.

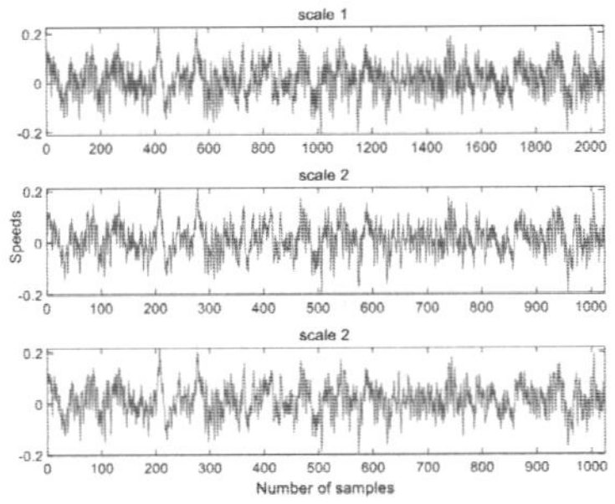

Figure 6. Multi-scale decomposition plots for scale parameters 1 and 2.

As the scale parameter increases, the number of sampling points decreases. Comparative experiments indicated that a scale parameter of 4 provided the best results, and thus it was selected in this study. With a scale parameter of 4, each state yields four dimensions of feature signals. The calculated composite multiscale entropy values are assembled into a matrix of size 300 × 12, which is subsequently reduced to a 300 × 8 matrix using a filtering method for feature selection, applying a correlation threshold of 0.15, as illustrated in Figure 7.

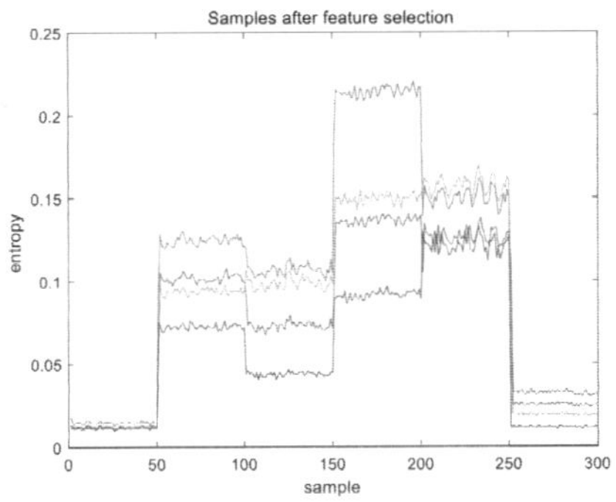

Figure 7. Samples after the filtration method.

Figure 7 illustrates the entropy values corresponding to various fault types. It is evident that the composite multi-scale permutation entropy effectively differentiates between each fault category. To further analyze the recognition rate, the faults are diagnosed using a fault recognition model. A support vector machine (SVM) model is established with the training and test datasets randomly split at a ratio of 0.2. A ten-fold cross-validation is performed on the SVM model, resulting in a recognition accuracy of 98.33% , as shown in Figure 8.

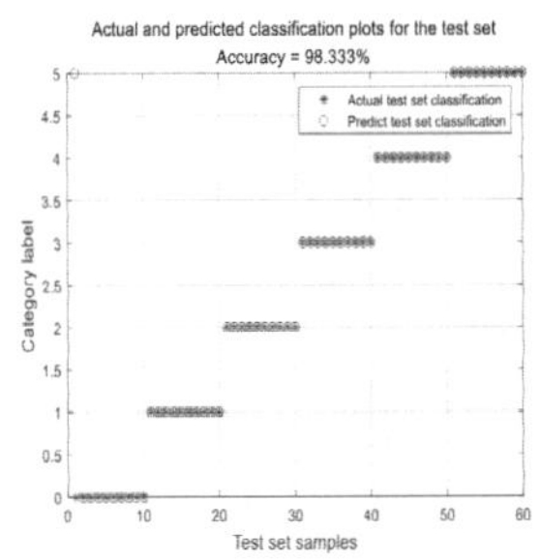

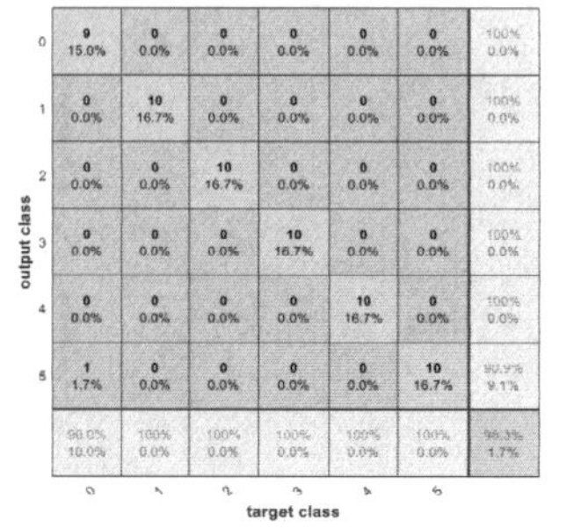

Figure 8. CME-PCC-10-fold SVM.

Figure 9 shows the graph of the recognition result of composite multiscale entropy-feature selection-random forest with 100% accuracy.

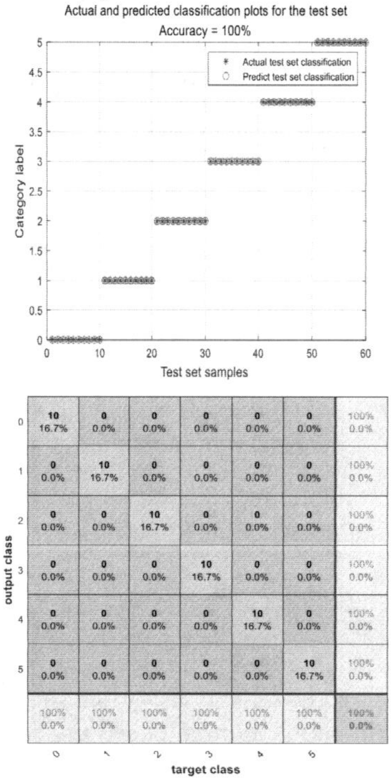

Figure 9. CME-PCC-RF.

To further validate the method proposed in this paper, the four methods are compared as in **Figure 10**.

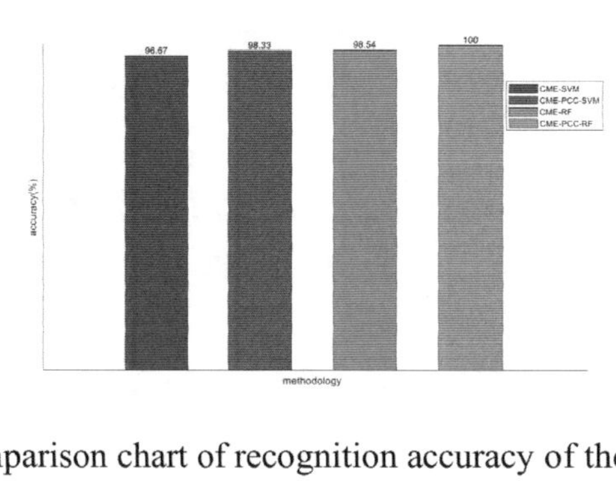

Figure 10. Comparison chart of recognition accuracy of the four methods.

Based on the accuracy comparison, it is evident that the proposed method outperforms the traditional ten-fold SVM approach by achieving a 1.67% higher recognition accuracy. Additionally, applying the filtering method further improves accuracy by 1.46%. These results validate the effectiveness of the feature selection process and demonstrate the superior fault diagnosis performance of the proposed approach.

7. Conclusion

This paper presents a rolling bearing fault diagnosis method that integrates composite multiscale entropy with a random forest model. The following conclusions are drawn in comparison to existing methods:

(1) CMPE revises the coarse-graining process for time series, effectively addressing the information loss that occurs in the MPE algorithm during single coarse-graining. Compared to MPE, the CMPE algorithm demonstrates greater stability.

(2) The comparison of recognition accuracy before and after feature selection demonstrates that the feature selection approach employed in this study is both efficient and accurate.

(3) This paper proposes a rolling bearing fault diagnosis method that combines composite multiscale entropy with a random forest model. Compared to traditional ten-fold SVM fault identification, the proposed method demonstrates improved accuracy and has certain advantages.

(4) When the distinctions between signal fault characteristics are subtle, the feature extraction method faces certain limitations, which can impact the accuracy of fault classification. Special processing of local feature signals is necessary, enhancing model stability as well as improving the accuracy of fault classification remain areas for further refinement in this paper.

References

[1] Xu Z, Li C, Yang Y. Fault diagnosis of rolling bearings using an improved multi-scale convolutional neural network with feature attention mechanism [J]. ISA transactions, 2021, 110: 379-393.

[2] Lu Y, Tang J. On time-frequency domain feature extraction of wave signals for structural health monitoring [J]. Measurement, 2018, 114: 51-59.

[3] Wang C. A sample entropy inspired affinity propagation method for bearing fault signal classification [J]. Digital Signal Processing, 2020, 102: 102740.

[4] Zhao S F, Liang L, Xu G H, et al. Quantitative diagnosis of a spall-like fault of a rolling element bearing by empirical mode decomposition and the approximate entropy method [J]. Mechanical Systems and Signal Processing, 2013, 40(1): 154-177.

[5] Noman K, Wang D, Peng Z, et al. Oscillation based permutation entropy calculation as a dynamic nonlinear feature for health monitoring of rolling element bearing [J]. Measurement, 2021, 172: 108891.

[6] Huang Y J, Liao A H, Hu D Y, et al. Multi-scale convolutional network with channel attention mechanism for rolling bearing fault diagnosis [J]. Measurement, 2022, 203: 111935.

[7] Zheng J , Pan H , Yang S ,et al.Generalized composite multiscale permutation entropy and Laplacian score based rolling bearing fault diagnosis [J].Mechanical Systems & Signal Processing, 2018, 99(jan.15): 229-243.DOI: 10.1016/j.ymssp.2017.06.011.

[8] Zheng J, Dong Z, Pan H, et al. Composite multi-scale weighted permutation entropy and extreme learning machine based intelligent fault diagnosis for rolling bearing [J]. Measurement, 2019, 143: 69-80.

[9] F. Wang, X. Wang, and S. Sun, 'A reinforcement learning level-based particle swarm optimization algorithm for large-scale optimization', Inf. Sci., vol. 602, pp. 298–312, Jul. 2022, doi: 10.1016/j.ins.2022.04.053.

[10] Qin L, Yang G, Sun Q. Maximum correlation Pearson correlation coefficient deconvolution and its application in fault diagnosis of rolling bearings [J]. Measurement, 2022, 205: 112162.

Research on the whole life cycle attribute decision method for gearbox product design based on WLS and TOPSIS

Yu Liu [1*], Juan Li [1], Jingying Li [1] and Ying Wang [1]

[1] China Productivity Center for Machinery Co., Ltd., Beijing 100044, China

*Corresponding author's e-mail: liuyulf@163.com

Abstract. Focusing on two different application scenarios and their entire lifecycle of gearboxes, covering various stages such as design, manufacturing, use, and disposal, this article focuses on exploring the impact of key material selection, manufacturing processes, and operation and maintenance management on the performance and sustainability of gearboxes. A comprehensive decision-making framework for the life cycle attributes of key components in gearboxes is proposed based on weighted least squares (WLS) and the ideal point method (TOPSIS). An evaluation index system for the entire life cycle attributes of gearbox product design is established. Through this method system, the optimal gear design scheme is determined in two different application scenarios: manufacturing accuracy and raw material selection. The effectiveness and feasibility of the proposed attribute decision-making method in practical applications are verified by using actual production data.

1. Introduction

Product design is considered to be the precursor of manufacturing, production, logistics, maintenance, and other stages in the product lifecycle, which significantly affects the improvement of resource productivity, engineering process reengineering of production and consumption profiles, and industry transformation [1]. Gearboxes have a wide range of application scenarios in electromechanical products, and their manufacturing processes are mostly discrete and in large quantities, covering various links such as gear processing technology, heat treatment, surface treatment, and assembly inspection. Each link has a significant impact on the performance, quality, and cost of gearboxes. By studying the decision-making of the entire lifecycle attributes of gearbox product design, we ensure the efficient operation and long service life of gearboxes in practical applications, and realize the important role of gearbox products in reducing greenhouse gas emissions, achieving green manufacturing, and promoting sustainable development in their raw materials, manufacturing, service, transportation, and recycling lifecycle attributes.

In the early stages of product development and design, a large number of decision-making factors need to be considered to determine product specifications [2]. However, in the design process of gearbox products, it is necessary to comprehensively consider the impact of various attributes of the lifecycle on the environment. By utilizing the common manufacturing methods of the vast majority of gearbox components and combining green features, the best design solution can be determined for designers to choose a more suitable and environmentally friendly basic component lifecycle model. In addition, in addition to considering the common characteristics of the above-mentioned gearbox components in design, manufacturing, assembly, and application, the decision-making of their lifecycle design scheme belongs to a multi-attribute and multi-objective decision-making problem, and there is also a phenomenon of deviation between the decision results and the actual design due to different evaluation scopes.

With the proposal of the "dual carbon" strategy, more "green information" will be injected into the product design stage, clearly expressing the environmental impact information of this product design [3]. In the research of lifecycle attribute decision-making methods, Feng et al. proposed a form matrix-based approach that considers fuzziness and uncertainty through intuitive fuzzy numbers, which can meet the sustainability requirements of the product concept stage [4]. Ng and Law proposed a method that combines fuzzy sets, COPRAS, and LCA to enable decision-makers to evaluate the effectiveness of design solutions in the early stages of product development [5]. Pacana et al. proposed a model that supports the improvement of product design processes during the lifecycle stage, enabling the development and analysis of alternative product design options during the design phase [6]. Ridvan and Badurdeen extended the product lifecycle stage by incorporating recycling strategies, achieving a sustainable product line design optimization model with economic and environmental performance goals, and applying it to toner cartridge products [7]. Ibn Mohammed et al. studied the use of normal accident theory and utility maximization framework theory to address environmental impact accidents when multiple targets conflict, to simplify and measure mitigation efforts [8]. In addition to the research on lifecycle decision-making methods, Zhang and Ming integrated customer product interaction into the lifecycle innovation service model, analyzed service influencing factors through fuzzy models, and achieved effective integration of service resources [9]. Luo et al. established a three-layer standard accumulation model for complex product systems, based on a subjective objective mixed fuzzy method using intuitionistic fuzzy sets, and obtained the optimal manufacturing resource allocation method by sorting the results [10].

After studying the decision-making methods for lifecycle attributes, most of the research has focused on the decision-making of the product lifecycle itself, but there has been no research on the decision-making methods for lifecycle attributes of product features in various application scenarios. Therefore, this article proposes a decision-making method for the lifecycle attributes of gearbox products based on WLS and TOPSIS, which combines two comprehensive decision evaluation methods, weighted least squares, and ideal point method, to analyze and determine the optimal component design scheme. This method can be applied to different component usage scenarios by focusing on different lifecycle stages, providing more suitable basic mechanical component design selection schemes for workers.

2. Establishment of a comprehensive life cycle attribute evaluation index system for gearbox product design

We establish a full lifecycle attribute evaluation index system for the characteristics of the gearbox products mentioned above and then determine the weights of each index from the target layer to the scheme layer through expert evaluation values and the weighted least squares method. We standardize the original data matrix after trending, construct a weighted standardization matrix, determine positive and negative ideal solutions, and calculate the distance. Finally, we determine the optimal gearbox design for the current application scenario by calculating the relative proximity.

The established evaluation index system for the full lifecycle attributes of gearbox products is shown in Figure 1, which includes a measure layer, a sub-criterion layer (evaluation index), a criterion layer (lifecycle attributes), and a target layer. The measured layer includes various types of components that meet the requirements of the service environment and performance; The sub-criteria layer (evaluation indicators) includes four types of parameters: resource consumption (*N*), energy consumption (*E*), greenhouse gas emissions (*G*), and cost (*C*); The criteria layer (lifecycle attributes) includes five stages: raw material acquisition stage (*M*), manufacturing stage (*P*), service stage (*S*), transportation stage (*T*), and recycling and disposal stage (*R*); The target layer is the optimal design scheme for the components of the gearbox product. The five stages of the product lifecycle mentioned have different scopes and requirements. Among them, the raw material acquisition stage ***M*** mainly considers whether the usage needs are met and the difficulty of processing the raw materials. The manufacturing stage ***P*** considers the environmental impact of the parts in the production and manufacturing process. The service stage ***S*** considers the maintainability, energy consumption, and environmental emissions of the parts. The transportation stage ***T*** considers the energy consumption and environmental emissions of the parts transported a certain distance after being packaged. The recycling and disposal stage ***R*** considers the reuse of the parts as a resource after service, mainly considering the recycling rate.

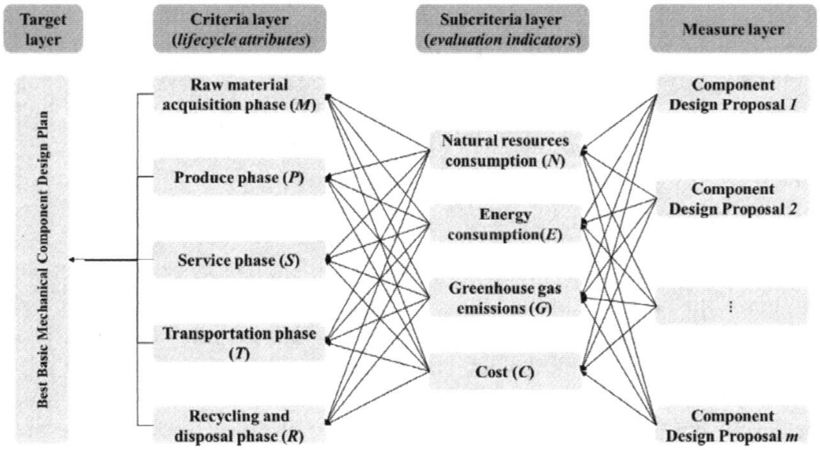

Figure 1. Evaluation index system for the full lifecycle attributes of gearbox products.

3. Decision-making method for the full lifecycle attributes of gearbox product design based on WLS and TOPSIS

Based on the attribute decision-making method for the entire lifecycle of gearbox product design constructed above, the WLS and TOPSIS comprehensive decision-making models are used for attribute decision-making. Among them, the Weighted Least Squares (WLS) method has many advantages compared to the traditional Ordinary Least Squares (OLS) method, especially when dealing with data with heteroscedasticity, which can apply variable weights to improve the robustness of the model; TOPSIS is a multi-criteria decision-making method proposed by Tzeng and Huang [11] for calculating the distance between the optimal solution and the positive or negative ideal values. The specific attribute evaluation process is shown in Figure 2. The basis for decision-making comes from the full lifecycle database established by using gas sensors and energy analysis instruments in manufacturing processes to collect data. Then, gear components that meet the requirements are preliminarily selected according to the application scenario requirements. Based on the constructed attribute evaluation index system, expert evaluation is combined with WLS to obtain a weighted evaluation matrix for each parameter index. The relative proximity is calculated by using the ideal point method based on the weights. Finally, the optimal design scheme for key components of the gearbox product is determined by the principle of maximum proximity.

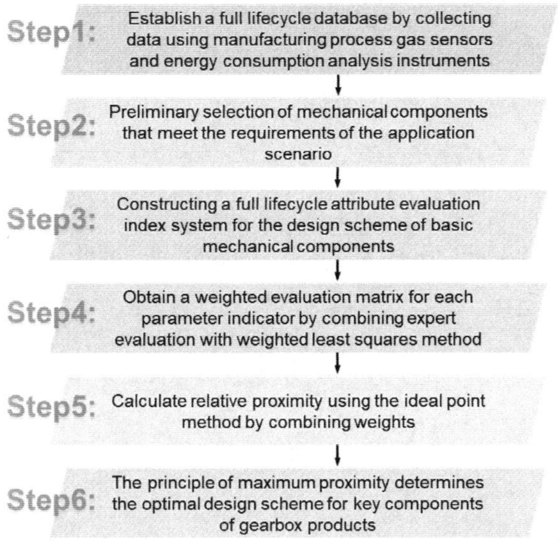

Figure 2. Decision-making and evaluation process for the full lifecycle attributes of gearbox product design.

The specific method execution steps are as follows:

Step 1: First, we invite m experts to provide relative importance scores (ranging from 1-10) for each indicator of different models of components in the criterion layer and sub-criterion layer and construct evaluation matrices for the criterion layer and sub-criterion layer;

The weight proportion vector of m experts is: $P=[p_1,\cdots,p_m]$.

The criterion layer evaluation matrix and sub-criterion layer evaluation matrix are shown in Equations (1) and (2):

$$A=\left(a_{ij}\right)_{m\times 5}=\begin{pmatrix} a_{11} & a_{12} & \cdots & a_{15} \\ \vdots & \vdots & \vdots & \vdots \\ a_{m1} & a_{m2} & \cdots & a_{m5} \end{pmatrix} \tag{1}$$

$$B=\left(b_{ij}\right)_{m\times 4}=\begin{pmatrix} b_{11} & b_{12} & \cdots & b_{14} \\ \vdots & \vdots & \vdots & \vdots \\ b_{m1} & b_{m2} & \cdots & b_{m4} \end{pmatrix} \tag{2}$$

where $a_{ij}\in[0,10]; b_{ij}\in[0,1]$.

Step 2: We input expert weights and solve for the score vectors of the criterion layer and sub criterion layer;

The criteria layer is $C_j=\sum_{i=1}^{m} p_j a_{ij}, j=1,\cdots,5$, where $C=[c_1,c_2,c_3,c_4,c_5]^T$;

Sub criteria layer is $D_j=\sum_{i=1}^{m} p_j b_{ij}, j=1,\cdots,4$, where $D=[d_1,d_2,d_3,d_4]^T$.

Step 3: We use Weighted Least Squares (WLS) to construct a weighted least squares model;

(i) Criteria layer:

The initial indicator weight vector is: $U=[u_1,u_2,u_3,u_4,u_5]^T$;

It is assumed that the true weight vector of the criterion layer indicators is: $W=[w_1,w_2,w_3,w_4,w_5]^T$;

The weighted least squares objective function of the criterion layer is shown in Equation (3):

$$f(u)=\sum_{j=1}^{5} w_j*\left(c_j-w_j\right)^2 \tag{3}$$

(ii) Sub-criteria layer:

The initial indicator weight vector is: $U'=\left[u'_1,u'_2,u'_3,u'_4,u'_5\right]^T$;

It is assumed that the true weight vector of the sub criterion layer indicators is: $V=\left[v_1,v_2,v_3,v_4\right]^T$;

The weighted least squares objective function of the sub criterion layer is shown in Equation (4):

$$f(v)=\sum_{j=1}^{4} w_j*\left(d_j-v_j\right)^2 \tag{4}$$

Step 4: We use gradient descent to solve for the weight vector that minimizes the objective function, and normalize it;

(i) Criteria layer:

The optimal weight vector is: $X=\left[x_1,x_2,x_3,x_4,x_5\right]^T$;

Normalized processing yields $X'_j=\frac{x_j}{\sum_{j=1}^{5}x_j}$, where $X'=\left[x'_1,x'_2,x'_3,x'_4,x'_5\right]^T$.

(ii) Sub-criteria layer:

The optimal weight vector is: $Y=\left[y_1,y_2,y_3,y_4\right]^T$

Normalized processing yields $Y'_j=\frac{y_j}{\sum_{j=1}^{4}y_j}$, where $Y'=\left[y'_1,y'_2,y'_3,y'_4\right]^T$.

Step 5: After using expert evaluation values combined with the weighted least squares method to determine the weights corresponding to each indicator in the criterion layer and sub criterion layer, we then use the ideal point method to calculate the corresponding scores (relative closeness) of each selected mechanical component. Firstly, the original data matrix composed of parameter data $\boldsymbol{r_{ij}}$ corresponding to each component in the underlying database is subjected to trend normalization and standardization processing to obtain the standardized matrix $\boldsymbol{Z}$;

(i) To ensure that all indicators change in the same direction, it is necessary to perform trend normalization on the parameter data $\boldsymbol{r_{ij}}$ of the sub-criterion layer. The lifecycle consists of five stages, each with four indicator values;

Standard 0-1 transformation: $a_{ij}=\frac{r_{ij}-a_j^{\min}}{a_j^{\max}-a_j^{\min}}$;

where $i=1, 2,\cdots, k$ (k is the number of selected components); $j=1,\cdots, 20$.

(ii) Nondimensionalization adopts the method of standardizing the original data matrix after trend normalization;

Standardization: $z_{ij}=\frac{a_{ij}}{\sqrt{\sum_{i=1}^{k}a_{ij}^2}}, j=1, 2,\cdots, 20$;

The standardized parameter matrix is shown in Equation (5):

$$Z=\begin{bmatrix} z_{11} & z_{12} & \cdots & z_{1n} \\ z_{21} & z_{22} & \cdots & z_{2n} \\ \vdots & & \vdots & \vdots \\ z_{k1} & z_{k2} & \cdots & z_{kn} \end{bmatrix}, n=1, 2, \cdots, 20 \tag{5}$$

Step 6: We calculate the weighted standardization matrix P from the weights of each indicator in the criterion layer and sub criterion layer, as well as the standardization parameter matrix, as shown in Equation (6):

$$P=\begin{bmatrix} w_1v_1y_{11}\cdots w_1v_4y_{14} & w_2v_1y_{15}\cdots w_2v_4y_{18} & w_3v_1y_{19}\cdots w_3v_4y_{1,12} & w_4v_1y_{1,13}\cdots w_4v_4y_{1,16} & w_5v_1y_{1,17}\cdots w_5v_4y_{1,20} \\ w_1v_1y_{21}\cdots w_1v_4y_{24} & w_2v_1y_{25}\cdots w_2v_4y_{28} & w_3v_1y_{29}\cdots w_3v_4y_{2,12} & w_4v_1y_{2,13}\cdots w_4v_4y_{2,16} & w_5v_1y_{2,17}\cdots w_5v_4y_{2,20} \\ \vdots & & \vdots & & \vdots \\ w_1v_1y_{31}\cdots w_1v_4y_{34} & w_2v_1y_{35}\cdots w_2v_4y_{38} & w_3v_1y_{39}\cdots w_3v_4y_{3,12} & w_4v_1y_{3,13}\cdots w_4v_4y_{3,16} & w_5v_1y_{3,17}\cdots w_5v_4y_{3,20} \end{bmatrix}$$

$$=\begin{bmatrix} P_{11} & P_{12} & \cdots & P_{1n} \\ P_{21} & P_{22} & \cdots & P_{2n} \\ \vdots & \vdots & & \vdots \\ P_{k1} & P_{k2} & \cdots & P_{kn} \end{bmatrix}, n=1, 2, \cdots, 20 \tag{6}$$

Step 7: We determine the positive ideal solution P_j^+ and negative ideal solution P_j^-, as shown in Equations (7)~(8), in which J^+ is the benefit-type index and J^- is the cost-type index.

Positive ideal solution: $P_j^+=\{(maxP_{ij}|j\in J^+), (minP_{ij}|j\in J^-)\}=(P_1^+, P_2^+, \cdots, P_{20}^+)$ (7);

Negative ideal solution: $P_j^-=\{(minP_{ij}|j\in J^+), (maxP_{ij}|j\in J^-)\}=(P_1^-, P_2^-, \cdots, P_{20}^-)$ (8).

Step 8: We calculate the total distance $\boldsymbol{S_i^+}$ and $\boldsymbol{S_i^-}$ between the various types of indicators of different welding processes and the positive and negative ideal solutions, respectively, as shown in Equations (9)~(10).

$$S_i^+=\sqrt{\sum_{j=1}^{20}(P_{ij}-P_j^+)^2}, i=1, 2, \cdots k \tag{9}$$

$$S_i^-=\sqrt{\sum_{j=1}^{20}(P_{ij}-P_j^-)^2}, i=1, 2,\cdots k \tag{10}$$

Step 9: We calculate the relative proximity $\boldsymbol{C_i^*}$ of different welding processes to obtain the relative proximity vector $\boldsymbol{C}$, as shown in Equations (11) to (12).

Relative proximity:

$$C_i^*=\frac{S_i^-}{S_i^+ + S_i^-}, i=1, 2,\cdots, n \tag{11}$$

Relative proximity vector:

$$C=(C_1^*, C_2^*, \cdots, C_5^*)^T \tag{12}$$

Step 10: Based on the principle of maximum relative proximity, we autonomously determine the key components of the gearbox corresponding to the maximum value of the elements in the relative proximity vector C as the optimal component design scheme for the current usage scenario.

4. Case study

The application case selects a certain type of gear as the large tooth gear in the first stage gear reducer and preliminarily selects three models required for this service scenario.

Gear 1: The gear weighs 10 kg and is made of cast steel. The manufacturing process involves grinding, normalizing, and quenching. It is transported by railway and needs to be recycled and reused;

Gear 2: The weight of the gear is 10 kg, made of cast iron, manufactured by milling and tempering, and transported by truck. It needs to be recycled and reused;

Gear 3: The gear weighs 10 kg and is made of stainless steel. The manufacturing process involves grinding, quenching, and tempering. It is transported by air and needs to be recycled.

When the lifecycle stages emphasized by usage scenario requirements are different, expert evaluation should focus on improving the corresponding stage scores, and the same consideration should be given to the four evaluation indicators. Based on the decision-making steps established in Figure 2 above, the specific calculation steps for the application scenario of the case are as follows:

Step 1: We use various analytical instruments and gas sensors to collect parameter data of various types of basic mechanical components at different stages of their lifecycle, calculate their cost values, and finally summarize them into a full lifecycle database of various types of basic mechanical components per unit weight.

Step 2: Based on factors such as the service environment and performance requirements of gearbox components, we preliminarily select all models of components that meet the conditions. Combined with the full life cycle database of various types of basic mechanical components per unit weight obtained in Step 1, the expected total data of various parameters of different models of components can be obtained, providing the underlying database for decision analysis in Step 3.

Step 3: We establish a full lifecycle attribute evaluation index system, and then use expert evaluation values and the weighted least squares method to determine the weights of each index from the target layer to the scheme layer. We standardize the original data matrix after trending, construct a weighted standardization matrix, determine positive and negative ideal solutions, and calculate the distance. Finally, we determine the optimal component design scheme for the current application scenario by calculating the relative proximity.

Step 4: We establish a full lifecycle attribute evaluation index system, including the measures layer, sub-criteria layer (evaluation indicators), criteria layer (lifecycle attributes), and target layer.

Step 5: We invite 3 experts to provide relative importance scores (ranging from 1-10) for each indicator of different models of components in the criterion layer and sub criterion layer, and construct evaluation matrices for the criterion layer and sub criterion layer. The weight ratio vector of the three experts is $P=\left[\frac{1}{3}, \frac{1}{3}, \frac{1}{3}\right]$.

(i) The first application scenario requires high precision and should focus on the manufacturing phase of the lifecycle, while also paying attention to energy consumption throughout the entire lifecycle.

Criteria level evaluation matrix: $A=\begin{pmatrix} 6.1 & 8.3 & 5.5 & 4.8 & 3.9 \\ 5.6 & 8.8 & 4.9 & 5.7 & 5.6 \\ 5.3 & 8.6 & 5.8 & 4.9 & 5.3 \end{pmatrix}$;

Sub criterion layer evaluation matrix: $B=\begin{pmatrix} 3.2 & 8.7 & 3.8 & 4.3 \\ 4.1 & 8.5 & 4.6 & 4.7 \\ 3.6 & 7.8 & 3.7 & 3.9 \end{pmatrix}$

(ii) The second application requirement scenario is high performance and good raw material quality, and we should focus on the raw material acquisition stage in the lifecycle while paying attention to the cost throughout the entire lifecycle.

Criteria level evaluation matrix: $A=\begin{pmatrix} 8.3 & 6.1 & 5.5 & 4.8 & 3.9 \\ 8.8 & 5.6 & 4.9 & 5.7 & 5.6 \\ 8.6 & 5.3 & 5.8 & 4.9 & 5.3 \end{pmatrix}$;

Sub criterion layer evaluation matrix: $B=\begin{pmatrix} 3.2 & 4.3 & 3.8 & 8.7 \\ 4.1 & 4.7 & 4.6 & 8.5 \\ 3.6 & 3.9 & 3.7 & 7.8 \end{pmatrix}$.

Step 6: We input expert weights and solve the score vectors for the criterion layer and sub-criterion layer.

(i) The first scenario:

Criteria layer: $C=[8.57, 5.67, 5.4, 5.13, 4.93]^T$;

Sub criteria layer: $D=[3.63, 8.33, 4.03, 4.3]^T$.

(ii) The second scenario:

Criteria layer: $C=[8.57, 5.67, 5.4, 5.13, 4.93]^T$;

Sub criteria layer: $D=[3.63, 4.3, 4.03, 8.33]^T$.

Step 7: We use the weighted least squares method to construct a weighted least squares model, with the same two scenarios;

(i) Criteria layer:

The initial indicator weight vector is: $U=[0.2, 0.2, 0.2, 0.2, 0.2]^T$;

It is assumed that the true weight vector of the criterion layer indicators is $W=[w_1, w_2, w_3, w_4, w_5]^T$;

The weighted least squares objective function of the criterion layer is shown in Equation (3).

(ii) Sub-criteria layer:

The initial indicator weight vector is: $U'=[0.25, 0.25, 0.25, 0.25]^T$;

It is assumed that the true weight vector of the sub-criterion layer indicators is $V=[v_1, v_2, v_3, v_4]^T$;

The weighted least squares objective function of the sub criterion layer is shown in Equation (4):

Step 8: We use gradient descent to solve for the weight vector that minimizes the objective function, and normalize it.

(i) The first scenario:

i) Criteria layer:

The optimal weight vector is: X=[5.67, 8.57, 5.4, 5.13, 4.93]^T

Normalization processing becomes: X' =[0.19, 0.29, 0.18, 0.17, 0.17]^T

ii) Sub criteria layer:

The optimal weight vector is: Y =[3.63, 8.33, 4.03, 4.3]^T

Normalization processing becomes: Y' =[0.18, 0.41, 0.2, 0.21]^T

(ii) The second scenario:

i) Criteria layer:

The optimal weight vector is: X=[8.57, 5.67, 5.4, 5.13, 4.93]^T

Normalization processing becomes: X' =[0.29, 0.19, 0.18, 0.17, 0.17]^T

ii) Sub criteria layer:

The optimal weight vector is: Y=[3.63, 4.3, 4.03, 8.33]^T

Normalization processing becomes: Y' =[0.18, 0.21, 0.2, 0.41]^T

Step 9: After using expert evaluation values combined with weighted least squares method to determine the corresponding weights of each indicator in the criterion layer and sub criterion layer, we then use the ideal point method to calculate the corresponding scores (relative closeness) of each selected mechanical component scheme. Firstly, the original data matrix composed of parameter data $\boldsymbol{r_{ij}}$ corresponding to each component in the underlying database is subjected to trend normalization and standardization processing to obtain the standardized matrix $\boldsymbol{Z}$;

We extract the raw data matrix $\boldsymbol{Q}$ from the database of basic mechanical components:

$$Q=\begin{pmatrix} 50 & 17.6 & 0.015 & 40 & 0.5 & 0.769 & 4.377 & 5.07 & 50 & 8107.78 & 259.4 & 100{,}000 & 5 & 0.032 & 0.051 & 0.8 & 0.375 & 0.577 & 3.283 & 3.80 \\ 50 & 12.6 & 1.813 & 5.0 & 0.5 & 0.474 & 4.440 & 6.32 & 50 & 8107.78 & 259.4 & 100{,}000 & 5 & 0.227 & 0.005 & 8.0 & 0.375 & 0.356 & 3.330 & 4.74 \\ 50 & 16.8 & 0.018 & 3.0 & 0.5 & 0.789 & 4.596 & 4.14 & 50 & 8107.78 & 259.4 & 100{,}000 & 5 & 3.449 & 1.554 & 1.2 & 0.375 & 0.592 & 3.447 & 3.11 \end{pmatrix}$$

Normalized matrix $\boldsymbol{Z}$ after normalization conversion:

$$Z=\begin{pmatrix} 0.577 & 0.642 & 0.008 & 0.990 & 0.577 & 0.641 & 0.565 & 0.557 & 0.577 & 0.577 & 0.577 & 0.577 & 0.577 & 0.009 & 0.033 & 0.098 & 0.577 & 0.641 & 0.565 & 0.557 \\ 0.577 & 0.460 & 1.000 & 0.124 & 0.577 & 0.395 & 0.573 & 0.695 & 0.577 & 0.577 & 0.577 & 0.577 & 0.577 & 0.066 & 0.003 & 0.984 & 0.577 & 0.395 & 0.573 & 0.695 \\ 0.577 & 0.613 & 0.010 & 0.074 & 0.577 & 0.658 & 0.593 & 0.455 & 0.577 & 0.577 & 0.577 & 0.577 & 0.577 & 0.998 & 0.999 & 0.148 & 0.577 & 0.658 & 0.593 & 0.456 \end{pmatrix}$$

Step 10: We calculate the weighted standardization matrix P based on the weights of each indicator in the criterion layer and sub-criterion layer, as well as the standardization parameter matrix;

(i) The first scenario:

$$Z=\begin{pmatrix} 0.0197 & 0.0500 & 0.0003 & 0.0395 & 0.0301 & 0.0762 & 0.0328 & 0.0339 & 0.0187 & 0.0426 & 0.0208 & 0.0218 & 0.0177 & 0.0006 & 0.0011 & 0.0035 & 0.0177 & 0.0447 & 0.0192 & 0.0199 \\ 0.0197 & 0.0358 & 0.0380 & 0.0049 & 0.0301 & 0.0470 & 0.0332 & 0.0423 & 0.0187 & 0.0426 & 0.0208 & 0.0218 & 0.0177 & 0.0046 & 0.0001 & 0.0351 & 0.0177 & 0.0276 & 0.0195 & 0.0248 \\ 0.0197 & 0.0478 & 0.0004 & 0.0030 & 0.0301 & 0.0782 & 0.0344 & 0.0277 & 0.0187 & 0.0426 & 0.0208 & 0.0218 & 0.0177 & 0.0695 & 0.0340 & 0.0053 & 00177 & 0.0458 & 0.0202 & 0.0163 \end{pmatrix}$$

(ii) The second scenario:

$$Z=\begin{pmatrix} 0.0301 & 0.0391 & 0.0005 & 0.1177 & 0.0301 & 0.0390 & 0.0328 & 0.0663 & 0.0187 & 0.0218 & 0.0208 & 0.0426 & 0.0177 & 0.0003 & 0.0011 & 0.0069 & 0.0177 & 0.0229 & 0.0192 & 0.0388 \\ 0.0301 & 0.0280 & 0.0580 & 0.0096 & 0.0301 & 0.0241 & 0.0332 & 0.0826 & 0.0187 & 0.0218 & 0.0208 & 0.0426 & 0.0177 & 0.0023 & 0.0001 & 0.0686 & 0.0177 & 0.0141 & 0.0195 & 0.0484 \\ 0.0301 & 0.0373 & 0.0004 & 0.0058 & 0.0301 & 0.0401 & 0.0344 & 0.0541 & 0.0187 & 0.0218 & 0.0208 & 0.0426 & 0.0177 & 0.0356 & 0.0340 & 0.0103 & 00177 & 0.0235 & 0.0202 & 0.0318 \end{pmatrix}$$

Step 11: We determine the positive ideal solution $P_j^{\ +}$ and the negative ideal solution P_j^-, where J^+ is a benefit-based indicator and J^- is a cost-based indicator;

(i) The first scenario:

Ideal solution:

P_j^+=(0.0197, 0.0500, 0.0380, 0.0395, 000301, 0.0782, 0.0344, 0.0423, 0.0187, 0.0426, 0.0208, 0.0218, 0.0177, 0.0695, 0.0340, 0.0351, 0.0177, 0.0458, 0.0202, 0.0248)

Negative ideal solution:

P_j^-=(0.0197, 0.0358, 0.0003, 0.0030, 0.0301, 0.0470, 0.0328, 0.0277, 0.0187, 0.0426, 0.0208, 0.0218, 0.0177, 0.0006, 0.0001, 0.0035, 0.0177, 0.0276, 0.0192, 0.0163)

(ii) The second scenario:

Ideal solution:

P_j^+=(0.0301, 0.0391, 0.0580, 0.1177, 0.0301, 0.0401, 0.0344, 0.0826, 0.0187, 0.0218, 0.0208, 0.0426, 0.0177, 0.0356, 0.0340, 0.0686, 0.0177, 0.0235, 0.0202, 0.0484)

Negative ideal solution:

P_j^-=(0.0301, 0.0280, 0.0004, 0.0058, 0.0301, 0.0241, 0.0328, 0.0541, 0.0187, 0.0218, 0.0208, 0.0426, 0.0177, 0.0003, 0.0001, 0.0069, 0.0177, 0.0141, 0.0192, 0.0318)

Step 12: We calculate the total distance between various indicators of different types of gear components and the positive and negative ideal solutions, denoted as S_i^+ and S_i^-, as well as the relative proximity, denoted as C_i^*;

(i) The first scenario:

The calculation results of the positive and negative ideal solutions for the first application scenario are shown in Table 1.

Table 1. The calculation results of positive and negative ideal solutions in the first application scenario.

Design scheme	S_i^+	S_i^-	C_i^*
Gear 1	0.0914	0.0523	0.3640
Gear 2	0.0898	0.0522	0.3675
Gear 3	0.0627	0.0857	0.5775

(ii) The second scenario:

The calculation results of the positive and negative ideal solutions for the second application scenario are shown in Table 2.

Table 2. The calculation results of positive and negative ideal solutions in the second application scenario.

Design scheme	S_i^+	S_i^-	C_i^*
Gear 1	0.0990	0.1146	0.5365
Gear 2	0.1200	0.0908	0.4308
Gear 3	0.1426	0.0533	0.2720

Step 13: We determine the optimal gear component design scheme;

(i) The first scenario:

Relative proximity vector: C_i^*=(0.3640, 0.3675, 0.5775). According to the principle of maximum relative proximity, gear 3 corresponding to C_i^*=0.5775 should be selected as the optimal reduction gear design scheme in the current scenario.

(ii) The second scenario:

Relative proximity vector: C_i^*=(0.5365, 0.4308, 0.2720). According to the principle of maximum relative proximity, gear 1 corresponding to C_i^*=0.5365 should be selected as the optimal reduction gear design scheme in the current scenario.

Based on the two different application scenarios set above, actual data calculations have verified that the design schemes of the reduction gearbox obtained in the two different application scenarios are also different, which also demonstrates the effectiveness and feasibility of the proposed comprehensive decision-making evaluation method for the full life cycle attributes of gearbox product design.

5. Conclusion

By analyzing the application scenarios and lifecycle of gearboxes from design, manufacturing, and use to disposal, with a focus on the selection, manufacturing, and operation stages of gearbox materials, and combining WLS and TOPSIS comprehensive decision-making evaluation methods, a decision-making method for the lifecycle attributes of key components of gearboxes is proposed. Based on evaluation indicators such as natural resource consumption, energy consumption, greenhouse gas emissions, and cost, the optimal reduction gear design scheme is analyzed and decided for two different application scenarios, namely the manufacturing stage and the raw material selection stage. This method can be flexibly applied to different component usage scenarios by focusing on different optimization objectives, providing more suitable gearbox product design and selection schemes for workers. However, this method still has a certain degree of subjectivity in the expert evaluation process. In the future, we will continue to improve the algorithm evaluation system and use technologies such as big data analysis to enrich the influencing factors and weight factors of product lifecycle attributes, making the evaluation results more referenceable.

References

[1] Chou J. R. (2024). An integrative review exploring the development of sustainable product design in the technological context of Industry 4.0. Advanced Engineering Informatics, 62, 102, 689.
[2] Ji X., Gao Q., and Wang H. A bilevel-optimization approach to determine product specifications during the early phases of product development: Increase customer value and reduce design risks [J]. Expert Systems with Applications, 2022, 188: 116, 012.
[3] Kong L., Wang L., Li F., Li J., Wang Y., Cai Z., ... & Wang G. (2023). Life cycle-oriented low-carbon product design based on the constraint satisfaction problem. Energy Conversion and Management, 286, 117, 069.
[4] Chunhua F., Shi H., & Guozhen B. (2020). A group decision-making method for sustainable design using intuitionistic fuzzy preference relations in the conceptual design stage. Journal of Cleaner Production, 243, 118, 640.
[5] Ng C. Y. & Law K. M. (2024). Eco-performance evaluation of product designs—an integrated fuzzy complex proportional assessment with life cycle assessment. Enterprise Information Systems, 18 (3), 2, 300, 995.
[6] Pacana A., Siwiec D., Bednárová L., & Petrovský J. (2023). Improving the Process of Product Design in a Phase of Life Cycle Assessment (LCA). Processes, 11 (9), 2, 579.
[7] Aydin R. & Badurdeen F. (2019). Sustainable product line design considering a multi-lifecycle approach. Resources, Conservation, and Recycling, 149, 727-737.
[8] Ibn-Mohammed T., Yamoah F. A., Acquaye A., Omoteso K., & Koh S. C. L. (2024). Enhancing life cycle product design decision-making processes: Insights from normal accident theory and the satisficing framework. Resources, Conservation, and Recycling, 205, 107, 523.
[9] Zhang X. & Ming X. (2022). Comprehensive understanding of smart product service system from multi-dimension and multi-perspective: an innovative service model for Customer-product Interaction Life Cycle (CILC). Advanced Engineering Informatics, 52, 101, 619.
[10] Luo X., Guo S., Du B., Guo J., Jiang P., & Tan T. (2023). Multi-criteria decision-making of manufacturing resource allocation for complex product systems based on intuitionistic fuzzy information entropy and TOPSIS. Complex & Intelligent Systems, 9 (5), 5, 013-5, 032.
[11] Tzeng G. H. & Huang J. J. (2011). Multiple attribute decision making: methods and applications. CRC press.

Influence of increasing temperature deformation on the removal uniformity of chemical mechanical polishing plate and improvement method

Guoqing Ye and Zhenqiang Yao *

School of Mechanical Engineering, Shanghai Jiao Tong University, Shanghai, China

* Corresponding author: zqyao@sjtu.edu.cn

Abstract—Chemical mechanical polishing is a process of precision machining of semiconductor silicon wafers. The removal uniformity affects the important inspection index total thickness variation (TTV) of silicon wafers. In this study, first, the geometric model of the circular arc deformation of the polishing plate and the calculation formula of the effect on the removal of nonuniformity is established. Second, the influences of the cooling water pressure, polishing pressure and thermal deformation of the polishing plate material on the deformation of the polishing plate are simulated, and the thermal expansion is found to be the most important factor. To reduce the thermal deformation of the polishing plate material, we developed a super invar material with a very low thermal expansion coefficient to achieve this goal. Finally, a silicon wafer polishing test verified that the removal uniformity of silicon wafer polishing can be improved.

1. Introduction

Chemical mechanical polishing (CMP) is a method for achieving high-precision planar processing in the semiconductor field [1][2]. The purpose of CMP polishing of silicon wafers is to remove the defects and damage layers left in the previous machining process under the combined action of mechanical and chemical methods [3] and obtain the required flatness and surface roughness to achieve the mirror effect to meet the process requirements in the field of integrated circuit (IC) manufacturing [4]. The most famous material removal model for chemical mechanical polishing is the Preston formula, which is MRR = KPV [5], where MRR (material removal rate) is the removal rate, K is the Preston constant, P is the polishing pressure applied to the effective polishing area and V is the relative running speed of the two interfaces. The shape of the polishing pad affects the pressure distribution and thus affects the uniformity of material removal during CMP. Therefore, many scholars have studied the polishing pad, as well as the temperature rise characteristics during the polishing process and the cooling effect of the polishing pad [6]. However, the polishing plate was heated from room temperature (23°C ± 3°C) to 30–40°C during processing. The increase in temperature causes deformation of the polishing plate, resulting in an uneven pressure distribution, which affects the MRR unevenly; that is, nonuniformity is removed [7], which leads to different wafer thicknesses and affects the total thickness variation (TTV) [8].

In this study, a curved geometric model of the deformation of a plate is established, a mathematical calculation of the removal uniformity is carried out, and a mathematical relation algorithm of the TTV

Journal of Physics: Conference Series **2992** (2025) 012032 doi:10.1088/1742-6596/2992/1/012032

change caused by the deformation of the polishing plate is obtained. By comparing the polishing plate because of the cooling water channel pressure, the four polishing head pressures and the polishing plate's temperature increase with respect to the three factors, the polishing plate because of the increase in temperature thermal deformation is the main factor affecting deformation. Compared with several commonly used polishing plate materials, such as stainless steel, cast iron and their combination, a lower thermal expansion coefficient significantly affects the thermal deformation of the polishing plate. Therefore, a super Invar [9] material suitable for polishing plates was developed, which can result in a low thermal expansion coefficient of 1.77E-6/°C. Experimental verification under the same process conditions revealed that the TTV of a high-temperature CMP silicon wafer polished with a temperature increase of 20 K was significantly improved.

2. Methodology

2.1 Principle of silicon wafer single-side CMP

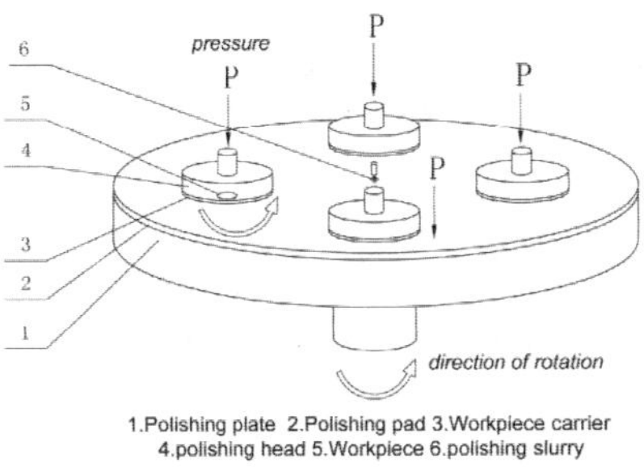

Figure 1. Single-side CMP schematic of a silicon wafer.

As shown in Figure 1, the polishing plate rotates approximately 60 rpm, and the polishing head (4 groups) rotates approximately 60 rpm in the same direction. The polishing head applies a set pressure to the workpiece (silicon wafer) above the workpiece carrier. The abrasive particles are SiO_2, and the alkaline slurry with KOH is sprayed on the polishing pad. The CMP process involves mechanical polishing removal after chemical corrosion softening. The polishing solution contains KOH, and its chemical reaction with the workpiece Si is as follows:

$$Si + 2OH^- + H_2O \rightarrow SiO_3^{2-} + 2H_2\uparrow$$
$$SiO_3^{2-} + H_2O \rightarrow H_2SiO_3 + 2OH^-$$

2.2 Geometric model and mathematical calculation of a deformable polishing plate

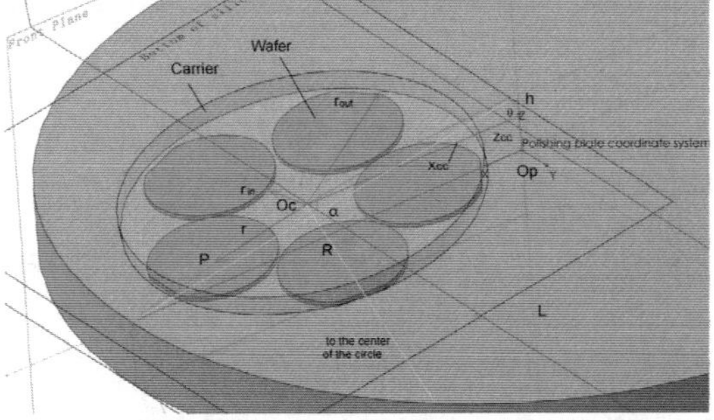

Figure 2. Geometric relationship diagram of the convex polishing plate and carrier.

The deformation of the polishing plate is regarded as a convex circular arc in the middle; if it is a concave middle case but the deformation direction is different, the principle is the same.

As shown in Figure 2, a three-dimensional Cartesian coordinate system is established with the center of the ideal datum plane of the polishing plate as the coordinate origin Op, where h represents the height of the thermal deformation projection, R is the radius of the polishing plate, and the connection angle between the axis OZ and the point (0,0, h) and the point (R, 0,0) is θ. The bus of the created polishing plate deformation arc geometric model passes through the points (0, 0, h) and (R, 0,0) and at the point (0, 0, h) tangent to the plane that passes through this point and is parallel to OXY so that the radius of the arc is L.

On the basis of the center of the carrier, the origin of the carrier coordinate system is Oc, and the bottom surface of the silicon wafer is taken as the plane polar coordinate. The bottom surface of the silicon wafer is tangent to the arc surface of the polishing plate, and the tangent point is Oc, where Oc is on the circular arc bus of the OXZ plane shown in Figure 2. P is any point on the silicon wafer, r is the distance from the carrier P point to Oc, r_{in} is the innermost distance from the silicon wafer to the center Oc, and r_{out} is the outermost distance from the silicon wafer to the center Oc. Therefore, the value range of r is $r_{in} \le r \le r_{out}$. α is defined as the angle between the horizontal plane at Oc and the line between the center of the circle of point Oc and the arc bus.

It is derived from the relationship between trigonometric functions:

$$tan\theta = \frac{R}{h}, L = \frac{R}{sin2\theta} \tag{1}$$

To evaluate the position of the P point rotated by the carrier, the Oc coordinate is set as (x_{cc}, y_{cc}, z_{cc}), and the coordinate expression in the graphical state is as follows:

$$x_{cc} = constant$$

$$y_{cc} = 0$$

From the right triangle Pythagorean theorem,

$$R^2 + (\mathrm{L} - \mathrm{h})^2 = x_{cc}{}^2 + [\mathrm{L} - (\mathrm{h} - \mathrm{z}_{cc})]^2 = L^2 \tag{2a}$$

Solving the formula, we can obtain:

$$z_{cc} = h - (\frac{h}{cos^2\theta} - \sqrt{\frac{h^2}{cos^4\theta} - 4x_{cc}{}^2})/2 \tag{2b}$$

We make t be the intermediate variable; then, the parameter vector formula during the rotation of any point on the silicon wafer is as follows:

$$\vec{p}(t) = \begin{cases} x_c(t) = rsin\alpha cost + x_{cc} \\ y_c(t) = sint \\ z_c(t) = -rcos\alpha cost + z_{cc} \end{cases} \tag{3}$$

Formula (3) yields the trigonometric function relationship:

$$cos\alpha = \frac{x_{cc}}{L}$$

$$sin\alpha = \frac{L - (h - z_{cc})}{L}$$

By substituting them and the expression of Formula (1) $L = \frac{R}{sin2\theta}$ into Formula (3), we can obtain Formula (4):

$$\vec{p}(t) = \begin{cases} x_c(t) = r\left[1 - \frac{(h - z_{cc})sin2\theta}{R}\right] cost + x_{cc} \\ y_c(t) = sint \\ z_c(t) = -r\frac{x_{cc}sin2\theta}{R} cost + z_{cc} \end{cases} \tag{4}$$

The normal vector of the Carrier is expressed as:

$$\vec{v}(t) = (cos\alpha, 0, sin\alpha)$$

$$=(cos\alpha = \frac{x_{cc}sin2\theta}{R}, 0{,}1 - \frac{(h - z_{cc})sin2\theta}{R}) \tag{5}$$

The vector function parallel to the normal vector and expressed by the analytic point P is $\vec{V}(m)$.

$$\vec{V}(m) = \begin{cases} x_p(m) = x_c + m\frac{x_{cc}sin2\theta}{R} \\ y_p(m) = y_c \\ z_p(m) = z_c + m(1 - \frac{(h-z_{cc})sin2\theta}{R}) \end{cases} \tag{6}$$

The spherical formula of the arc is expressed as follows:

The centroid coordinate (0, 0, h-L) is (0, 0, h-Rsin2θ). Therefore, the spherical formula is as follows:

$$x^2 + y^2 + (z - \text{h} + \frac{R}{sin2\theta})^2 = \frac{R^2}{sin^2 2\theta} \tag{7}$$

Substituting Formula (6) into Formula (7), we can obtain:

$$(x_c + m\frac{x_{cc}sin2\theta}{R})^2 + {y_c}^2 + (z_c + m(1 - \frac{(h - z_{cc})sin2\theta}{R}) - \text{h} + \frac{R}{sin2\theta})^2 = \frac{R^2}{sin^2 2\theta} \tag{8}$$

Formula (9) is obtained by converting Formula (8) into a quadratic formula with one variable of approximately m.

$$\frac{{x_{cc}}^2 sin^2 2\theta + [R - (h - z_{cc})sin2\theta]^2}{R^2} m^2 + \frac{2\left\{x_c x_{cc} sin2\theta + (z_c - \text{h} + \frac{R}{sin2\theta})[(R - (h - z_{cc})sin2\theta)]\right\}}{R} m + {x_c}^2 + {y_c}^2 + (z_c - h)^2 + 2(z_c - h)\frac{R}{sin2\theta} = 0 \tag{9}$$

The coefficients of this quadratic formula with respect to m for one variable are as follows:

$$a = \frac{{x_{cc}}^2 sin^2 2\theta + [R - (h - z_{cc})sin2\theta]^2}{R^2}$$

$$b = \frac{2\left\{x_c x_{cc} sin2\theta + (z_c - \text{h} + \frac{R}{sin2\theta})[(R - (h - z_{cc})sin2\theta)]\right\}}{R}$$

$$c = {x_c}^2 + {y_c}^2 + (z_c - h)^2 + 2(z_c - h)\frac{R}{sin2\theta}$$

Therefore, the solution form is:

$$m_d = \frac{-b+\sqrt{b^2-4ac}}{2a} \tag{10}$$

Therefore, the distance from point p to the surface is:

$$d_{cv} = \sqrt{(x_d - x_c)^2 + (y_d - z_c)^2 + (z_d - z_c)^2} = |m_d| \tag{11}$$

Therefore, it affects the thickness of the silicon wafer.

$$\delta_{cv} = |m_d|_{max} - |m_d| \tag{12}$$

By substituting R=457, δ_{cv} is 0.04, 0.07 and 0.14 when the deformation h is 0.3, 0.5 and 1, respectively. Therefore, the smaller the deformation h value is, the smaller the impact is on the TTV.

2.3 Comparison of the influence of three factors on the deformation h of the polishing plate

The internal structure of the polishing plate of the single-sided polishing machine adopts a radial cooling channel structure. The deformation of the polishing plate is studied from three aspects. The first aspect is the deformation caused by the internal cooling water, the second aspect is the deformation caused by the pressure of the four polishing heads, and the third aspect is the thermal expansion deformation caused by the heat generated by the material itself during the processing process.

To compare the deformation sizes of several commonly used polishing plate materials, the following three simulation comparisons are made. The values of Young's modulus E of several commonly used polishing plate materials are shown in Table 1.

Journal of Physics: Conference Series **2992** (2025) 012032 doi:10.1088/1742-6596/2992/1/012032

Table 1. Young's modulus of several commonly used materials for polishing plates.

	SUS316L	Invar	low expansion coefficient cast iron	SUS410
Young's modulus E (GPa)	193	123	130	200

Because the difference in Young's modulus E of the four materials is 1.62 times, which is not on the order of magnitude, the simulation calculation is carried out according to Young's modulus E of 200 GPa.

Figure 3. Simulation of the influence of 0.3 MPa pressure of cooling water on the flatness of the polishing plate.

As shown in Figure 3, under a cooling water pressure of 0.3 MPa, the flatness deformation of the polishing plate is simulated to be 0.17 μm. Owing to the different Young's moduli E of the above materials, the deformation range caused by cooling water deformation is between 0.17 and 0.28 μm.

As shown in Figure 4, the polishing plate is simulated under the pressure of the polishing head. When the four polishing heads are pressurized by 150 kgf, the deformation of the polishing plate is 0.5 μm. Owing to the different Young's moduli E of the above materials, the deformation range caused by the pressure deformation of the polishing head is between 0.5 and 0.89 μm.

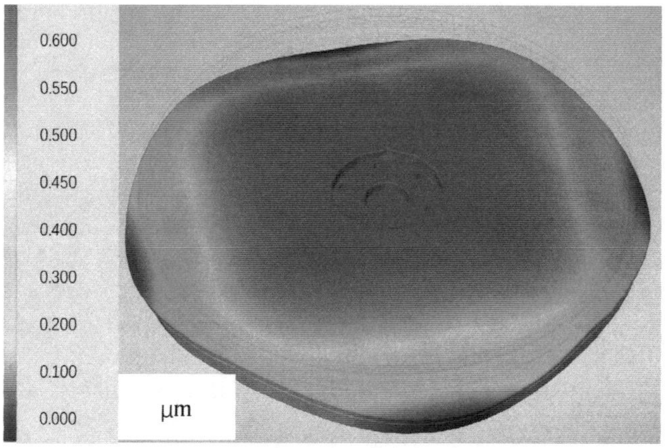

Figure 4. Deformation of the polishing plate under static pressure at the four polishing heads.

Table 2. Deformation of the polishing plates of several materials with different thermal expansion coefficients.

Recipe No.	Water-cooled plate material /expansion coefficient (E-6/°C)	Panel materials /expansion coefficient (E-6/°C)	Maximum deformation of polishing plate (μm)
1	Low thermal expansion cast iron/4.5	SUS316/15.9	83
2	Low thermal expansion cast iron/4.5	SUS410/10	44
3	SUS410/10	SUS410/10	31
4	Invar/1.5	Invar/1.5	3.3

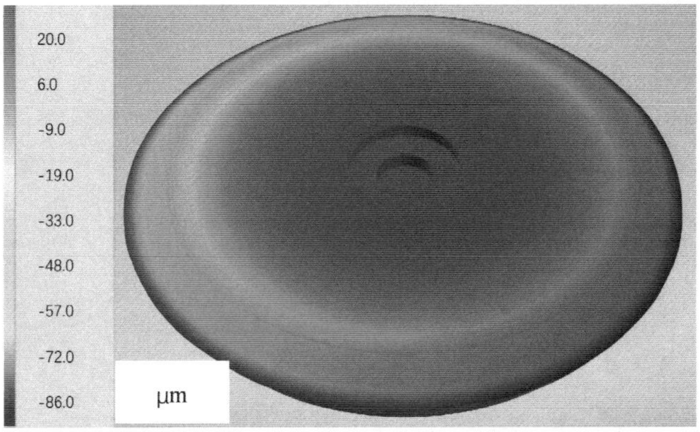

Figure 5. Deformation simulation under a temperature increase of 20 K.

As shown in Figure 5, as the temperature increased from 20°C to 40°C, the polishing plate changed into a convex shape in the middle. Although the thermal expansion coefficients of different materials are different, due to the same working conditions, their deformation modes should be the same, and there should be only a difference in the magnitude of deformation. Materials with different thermal expansion coefficients are simulated and compared, and the deformation is obtained, as shown in Table 2.

Table 2 shows that panels and water-cooled plates with different materials have different coefficients of thermal expansion, resulting in significantly greater deformation than the same material does. Compared with Recipe 3, Recipe 2 also uses a material with a lower thermal expansion coefficient, but the maximum thermal deformation is greater than that of Recipe 3. Therefore, it is better to use the same material in recipe selection.

Additionally, owing to the interchangeability of the polishing plate installation and the size limitations of the mechanical design, all the simulations were performed with the same material thickness and part size. If the thickness of the same material is increased, the thermal deformation improvement effect is not obvious after simulation, so a newly developed super invar material is adopted to improve the thermal deformation.

Table 2 also shows that among the four recipes, the first recipe uses a water-cooled plate as the cast iron with a low expansion coefficient, and the panel uses a SUS316 polishing plate to produce a maximum thermal deformation of 83 μm, which is much greater than that of 3.3 μm of Invar's material.

Table 3. Comparison of the influences of the three factors on the deformation of the polishing plate.

	Influence of cooling water	Influence of polishing head pressure	Material temperature rise 20 k thermal deformation
Maximum deformation of polishing plate (μm)	0.17-0.28	0.5-0.89	3.3-83

Journal of Physics: Conference Series **2992** (2025) 012032 doi:10.1088/1742-6596/2992/1/012032

From the influence of the above cooling water on the deformation of the polishing plate, the influence of the polishing head pressure on the polishing plate and the influence of the thermal deformation of the material (see Table 3), the main factor affecting the deformation is the thermal deformation of the material.

Because thermal deformation has the greatest influence on the polishing plate, in this work, a new polishing plate material with a low thermal expansion coefficient and chemical corrosivity is developed. Through experiments, a super Invar material was developed, which consists of a ternary alloy of 28.08% nickel, 7.15% cobalt, and 64.77% iron. Using a casting process, a material with an expansion coefficient of 1.7E-6/°C can be obtained. The deformation and thermal expansion coefficient of the material are shown in Figure 6.

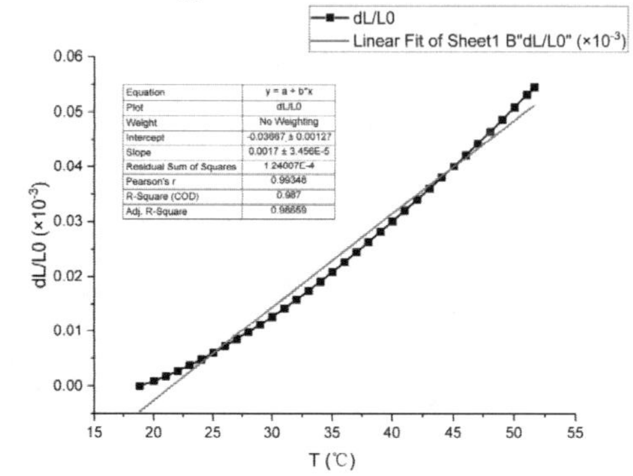

Figure 6. The thermal deformation and expansion coefficient of the new material.

2.4 Silicon wafer polishing test

Polishing plates made of two different materials were installed on the polishing machine for comparative testing. See Figure 7. The test process conditions and environmental conditions are the same. The data of the processed silicon wafer are monitored, and the TTV data are compared. The test conditions are shown in Table 4.

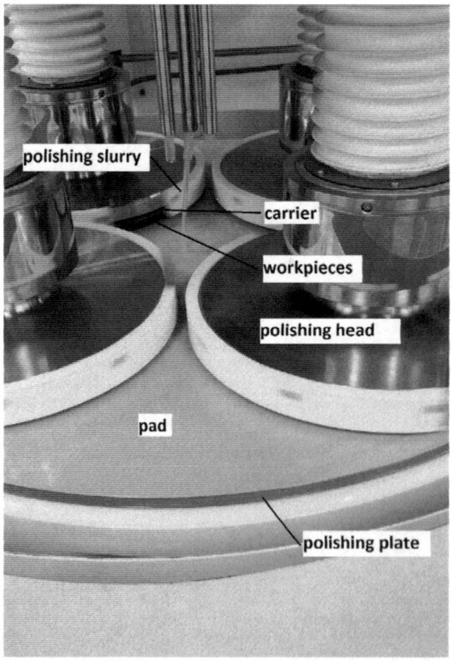

Figure 7. Silicon wafer polishing test.

Table 4. Experimental conditions.

Pad	MH™-S15S (NITTA DuPont INCORPORATE) Diameter: 914 mm
Polishing slurry	AJ-7120 colloidal silica, Fujimi Chemical additive: KOH, PH value of polishing solution: PH11 Flow rate: 2 L/min
Carrier	SiC ceramic (Waxing with silicon wafers), 4 pcs
workpiece	Material: Silicon wafer Diameter: 125 mm, Thickness:540 μm Quantity: 4×5=20
Rotation condition	Polishing plate: 60 rpm Polishing head: 60 rpm
Applied load	250 N (15 sec)→1050 N (29 min)→250 N (75 sec)
Rotational speed of polishing plate/head	60 rpm/60 rpm

3. Results and Discussions

Under the conditions of the four different polishing plate materials in Table 2, the same process recipe in Table 4 was used for comparative testing. Each type of polishing plate is tested 5 times, each time with a full load of 20 silicon wafers. The average TTV of 20 silicon wafers in each test was calculated, and a comparison chart was drawn, as shown in Figure 8.

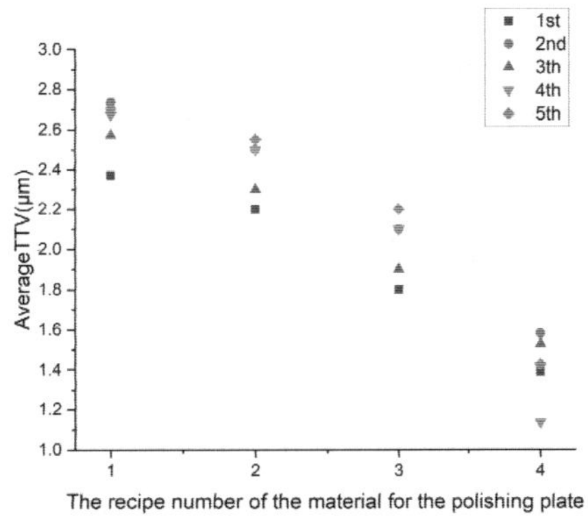

Figure 8. TTV comparison diagram of the processed silicon wafers with different polishing plates.

Figure 8 shows the average TTV values of the silicon wafers measured by the polishing experiments under the different polishing plate material recipes in Table 2. The TTV of Recipe 1 is the largest, and that of recipe 4 is the smallest, which is related to their coefficient of thermal expansion, indicating that the thermal expansion coefficient of the invar material is the smallest, and the TTV data of the silicon wafer obtained are also the best among the four kinds.

To determine why the thermal deformation of the polishing plate affects the TTV, the model carrier shown in Figure 2 was dissected, as shown in Figure 9.

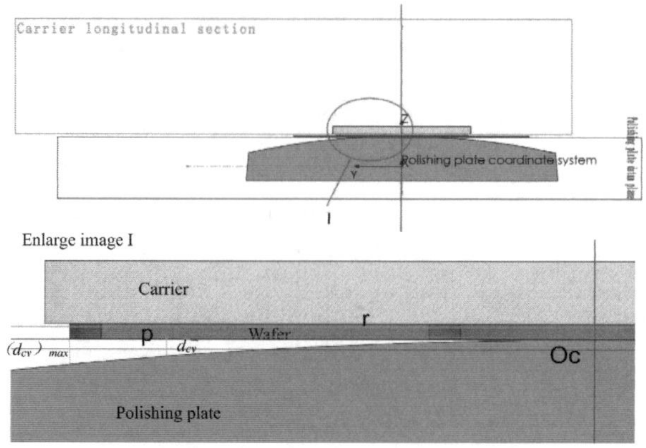

Figure 9. Carrier longitudinal profile.

$$\Delta L = \alpha L \Delta t \tag{13}$$

where ΔL is the amount of deformation, α is the coefficient of thermal expansion, L is the dimension of the deformation direction, and Δt is the temperature change.

In this model, the sizes of the different polishing plates and the temperature increase during processing are the same; however, the difference is that the coefficient of thermal expansion α is that the lower α is, the lower the degree of thermal deformation is, that is, the lower the $(d_{cv})max$ in Figure 9 is.

$$TTV = Max(t_1, t_2, t_3, t_4, t_5) - Min(t_1, t_2, t_3, t_4, t_5) \tag{14}$$

where t_1~t_5 are the thicknesses measured from the center of the silicon wafer and the other 4 positions near the edge.

Different gaps d_{cv} lead to different pressures P, and it can be seen from the Preston formula that MRR=KPV results in different MRRs. The difference in MRR affects the thickness of the silicon wafers. Formula (14) shows that if the thickness difference is small, the TTV level will be improved. This is the mechanism by which a reduction in the thermal expansion coefficient of the polishing plate material can improve the TTV of the silicon wafer.

In addition, different materials with different thermal expansion coefficients lead to greater thermal deformation, so it is better to use the same material. At present, we have not performed experiments in the form of polishing plates composed of Invar materials and other materials.

4. Conclusion

By establishing a geometric model and mathematical formula for polishing plate deformation, the influence of polishing plate deformation on removal uniformity can be calculated. A comparison of the three aspects of the influence of cooling water on the polishing plate, the pressure of the polishing head and the increase in temperature revealed that thermal expansion deformation was the main factor. Compared with several commonly used polishing plate materials, a material with a formula of 28.08% nickel, 7.15% cobalt and 64.77% iron was developed, and the thermal expansion coefficient was measured to be 1.7E-6/°C.

Through experiments, under the same temperature rise and process conditions, the best TTV level can be obtained for the polishing plate of the invar material. After the experimental results were obtained, the internal mechanism between them was analysed. The difference in deformation caused by the different thermal expansion coefficients of the polishing plate material led to the difference in pressure P. According to the Preston formula, MRR=KPV, which results in differences in the uniformity of polishing removal. Therefore, the newly developed Invar material with a low coefficient of thermal expansion can improve the TTV level of polished silicon wafers. In addition, the newly developed invar material has corrosion resistance and meets the requirements of sensitive elements for chemical and mechanical polishing of silicon wafers, such as copper and other elements, which have also been confirmed by alkaline slurry experiments to meet the requirements.

Acknowledgement
This research was funded by the Key R&D Program of Shandong Province, China, grant number 2021CXGC010205.

References

[1] Babu Suryadevara. Advances in Chemical Mechanical Planarization (CMP), Second Edition. [M] Elsevier. 2022, p. 3
[2] Debottam Datta, Himanshu Rai, Swarnima Singh, Meenakshi Srivastava, Rajesh Kumar Sharma, Nitya Nand Gosvami, Nanoscale tribological aspects of chemical mechanical polishing: A review, Applied Surface Science Advances, Volume 11, 2022.
[3] Fanning Meng, Zhenyu Zhang, Zinuo Zeng, Cheng Fan, Yang Gu, Fei Gao, Hao Luo, Jianmei Wang, Guo Li, Review on polishing of atomic-level surfaces for potassium dihydrogen phosphate, Tribology International, Volume 198, 2024.
[4] Pandey, Kheelraj, Ashwani Sharma, and Ayush Kumar Singh. "Silicon Wafers; Its Manufacturing Processes and Finishing Techniques: An Overview. " SILICON 14. 18 (2022): 12031–12047. Print.
[5] Zhun Luo, Zhenyu Zhang, Feng Zhao, Cheng Fan, Junyuan Feng, Hongxiu Zhou, Fanning Meng, Xuye Zhuang, Jianmei Wang, Advanced polishing methods for atomic-scale surfaces: A review, Materials Today, Sustainability, Volume 27, 2024
[6] K. Park, J. Park, B. Park, and H. Jeong, Correlation between break in characteristics and pad surface conditions in silicon wafer polishing. [J]. Mater. Process. Tech. 205(1)-(3), 360–365 (2008).
[7] M. Tsujimura, 16 - Chemical mechanical polishing (CMP) removal rate uniformity and role of carrier parameters, Editor(s): Suryadevara Babu, In Woodhead Publishing Series in Electronic and Optical Materials, Advances in Chemical Mechanical Planarization (CMP) (Second Edition), Woodhead Publishing, 2022, Pages 451-467.
[8] Liu, Yuanhang et al. An Investigation on the Total Thickness Variation Control and Optimization in the Wafer Backside Grinding Process. Materials 15. 12 (2022): 4230. Print.
[9] Renjie Zhao, Yakai Zhao, Meng Qin, Yubo Jia, Kai Feng, Zhuguo Li, Upadrasta Ramamurty, Renbiao Xie, Influence of process parameters on properties of Super Invar alloy fabricated by laser powder bed fusion for semiconductor equipment, Additive Manufacturing, Volume 92, 2024.

Research on stress distribution and dispersion of cylinder head

Xuanyi Zhao, Weiqing Huang, Peirong Ren *, Pengyu Xia, Jiayu Kuang and Jiaheng Tang

School of Mechanical Engineering, Beijing Institute of Technology, Beijing 100081, China

*Corresponding author's e-mail: renpr@bit.edu.cn (Peirong Ren)

Abstract: The increasing power density and lightweight requirements of engines pose higher challenges to cylinder head reliability, while the actual stress dispersion in the cylinder head poses an obstacle to cylinder head reliability checking. This study investigates the stress distribution and dispersion of a specific type of cylinder head through simulation. First, heat transfer calculations are conducted by incorporating flow-solid coupling, comparing temperature results derived from various turbulence models. The model with the highest accuracy is selected to provide precise thermal load inputs for stress calculations. Subsequently, the stress of the cylinder head is calculated while considering heat-mechanical coupling. The critical location of the cylinder head is then identified through fatigue analysis, utilizing the stress simulation results as input. Finally, the stress dispersion at the critical location is analyzed using the Monte Carlo method, accounting for dispersion in preload, material parameters, and burst pressure. The results indicate that temperature calculations using the standard k−ε model yield the highest accuracy, with an error margin of 0.221%. Additionally, the stress dispersion at the critical location follows a normal distribution, with a standard deviation of 2.48 MPa. Among the factors influencing stress dispersion, the dispersion in the material's elastic modulus, burst pressure, and coefficient of linear expansion significantly impact the stress dispersion, contributing 25.5%, 23.5%, and 20.9%, respectively. This study provides a valuable reference for the simulation and evaluation of the structural strength of cylinder heads.

1. Introduction

The cylinder head, as a key component of the combustion chamber in a diesel engine, features a complex structure. During operation, it endures the combined effects of mechanical loads such as gas explosion pressure and bolt pre-tightening force, along with thermal loads generated by combustion. So, it is one of the critical factors affecting the operational reliability of the diesel engine. The inner surface of the cylinder head is situated directly in front of the high-temperature combustion flame, leading to the highest heat load in the engine. Concurrently, the flow of coolant removes heat from the cylinder head. The high density of heat flux creates a significant temperature gradient within the cylinder head, resulting in substantial thermal strain. Additionally, the high pressure generated by combustion exerts considerable mechanical loads on the cylinder head. These factors contribute to the cylinder head being one of the most frequently failing components in engine systems [1]. Therefore, it is essential to analyze the working conditions of the cylinder head.

In the past, experimental measurement was the primary method for analyzing the working conditions of the cylinder head due to limitations in computer software and hardware technology. Rakopoulos et al. [2] conducted an experimental analysis of the transient heat flux at the walls of the combustion chamber during the engine cycle on a four-stroke direct-injection air-cooled diesel engine. With advancements in hardware performance and finite element simulation technology, an increasing number of researchers are turning to finite element analysis to assess the working conditions of the cylinder head. This technique for analyzing the cylinder head's operational state is becoming increasingly refined. The finite element simulation of the cylinder head consists of three main components: input, calculation, and output. Accurate input of material parameters and boundary conditions enhances the precision of finite element calculations. Many researchers have focused on improving the accuracy of these inputs. Farrahi et al. [3] successfully predicted the mechanical properties of light alloy LCF/TMF through numerical simulation with high accuracy, providing a solid foundation for the input of material parameters in the cylinder head simulation. The simulation of the cylinder head involves multi-field coupling, including flow fields, combustion, and static loads, making it challenging to determine the boundary conditions at the coupling interfaces. Some researchers derive the input for each boundary condition of the cylinder head by calculating different physical fields separately. Fontanesi et al. [4] applied a combined CFD and FEM analysis approach, implementing a decoupled analysis of CFD and FEM to accurately estimate the fatigue strength of the cylinder head in a V6 turbocharged diesel engine. Yang et al. [5] obtained the stress-strain field through coupled thermo-mechanical simulation, predicting the fatigue life of the cylinder head using FEMFAT, with results validated through experiments. Sharif et al. [6] employed a decoupling assumption in their structural analysis of the cylinder head, extracting the coolant heat load on the water jacket using three-dimensional CFD analysis. This calculated heat load was then applied to the cylinder head during the stress analysis phase to obtain heat-mechanical coupling results. The accuracy of the computational model significantly impacts the simulation results, particularly regarding the heat transfer model, which is a critical and complex component of the cylinder head calculation. Numerous researchers have conducted in-depth studies on this aspect. Hu et al. [7] calibrated the performance of the diesel engine's heat engine coupling that included the cylinder head and optimized the cylinder head through experimental design. Ostapski [8] used a simplified mathematical model, CAE simulations, and test bench assessments to analyze the impact of thermal and thermomechanical loads on the stress-strain state of an air-cooled piston aero-engine. The results provided insights for proposing technically feasible structural improvements to mitigate uneven heating of the cylinder head and reduce deformation of the cylinder liner near the upper stop. Tan and Yuan [9] optimized the coolant velocity distribution based on simulation analyses of the engine cooling system. Additionally, Tan further improved the coolant velocity distribution through simulation analysis of the engine coolant flow in a single-cylinder water jacket to enhance the cooling effect [10]. Mohsen [11] concluded from simulation analyses that reducing the thickness of the ferrous baffle block could decrease fuel consumption and emissions during the warm-up period.

The evaluation of cylinder head performance through simulation has been extensively researched. However, current simulation studies typically treat material parameters and boundary conditions as constants. In reality, due to microstructural variations arising from the casting process and fluctuations in external loads, the material properties of the cylinder head and the applied loads exhibit a degree of dispersion. This dispersion is a significant factor influencing the performance of the cylinder head. To conduct a more comprehensive simulation assessment of the structural strength of the cylinder head, this paper aims to investigate the stress dispersion at the critical location of a specific type of cylinder head through simulation.

2. Research methodology

The cylinder head operates under both thermal and mechanical loads, necessitating a simulation calculation that accounts for multi-field coupling, specifically involving the interactions of cooling water, high-temperature gas, burst pressure, and bolt preload. The consideration of these multi-physical field couplings increases the complexity of the calculations and can reduce the accuracy

Journal of Physics: Conference Series **2992** (2025) 012033 doi:10.1088/1742-6596/2992/1/012033

of the results. Therefore, this paper simplifies the model used for calculations: for the cylinder head, the mutual influence of the flow field and mechanical loads is relatively minimal. Consequently, decoupling is applied to the flow field and mechanical loads, transforming the flow-solid-thermal coupling calculations into separate flow-solid coupled heat transfer calculations and heat-machine coupling stress calculations. In the flow-solid coupled heat transfer calculation stage, a suitable turbulence model is selected to enhance calculation accuracy and provide precise input for the subsequent heat-machine coupled stress analysis. The stress simulation results, derived from the heat-machine coupling, are then analyzed to identify the critical location within the cylinder head through fatigue calculations. Finally, material and load dispersion are incorporated using the Monte Carlo method during stress simulations to assess the dispersion characteristics of stress at the critical location. This approach also identifies sensitive factors affecting stress dispersion at the location.

2.1. Calculation of flow-solid coupled heat transfer

In the flow-solid coupling analysis, the factors considered include the effects of cooling water and high-temperature gas. The influence of high-temperature gas is incorporated into the calculations through imposed thermal boundary conditions, while the effect of cooling water is integrated through flow-solid coupling, allowing for the calculation of the temperature field of the cylinder head under the combined influence of both cooling water and high-temperature gas. Since the Reynolds number of the cooling water flow exceeds the critical threshold, it can be concluded that the flow is turbulent. Turbulence models are highly empirical, which limits their scope of applicability; therefore, the specific choice of turbulence model must be based on a detailed analysis of the object in question. To select the turbulence model appropriately, this paper employs experimental temperature measurements from cylinder head A of a specific type for verifying the accuracy of turbulence model calculations. Heat transfer calculations for cylinder head A are conducted using standard $k-\varepsilon$, RNG $k-\varepsilon$, and realizable $k-\varepsilon$ turbulence models. Based on comparisons between the temperature simulation results and experimental data, as well as analyses of the flow field simulation results, a suitable turbulence model is chosen for the flow-solid coupling calculations of cylinder head B, which is of the same type. Through these flow-solid coupling heat transfer calculations, the temperature distribution results for cylinder head B can be obtained.

2.1.1. Cylinder head fluid-solid coupling system composition. The geometric model of cylinder head B is shown in Figure 1, and finite element meshing was performed based on the geometric model. To balance computational accuracy and time efficiency, this paper conducts a convergence analysis of the cylinder head mesh size, ultimately yielding a converged mesh, as illustrated in Figure 2.

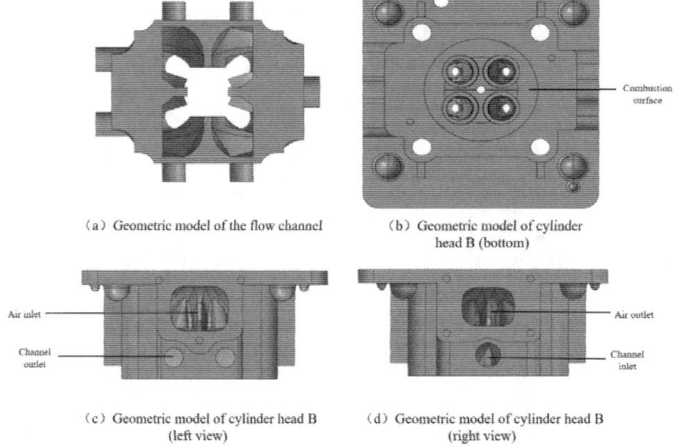

Figure 1. Geometric model of cylinder head B.

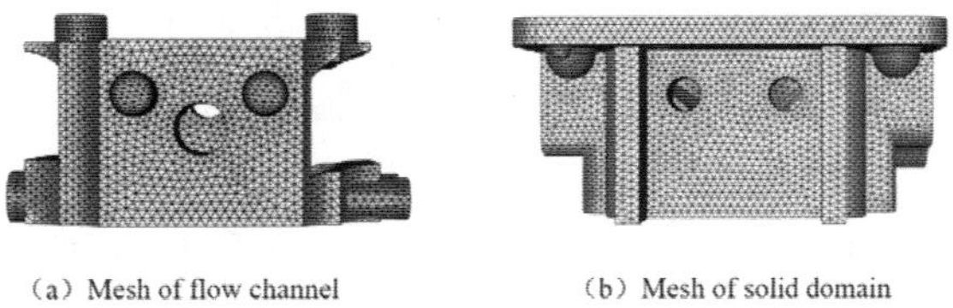

Figure 2. Finite element mesh of cylinder head B.

2.2. Thermo-mechanical coupling stress calculation

In the thermal-mechanical coupling analysis, the influence of the temperature field on solid stress is significantly greater than that of solid stress on the temperature field. Therefore, the thermal-mechanical coupling calculation focuses exclusively on the effect of the temperature field on solid stress, utilizing a one-way coupling method. In this approach, the thermal load of the cylinder head, along with burst pressure and bolt preload, is applied as the loading boundary conditions to obtain the stress field resulting from the combined effects of cooling water, high-temperature gas, burst pressure, and bolt preload. For the specific type of cylinder head, the peaks and valleys of the cyclic loads correspond to working conditions with and without burst pressure. Consequently, this paper comprehensively considers both working conditions in the thermal-mechanical coupling stress calculation. Through this analysis, the stress distribution of the cylinder head under these two conditions can be determined, providing critical data for fatigue calculations.

2.2.1. Thermo-mechanical coupling system composition. The strength of the cylinder head is significantly influenced by both mechanical and thermal loads, necessitating the establishment of a thermo-mechanical coupling system. Given that contact plays a crucial role in stress calculations, the model employed for the thermal-mechanical coupling is an assembly model that includes the cylinder head, bolts, and cylinder block, as illustrated in Figure 3:

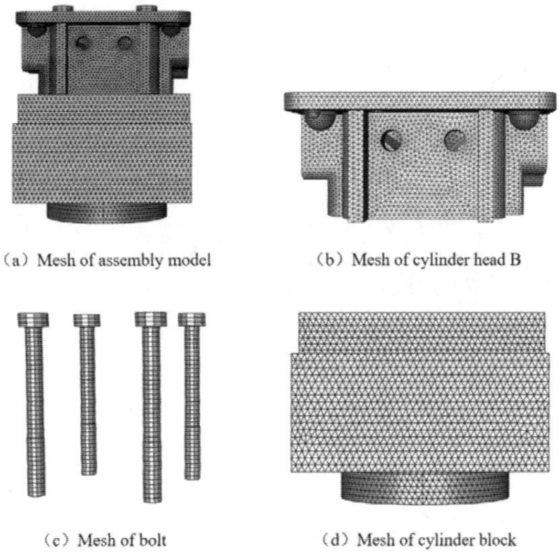

Figure 3. FE Mesh used for thermal-mechanical coupling.

The finite element mesh of the cylinder head used for thermal-mechanical coupling is the C3D4 type mesh in Abaqus. The number of elements was determined following a mesh independence analysis, resulting in a final mesh count of approximately 1.3 million for the cylinder head.

2.2.2. Material parameters and boundary conditions. The material used for cylinder head B is peristaltic cast iron and its material parameters are shown in Table 1:

Table 1. Material parameters of cylinder head B.

Young's Modulus (MPa)	150000
Poisson's Ratio	0.26
Density/(kg/m^3)	7800
Conductivity/(W/(m·K))	41.5
Expansion/(1/K)	1.1E-05

The application of boundary conditions includes displacement boundary conditions, mechanical load boundary conditions, and thermal load boundary conditions.

(1) Displacement boundary conditions: Based on the assembly relationships among the cylinder head, bolts, and cylinder block, the corresponding contact pairs are established to simulate these relationships. Additionally, full constraints are applied to the lower part of the cylinder block in all three translational and rotational directions.

(2) Mechanical load boundary conditions: These include the burst pressure on the combustion surface and the bolt preload.

(3) Thermal load boundary conditions: Utilizing the one-way coupling method, the temperature field results from the flow-solid coupling calculations are incorporated into the thermal-mechanical coupling calculation process as thermal load boundaries.

In this study, the aforementioned conditions serve as inputs for calculating the stresses under thermal-mechanical coupling for the specific type of cylinder head.

2.3. Calculation of fatigue safety factor

Fatigue is the most common form of failure in cylinder heads. Therefore, this paper identifies the critical location based on the fatigue calculation results. The analysis incorporates average stress correction using Goodman's equation for fatigue calculations specific to this type of cylinder head.

The S-N curve high-cycle fatigue model is widely utilized, as it is convenient, accounts for multiple influencing parameters, and is suitable for the strength assessment and life prediction of engineering structures. The S-N curve is based on the assumption of constant amplitude stress. However, since the cyclic load on the cylinder head is not a constant amplitude load, it requires correction by considering the effects of average stress.

Goodman's equation is employed to convert the stress amplitude that includes average stress into an equivalent stress amplitude with an average stress of zero. This average stress correction is applied while maintaining the advantages of the high-cycle fatigue model represented by the S-N curve. Consequently, this paper adopts Goodman's equation for fatigue calculations in the following form:

$$\frac{S_a}{S_a(R=-1)}+\frac{S_m}{S_{UTS}}=1 \tag{1}$$

where S_a and S_m are the actual material stress amplitude and average stress, S_{UTS} is the ultimate tensile strength of the material, and $S_a(R=-1)$ is the stress amplitude when R=-1. The life corresponding to the stress amplitude containing the average stress is obtained by substituting $S_a(R=-1)$ into the S-N curve.

The minimum fatigue safety factor is used to characterize the fatigue performance of the cylinder head, defining the fatigue safety factor as:

$$FOS=\frac{\Delta P_0}{\Delta P} \tag{2}$$

where ΔP_0 is the equivalent load amplitude for the specified life, and ΔP is the actual load amplitude of the cylinder head.

2.4. Stress dispersion analysis

In structural simulation evaluations, the material parameters and boundary conditions of components are typically fixed. However, in practice, factors such as manufacturing dimensional errors, variations in operating cycles, and differences in material production can lead to a certain degree of dispersion in the material parameters and loads, resulting in stress dispersion. This stress dispersion is a significant factor affecting the performance of the cylinder head.

In most cases, the material parameters of the cylinder head and the load can be approximated as following a normal distribution. This paper identifies the factors influencing stress dispersion in the cylinder head and samples these factors according to a normal distribution using the Monte Carlo method to generate several examples. Each sampled example is then analyzed. The results are used to fit the stress distribution curve at the fatigue-critical location of the cylinder head. Additionally, the sensitivity factors affecting stress dispersion are analyzed based on the results of the stress dispersion calculations, providing a precise basis for the simulation and evaluation of the structural strength of the cylinder head.

Based on the factors influencing stress in the cylinder head, the parameters considered for dispersion in the strength simulation assessment include: modulus of elasticity, coefficient of thermal expansion, bolt preload, and burst pressure. In stochastic structural analysis, the load is generally modeled as following a normal or lognormal distribution, while the mechanical property parameters of the material typically adhere to a normal distribution. Thus, in this paper, the parameters considered for dispersion are assumed to conform to a normal distribution, the density function and distribution function of the normally distributed random variable x are as follows [12]:

$$f(x)=\frac{1}{\sqrt{2\pi}\sigma_x}\exp\left[-\frac{1}{2}\left(\frac{x-\mu_x}{\sigma_x}\right)^2\right] \tag{3}$$

$$F(x)=\frac{1}{\sqrt{2\pi}\sigma_x}\int_{-\infty}^{x}\exp\left[-\frac{1}{2}\left(\frac{x-\mu_x}{\sigma_x}\right)^2\right]dx \tag{4}$$

where μ_x and σ_x are the mean and standard deviation of the random variable x, respectively.

The statistical characteristics of the stochastic parameters are determined based on engineering experience and data provided in the literature. The statistical characteristics of the cylinder head parameters are shown in Table 2:

Table 2. Statistical characteristics of cylinder head parameters.

Cylinder head parameters	Distribution type	Coefficient of variation
Young's Modulus/MPa	Normal	2%
Expansion/(1/°C)	Normal	2%
Preload of bolt1(N)	Normal	2%
Preload of bolt2(N)	Normal	2%
Preload of bolt3(N)	Normal	2%
Preload of bolt4(N)	Normal	2%
Burst pressure on combustion surface (MPa)	Normal	2%
Burst pressure on intake valve seat ring (MPa)	Normal	2%
Burst pressure on exhaust valve seat ring (MPa)	Normal	2%

Simple random sampling is conducted for parameters such as preload, material properties, and burst pressure. The values obtained from this sampling serve as inputs, leading to different combinations of parameters and thereby generating various examples. Finite element calculation software is utilized through Isight to perform fluid-solid-thermal coupled finite element analysis for each example, allowing for the determination of stress at the critical location of the cylinder head under each parameter set. Using the obtained stress values, the stress distribution characteristics at the

Journal of Physics: Conference Series **2992** (2025) 012033 doi:10.1088/1742-6596/2992/1/012033

hazardous location are fitted. The sensitivity factors related to stress dispersion at this location are analyzed using the significance analysis module in Isight.

3. Results and discussion

3.1. Flow-solid coupling calculation results

The combustion surface temperatures of cylinder head A, obtained using the standard $k-\varepsilon$, RNG $k-\varepsilon$, realizable $k-\varepsilon$ turbulence models, are presented in Figure 4:

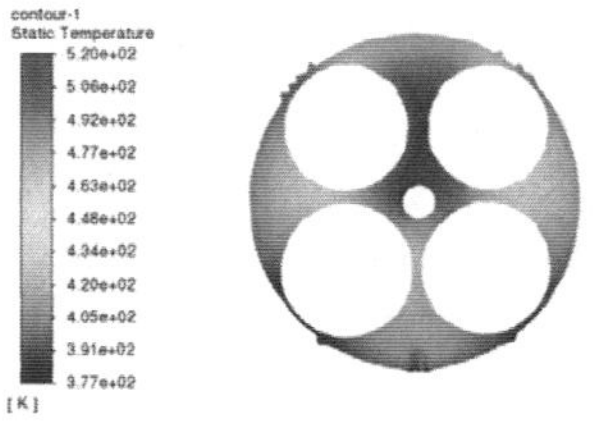

(a) Result under Standard k−ε model

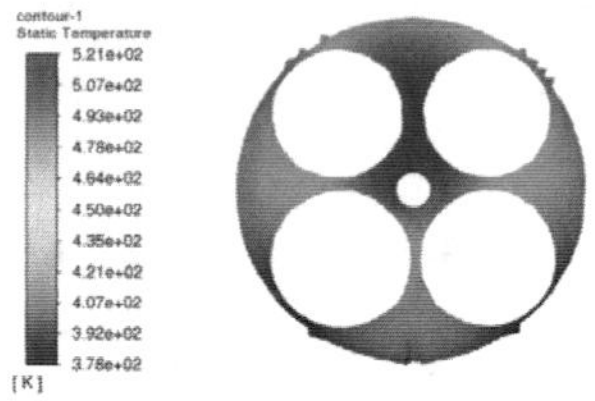

(b) Result under RNG k−ε model

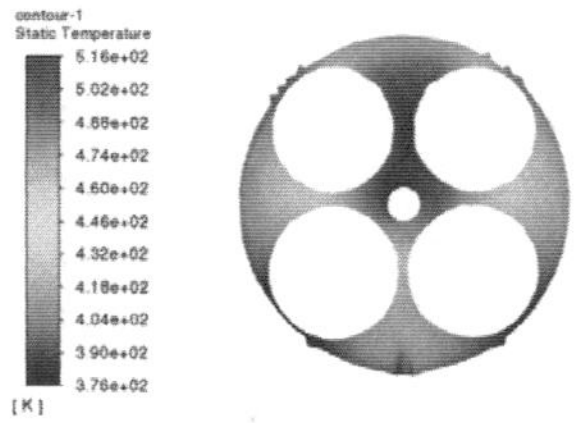

(c) Result under Realizable k−εmodel

Figure 4. Calculation result of combustion surface temperature of cylinder head A.

We drill thermocouple mounting holes at appropriate locations on the combustion surface and install thermocouples to measure the actual temperature of the cylinder head combustion surface. The distribution of the thermocouple measurement points is illustrated in Figure 5:

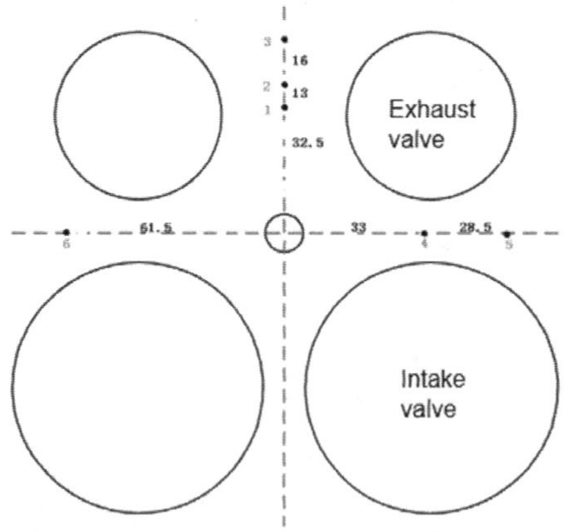

Figure 5. Temperature test points of cylinder head A.

The simulation calculations using the standard $k-\varepsilon$, RNG $k-\varepsilon$, and realizable $k-\varepsilon$ models correspond to Group 1, Group 2, and Group 3, respectively. A comparison of the temperature simulation results with the experimental results for each group at the measurement points shown in Figure 5 is presented in Table 3.

Table 3. Comparison of simulation and experimental results (K) of temperature at each measurement point.

Measurement point	Group 1 simulation results	Group 2 simulation results	Group 3 simulation results	Experimental results
Point 1	520.3K	521.2K	516.0K	520.6K
Point 2	518.7K	519.7K	514.7K	517.9K
Point 3	485.9K	487.4K	483.2K	484.3K
Point 4	470.5K	473.0K	469.0K	472.8K
Point 5	437.3K	439.0K	436.0K	436.2K
Point 6	432.8K	435.1K	431.5K	433.0K

According to the experimental measurement results, the highest temperature on the combustion surface of cylinder head A occurs at the measurement point, reaching 520.6 K. For the six measurement point locations, the average errors in calculations are as follows: 0.221% for the standard $k-\varepsilon$ model, 0.379% for the RNG $k-\varepsilon$ model, and 0.488% for the realizable $k-\varepsilon$ model.

A cross-section analysis was performed at the water chamber beneath cylinder head A, with the resulting cooling water flow velocity vector diagrams under standard $k-\varepsilon$, RNG $k-\varepsilon$, and realizable $k-\varepsilon$ models presented in Figures 6, 7, and 8, respectively. The flow velocity at the interface between the cooling runner and the combustion surface significantly influences heat transfer from the combustion surface. It can be observed that under all three turbulence models, the flow state of the cooling water is largely similar, with a relatively uniform flow velocity distribution at the contact area, characterized by a small velocity gradient and few eddy features.

Consequently, the accuracy of the RNG $k-\varepsilon$ model, which includes velocity gradient correction, does not show a clear advantage over the standard $k-\varepsilon$ model, while the accuracy of the realizable $k-\varepsilon$ model, which incorporates eddy correction, is lower than that of the standard $k-\varepsilon$ model. Additionally, the mathematical formulation of RNG $k-\varepsilon$ model is more complex than that of the standard $k-\varepsilon$ model, necessitating stricter mesh quality for calculations, resulting in greater difficulty in convergence compared to the standard $k-\varepsilon$ model.

A sudden temperature change is observed at the top of the cylinder head under the RNG $k-\varepsilon$ model, as illustrated in Figure 9. This issue can be mitigated by enhancing mesh quality, though it may require significant time and resources. Overall, for cylinder head A, the standard $k-\varepsilon$ model is determined to be the most suitable turbulence model, considering both simulation accuracy and efficiency.

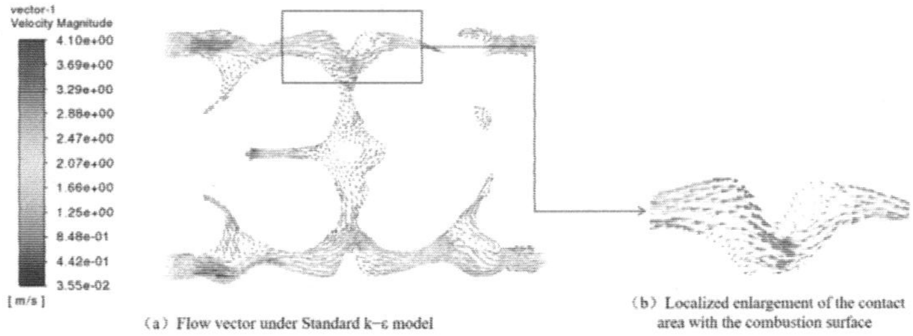

Figure 6. Flow vector diagram of cylinder head cooling water under standard $k-\varepsilon$ model.

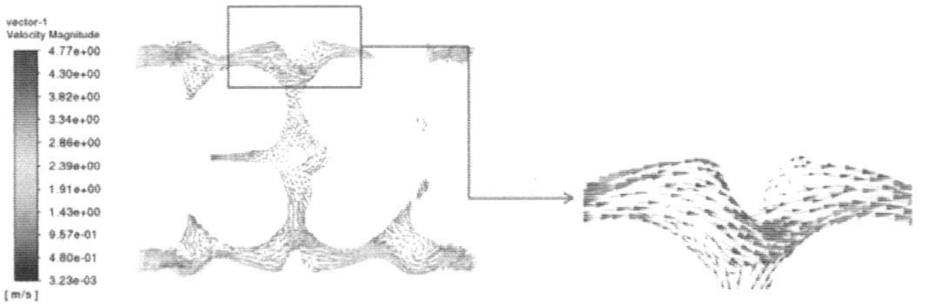

(a) Flow vector under RNG k-ε model

(b) Localized enlargement of the contact area with the combustion surface

Figure 7. Flow vector diagram of cylinder head cooling water under RNG $k-\varepsilon$ model.

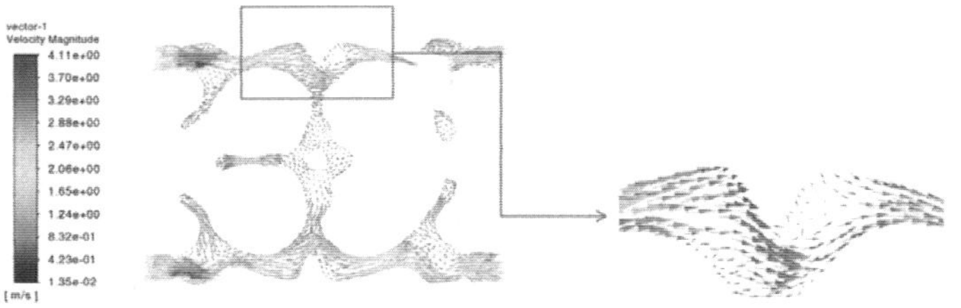

(a) Flow vector under Realizable k-ε model

(b) Localized enlargement of the contact area with the combustion surface

Figure 8. Flow vector diagram of cylinder head cooling water under a realizable $k-\varepsilon$ model.

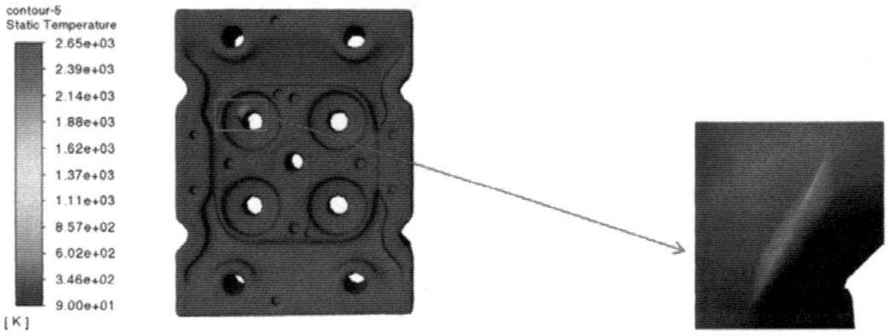

(a) temperature mutation location

(b) Localized amplification of temperature mutation location

Figure 9. Temperature mutation of cylinder head A under the RNG $k-\varepsilon$ model.

Through the calibration of the turbulence model on cylinder head A, it is observed that the standard model is the most suitable turbulence model when the flow at the location where the cooling runner contacts the combustion surface is relatively uniform, the velocity gradient is small, and there are no pronounced eddy features. A cross-section is made in the lower water chamber of cylinder head B under the same model. The cooling water flow velocity vector diagram at this cross-section location is shown in Figure 10. Although eddy features are present near the inlet and outlet, the cooling water flow at the contact point between the cooling runner and the combustion surface exhibits gentle flow direction changes, a small velocity gradient, and inconspicuous vortex features. Therefore, the standard $k-\varepsilon$ model is selected as the turbulence model for the fluid-structure coupling simulation of cylinder head B.

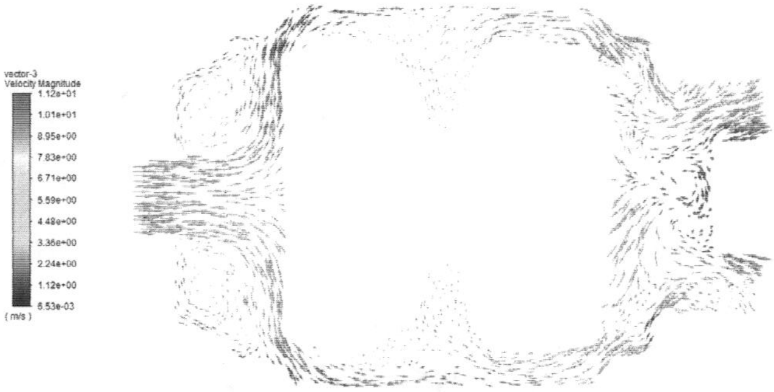

Figure 10. Vector diagram of cooling water flow rate for cylinder head B.

The temperature field of cylinder head B is calculated using the established fluid-solid coupling model. The temperature distribution is illustrated in Figure 11, where the highest temperature recorded is 773 K, located at the nose area of the combustion surface between the exhaust channels. Elevated temperature regions are found near the injector hole, in the nose area of the combustion surface favoring the exhaust side, and within the wall area of the exhaust channel. Among these, the nose area near the exhaust gate exhibits the highest temperature.

The elevated temperatures in these regions result from the heating of the exhaust channel wall surface by high-temperature gas, which raises the wall surface temperature to a maximum of 719 K. The area near the injector hole can reach up to 765 K due to the heating effects of high-temperature and high-pressure gas. Additionally, the nose area of the combustion surface near the exhaust door experiences heating from high-temperature gas and the impact of pressure fluctuations, reaching a maximum temperature of 773 K.

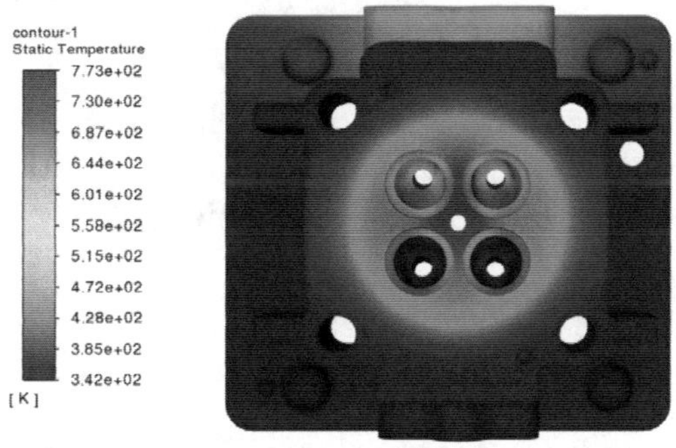

Figure 11. Temperature distribution of the cylinder head B.

3.2. Thermo-mechanical coupling calculation results

With both mechanical and thermal loads applied, the maximum Mises equivalent stress in the cylinder head occurs around the bolt holes, measuring 324.9 MPa. Additionally, the maximum Mises equivalent stress in the combustion face reaches 324.6 MPa. When only mechanical loads are applied, the maximum Mises equivalent stress in the cylinder head around the bolt holes increases to 330.7 MPa, while the maximum Mises equivalent stress on the combustion face is only 98.8 MPa. This observation shows that the thermal load increases stress on the combustion face while reducing stress at the bolt-hole locations. This effect significantly influences the overall stress distribution, underscoring the necessity of considering thermal-mechanical coupling in the strength assessment of the cylinder head. Under conditions without applied burst pressure, the maximum Mises equivalent stress of the cylinder head is found around the bolt holes, measuring 315.1 MPa. The results are shown in Figure 12.

Journal of Physics: Conference Series **2992** (2025) 012033 doi:10.1088/1742-6596/2992/1/012033

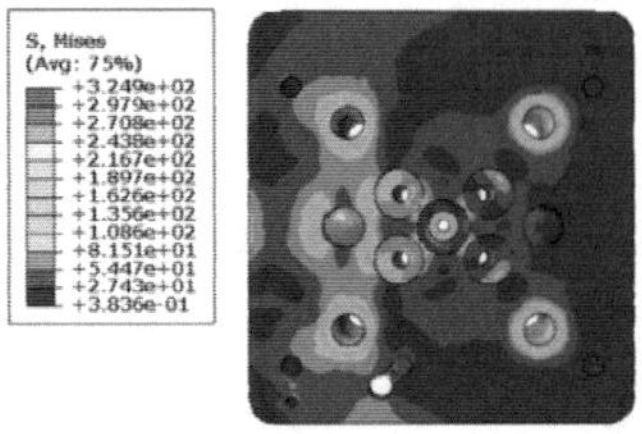

(a) Result of applying thermal load and mechanical load at the same time

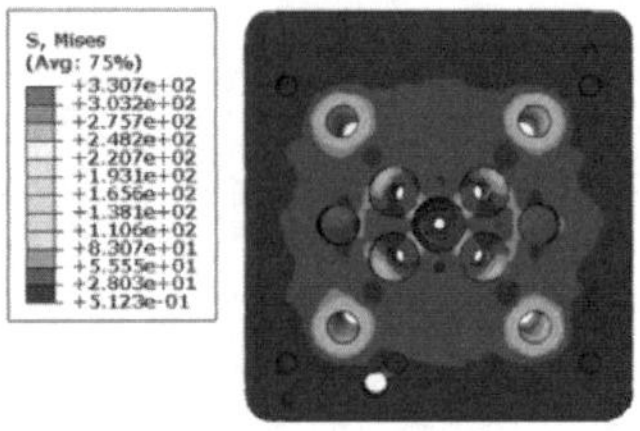

(b) Result of applying mechanical load

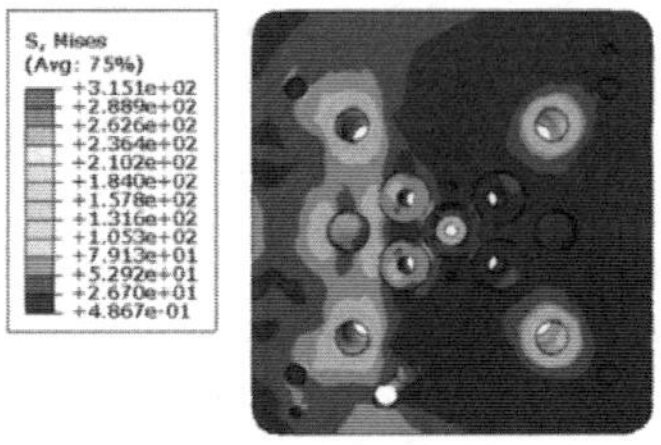

(c) Result of applying thermal load and mechanical load at the same time without explosion pressure

Figure 12. Mises equivalent stress nephogram of cylinder head B.

3.3. *Fatigue safety factor calculation results*

The Cloud image of the fatigue safety factor of cylinder head B is shown in Figure 13:

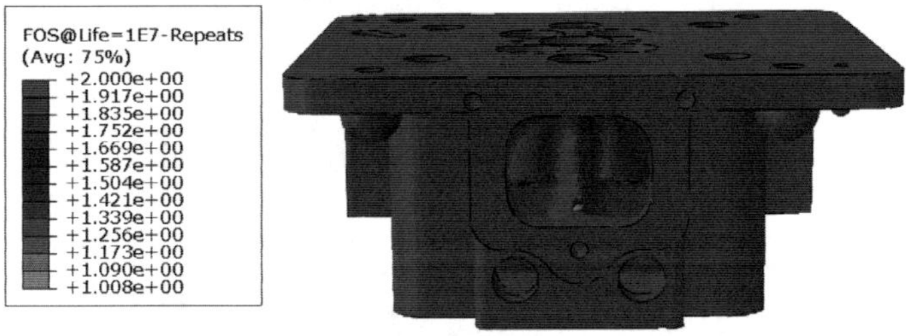

Figure 13. Cloud image of the fatigue safety factor of cylinder head B.

As shown in Figure 13, the most critical position of the cylinder head, indicated by the lowest fatigue safety factor, is located at the inlet channel pinch wall, with a fatigue safety factor of 1.008. When comparing the stress results under burst pressure conditions with those without burst pressure, it is evident that the stress at the inlet channel pinch wall varies significantly. This variation leads to a reduced fatigue safety factor, making the inlet channel pinch wall a particularly vulnerable location for fatigue failure in the cylinder head.

3.4. *Calculation of stress dispersion at the critical location of the cylinder head*

Parameters to be considered for dispersion—factors influencing stress variation—were established with normal distributions using Monte Carlo algorithms via Insight software. The resulting stress distribution pattern at the critical location of the cylinder head is presented in Figure 14:

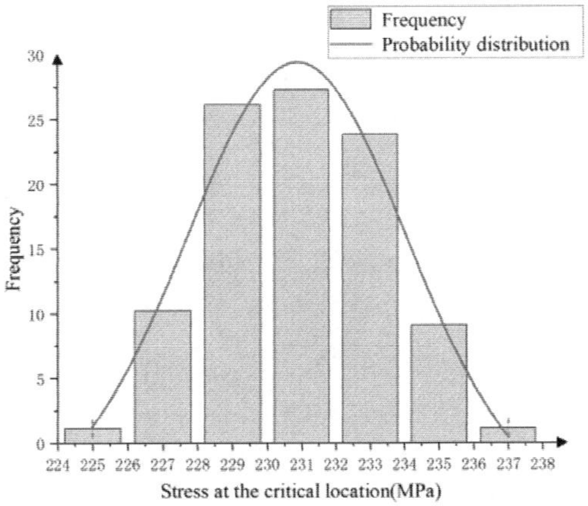

Figure 14. Stress dispersion at the critical location of cylinder head B.

The fitting results in a stress distribution at the critical location of the cylinder head that follows a normal distribution, with a standard deviation of 2.48. Significance analysis indicates that the dispersion of the modulus of elasticity, burst pressure at the fire surface, and coefficient of expansion have a significant impact on stress variation, with influence degrees of 25.5%, 23.5%, and 20.9%, respectively.

4. Conclusions

This study investigates the stress distribution and dispersion of a cylinder head through the four steps of flow-solid coupling heat transfer calculation, heat-machine coupling stress calculation, fatigue calculation, and stress dispersion analysis, and obtains the following conclusions:

1. In the heat transfer calculation of the cylinder head, the error between the temperature calculation results and the experimental data using the standard $k-\varepsilon$ model is the smallest, at only 0.221%. This model demonstrates the best convergence.

2. The critical location of the cylinder head is found at the inlet channel pinch wall, where the stress dispersion follows a normal distribution with a standard deviation of 2.48.

3. The dispersion of the modulus of elasticity, burst pressure at the combustion surface, and expansion significantly impact stress dispersion, with degrees of influence of 25.5%, 23.5%, and 20.9%, respectively.

References

[1] Luo X, Zeng X, Zou P, et al. A finite element analysis-computational fluid dynamics coupled analysis on thermal-mechanical fatigue of cylinder head of a turbo-charged diesel engine [J]. Proceedings of the Institution of Mechanical Engineers, Part D: Journal of Automobile Engineering, 2020, 234(6): 1634-1643.

[2] Rakopoulos C, Mavropoulos G. Experimental instantaneous heat fluxes in the cylinder head and exhaust manifold of an air-cooled diesel engine [J]. Energy Conversion and Management, 2000, 41(12): 1265-1281.

[3] Farrahi GH, Shamloo A, Felfeli M, et al. Numerical simulations of cyclic behaviors in light alloys under isothermal and thermo-mechanical fatigue loadings [J]. Materials and Design, 2013, 56245-253.

[4] Fontanesi S, Giacopini M. Multiphase CFD–CHT optimization of the cooling jacket and FEM analysis of the engine head of a V6 diesel engine [J]. Applied Thermal Engineering, 2013, 52(2): 293-303.

[5] Yang WJ, Pang JC, Kang X, et al. Thermo-mechanical fatigue life prediction based on the simulated component of cylinder head [J]. Engineering Failure Analysis, 2022, 135

[6] Sharifi H M S, Googarchin S H, Forouzesh F. Three dimensional analysis of low cycle fatigue failure in engine part subjected to multi-axial variable amplitude thermo-mechanical load [J]. Engineering Failure Analysis, 2016, 62128-141.
[7] Hu L, Yang J, Yu Y, et al. Analysis and Optimisation of Thermo-Mechanical Coupling Load of Cylinder Head Considering Fluid-Structure Interaction for a Marine High-Power Diesel Engine [J]. Energies, 2020, 13(14).
[8] Ostapski W. Analysis of Thermo-Mechanical Response in an Aircraft Piston Engine by Analytical, FEM, and Test-Stand Investigations [J]. Journal of Thermal Stresses, 2011, 34(3): 285-312.
[9] Tan LB, Yuan YJ. Numerical simulation on flow field characteristics of the four cylinder engine cooling system and its performance improvement [J]. Journal of Mechanical Science and Technology, 2023, 37(1): 487-500.
[10] Tan LB, Yuan YJ, Huang C. CFD modelling on flow field characteristics of engine cooling water jacket and its cooling performance improvement based on coolant transport path analysis method [J]. Proceedings of the Institution of Mechanical Engineers, Part A: Journal of Power and Energy, 2023, 237(2): 385-401.
[11] Mohsen N. Investigation of thermal transient behavior of block and coolant in an internal combustion engine after shutdown [J]. Journal of Thermal Analysis and Calorimetry, 2022, 148(5): 2119-2128.
[12] Khoukhi E D, Morel F, Saintier N, et al. Experimental investigation of the size effect in high cycle fatigue: Role of the defect population in cast aluminium alloys [J]. International Journal of Fatigue, 2019, 129105222-105222.

Journal of Physics: Conference Series **2992** (2025) 012034 doi:10.1088/1742-6596/2992/1/012034

Investigation of floating ball mandrel application in tube free bending

Zhenhua Xu, Jing Huang, Hua Wang * and Rongjing Hong

College of Mechanical and Power Engineering, Nanjing Tech University, Nanjing 211816, China

*Corresponding author's e-mail: wanghua@njtech.edu.cn (Hua Wang)

Abstract: Cross-section distortion and axial instability are inherent challenges in the tube bending process. This paper investigates the use of a floating ball within the tube to mitigate ovality. Initially, the tube bending process with the floating ball is analyzed to elucidate the floating ball's role during tube bending. Subsequently, a finite element model without a floating ball was developed and validated through simulations. Finally, the influence of the floating ball's position and the gap between the floating ball-tube on the formability of the tube was analyzed by using finite element analysis (FEA). Optimized parameters for the floating ball were identified, enhancing tube bending quality and offering practical guidelines for implementing the floating ball in tube free bending.

1. Introduction

Hollow tubes are extensively utilized in aerospace, construction, automobile, and fluid transport industries owing to the characteristics of high strength and excellent processability [1, 2]. Cold-bended seamless tubular components integrated excellent stability and better load-bearing properties than welded tubes [3]. However, traditional tube bending technologies, push bending, rotary draw bending, and press bending, are increasingly unable to meet the growing demand of the market for tube components with high degrees of freedom and complex shapes [4, 5]. Consequently, there is an urgent need for tube-bending technologies to meet evolving market requirements. The free-bending technology developed by Murata et al. [6] addresses these demands by offering simple structures and eliminating the need to change bending dies for varying tube bending radii. Moreover, this technology reduces the time and cost of mold development, enabling the fabrication of tubular components with greater flexibility compared to traditional bending methods [7].

Springback, wrinkles, and cross-section distortion are common defects in tube bending [8], with cross-sectional distortion significantly affecting the suitability of tubes for oil transport and aeronautical applications [9]. Numerous studies have explored the use of internal fillings to provide support during tube bending. Guo et al. [10] researched the effects of various internal mandrels in tubes on bending quality. Beulich et al. [11] established a finite element model incorporating a spherical mandrel within the tube, validated through tube bending experiments. Song et al. [1] proposed a granular media filling method for tubes, analyzing the effects of particle diameter, particle system, and counterpressure on internal pressure distribution, and demonstrated that granular media substantially enhanced bending quality. Sedighi et al. [12] investigated the influence of rubber and disposable mandrels, revealing that rubber was unsuitable as a filler material, whereas disposable lead and tin mandrels achieved superior bending quality. Kajikawa et al. [13] explored the influence of mandrel diameter and positioning on tube quality, predicting defect occurrence under specific conditions. Cheng et al. [14] analyzed mandrel location, mandrel ball number, and the gap between mandrel and tube using numerical simulations. The findings indicated that the optimized mandrel parameters substantially enhance the quality of tube bending. Koyama [15] introduced the mechanics of the floating ball and confirmed its application in tube diameter expansion.

The mandrel of the floating ball proposed by H. Koyama simplifies the bending parameters compared to the tandem ball mandrel. However, the effects of the floating ball on the tube free bending are ambiguous and require further investigation. Hence, this study investigates the application of a floating ball in tube bending. Initially, the tube bending process was analyzed. Subsequently, a finite element model was developed and validated through tube bending experiments. Finally, the influence of the floating ball position and the clearance between the tube and the floating ball on the ovality and thickness of the tube was studied by using the finite element model (FEM).

2. Tube-free bending process and mandrel structures

Figure 1 shows the components of tube-free bending technology with a mandrel including bearings, bending dies, guide, tubes, and mandrel. During tube bending, the tube is pushed due to the push force of P_L. The bearing deviates from the equilibrium position due to the force of P_U. The tube is curved to the specified bending radius and angle by the combined influence of P_L and P_U. During the free-bending process, the floating ball stays close to the tube's inner wall. Bending moment M can be noted as:

$$M = P_U \times A + P_L \times U \tag{1}$$

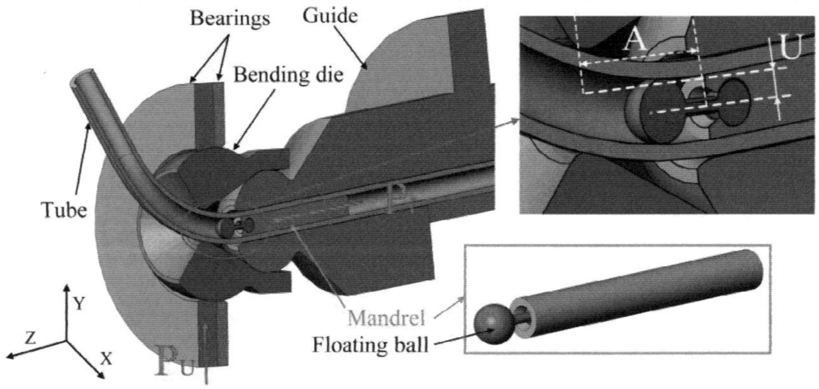

Figure 1. Tube bending with mandrel.

3. FEM of tube bending

The Abaqus/Explicit algorithm was employed to simulate the tube's free-bending process. Material parameters are shown in Table 1. Furthermore, the mandrel was simplified to a floating ball, and the finite element model is exhibited in Figure 2. The FEM consists of a tube, bearings, bending die, guide, and floating ball. The tube was modeled as a deformable solid by using C3D8R. Other components were modeled as rigid solids by using the same element type (C3D8R). A velocity V was applied along the Z-axis at the tube's end, while a displacement U was applied to constrain the bearing's motion in the XY plane. The floating ball's position varies with the tube's deformation. The friction coefficient was set at 0.05, as the parts are fully lubricated [16].

Table 1. Material parameters.

Material	Density ρ (g/cm^3)	Yield stress (MPa)	Passion ratio μ	Strength coefficient K (MPa)	Elastic modulus E (MPa)
304 stainless steel	7.93	260.01	0.3	966.1	207.97

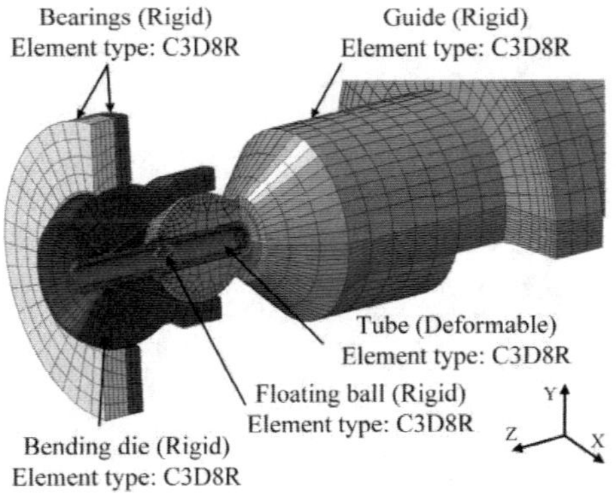

Figure 2. FEM after meshed.

4. Results and discussion

4.1 FEM validation

Cheng et al. [14] indicate that the strain-neutral surface moves outward of the tube bend. Besides, the displacement of the strain-neutral surface is more pronounced with a mandrel than without one. Therefore, the validity of the coreless bending model needs to be verified first. Subsequently, the validity of the tube bending model with a mandrel is verified by finite element simulation based on the above conclusion. The floating ball was removed according to the finite element model presented in Section 3, and a mandrel-less model was obtained. Tube bending experiments were carried out with a tube bending machine in our laboratory, and tube bending simulations were performed by using the mandrel-less FEM. The simulation and experiment results are shown in Figure 3. The findings indicate that the largest discrepancy between the experiment and simulation is 3.56%, and this discrepancy is primarily attributed to mesh partitioning. These findings demonstrate that the results of the tube bending experiment conducted without a mandrel align well with the simulation results, thereby validating the FEM without mandrel.

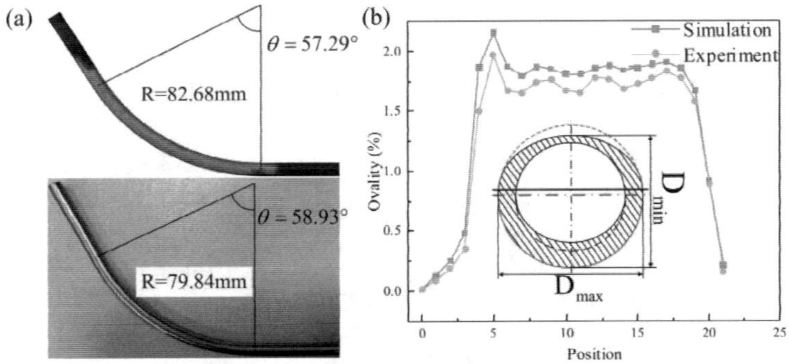

Figure 3. Tube free bending results comparison between simulation and experiment (without mandrel): (a) Tube shape; (b) Ovality.

Based on the aforementioned results, tube bending simulations were performed both with and without a floating ball, using identical bending parameters. As shown in Figure 4, the findings express that the displacement of the strain-neutral layer with a floating ball exceeds that without a floating ball. The validity of the FEM for tube bending with a floating ball has been verified based on the conclusion of Cheng et al. [14].

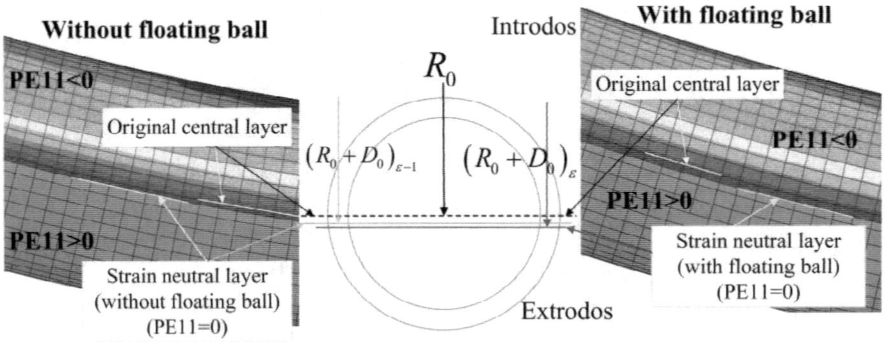

Figure 4. Displacements of the strain-neutral surface.

4.2 Effects of floating ball location on the quality of the tube bending

The analysis in Section 2 indicates that the plastic deformation initiates at the guide's front and ends at the contact point of the bending die and the tube, located at a distance A from the guide's front end. Consequently, positioning the floating ball within the region of tube plastic deformation (0A-1A) effectively reduces the section deformation of the tube. Nevertheless, the floating ball significantly influences the tube quality at different locations. Therefore, it is essential to examine the impact of the floating ball's position on the quality of the tube. As shown in Figure 5, the floating ball's impact on the tube's quality at 0A, 0.5A, and 1A was investigated by using the FE method.

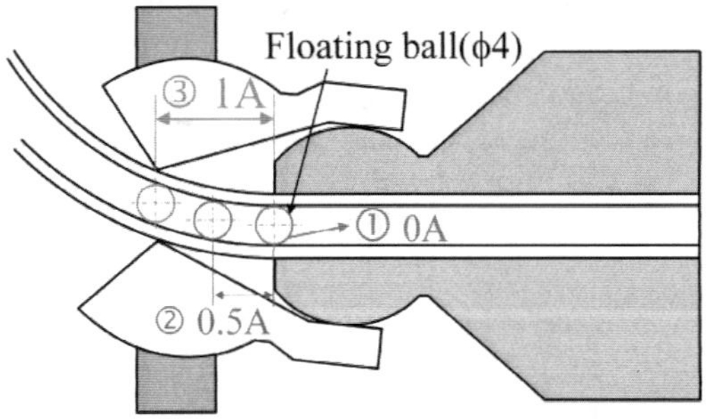

Figure 5. Schematic diagram of floating ball position

Figures 6 (a) and (b) illustrate the variations in ovality and thickness of the tube at various locations of the floating ball. The findings indicate that the ovality and thickness of the tube decrease as the distance between the floating ball and the front end of the guide increases, indicating that positioning the floating ball at the termination of the distortion zone can effectively mitigate the ovality of the tube. In Figure 6 (a), a peak is observed at 0A and 0.5A at Position 5, attributable to stress concentration resulting from the bearing's displacement from its original location. Nevertheless, the ovality of 1A at Position 5 is eliminated as the floating ball provides sufficient internal support at the stress concentration site. Figure 6 (b) indicates that 1A shows greater thickness on the outside and smaller thickness on the inner side of the pipe bend. Moreover, the thickness of 1A varies within a small range without extreme values, indicating higher bending quality. The investigation indicates that 1A shows more stable variation in thickness and ovality, without peak points. Consequently, superior tube bending quality is achieved when the floating ball is positioned at 1A.

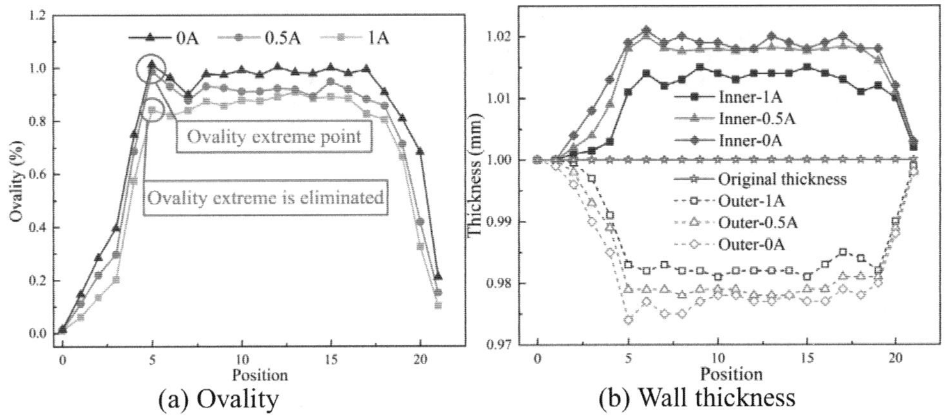

Figure 6. Tube bending results in floating balls in different positions: (a) Ovality; (b) Wall thickness.

4.3 Influence of the floating ball-tube gap on the formability of the tube bending

Excessive clearance will inhibit the floating ball from providing sufficient internal support, increasing the ovality of the tube. Smaller clearances will increase friction between the tube and the floating ball, which may hinder the tube's successful bending. The floating ball was placed at 1A based on the analysis in Section 4.2, and the gap between the tube and the floating ball was set as 0.05 mm, 0.1 mm, and 0.15 mm.

Figure 7 (a) indicates the ovality results of tube bending with varying clearances. The results indicate that the ovality increases with increasing clearance. The distortion of the tube results from radial stresses, which are mitigated by the internal support from the floating ball. Figure 7 (b) shows the thickness results of pipe bending under different clearances. The results reveal that the thickness on the inner side of the pipe bend decreases while increasing on the outer side of the pipe bend as the clearance increases. The smaller the clearance is, the larger the distance between the tube deformation area and the strain-neutral surface is. Consequently, the tube exhibits a reduced thickness on the outer side of the pipe bend and an increased thickness on the inner side of the pipe bend. Consequently, a clearance of 0.1 mm is optimal, considering both wall thickness and ovality.

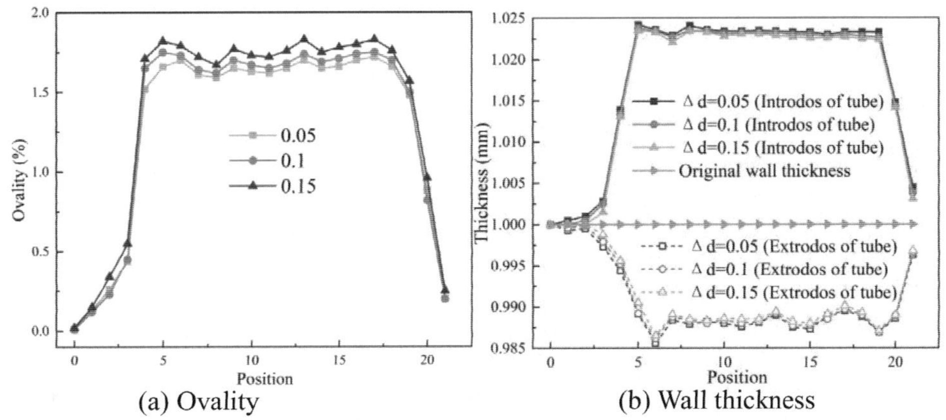

Figure 7. Finite element analysis results of the tube under various clearances: (a) Ovality; (b) Wall thickness.

5. Conclusion

This research provides scientific guidance for the application of a floating ball on tube-free bending. The effects of floating ball position and clearance on tube bending quality were investigated.

Compared to coreless bending, the application of floating balls significantly improves tube bending quality. The findings are concluded as:

(1) The floating ball mandrel provides effective support for tube-free bending and minimizes ovality.
(2) The positioning of the floating ball has been optimized, resulting in reduced tube ovality and resolving the issue of peak points in ovality.
(3) An optimal clearance for the tube-floating ball, considering both ovality and wall thickness, has been obtained, thus improving the floating ball's support in tube-free bending.

Acknowledgment
This research was funded by the Jiangsu Key Research and Development Program (Grant No. BE2019007-3).

Reference

[1] Song H. W., Xie W., Zhang S. H., Jiang W., Lăzărescu L., and Banabic D. Granular media filler assisted push bending method of thin-walled tubes with small bending radius (2021). International Journal of Mechanical Sciences, 198.
[2] Li H., Ma J., Liu B. Y., Gu R. J., and Li G. J. An insight into neutral layer shifting in tube bending (2018). International Journal of Machine Tools and Manufacture, 126: 51-70.
[3] Cheng C., Pan C., Bai X., Liu C., and Guo X. Investigation on the influence of weld position on the deformation behavior of welded tube during the free bending process (2022). The International Journal of Advanced Manufacturing Technology, 120: 2, 201-2, 215.
[4] Zhu Y. X., Chen W., Tu W. B., Guo Y., and Chen L. Three-dimensional finite element modeling of rotary-draw bending of copper-titanium composite tube (2019).The International Journal of Advanced Manufacturing Technology, 106: 2, 377-2, 389.
[5] Lukassen T. V., Gunnarsson E., Krenk S., Glejbøl K., Lyckegaard A., and Berggreen C. Tension-bending analysis of flexible pipe by a repeated unit cell finite element model (2019). Marine Structures, 64: 401-420.
[6] Murata M., Uemura Y., and Suzuki H. New flexible bending penetration of a tube, in: Proceeding of the 3rd International Conference on Technology of Plasticity, Advanced Technology of Plasticity, Tokyo, Japan, 1989, pp. 1, 573-1, 578.
[7] Gantner P., Harrison D. K., De Silva A. K., and Bauer H. The Development of a Simulation Model and the Determination of the Die Control Data for the Free-Bending Technique (2007). Proceedings of the Institution of Mechanical Engineers, Part B: Journal of Engineering Manufacture, 221: 163-171.
[8] Wang Z., Wang C., Zhang S., Qiu L., Lin Y., Tan J., and Sun C. Towards high-accuracy axial spring back: Mesh-based simulation of metal tube bending via geometry/process-integrated graph neural networks (2024). Expert Systems with Applications, 255.
[9] Yang Q., Liu C., Cheng C., Guo X., Tao J., and Ma Z. Effect mechanism of torsion on flat oval profile free bending (2023). International Journal of Mechanical Sciences, 250.
[10] Guo X., Cheng X., Xu Y., Tao J., Abd El-Aty A., and Liu H. Finite element modeling and experimental investigation of the impact of filling different materials in copper tubes during the 3D free bending process (2020). Chinese Journal of Aeronautics, 33: 721-729.
[11] Beulich N., Craighero P., and Volk W. FEA Simulation of Free-Bending – a Preforming Step in the Hydroforming Process Chain (2017). Journal of Physics: Conference Series, 896.
[12] Sedighi M., Kahnamouei J. T. J. J. O. T., and Mechanics A. Role of filling material on defects of the thin-walled tube bending process (2014). Journal of Theoretical and Applied Mechanics, 52.
[13] Kajikawa S., Wang G., Kuboki T., Watanabe M., and Tsuichiya A. J. P. M. Prevention of defects by optimizing mandrel position and shape in rotary draw bending of copper tube with a thin wall (2018). Procedia Manufacturing, 15: 828-835.

[14] Cheng C., Chen H., Guo J., Guo X., and Shi Y. Investigation on the influence of mandrel on the forming quality of thin-walled tube during the free bending process (2021). Journal of Manufacturing Processes, 72: 215-226.
[15] Koyama H. Bending of tubes and pipes (2019). The Japan Society for Technology of Plasticity, 2: 204-208.
[16] Guo X., Ma Y., Chen W., Xiong H., Xu Y., El-Aty A. A., and Jin K. Simulation and experimental research of the free bending process of a spatial tube (2018). Journal of Materials Processing Technology, 255: 137-149.

Analysis of thermal shock failures in ceramic materials under varying temperature conditions based on phase field model

Qiang Shen [1*], Haoyu Bai [1] and Yu Xue [1]

[1] School of Applied Science, Taiyuan University of Science and Technology, Taiyuan, Shanxi 030024, China

[*]Corresponding author's e-mail: shenq@tyust.edu.cn

Abstract: Ceramics, particularly alumina, are renowned for their high-temperature resistance and chemical stability but are susceptible to thermal shock failures due to brittleness and thermal expansion. This study employs phase field theory to investigate the thermal shock failures of alumina ceramics, evaluating their resistance under varying initial temperatures and heat transfer coefficients. The results indicate that different initial temperatures and heat transfer coefficients significantly affect the number, morphology, and propagation process of cracks in ceramic materials. The findings enhance the understanding of thermal shock phenomena and provide valuable insights for optimizing ceramic materials in challenging thermal environments.

1. Introduction

Ceramic materials are renowned for their excellent high-temperature stability, allowing them to maintain structural and performance integrity in demanding environments [1]. This capability has resulted in extensive applications in aerospace and other fields, garnering significant attention [2]. Additionally, their high hardness and wear resistance make them ideal for cutting tools, bearings, and abrasive materials [3]. Despite their unique advantages in various applications [4], ceramic materials face significant challenges in practical use, particularly regarding thermal shock resistance [5]. Their inherent brittleness makes them susceptible to cracking under thermal stresses when exposed to complex thermal environments with wide temperature variations. This susceptibility can lead to degradation and failure of performance [6]. Additionally, the thermophysical properties and damage criteria of ceramics are heavily influenced by temperature fluctuations, resulting in considerable variability in performance across different thermal conditions. Accurately simulating and evaluating thermal shock behavior in complex environments is further complicated by factors such as crack surface roughness, extension instability, and potential bifurcation phenomena [7].

Regarding experimental investigations, Jiang et al. [8]and Liu et al. [9] employed circular and thin rectangular specimens of Al_2O_3 ceramic to investigate the principal characteristics of thermal shock-induced cracking. Notably, the foregoing investigation was conducted through a combination of water-quenching experiments and numerical methods. On the same note, Li et al. [10], meticulously controlled the water entry orientation of Al_2O_3 rectangular rod specimens in quenching trials, thereby revealing a prominent influence exerted by the entry orientation on thermal shock fracture. Apart from observing the multi-faceted cracking process triggered by thermal shock, Shao et al. [11] utilized high-speed camera imaging technology to monitor the propagation velocity of cracks in thin ceramic strips.

There has been extensive research on the phenomenon of crack propagation and crack length hierarchy in thermal shock, including energy minimization [12], energy release rate [13], meso-damage mechanics [14], and the variational model [15], etc. The phase field method is a widely utilized technique for calculating microscale crack evolution, particularly effective in complex scenarios with well-defined boundary conditions [16]. These works mainly focus on the effect of boundary temperature on thermal shock failure of Al_2O_3 ceramic. However, the influence of the heat transfer coefficient is not considered. This work mainly investigates the thermal shock failure of Al_2O_3 ceramic under different temperature conditions.

2. Governing equations of thermal shock failure

At first, this research assumes that there is a cracked solid, denoted as $\Omega \subset \mathbb{R}^{n_{\dim}}$ ($n_{\dim}$=1, 2, 3), with a sharp crack set $\mathcal{S} \subset \mathbb{R}^{n_{\dim}-1}$ serving as the reference configuration. Notably, its outer boundary is represented by $\partial\Omega \subset \mathbb{R}^{n_{\dim}-1}$. Besides, the body forces $\boldsymbol{b}$ is distributed throughout the entire domain Ω, whereas a prescribed traction force t^* is applied to a portion of the boundary $\partial\Omega_t = \partial\Omega_t \subset \partial\Omega$. The deformations of the solid resulting from these conditions are characterized by the displacement field $(\boldsymbol{x})$ and the (infinitesimal) strain field $\varepsilon(x) := \nabla^{\text{sym}}\boldsymbol{u}(x)$.

Within the PF-CZM framework [17], under the irreversibility condition that the crack phase field or damage field $d(x): \mathcal{B} \to [0,1]$ satisfies $\dot{d}(\boldsymbol{x}) \geq 0$, a sharp crack $\mathcal{S}$ is regularized through the localization band $\mathcal{B} \subseteq \Omega$. In classical continuum mechanics, $d(x) = 0$ denotes the material's integrity and soundness, whereas $d(x) = 1$ signifies complete material fracture. The phase-field length scale $b > 0$ is introduced to measure the bandwidth; when it approaches 0, it indicates the restoration of sharp cracks.

Concerning the phase-field regularized cohesive zone model for pure mechanics problems, it is governed by the following damage-displacement coupling equation:

$$\begin{cases} \nabla \cdot \boldsymbol{\sigma} + \boldsymbol{b} = 0 & \text{in } \Omega \\ \boldsymbol{\sigma} \cdot \boldsymbol{n} = \boldsymbol{t}^* & \text{on } \partial\Omega_t \end{cases} \tag{1}$$

$$\begin{cases} \nabla \cdot \boldsymbol{q} + Q \leq 0 & \text{in } \mathcal{B} \\ \boldsymbol{q} \cdot n_{\mathcal{B}} \geq 0 & \text{on } \partial\mathcal{B} \end{cases} \tag{2}$$

where $\boldsymbol{\sigma}$ denotes the stress tensor, and the flux vector $\boldsymbol{q}$ denotes the conjugate of the damage gradient d, with its divergence being balanced by the source term Q, which is given by:

$$\begin{cases} \boldsymbol{q} = \dfrac{2b}{c_\alpha} G_{\mathrm{f}} \nabla d \\ Q = -\omega'(d)\bar{Y} - \alpha'(d)\dfrac{G_{\mathrm{f}}}{c_\alpha b} \end{cases} \tag{3}$$

where G_{f} represents fracture energy (i.e., energy dissipated per unit area), and $\bar{Y}$ represents the effective crack driving force. Concurrently, its mechanical process involves a strong coupling between the crack phase fields.

The governing equations mentioned above are characterized by a monotonically increasing geometric crack function $\alpha(d) \in [0,1]$ as well as a monotonically decreasing energy degradation function $\omega(d) \in [0,1]$, with their first derivatives being $\alpha'(d) := \partial\alpha / \partial d$ and $\omega'(d) := \partial\omega / \partial d$, respectively.

The damage field $d(x)$ must satisfy the irreversible condition $\dot{d}(\boldsymbol{x}) \geq 0$ in the fracture phase-field model, so that the effective fracture driving force $\bar{Y}$ in the phase-field source term can be replaced by its maximum value denoted as $\mathcal{H}$, which is determined as:

$$Q = -\omega'(d)\mathcal{H} - \alpha'(d)\frac{G_{\mathrm{f}}}{c_\alpha b} \quad \text{with} \quad \mathcal{H} = \max(\bar{Y}_0, \max_{n \in [0,T]} \bar{Y}_n) \tag{4}$$

Given the simplest isotropic relationship, the constitutive relationship between stress σ and the effective damage-driving force $\bar{Y}$ must be integrated into the damage-mechanical coupling control equation, which is defined as:

$$\boldsymbol{\sigma} = \omega(d)\bar{\boldsymbol{\sigma}} = \omega(d)\mathbb{E}_0 : \boldsymbol{\varepsilon} \tag{5}$$

Based on the preceding research by Wu et al., the following characteristic functions are identified:

$$\begin{cases} \alpha(d) = 2d - d^2 & c_\alpha = 4\int_0^1 \sqrt{\alpha(\beta)}\mathrm{d}\beta = \pi \\ \omega(d) = \dfrac{(1-d)^p}{(1-d)^p + a_1 d \cdot P(d)} & P(d) = 1 + a_2 d + a_3 d^2 \end{cases} \tag{6}$$

Building upon the foregoing discussion, phase field theory is coupled with the heat conduction equation while being extended to incorporate thermo-mechanical fracture. Introducing the temperature field denoted as $T(x)$, the ensuing temperature induces thermal strain that leads to deformation and then generates the initial driving force $\bar{Y}$ to promote the evolution of damage, which ultimately degrades the material's performance. Furthermore, the external boundary is divided into two parts, one of which is $\partial\Omega_h$ with a specified heat flux h^*, while the other is $\partial\Omega_T = \Omega/\Omega_h$ with a specified temperature T^*.

The heat transfer process is determined by the following system of partial differential equations:

$$\begin{cases} \rho c\dot{T} + \nabla \cdot \boldsymbol{h} = \Theta^* & \text{in } \Omega \\ h \cdot n = h^* & \text{on } \partial\Omega_h \end{cases} \tag{7}$$

Due to the shrinkage caused by thermal expansion and hydration, the total strain is decomposed to:

$$\varepsilon^{\mathrm{t}} = \eta_t (T - T_0)\mathbf{1} \tag{8}$$

3. Numerical simulation

The finite element geometric model of the ceramic plate is established based on the water quenching experiment. As shown in Figure 1, the plate measures 50 mm in length and 25 mm in width. It is not subjected to external loads or additional constraints at its four boundaries. The bottom and right sides of the plate are set as insulating conditions, preventing any heat exchange, while the upper and lower surfaces are designated as cold boundary conditions, creating thermal shock within the plate. As the temperature boundary approaches a fixed value (i.e., with an infinite heat transfer coefficient), temperature changes will occur over time, leading to expansion or contraction of the material. These boundary conditions limit deformation, resulting in temperature-induced stresses. To accurately model the ceramic plate, symmetric boundary conditions were employed. A 2D quadrilateral quadratic Lagrangian mesh with a cell size of 0.15 mm was used. The initial temperature was uniformly set at 20°C.

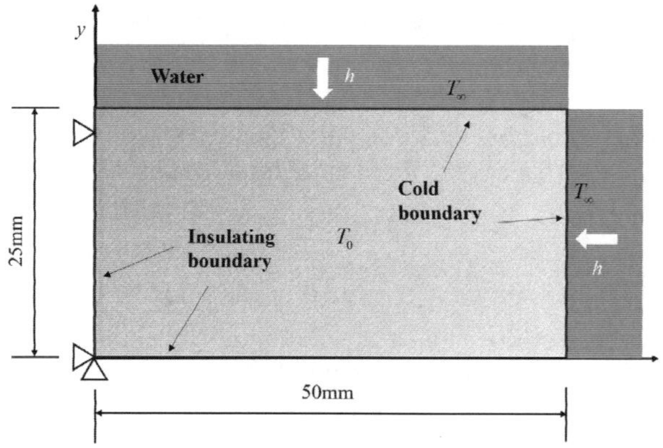

Figure 1. Schematic diagram of a two-dimensional plate subjected to cold shock.

Semi-transparent ceramic materials, which were synthesized by utilizing 99.5 % high-purity alumina (i.e., Al_2O_3) powder with a particle size of 0.5 mm, were processed by tape casting and subsequently sintered at 1850°C for two hours, thereby generating an approximate bulk density of 3960 kg/m³ and an average surface grain size of approximately 18.9 mm. Required material parameters are detailed in Table. 1. Regardless of the fact that ceramics are typically characterized as brittle solids with elastic behavior, they demonstrate unstable failure, encompassing instances of quasi-brittle behavior and nonlinear mechanical characteristics.

Table. 1 Aluminum plate sample part material parameters.

Material parameters	Value	Unit
Elastic modulus	373	GPa
Density	2500	g/cm³
Poisson's ratio	0.22	-
Thermal conductivity	20	W/m/K
Thermal expansion	6.8×10^{-6}	1/K
Specific heat	880	J/kg/K

To investigate the influence of varying temperature conditions on the fracture process of the material, a series of simulations were conducted. Using the model and boundary conditions shown in Figure 1, the initial temperature was respectively set to 300°C, 400°C, 500°C, and 600°C, with the water temperature at 20°C. The heat transfer coefficients were varied at levels of 10, 000 W/m²/K, 20, 000 W/m²/K, 30, 000 W/m²/K, 40, 000 W/m²/K, 50, 000 W/m²/K, and 60, 000 W/m²/K. To isolate the effects of heat transfer coefficients, all other variables were controlled. Each of the six models employed a linear quadrilateral mesh comprising a total of 14, 028 domain cells and 502 boundary cells. The simulations analyzed crack extension, stress distribution, and temperature variation to assess the impact of different heat transfer coefficients on the crack propagation process.

4. Results and discussion

This section centers on the investigation of the influence exerted by heat transfer coefficient and initial temperature on the thermal shock fracture behavior of ceramic plates in quenching experiments, primarily encompassing thermal shock crack patterns, energy evolution, and thermal crack propagation rate.

Figure 2 shows the crack expansion at t = 1 s for different heat transfer coefficients. The four figures illustrate how varying coefficients influence the crack expansion rate. At a heat transfer coefficient of 10, 000 W/m²/K, few cracks form at the material boundary, with the direction of expansion predominantly deflected from a straight line. Increasing the coefficient to 20, 000 W/m²/K results in more cracks and fewer deflected paths. At 30, 000 W/m²/K, the number of cracks further increases, with only a few deflections. As the heat transfer coefficient rises, the crack density increases, the spacing narrows, and cracks begin to expand more linearly. This indicates that higher heat transfer coefficients lead to faster crack expansion. A higher coefficient signifies better thermal conductivity, allowing heat to dissipate quickly from the crack area, resulting in lower temperatures around the cracks. This reduction slows crack growth. Additionally, enhanced thermal conductivity promotes a more uniform temperature distribution, further decelerating crack expansion by minimizing temperature gradients. Consequently, changes in the heat transfer coefficient can influence the selection of crack extension paths.

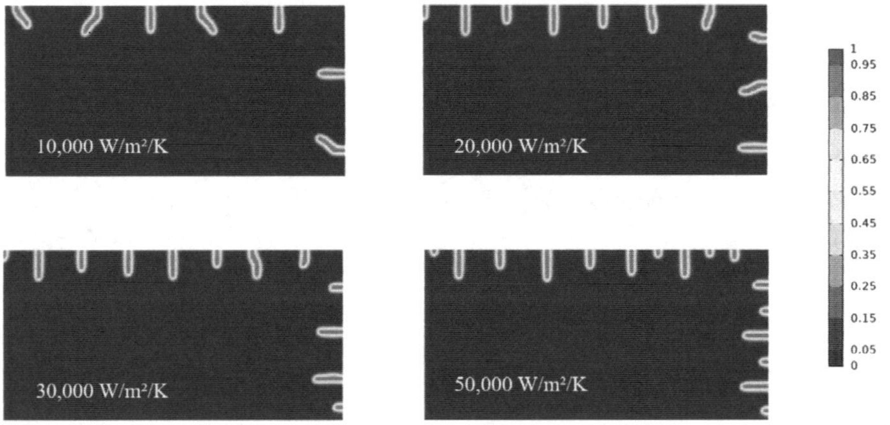

Figure 2. Crack propagation at t=1s with different heat transfer coefficients: (a) 10, 000 W/m²/K, (b) 20, 000 W/m²/K, (c) 30, 000 W/m²/K, and (d) 50, 000 W/m²/K.

Figure 3 presents a simulation of crack extension under different heat transfer coefficients, this comparison highlights the impact of various coefficients on the number of cracks, crack morphology, and extension processes. At a heat transfer coefficient of 10, 000 W/m²/K, the number of cracks generated by thermal stress at the material boundary is low, with significant spacing between them. Long and short cracks are distinctly layered, and long cracks extend obliquely rather than in a straight line. At 20, 000 W/m²/K, the number of cracks increases, with reduced spacing and the emergence of long, straight cracks. However, many cracks still do not extend linearly. At 50, 000 W/m²/K, the crack density continues to rise, spacing decreases, and delamination becomes more pronounced, with all cracks expanding in a straight direction without deflection, demonstrating the significant influence of heat transfer coefficients. Specifically, higher coefficients lead to short cracks nucleating within long cracks and expanding into the material, creating multi-level crack structures. This phenomenon is attributed to inhomogeneous local temperature and stress distributions. High thermal conductivity materials distribute heat more evenly, reducing thermal stress and leading to slower crack growth. In contrast, low-conductivity materials create steeper temperature gradients, increasing thermal stress at the crack tip and accelerating crack initiation. Additionally, cracks in low-conductivity materials often follow more irregular paths due to varying stress fields. These factors highlight the role of thermal conductivity in both the rate and morphology of crack propagation, emphasizing the importance of thermal management in preventing material failure under thermal shock.

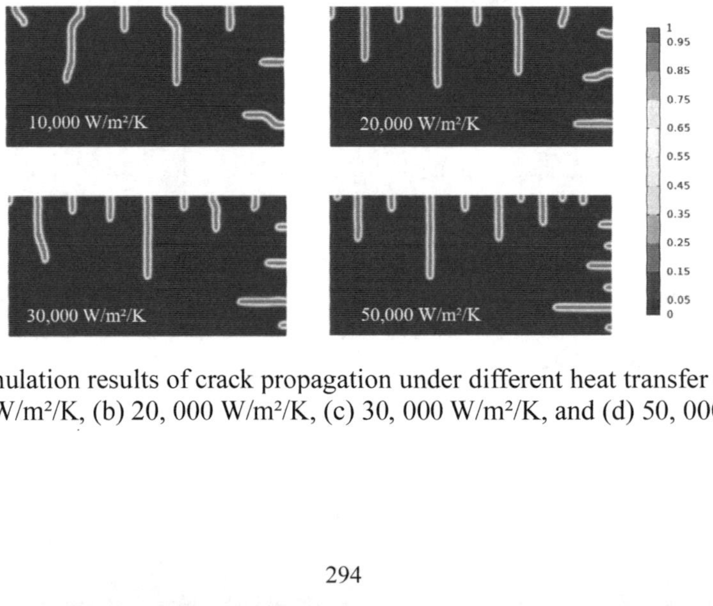

Figure 3. Simulation results of crack propagation under different heat transfer coefficients: (a) 10, 000 W/m²/K, (b) 20, 000 W/m²/K, (c) 30, 000 W/m²/K, and (d) 50, 000 W/m²/K.

Figure 4 shows different initial temperatures significantly impact the material's behavior during thermal shock. It can be observed that when the initial temperature is 300°C, the cracks exhibit a more dispersed distribution, a larger spacing, and a lack of accordance with the crack straight line expansion. Additionally, a crack layering phenomenon is evident, with the cracks divided into short, medium, and long categories. The number of cracks was found to increase significantly when the initial temperature was 400°C. The spacing was also found to be significantly reduced, with an increase in the number of short, longer, and three-level cracks. At an initial temperature of 600°C, the ceramic plate's cold exchange boundary is covered with various types of cracks. The boundary is almost completely destroyed with the increase in the number of cracks and the extension of long cracks compared to the initial temperature. The initial temperature of a material significantly affects its thermal shock behavior by influencing the temperature gradient, heat exchange rate, and stress concentration during thermal loading. Higher initial temperatures create larger temperature gradients and faster heat exchange rates, resulting in increased stress concentrations. In contrast, lower initial temperatures slow down temperature changes, allowing the system to better absorb or release heat and reducing stress concentration formation. As a result, higher initial temperatures generally lead to greater thermal stresses, accelerating crack propagation. This is reflected in crack extension maps, where higher temperatures produce a denser distribution of cracks with smaller spacing, while lower temperatures yield a more dispersed crack distribution with larger spacing. Therefore, variations in initial temperature directly influence thermal stress distribution, affecting crack extension behavior and material stability. These factors highlight the critical role of initial temperature in determining thermal shock resistance, suggesting that controlling the initial temperature is key to optimizing material performance during thermal cycling.

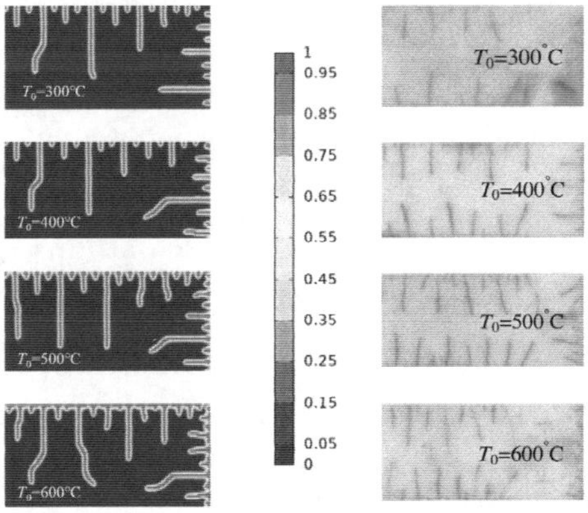

Figure 4. Comparison of simulation and experimental results [8] of crack growth withdifferent initial temperatures: (a) 300°C, (b) 400°C, (c) 500°C, and (d) 600°C

5. Conclusions

In this paper, the study demonstrates that varying heat transfer coefficients significantly influence crack morphology in ceramic materials, particularly in terms of expansion rate, direction, morphology, and path selection. Higher heat transfer coefficients enhance the material's capacity to dissipate heat, resulting in lower temperatures around cracks. These lower temperatures typically increase the material's toughness and tensile resistance, thereby slowing crack extension. Furthermore, improved thermal conductivity associated with higher heat transfer coefficients facilitates a more uniform temperature distribution around cracks. This uniformity is essential for mitigating crack propagation, as it ensures an even distribution of thermal stresses, reducing the likelihood of further crack growth. Overall, the findings underscore the critical role of heat transfer coefficients in shaping the thermal shock behavior of ceramic materials, with important implications for their design and application in high-temperature environments.

Initial temperatures significantly influence material behavior during thermal shock. Higher temperatures result in larger temperature gradients and faster heat exchange, leading to increased stress concentrations. Conversely, lower temperatures slow temperature changes, allowing better heat absorption or release, which reduces stress formation. Consequently, higher initial temperatures generally cause greater thermal stresses, accelerating crack propagation. This is evident in crack extension maps, where higher temperatures produce a denser distribution of cracks with smaller spacing, while lower temperatures lead to a more dispersed crack distribution with larger spacing. Thus, variations in initial temperature directly affect thermal stress distribution and material stability.

This study has important implications for the design of materials in environments subject to thermal shock, particularly in high-performance applications like aerospace, electronics, and power generation. The research provides a deeper understanding of how thermal conductivity and initial temperature influence crack propagation in alumina ceramics. By optimizing these factors, engineers can enhance the thermal shock resistance of ceramics, improving their performance in extreme temperature conditions. These insights can help guide the selection and design of materials for applications requiring high thermal stability, such as aerospace components and industrial machinery, ultimately improving reliability and lifespan.

Reference

[1] Opeka M M, Talmy I G, Wuchina E J, Zaykoski J A and Causey S J 1999 Mechanical, thermal, and oxidation properties of refractory hafnium and zirconium compounds. J. Eur. Ceram. Soc. 19(13–14): 2405–14.

[2] Bobaru F and Zhang G. 2015 Why do cracks branch? A peridynamic investigation of dynamic brittle fracture. *Int. J. Fract.* 196: 59–98.

[3] Menouillard T and Belytschko T 2011 Analysis and computations of oscillating crack propagation in a heated strip. *Int. J. Fract.* 167: 57–70.

[4] Gasch M, Ellerby D, Irby E, Beckman S, Gusman M and Johnson S 2004 Processing, properties and arc jet oxidation of hafnium diboride/silicon carbide ultra high temperature ceramics. *J. Mater. Sci.* 39(19): 5925–37.

[5] Wu J Y. 2017 A unified phase-field theory for the mechanics of damage and quasi-brittle failure. *J. Mech. Phys. Solids* 103: 72–99.

[6] Li D Y, Wang R Z, Wang X R and Li W G 2020 Simulation of the thermal shock cracking behaviors of ceramics under water quenching for 3-dimension conditions. *Eur. J. Mech. A Solids* 84: 104080.

[7] Zeng X, Ning Y and Yang Y. 2023 A multiscale damage prediction model for the thermal shock of carbon fiber-reinforced silicon carbide matrix composites. Mech. Adv. Mater. Struct. 30: 1769–89.

[8] Jiang C P, Wu X F, and Li J 2012 A study of the mechanism of formation and numerical simulations of crack patterns in ceramics subjected to thermal shock. Acta Mater. 60: 4540–50.

[9] Liu Y, Wu X, Guo Q, Jiang C, Song F and Li J. 2015 Experiments and numerical simulations of thermal shock crack patterns in thin circular ceramic specimens. Ceram. Int. 41: 1107–14.

[10] Li D, Li W, Wang R and Fang D 2016 The effects of water entry postures on the thermal shock behavior of alumina. Int. J. Appl. Ceram. Tech. 13: 56–60.

[11] Shao Y F, Liu B, Wang X, Li L, Wei J and Song, F 2018 Crack propagation speed in ceramic during quenching. J. Eur. Ceram. Soc. 38(7): 2879–85.

[12] Jenkins D R 2005 Optimal spacing and penetration of cracks in a shrinking slab. Phys Rev E. 71:056117, 8pp.

[13] Bahr H A, Weiss H J, Maschke H G and Meissner F. 1988 Multiple crack propagation in a strip caused by thermal shock. Theor. Appl. Fract. Mech. 10: 219–26.

[14] Tang S B, Zhang H, Tang C A, and Liu H Y 2016 Numerical model for the cracking behavior of heterogeneous brittle solids subjected to thermal shock. Int. J. Solids Struct. 80: 520–31.

Journal of Physics: Conference Series **2992** (2025) 012035 doi:10.1088/1742-6596/2992/1/012035

[15] Bourdin B, Marigo J J, Maurini C and Sicsic P. 2014 Morphogenesis and propagation of complex cracks induced by thermal shocks. Phys. Rev. Lett. 112: 014301.
[16] Miehe C, Hofacker M, and Welschinger F. 2010 A phase field model for rate-independent crack propagation: Robust algorithmic implementation based on operator splits. Comput. Methods Appl. Mech. Eng. 199: 2765–78.
[17] Wu J Y and Chen W X. 2022 On the phase-field modeling of fully coupled chemo-mechanical deterioration and fracture in calcium leached cementitious solids. Int. J. Solids Struct.; 238: 111380.

Research on the physical characteristics and rock-breaking effect of high-energy plasma drill bit

Yancong Han[1], Chao Zheng[2], Yonghong Liu[3*], Yuming Liu[4] and Wenhao Zhao[5]

[1,2,3,4,5]College of Mechanical and Electronic Engineering, China University of Petroleum East China, Qingdao, China

[*]Corresponding author's e-mail: liuyhupc@163.com

Abstract—High-energy plasma jet rock-breaking is a highly promising new drilling method with advantages such as high energy density, high rock breaking efficiency, and no mechanical wear. The internal physical characteristics of plasma drill bit significantly affect both the bit lifespan and rock-breaking effect, warranting detailed study. This paper establishes a discharge model of the plasma drill bit through numerical simulation, deeply analyzing the distribution characteristics of temperature and velocity and discussing the influence of plasma current on the internal physical properties of the bit. Experimentally, the rock-breaking effects of different rock types under the action of the plasma jet are compared. The findings reveal that the maximum temperature of the arc plasma can reach 21,200 K; the current significantly impacts the velocity, reaching 1,730 m/s at 300 A. The effectiveness of the plasma drill in breaking rock is independent of the type of rock.

1. Introduction

Currently, traditional mechanical drilling technology faces challenges such as low drilling speed and severe bit wear when drilling for deeper oil and gas resources. Costly tool replacement not only incurs significant expenses but also extends unproductive time due to frequent trips to and from the well [1]. In contrast, high-energy plasma arc drilling technology has advantages of high energy density, including high rock-breaking efficiency, stability, easy control, and no mechanical wear during operation. As a new drilling technology with great development potential, it is becoming a focal point of research [2].

Non-transferred arc plasma is generated between the anode nozzle and the cathode electrode. The fluctuating behavior of the arc within the torch can lead to the plasma jet instability and accelerate the ablation rate of the electrode. This process directly impacts the drill bit's operational lifespan and the efficiency of rock breaking. The simulations account for complex interactions among electromagnetic fields, ionization processes, conductive fluids, and heat and mass transfer within the plasma torch. Modirkhazeni et al. [3] established a model that transitions from non-equilibrium plasma laminar flow to turbulent flow. Perambadur et al. [4] developed 2D and 3D transient models of the torch, focusing on the effects of process parameters on the fluctuations of the anode arc root. Sun et al. [5] explored the effects of electrical and structural parameters of the plasma drill bit on rock-breaking effectiveness through experiments. Their findings indicated that the effect of plasma jet rock-breaking has been significantly improved.

In summary, while numerous models for DC non-transfer arc plasma torches have been simulated, the internal characteristics of a plasma bit proposed in this paper for drilling applications remain underexplored. This study addresses this gap by resolving the internal discharge model of the plasma drill bit. Additionally, the anode nozzle ablation and rock-breaking effect under different plasma currents were investigated experimentally. The results can provide theoretical guidance for optimizing the plasma drill bit design and improving the rock-breaking effect.

2. Plasma drill bit model

2.1 Underlying assumptions

Arc plasma generation is a complex physical process involving strong coupling among electric fields, magnetic fields, flow fields, and thermal fields. To simplify the mathematical model of arc plasma and facilitate easier numerical simulation, the following basic assumptions are made in the study regarding the heat transfer and flow dynamics of non-transferring plasma torch [6]:

- Arc plasma is modeled as a two-dimensional, axially symmetric, steady-state laminar flow.
- The arc plasma is considered compressible and in local thermodynamic equilibrium (LTE).
- The arc plasma is considered optically thin, ignoring the reabsorption component of the total radiation loss. The gravity effects, viscous dissipation, pressure workers and induced electric fields are neglected.

2.2 Governing equations

Within the plasma drill bit, the arc plasma can be treated as a continuous fluid medium and described using the mass conservation, energy conservation, and momentum conservation equations of hydrodynamics. Additionally, in the presence of an applied electric field and self-magnetic induction, it becomes necessary to incorporate Maxwell's electromagnetic equations. The magnetohydrodynamics (MHD) equations used in the model are as follows [7]:

$$\frac{\partial}{\partial z}(\rho v_z)+\frac{1}{r}\frac{\partial}{\partial r}(r\rho v_r)+\frac{1}{r}\frac{\partial}{\partial \theta}(\rho v_\theta)=0 \tag{1}$$

$$\begin{aligned}&\frac{\partial}{\partial z}(\rho v_z^2)+\frac{1}{r}\frac{\partial}{\partial r}(r\rho v_r v_z)+\frac{1}{r}\frac{\partial}{\partial \theta}(\rho v_\theta v_z)=-\frac{\partial p}{\partial z}+\frac{\partial}{\partial z}\left(2\mu\frac{\partial v_z}{\partial z}\right)\\&+\frac{1}{r}\frac{\partial}{\partial r}\left[r\mu\left(\frac{\partial v_r}{\partial r}+\frac{\partial v_z}{\partial z}\right)\right]+\frac{1}{r}\frac{\partial}{\partial \theta}\left[\mu\left(\frac{1}{r}\frac{\partial v_\theta}{\partial \theta}+\frac{\partial v_z}{\partial z}\right)\right]+j_r B_\theta-j_\theta B_r\end{aligned} \tag{2}$$

$$\begin{aligned}&c_p\frac{\partial}{\partial z}(\rho v_z T)+c_p\frac{1}{r}\frac{\partial}{\partial r}(r\rho v_r T)+c_p\frac{1}{r}\frac{\partial}{\partial \theta}(\rho v_\theta T)=\\&\frac{\partial}{\partial z}\left(k\frac{\partial T}{\partial z}\right)+\frac{1}{r}\frac{\partial}{\partial r}\left(rk\frac{\partial T}{\partial r}\right)+\frac{1}{r^2}\frac{\partial}{\partial \theta}\left(k\frac{\partial T}{\partial \theta}\right)\\&+\frac{5}{2}\frac{k_B}{e}\left(j_z\frac{\partial T}{\partial z}+j_r\frac{\partial T}{\partial r}+j_\theta\frac{1}{r}\frac{\partial T}{\partial \theta}\right)+\frac{j_z^2+j_r^2+j_\theta^2}{\sigma}-Q_r\end{aligned} \tag{3}$$

$$\frac{\partial}{\partial z}\left(\sigma\frac{\partial \phi}{\partial z}\right)+\frac{1}{r}\frac{\partial}{\partial r}\left(r\sigma\frac{\partial \phi}{\partial r}\right)+\frac{1}{r^2}\frac{\partial}{\partial \theta}\left(\sigma\frac{\partial \phi}{\partial \theta}\right)=0 \tag{4}$$

where v_z, v_r, and v_θ represent the axial, radial, and tangential components of velocity, respectively, m/s; ρ denotes the mass density of arc, kg/m^3; c_p represents the specific heat capacity, J·kg^{-1}·K^{-1}; T indicates the thermodynamic temperature, K; k denotes the thermal conductivity, W·m^{-1}·K^{-1}; k_B is Boltzmann's constant, valued at 1.38×10^{-23} J·K^{-1}; e is the basic charge, with a value of 1.6×10^{-19} C; Q_r is the radiation heat loss of arc plasma, J/m^3, φ is the electric potential, V.

2.3 Computational domain and boundary conditions

This paper presents a two-dimensional axisymmetric model of plasma drill bit discharge in the cylindrical coordinate system, using COMSOL Multiphysics (Version 6.2). The geometric model and corresponding mesh division are shown in Figure 1. The discharge region, which is critical for the computational region, employs a free triangular mesh. Mesh densification and refinement are specifically implemented near the electrode wall boundary in the radial direction to enhance accuracy. The model consists of 67,751 total mesh elements, including 62,401 triangular mesh units and 5,350 quadrilateral mesh units. The mesh quality was rigorously evaluated, achieving an average mesh quality of 0.8434, which surpasses the requirements for computational accuracy. The boundary conditions of the plasma drill bit internal discharge model are shown in Table 1.

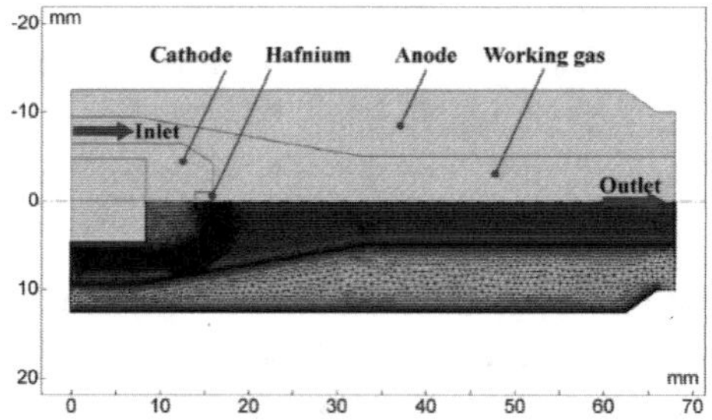

Figure 1. Models and mesh.

Table 1. Boundary Condition of the Model.

Boundary type	Inlet	Outlet	Cathode	Anode
Temperature	300 K	$\partial_n T = 0$	$T(r)$	$-k\partial_n T = h(T - T_f)$
Velocity	V_{in}	$\partial_n V = 0$	0	0
Pressure	$\partial_n P = 0$	0	$\partial_n P = 0$	$\partial_n P = 0$
Electric potential	$\partial_n \phi = 0$	$\partial_n \phi = 0$	$-\sigma\partial_n \phi = J(r)$	0
Magnetic vector potential	0	$\partial_n A = 0$	$\partial_n A = 0$	$\partial_n A = 0$

3. Simulation result analysis

3.1 Arc plasma thermal characteristics

Figure 2 shows the temperature distribution cloud diagram for the arc plasma at a current of 300 A. The temperature of the arc plasma peaks near the center of the cathode end. As the flows axially toward the plasma torch exit, its temperature gradually decreases, with the reduction becoming more pronounced at the exit. The reason for this phenomenon is that the current density of the arc plasma gradually decreases along the central axis, reaching its maximum near the cathode end, which generates a large amount of Joule heat, causing the arc plasma to heat up rapidly. Radially, the temperature of the arc plasma is highest along the central axis and decreases progressively towards the sides. This cooling effect is attributed to the direct water cooling of the anode. A non-ionized gas jacket forms near the anode's inner wall. This gas layer compels the charged particles to migrate towards areas of higher ionization in the center of the arc column, thereby increasing the current density at the center.

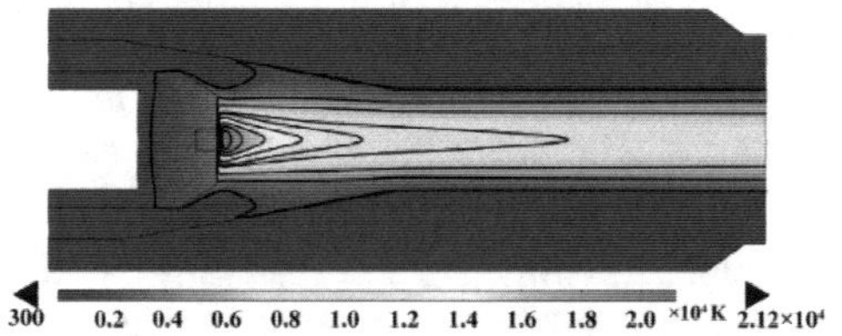

Figure 2. Temperature distribution of arc plasma.

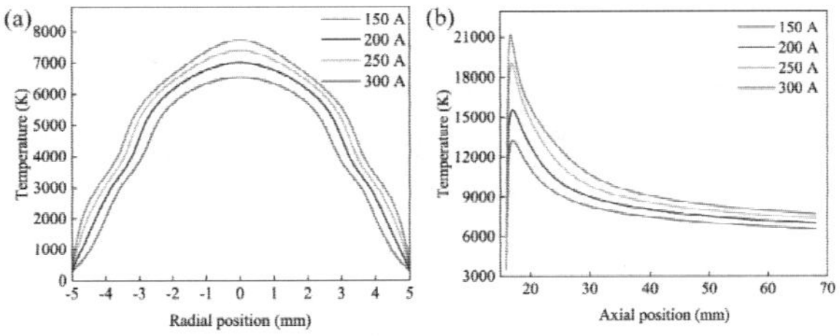

Figure 3. Radial (a) and axial (b) temperature distribution curves of arc plasma under different working currents.

Within the plasma drill, two line segments were defined to assess temperature distribution: one radially across the outlet (r = -5 to 5 mm, z = 68 mm) and the other along the center axis (z = 16 to 68 mm, r = 0 mm). Temperature distribution curves were then obtained for each node on these segments, as depicted in Figure 3. As the operating current increases, the temperature at each node in the radial and center-axis upward direction of the outlet inside the plasma drill bit shows an increasing trend, which is consistent with the Gaussian distribution. From Figure 3(b), the temperature of the arc plasma initially increases and then decreases along the center axis toward the exit. Furthermore, the rate of temperature increase and subsequent decay accelerates with higher operating currents. The temperature peaks near the cathode end (z=16 mm).

3.2 Arc plasma temperature distribution characteristics

Figure 4(a) displays the velocity distribution cloud diagram of arc plasma in the plasma drill bit. The velocity of the arc plasma increases rapidly and significantly near the cathode end. This increase is due to the working gas entering the plasma torch from the inlet, where it is continuously heated, ionized, and expanded. However, as the plasma moves toward the outlet, the current density and the forces acting on the plasma—namely the Lorentz force and thermal expansion force—begin to decrease, leading to a gradual reduction in velocity along the central axis. Figures 4(b) and (c) show the velocity streamline results within the drill bit under two different air intake modes: axial direct flow and tangential swirling flow, respectively. Under axial direct flow, the velocity streamline distribution of the arc plasma is primarily influenced by the geometry of the outer surface of the cathode and the inner surface of the anode nozzle. The streamlines in the gas chamber and compression channel remain mostly parallel to the central axis of the plasma torch. In contrast, when tangential swirling air intake is employed, the working gas enters the air chamber and follows a spiral path along the inner wall of the anode, entering the compression channel with a pronounced rotational movement.

Figure 5(a) shows the radial velocity distribution curves at the exit of the plasma drill bit under different plasma currents, with the data interception position corresponding to those in Figure 3. As the operating current increases, there is a noticeable rise in velocities at both the exit radial and center-axis upward nodes within the plasma torch. This trend is attributed to the increase in working current, which amplifies the current density at each node within the plasma torch. Consequently, more Joule heat is generated. This increase in temperature causes a decrease in the density of the working gas and a rapid expansion in volume, thereby boosting the velocity of the arc plasma. Similarly, the distribution of velocities follows the law of Gaussian distribution. From Figure 5(b), it is observed that the velocity of the arc plasma initially increases, then decreases along the central axis toward the exit, peaking at a point significantly distant from the cathode end. The Lorentz force and thermal expansion force contribute to this velocity pattern, exhibiting a certain lag before influencing the increase in velocity.

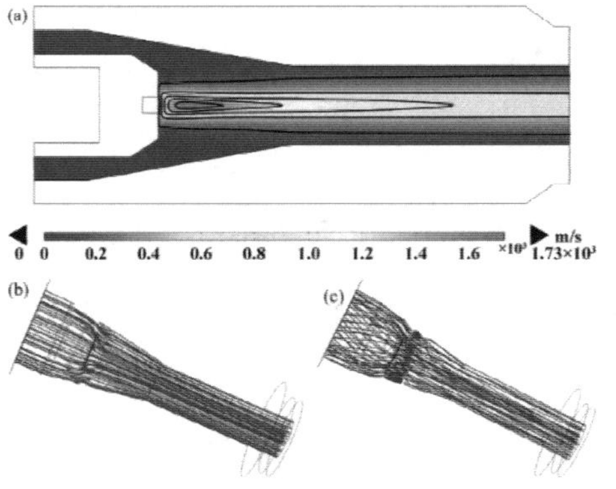

Figure 4. Velocity distribution of arc plasma (a), velocity streamline distribution of axial direct flow intake (b) and tangential swirling flow intake (c).

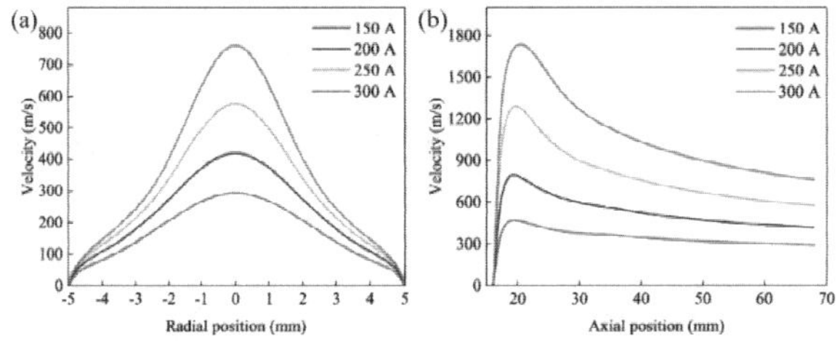

Figure 5. Radial (a) and axial (b) velocity distribution curves of arc plasma under different working currents.

4. Experimental research

4.1 Experimental principles and apparatus

To deepen the understanding of the discharge process and rock-breaking capabilities of the plasma drill bit, as well as to provide practical insights for optimizing the drill bit structure, the experiments were conducted. Figure 6 shows the experimental schematic diagram. The anode loss experiments were carried out under an airflow of 40 slm and a plasma current of 150 A–300 A. The granite, marble, concrete and limestone were broken under a current of 300 A and a breaking distance of 40 mm. Under high-frequency and high-voltage conditions facilitated by the arc ignition circuit, the working gas between the cathode and anode is ionized, forming a plasma channel. Once the arc is initiated, the arc-inducing circuit is deactivated and the arc-dimensioning circuit is activated. The sustained high current through the arc-dimensioning circuit maintains the arc plasma, which is expelled from the anode nozzle as high-temperature jets. These jets continuously erode the rock, causing it to melt and fracture.

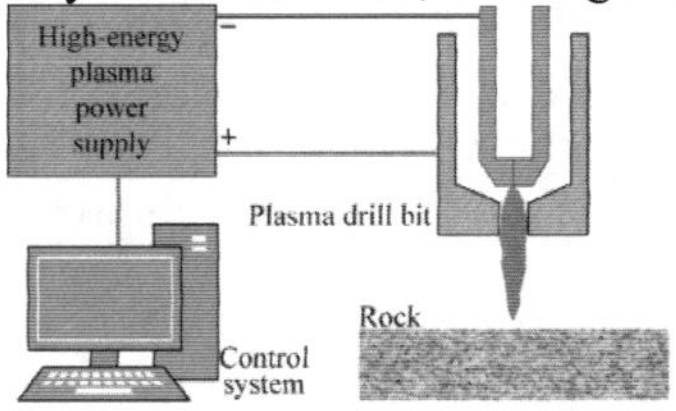

Figure 6. Schematic diagram of rock-breaking experiment.

4.2 Analysis of anode nozzle loss results

To quantify the wear on the anode nozzle, the mass difference before and after experimentation was defined as absolute mass loss, serving as a measure of the anode nozzle loss rate, as shown in Figure 7. As the operating current increases, the number of arc roots of the arc plasma also increases, enhancing the formation of corrosion points on the anode nozzle and accelerating its wear. Figure 8 illustrates how the arc plasma intensifies with increased current; it elongates, becomes stiffer, and stabilizes. This stabilization leads to a reduction in arc root movement and prolongs contact at specific points, thus forming distinct arc corrosion points. Once the arc corrosion point is formed, the anode arc root tends to stay at the corrosion point. The prolonged contact enlarges the corrosion point, creating a feedback loop that further increases the nozzle's wear rate. According to the law of energy conservation, as the operating current of the plasma torch increases, so does the energy and heat transferred to the anode nozzle, accelerating the loss rate.

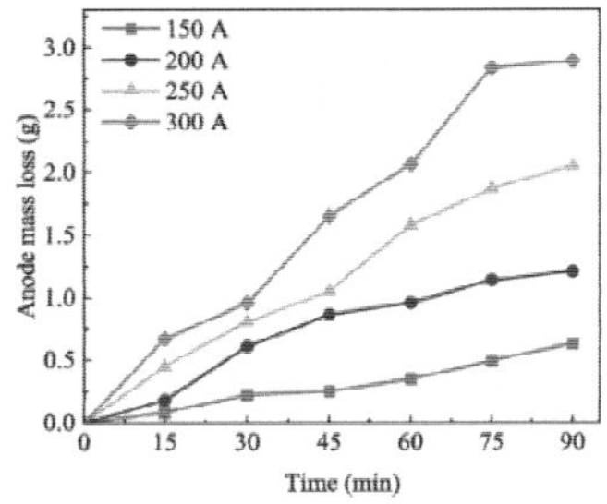

Figure 7. Experimental curve of the anode nozzle loss under different plasma currents.

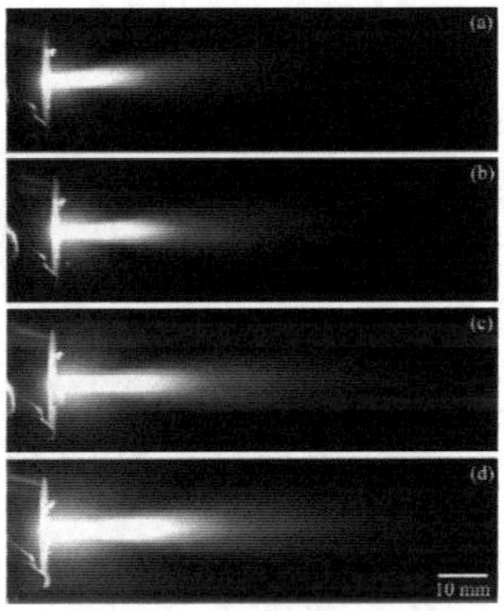

Figure 8. Appearance of arc plasma under different plasma currents.

Figure 9 shows the rock-breaking effects of high-energy plasma jets on four different types of rocks at the breaking time of 30 s. Due to its high energy density, plasma jets can quickly and efficiently remove rocks through fracture and melting. The resulting holes produced in all four rock types are similar in size and depth. This consistency demonstrates that the rock-breaking efficacy of the high-energy plasma drill is largely independent of rock hardness. The maximum rate of penetration is up to 6 m/h and the minimum specific energy is 7 kJ/g. The rock-breaking effect is superior to other non-contact rock-breaking methods such as water jets and laser beams.

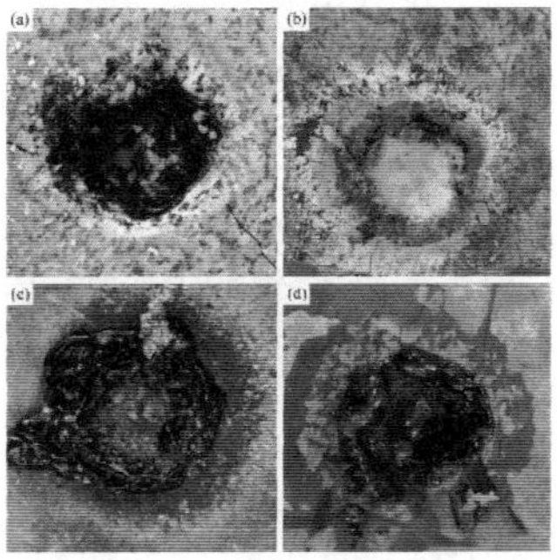

Figure 9. Rock-breaking effect pictures of different rocks: (a) granite, (b) marble, (c) concrete, (d) limestone.

5. Conclusions

This paper details the mathematical description of a two-dimensional axisymmetric steady-state model of the non-transferring arc plasma torch, assuming local thermodynamic equilibrium. The corresponding magnetohydrodynamic equations, computational domains, and boundary conditions were established and subsequently solved using COMSOL Multiphysics. This approach enabled the elucidation of the laws governing the temperature and velocity distribution characteristics of the arc plasma within the high-energy plasma drill. Using self-designed high-energy arc plasma rock-breaking equipment, we conducted experiments to evaluate anode nozzle loss and plasma jet rock-breaking efficiency. The experimental findings reveal that under conditions of a very high-temperature output, the mass loss of the anode nozzle is minimal.

Acknowledgment

This work was supported by the Equipment Pre-Research Collaborative Foundation for Innovation Team from the Ministry of Education of the People's Republic of China (No. 8091B042209) and the Taishan Scholars Project Special Fund (No. tsqn202211085).

References

[1] Y. Li, J. Peng, G. Zhang, Z. Yang, and K. Li, "Study on perforation performance of abrasive water jet enhanced by percussive drilling," J Pet Sci Eng, vol. 192, no. March, p. 107259, 2020.
[2] F. Yang, W. Liu, X. Zhu, and C. Xiang, "The Rock-Breaking Mechanism of Thermal Spalling-Assisted Rock Cutting by PDC Cutter," Rock Mech Rock Eng, vol. 57, no. 2, pp. 993–1012, Feb. 2024.
[3] S. M. Modirkhazeni and J. P. Trelles, "Non-transferred Arc Torch Simulation by a Non-equilibrium Plasma Laminar-to-Turbulent Flow Model," Journal of Thermal Spray Technology, vol. 27, no. 8, p. 1447-1464, 2018.
[4] J. Perambadur, P. Shukla, A. Y. Klimenko, A. B. Murphy, and V. Rudolph, "Arc dynamics in a vortex-stabilized non-transferred plasma torch with a tangential gas feed," J Phys D Appl Phys, vol. 54, no. 49, p. 495501, 2021.
[5] Q. Sun, Y. Liu, Z. Shen, Y. Han, Q. Wang, and P. Liu, "A Combined Thermal Spallation and Melting Technology by Plasma Jet for Deep and Hard Rock Reservoirs," SPE Journal, vol. 28, no. 1, p. 49-63, 2023.
[6] I. A. Alshunaifi, S. Elaissi, I. Ghiloufi, S. S. Alterary, and A. A. Alharbi, "Modelling of a non-transferred plasma torch used for nano-silica powders production," Applied Sciences (Switzerland), vol. 11, no. 21, p. 9842, 2021.
[7] H. P. Li, E. Pfender, and X. Chen, "Application of Steenbeck's minimum principle for three-dimensional modelling of DC arc plasma torches," J Phys D Appl Phys, vol. 36, no. 9, p. 1084, 2003.

Journal of Physics: Conference Series **2992** (2025) 012037 doi:10.1088/1742-6596/2992/1/012037

Research on the variation law and characterization model of kinetic energy block penetration overload

Zhe Yang [1] and Fei Shang [1*]

[1] Nanjing University of Science and Technology, Nanjing, Jiangsu, 210094, China

*Corresponding author's e-mail: shangfei23@126.com

Abstract: During the penetration process of kinetic blocks, they experience penetration overloads of tens of thousands of g's, which poses significant challenges to the structural integrity of the kinetic blocks and the functionality of their internal components. Accurate assessment of penetration overload is crucial for the optimization of kinetic block design. Existing studies on penetration overload are mostly focused on qualitative research and single-factor quantitative research, lacking a unified mathematical representation model that accounts for multiple factors. In this paper, we established a simulation model using finite element analysis software, obtaining penetration overload data under different conditions of kinetic block shapes, penetration speeds, and target thicknesses. We proposed a rigid body filtering method for penetration overload signals and analyzed the trends of penetration overload changes with different factors, providing a theoretical basis and technical support for improving the penetration performance of kinetic blocks. We constructed a unified mathematical representation model for penetration overload that varies with multiple factors, with a model fitting degree of 0.94, offering new perspectives and tools for subsequent research.

1 Introduction

During the penetration of a target by a kinetic energy block, penetration overloads of tens of thousands of g's are generated, posing a severe challenge to the structural integrity and functionality of its internal components. An accurate assessment of penetration overload is crucial for the design optimization of the kinetic energy block.

In recent years, significant progress has been made in the study of penetration overload, with numerous scholars approaching the topic from various angles and employing simulation, experimentation, and data analysis to reveal the nature and influencing factors of penetration overload. For example, Ren et al. [1] combined dimensional theory with experimental research, successfully deriving a set of model formulas suitable for engineering calculations through regression analysis of acceleration curves during the penetration of concrete. They validated the accuracy of these formulas using measured overload characteristic parameters. Peng et al. [2] conducted meticulous measurements of deceleration and structural vibrations during the penetration process through experimental means, proposing a theoretical model to describe the deceleration time during penetration or perforation, with highly consistent prediction results compared to actual test data.

In-depth research on penetration overload and its influencing factors has also yielded significant results in the academic community. The study by Xiaodong et al. [3] found that parameters such as velocity, mass, diameter, and length-to-diameter ratio (CRH) significantly affect static resistance, leading to the proposal of an empirical formula for predicting static resistance. Zhang et al. [4] demonstrated, through an analysis of the relationship model between mass wear and CRH, combined with experimental data, that the influence of CRH on penetration depth gradually increases with rising velocity. Zhou et al. [5] and Hui et al. [6] explored the effects of different nose shapes on penetration overload using numerical simulation and finite element analysis, respectively. Ma et al. [7] constructed a simulation model for the penetration process of multi-layer steel targets, analyzing the complex relationship between the size parameters of semiconductor bridge explosives and target plate configurations. Yan et al. [8] proposed a new method for non-proportional scaling design, experimentally validating that this method effectively simulates the rigid body deceleration behavior of prototypes during the penetration process.

In summary, existing studies on penetration overload have largely focused on qualitative research and single-factor quantitative analysis, lacking a unified mathematical representation model that accounts for multiple factors. This paper, through numerical simulations, has obtained penetration overload data under various conditions of kinetic energy block shapes, penetration velocities, and target thicknesses. A rigid body filtering method for penetration overload signals was proposed, analyzing the trends of penetration overload with different influencing factors. Consequently, a unified mathematical model representing penetration overload influenced by various factors was constructed, providing new perspectives and tools for subsequent research.

2 Numerical simulation of the penetration of kinetic blocks into steel plate targets

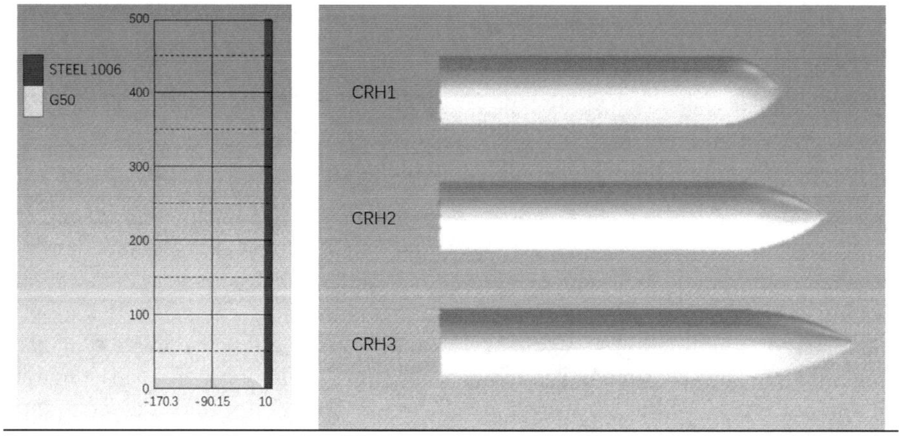

(a) (b)

Figure 1 Two-dimensional simulation model of kinetic blocks penetration into a steel plate target.

Due to the lengthy duration of penetration tests and the lack of reliable experimental methods for ultra-high overloads in penetration steel plates, this study employs simulation methods to obtain penetration overload data under various conditions. The research establishes a two-dimensional simulation model of a specific type of kinetic energy block penetrating a steel plate target using finite element analysis software, as illustrated in Figure 1. The kinetic energy block has a diameter of 30 mm, with a shape factor (CRH) ranging from 1 to 3, and a total weight of 740 grams. The material of the kinetic energy block is G50, which possesses sufficiently high strength, fracture toughness, and impact load resistance [9], making it widely used as the casing material for a new generation of penetrative warheads. The target has a diameter of 1000 mm and a thickness ranging from 10 mm to 30 mm, with a thickness increment of 5 mm. The target material is STEEL1006. The average acceleration of the projectile is used to represent the overall penetration overload. Both the kinetic energy block and the target plate are modeled using the Lagrange algorithm, which is suitable for solid mechanics problems. This algorithm effectively handles complex phenomena such as contact and separation. It can achieve relatively accurate results within a reasonable computation time when addressing large deformation issues. It is particularly applicable for simulations of processes such as penetration, impact, and explosion.

Table 1 The Johnson-Cook model parameters.

Material	G50	STEEL1006
Yield strength A/MPa	1390	350
Strain hardening coefficient B/MPa	1226	275
Strain rate sensitivity coefficient C	0.005	0.022
Strain hardening coefficient n	0.356	0.36
Temperature sensitivity coefficient m	1.12	1

The stress-strain relationship of the material follows the classical Johnson-Cook constitutive model (material parameters are shown in Table 1 [10]). This constitutive model characterizes the effects of strain, strain rate, and temperature on the mechanical properties of the material. Its equation is expressed as follows:

$$\sigma = (A + B\varepsilon^{n}) \cdot (1 + C \cdot \ln\frac{\dot{\varepsilon}}{\dot{\varepsilon}_0}) \cdot (1 - \frac{T - T_0}{T_m - T_0}) \tag{1}$$

In this equation, σ represents the stress of the material; ε denotes the strain; $\dot{\varepsilon}$ and $\dot{\varepsilon}_0$ are the strain rate and quasi-static strain rate, respectively; T is the actual temperature; T_0 is the reference temperature T_m is the melting point temperature of the material. A, B, C, m, and n are material parameters, where A is the yield strength at the reference temperature and strain rate, B and n are strain hardening coefficients, C is the strain rate sensitivity coefficient, and m is the temperature sensitivity coefficient.

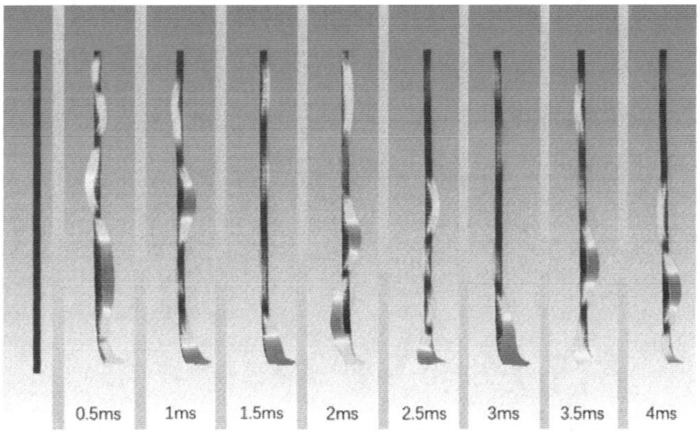

Figure 2 illustrates the variation process of the surface velocity signal of the target.

The colors in Figure 2 represent the direction of the velocity of the target. At the initial moment (0 milliseconds), the velocity direction of the target surface is not exciting. The target exhibits a uniform color, indicating that it is at rest. At 0.5 milliseconds, the velocity direction of the target surface changes, and the target shows a noticeable vibrational response due to the impact. By 2 milliseconds, the color distribution on the target surface gradually stabilizes, and the rate of change in velocity direction slows down, entering a relatively stable phase.

3 Analysis of the variation law of penetration overload

In the penetration characterization method proposed in this paper, it is necessary to filter the signals generated by the simulation to eliminate the influence of different kinetic block structures on the characterization results, in order to obtain the rigid body overload of the kinetic block[11].

Chen et al. [12] contend that low-pass filtering, as a traditional method for obtaining rigid body overload, is straightforward and practical. However, it exhibits poorer adaptability compared to other signal processing methods. To mitigate the influence of human factors on the evaluation of penetration overload, this study combines the approach proposed by Wang et al. [13] which involves integrating the original signals and selecting the point of abrupt change in penetration speed as the filtering frequency. The filtering principle applied in this paper is based on the condition that the deviation of the velocity change obtained from the filtered integration compared to that from the original overload integration exceeds 1%. This approach allows for the determination of the rigid body penetration overload of the kinetic block while avoiding human interference.

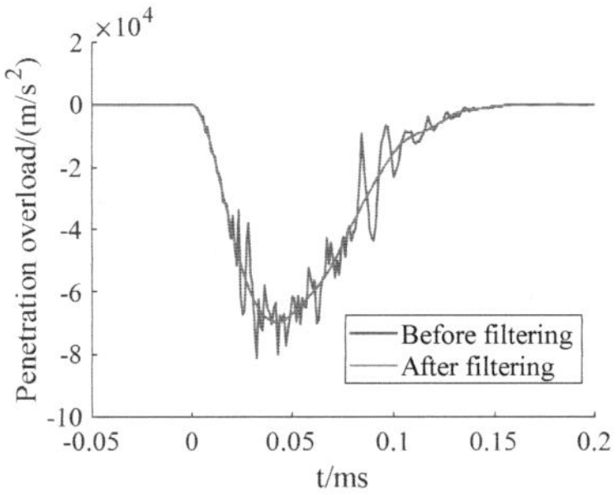

Figure 3 Comparison of penetration overload curves before and after filtering.

Figure 3 presents a comparison of the penetration overload curves before and after filtering. The unfiltered data exhibits significant fluctuations and displays distinct vibrational characteristics, failing to meet the unimodal requirement of rigid body penetration overload. In contrast, the curve processed by the aforementioned algorithm appears smoother and more stable, indicating that the proposed rigid body filtering algorithm effectively removes unnecessary interference components and accurately captures the rigid body overload during the penetration process.

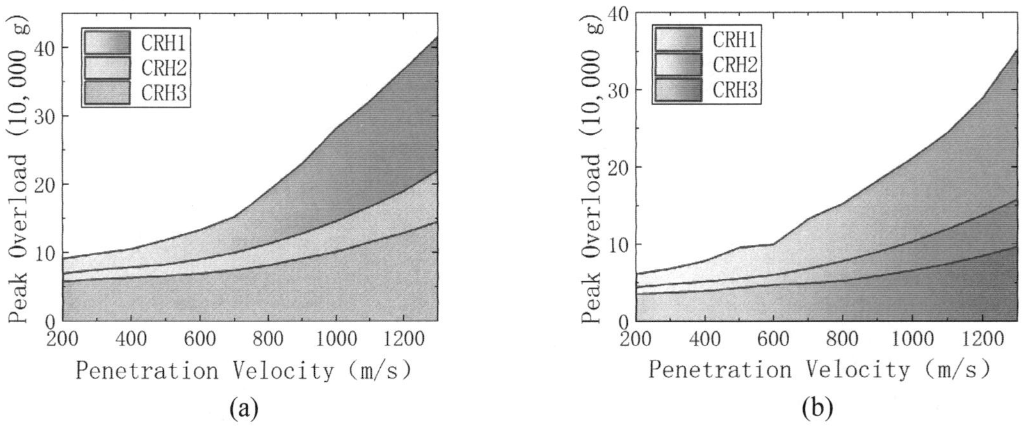

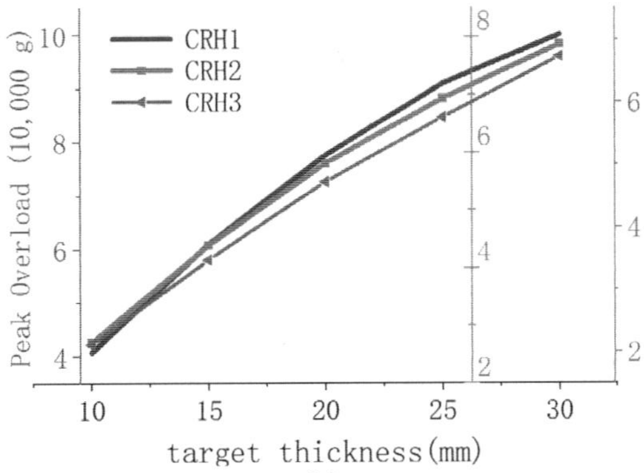

(c)

Figure 4 The penetration overload variation law.

Figures 4(a) and (b) illustrate the changes in penetration overload curves of kinetic blocks with different velocities and aspect ratios against targets of thicknesses 15 mm and 25 mm. It can be observed that for the same type of kinetic block, as the penetration velocity increases, the peak penetration overload gradually rises. Under the same velocity conditions, within the simulation range, the penetration overload decreases with an increase in the aspect ratio. Figure 4(c) presents the variation of penetration overload under the same velocity conditions for different target thicknesses and kinetic block shapes. The trend of penetration overload changes for different kinetic blocks is generally consistent with the trend of target thickness. At smaller thicknesses, the rate of increase in penetration overload for different kinetic blocks is relatively similar; however, as the thickness continues to increase, the differences between them become increasingly pronounced.

4 Mathematical representation model of penetration overload

Based on simulations under different conditions, the rigid body overload of the kinetic block was obtained after processing through the aforementioned filtering algorithm, and the peak value of the penetration overload was determined. The relationship between the penetration overload and factors such as the shape coefficient, target thickness, and penetration velocity was investigated. Assuming the kinetic block is penetrating a steel plate target, the penetration overload parameter T of the kinetic block is related to, but not limited to the following factors: M (mass of the kinetic block), D (diameter of the kinetic block), N (shape of the kinetic block's head), Y (strength parameter of the steel plate target), H (thickness parameter of the steel plate target), V (penetration velocity of the kinetic block), and g (acceleration due to gravity). Thus, the relationship between the penetration overload parameter T and the aforementioned factors is established as follows:

$$\frac{A_{\max}}{g} = \alpha N^{\beta} \left(\frac{YD^2}{Mg}\right)^{\varepsilon} \left(\frac{H}{D}\right)^{\delta} \left(\frac{V}{\sqrt{gD}}\right)^{\zeta} \tag{2}$$

For the sake of comparison in results, M (mass of the kinetic block), D (diameter of the kinetic block), and Y (strength parameter of the steel plate target) are treated as constants, simplifying the model to:

$$\frac{A_{\max}}{g} = \alpha N^{\beta} H^{\theta} V^{\zeta} \tag{3}$$

Using the previously mentioned finite element analysis software, numerical simulations are conducted where different kinetic blocks penetrate various steel plates at different velocities. The penetration overload data obtained under these varying conditions are processed through the aforementioned filtering algorithm to yield the rigid body overload curve and peak values during the penetration process. By applying multivariable function fitting to the resulting data, we obtain a = 0.0024, β = -0.9186, θ = 0.5024, ζ = 1.1250, and R^2 = 0.9404:

$$\frac{A_{max}}{g} = 0.0024 \times N^{-0.9186} H^{0.5024} V^{1.1250} \tag{4}$$

The expression of Equation (6) has a fitting degree of 0.94, indicating that the model can calculate penetration overload within an acceptable error range. The signs and magnitudes of the exponents of each influencing factor in the formula are consistent with the patterns analyzed previously, which validates the reliability of the previously discussed patterns and the mathematical model.

5 Summary

This study addresses the phenomenon of high overload generated during the penetration of kinetic blocks. A two-dimensional simulation model for the penetration of kinetic blocks into steel plate targets was established using finite element analysis software, allowing for an in-depth investigation of the variation patterns of penetration overload and its influencing factors. This research contributes to the optimization of kinetic block penetration performance and lays the groundwork for the precise measurement of penetration overload and the development of signal processing technologies. The main conclusions are as follows:

A signal processing algorithm for penetration overload based on low-pass filtering and integration has been designed. By setting a threshold for the rate of change of the filtering frequency, high-frequency noise in the signal has been effectively removed, resulting in a smooth and stable rigid body overload curve. This provides a reliable technical means for the accurate measurement of penetration overload.

By analyzing simulation data regarding various factors such as the shape of kinetic blocks, penetration speed, and target thickness, it was found that the penetration overload increases with the rise in penetration speed. Furthermore, under the same speed conditions, the penetration overload decreases as the aspect ratio of the kinetic block increases. The variation trend of penetration overload remains consistent across different target thicknesses; however, in the case of thicker targets, the differences in penetration overload among various kinetic blocks become more pronounced.

A dimensionless expression for the peak penetration overload has been established. The model has been further simplified through dimensional analysis. Based on simulation results, a mathematical expression has been constructed to describe how the peak penetration overload varies with multiple factors. This model effectively reflects the influence of various factors, including the shape of the kinetic block, target thickness, and penetration velocity, providing a theoretical basis for the design and performance evaluation of kinetic blocks.

Unfortunately, this paper does not conduct an in-depth study of the material parameters related to penetration overload, including the kinetic block and target material properties, as well as the diameter and mass parameters of the kinetic block and the angle of penetration. Future research could further expand to investigate penetration scenarios under these more complex conditions, developing a more comprehensive characterization model that would provide more robust support for the design of kinetic blocks.

Acknowledgments

This work was supported by funds from the Postgraduate Research & Practice Innovation Program of Jiangsu Province (KYCX23_0447).

Reference

[1] Ren Huiqi, He Xiang, Liu Ruichao, et al. Study on the Overload Characteristics of Projectile Penetration into Concrete. Journal of Civil Engineering, 2005, 38(1): 110-116.

[2] Peng Y, Wu H, Fang Q, et al. Deceleration Time of Projectile Penetration/Perforation into a Concrete Target: Experiment and Discussions. Advances in Structural Engineering,

[3] Xiaodong N, Xiangyun W, Zhi Y, et al. Discussion on the Static Resistance of Granite under Penetration. Materials (Basel, Switzerland), 2023, 16(9).
[4] Zhang Xuelun, Liu Zongwei. Study on the Influence of Projectile CRH Value on Penetration Depth into Concrete. Journal of Weapon Equipment Engineering, 2016, 37(10): 31-34.
[5] Zhou Dong, He Chun. Impact of Typical Nose Shapes of a Certain Type of Kinetic Penetration Warhead on Penetration Performance. Tactical Missile Technology, 2013, (01): 106-110.
[6] Hui Jianghai, Shang Yaling, Xu Haowen. Variation Law of High-Speed Penetration Overload of Warheads with Changes in Shape. Modern Defense Technology, 2015, 43(05): 63-69 + 93.
[7] Ma Hongliang, Li Jun, Fu Dongxiao, et al. Simulation Study on the Overload Response of Semiconductor Bridge Explosive Devices under Penetration of Multi-Layer Steel Targets. Explosives, 2019, (05): 25-28.
[8] Amin Y, Aiguo P, He Y, et al. Study on the Equivalence of Penetration Overloading for Projectile-Borne Components in Nonproportional Penetrators. Shock and Vibration, 2022.
[9] Shi Junfei. Discussion on Two Process Routes for Cobalt-Free Ultra-High Strength Steel G50. Special Steel Technology, 2006, 11(2): 14-16.
[10] Kong Q., Shen F., Xing Y., et al. Comparative Experimental Study on the Dynamic Mechanical Properties of G50 Steel and G31 Steel. Journal of High Pressure Physics, 2021, 35(1): 68-74.
[11] Ronald G. Lundgren. Method and Signal Processing Means for Detecting and Discriminating Between Structural Configurations and Geological Gradients Encountered by Kinetic Energy Subterranean Terra-Dynamic Crafts: US11953562 [P]. 2010-05-18.
[12] Chen B., Zhao S., Zou H., et al. Research Progress on the Characteristics and Processing Techniques of Penetration Projectile Overload Signals. Journal of Ordnance Engineering, 2024, 45(9): 2906-2928.
[13] Wang C., Chen P., Xu X. Filtering of Measured Data on Penetration Overload and Determination of Rigid Body Overload During Projectile Penetration. Explosion and Shock, 2007, 27(5): 416-419.

Numerical analysis of bursting pressure of a brazed plate heat exchanger

Wei Wang[1], Yadong Zhao[2,*] and Dao Chi[3]

[1]School of Automotive Engineering, Wuhan university of technology, Wuhan, Hubei, China
[2]School of Automotive Engineering, Wuhan university of technology, Wuhan, Hubei, China
[3]Valeo Automotive Air Conditioner Hubei Co., Ltd, Jingzhou, Hubei, China

[*]Corresponding author's e-mail: 334025@whut.edu.cn

Abstract—In this paper, the bursting pressure of a brazed plate heat exchanger (BPHE) was studied through numerical and experimental analysis. A numerical model of BPHE in FEM (Finite element method) was established. The bursting test of three plate heat exchanger samples was carried out. The results show that the gap in the bursting pressure between the simulation and experiment is 11.50%. Strain gage tests were carried out during one of the bursting tests. The results show that the gap of von Mises stresses between simulation and experiment is minimized to 1.84%. CT scanning was also carried out to obtain the BPHE deformation after the bursting test, which was consistent with the simulation.

1. Introduction

Plate heat exchangers are widely used in petrochemicals, ships, food and other fields because of their compact structure and high heat transfer coefficient. As heat exchange equipment, it is also pressure equipment. With the development of industry, the operating pressure of plate heat exchangers is getting higher and higher. If the bearing capacity is insufficient, the plates will be deformed and the heat transfer will be affected. In serious cases, it causes the plate to break, resulting in media leakage, and then a safety accident. Therefore, the structural strength study of plate heat exchangers is very important[1].

Some scholars have studied the stress characteristics and bearing capacity of plate heat exchangers under pressure load by using the finite element method. Liu et al.[2] established a global model to evaluate mechanical stresses and estimate the limit load in the plate heat exchanger. Plate heat exchangers have many repeated flow paths. Some scholars simplified the finite element model to only a few flow paths, ignoring the effect of the number of layers on the stress characteristics. Ma et al. [3][4] proposed a local model of four flow paths with five plates to study the LNG heat exchanger, and the results show that the equivalent stress reaches the peak value in the brazed joint. Wang et al.[5] also established a simplified model of four flow paths to study the printed circuit heat exchanger, and the results indicate that the highest stress appears in the middle of the semicircular arc of the channel. Some scholars also have studied the effect of structural parameters on the structural strength of plate heat exchangers by using the finite element method. Gürler[6] studied the effect of the chevron angle on the mechanical properties of BPHEs. Ma et al. [7] studied the effect of curvature radius on the brazing arc on stress characteristics. These studies demonstrate that the finite element method is an important tool in the study of structural analysis of plate heat exchangers, and it can help to get the appropriate structural parameters.

Some scholars have carried out consistency studies of simulation and experiments. Zibart et al.[8] proposed an approach in which the bursting pressure of pillow plates is determined numerically based on the finite element methods, and the gap between simulation and experiment is 11%. Santiago et al. [9] established a finite element model of a plate and shell heat exchanger and conducted static analysis, revealing that the highest stress occurred along the corrugation top at contact regions, and the numerical results were consistent with the experimental results of Martins[10]. These studies show the consistency of simulation and experiment, thus showing that the finite element method is an effective method for the structural analysis of plate heat exchangers.

In this paper, the bursting pressure of a BPHE was studied. For this purpose, the global finite element model of BPHE was established for static structural analysis. The bursting pressure and fracture position of BPHE were predicted by numerical analysis. In order to verify the validity of the numerical model, the bursting pressure test, strain gage test and CT scanning were carried out. The results showed that the numerical model agrees with the experimental results.

2. Numerical analysis

2.1 BPHE structure

The BPHE used in this study is a part of the thermal management system of an electric vehicle, provided by Valeo Hubei Automotive Air Conditioning Co., LTD. It functions as a coupling between the battery cooling circuit and the cabin air conditioning circuit. The global three-dimensional geometric model is shown in Figure 1, consisting of a refrigerant side block, a top plate, a bottom plate, an installing bracket and many heat exchange plates. The block has an inlet and an outlet for the refrigerant. The components are all made of aluminum alloy and connected by high-temperature brazing. Different from the common chevron heat exchange plates, the heat exchange plates in this BPHE adopt the structural design of a dimple, as shown in Figure 2.

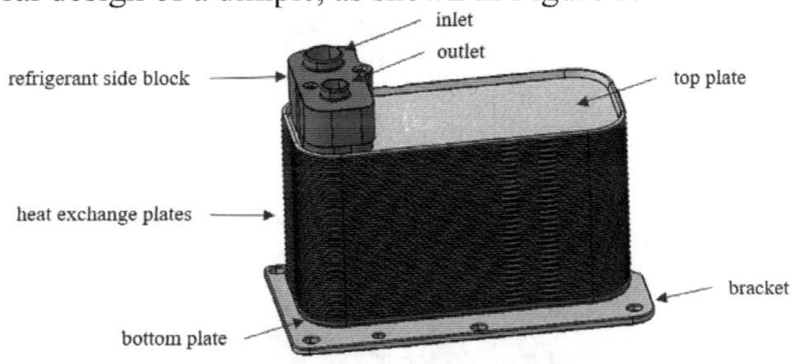

Figure 1. BPHE model.

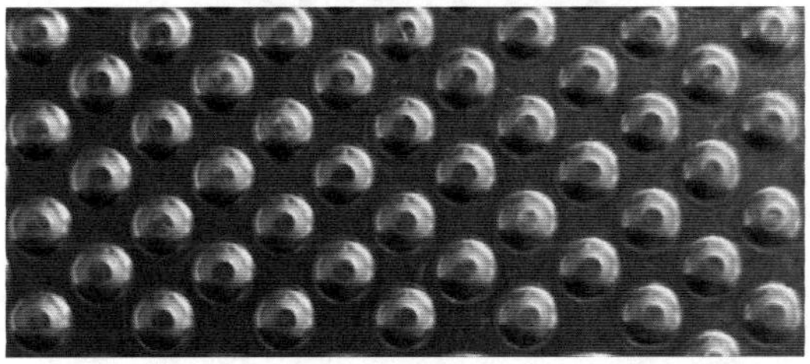

Figure 2. Dimple heat exchange plate.

2.2 Meshing

Heat exchange plates are the most important parts of the BPHE, and it is taken as an example to illustrate the meshing. The heat exchanger plates are identical, stacked in parallel to form a refrigerant flow path and a coolant flow path. The two flow paths are adjacent, and the pressure on the refrigerant side is much greater than the coolant side in the actual working process, so the refrigerant side is pressurized and the coolant side is not pressurized in the bursting test. To compare with the test, the BPHE model was established, the pressure was applied only on the refrigerant side, and no pressure was applied on the coolant side. The two heat exchange plates forming the refrigerant side flow path were modeled as a whole and then arrayed. The local meshing is shown in Figure 3. It was divided by hexahedral elements, two layers of mesh in the thickness direction, and the brazing radius was taken into account. The welding foot areas are usually prone to stress concentration, so mesh refinements are carried out on them. The two heat exchange plates forming the coolant side flow path were contacted by tie constrain.

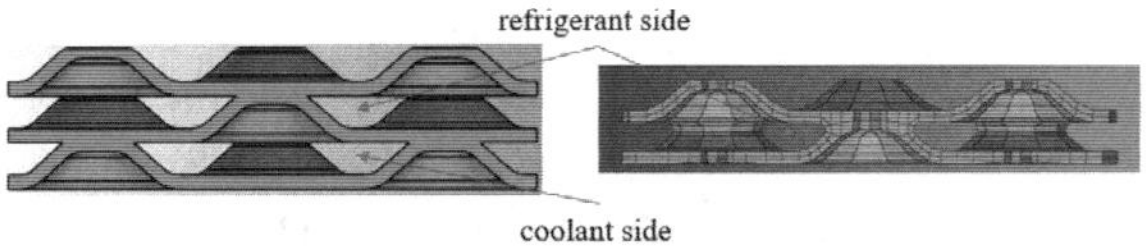

Figure 3. Meshing method.

2.3 *Material properties*

The components are all made of aluminum alloy. Elastic-plastic material constitutive relation was used, so geometric nonlinearity was chosen. Density is 2.7 g/cm^3, Young's modulus is 65 GPa, and Poisson's ratio is 0.33.

2.4 *Loads and boundary conditions*

Pressure P was applied to the inner surface of the refrigerant side flow path, and counter pressure $P_{counter}$ was applied to the inlet and outlet of the refrigerant side block. The relationship between $P_{counter}$ and P is shown in Equation (1), where R is the outer circle radius of the inlet and outlet, and r is the inner circle radius of the inlet and outlet.

$$P_{counter} = \frac{Pr^2}{R^2 - r^2} \tag{1}$$

According to the Saint-Venant principle, the specific distribution of loads only affects the stress distribution near the load area. In order to improve the convergence, the constraints shown in Figure 4 were applied to the four sides of the middle heat exchange plate as boundary conditions. The reason for choosing the middle plate is that according to the actual test results, the middle plate has almost no displacement during pressure loading.

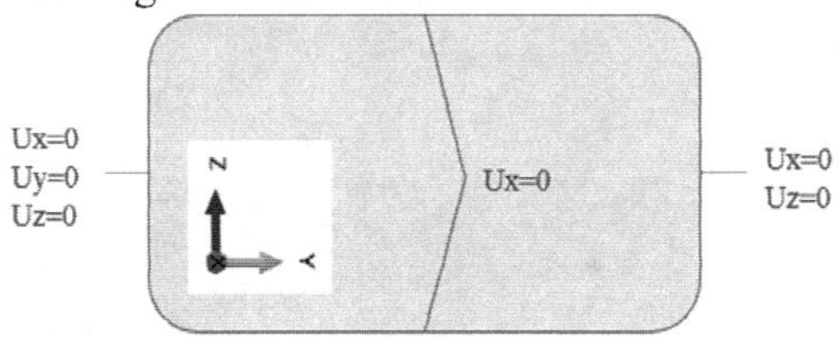

Figure 4. Boundary conditions.

3. Validation

3.1 *Bursting test*

To obtain the bursting pressure of the BPHEs, a bursting test bench capable of producing loads of up to 15 MPa was used. The medium used in the equipment is hydraulic oil. The pressure boost rate ranges from 0.02-10 bar/s, and the pressure sensor is placed at the medium inlet. The bursting pressure test of three BPHEs from the same production batch was carried out, and the pressure was increased at a rate of 0.25 bar/s until the BPHEs burst.

As shown in Figure 5, the strain gage test was carried out on the top plate during one of the bursting tests. The reason for choosing this measuring point is that numerical analysis shows a large strain at this position, and it has enough area for easy to paste 45°-3 strain rosettes. The strain measuring system consisted of four parts: strain gage, data acquisition card, signal processor and upper computer. The strain gage used the Kyowa KFGS series. The gage factor is 2.1, the resistance is 120 Ω, the sensitive grid size is 1 mm and the adoptable thermal expansion is 23.4×10^{-6}/°C. The data acquisition card used *National Instruments*, NI 9235 and the signal processor used *National Instruments*, NI cDAQ-9178. The principal strain direction was unknown, a 45°-3 strain rosette was used. The data collected in the strain gage test are in the time domain, and the amount of data is not only huge, but also some of the data will have certain interference. Therefore, this test applied the pressure load in the step loading mode several times, and collected the data for analysis in the stable stage. In order to facilitate the reading of the strain value corresponding to the pressure, the pressure was held once per P_1 bar starting from $2P_1$ during the pressurization. The pressure was held for 25 s per P_1 bar, and the sampling frequency was 10 Hz.

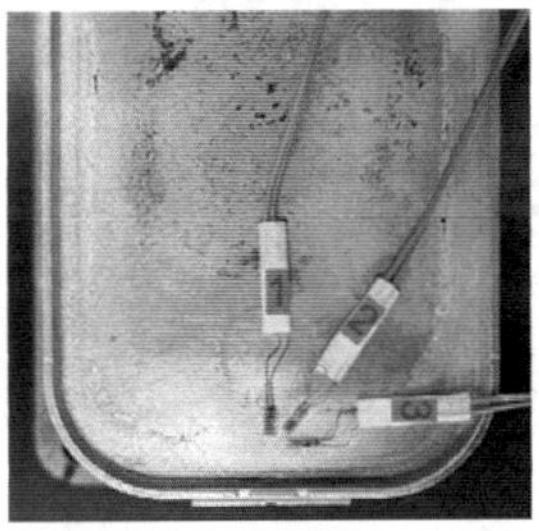

Figure 5. Positioning of strain gages.

3.2 CT scanning

Because the heat exchange plates are connected by brazing, it is difficult to obtain the deformation inside the BPHE after pressure loading. Traditionally, identifying defects, voids, and cracks within an object has required destructive testing. CT scanning can detect internal features and defects without damaging the part and display this information in three-dimensional form. For this purpose, the Nikon XT H 225 industrial CT scanning device was used to obtain the deformation inside the BPHE, it can perform high resolution detection and can clearly display the internal structure of the BPHE. The X-ray source is an open-tube microfocus type with a maximum energy of 225 KV, a maximum power of 450 W and a minimum focal point of 1 μm.

4. Results

4.1 Bursting pressure comparison

The limit load of the BPHE is determined by the maximum equivalent plastic strain (PEEQ). According to the equation below, when PEEQ is greater than material elongation, the BPHE is considered to be damaged.

$$\mathrm{PEEQ} \geq \varepsilon_{\max}^{\mathrm{P}} \tag{2}$$

According to the numerical analysis, when the internal pressure of the refrigerant side reaches P_0 bar, PEEQ reaches the failure strain value of the material, so P_0 is the bursting pressure of the BPHE predicted by the numerical analysis. As shown in Figure 6, the maximum PEEQ is located at the welding foot at the outlet of the first refrigerant flow path.

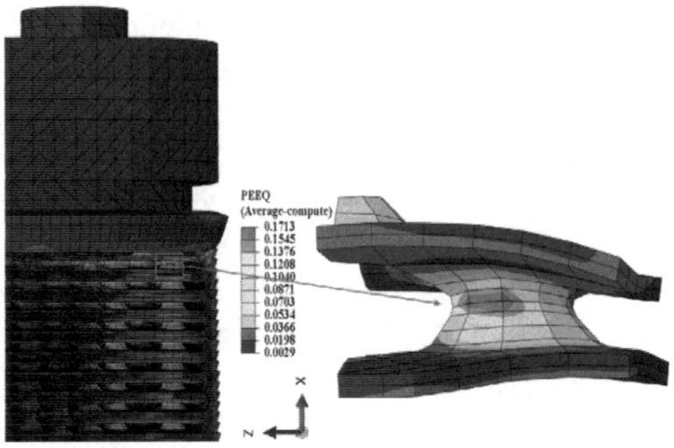

Figure 6. Maximum PEEQ position.

Figure 7 shows the bursting pressure of the three BPHEs obtained from the bursting test, the average is 113%P_0 bar, the gap between predicted and actual bursting pressure is 11.50%. The reason that the predicted pressure is lower than the actual one may be related to the material properties. The material properties of the brazed area will be enhanced[9], but the brazed area and the non-brazed area were not divided in this study. Therefore, the result of the simulation is more conservative than that of the experiment.

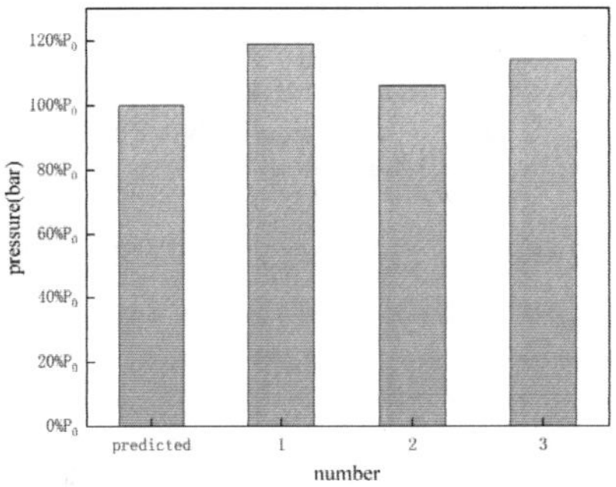

Figure 7. Bursting pressure comparison.

4.2 *Stress comparison*

As shown in Figure 8, the strain at the measured point increases during pressurization. At $4P_1$ bar, the strain exceeds the material yield strain, and it can be judged that the measured point has entered the plastic deformation stage.

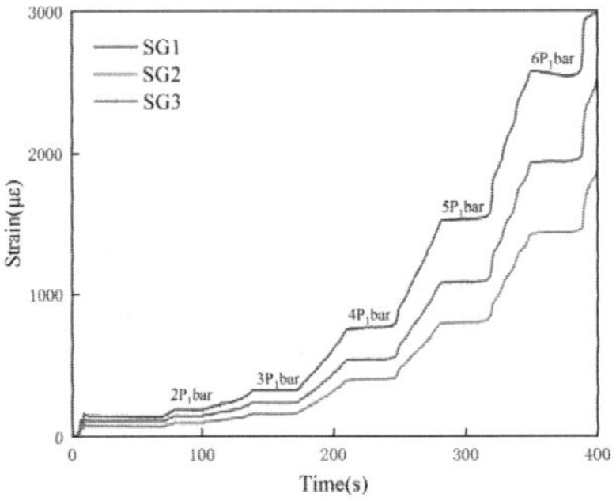

Figure 8. Strain-time curve.

At $2P_1$ bar and $3P_1$ bar, the measured point is in the elastic region, and the strain value corresponding to the pressure can be obtained. 45°-3 strain rosettes require extended forms of Hooke's law, and the local stress state of the measured point can be obtained by Equation (3).

$$\sigma_{1,2} = \frac{E}{2}\left[\frac{\varepsilon_0 + \varepsilon_{90}}{1-\nu} \pm \frac{\sqrt{2}}{1+\nu}\sqrt{(\varepsilon_0 - \varepsilon_{90})^2 + (\varepsilon_{45} - \varepsilon_{90})^2}\right] \quad (3)$$

where ν is the Poisson coefficient and E is Young's modulus. The subscripts 0, 45, and 90 refer to the directions of the strains, while the "1" and "2" refer to the main stresses of the measuring point.

The multiaxial von Mises criterion is used to determine the von Mises equivalent stress, according to the equation below.

$$\sigma_{Mises} = \sqrt{\frac{\sigma_1^2 + \sigma_2^2 + (\sigma_1 + \sigma_2)^2}{2}} \quad (4)$$

As shown in Figure 9, the von Mises stress of the experiment and simulation is compared. The relative gap is minimized to 1.84% at $3P_1$ bar. The gap between the experiment and simulation is within a reasonable range, which further verifies the validity of the finite element model.

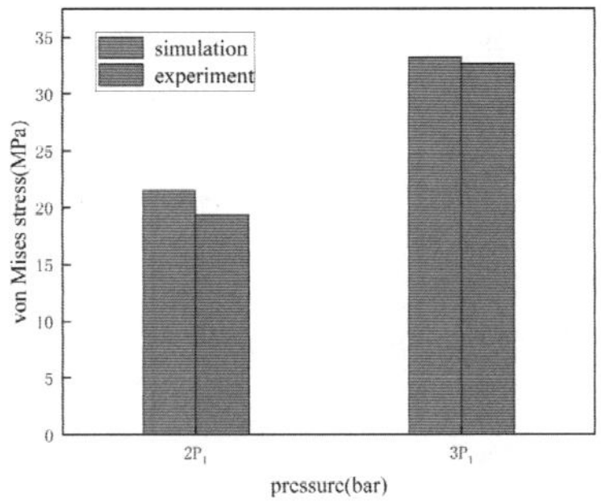

Figure 9. von Mises stress comparison.

4.3 *Deformation comparison*

The XY cross section of the maximum PEEQ position was used for deformation comparison between the CT scan and simulation, as shown in Figure 10. It is obvious that the deformation of the first refrigerant flow path is the largest. The result of CT scanning is generally consistent with that of simulation.

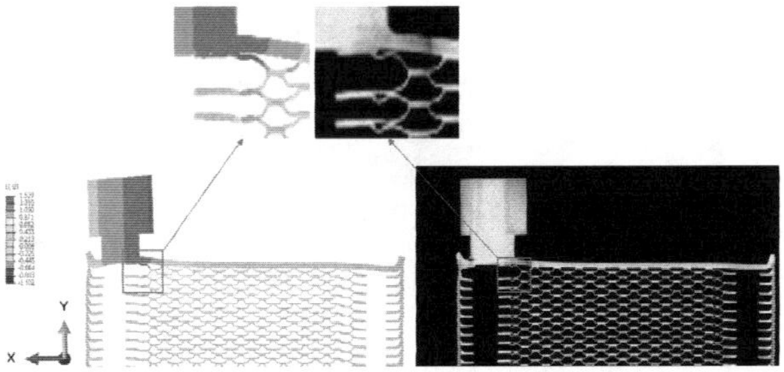

Figure 10. Deformation comparison.

5. Conclusion

Based on the results and discussions presented above, the conclusions are obtained as below:

(1) The bursting pressure of BPHE can be predicted by numerical analysis, and the numerical result is more conservative than the experimental result. FEM is an effective method to study the bursting pressure of BPHEs.

(2) Strain gage tests show that the gap of von Mises stresses between simulation and experiment is small, and a CT scan of the BPHE after the bursting test shows that the deformation is consistent with the numerical analysis, which proves the validity of the simulation model.

References

[1] Martins G.S.M., Zanzi M.S., Paiva K.V., Oliveira A.A.M., Oliveira J.L.G. (2023). Mechanical stress analysis of various GPHE corrugated plates during the assembly and in working conditions. International Journal of Pressure Vessels and Piping, 206, 105013.
[2] Liu Z., Li H., Qian C. (2021). Strength Analysis of a Plate Heat Exchanger with Two Design-By-Analysis Methods. Mechanical Engineering and Materials, 143-154.
[3] Ma H., Cai W., Zheng W., Chen J., Yao Y., Jiang Y. (2014). Stress characteristics of plate-fin structures in the cool-down process of LNG heat exchanger. Journal of Natural Gas Science and Engineering, 21, 1113-1126.
[4] Ma H., He B., Lan S., Liu Y., Gao X., Xue X., Hou C., Li C. (2018). The stress characteristics of plate-fin structures at the different operation parameters of LNG heat exchanger. Oil & Gas Science and Technology, 73, 13.
[5] Wang J., Yan X., Lu M., Sun Y., Wang J. (2022). Structural Assessment of Printed Circuit Heat Exchangers in Supercritical CO2 Waste Heat Recovery Systems for Ship Applications. Journal of Thermal Science, 31(3), 689-700.
[6] Gürler Y. (2018). Design and Mechanical Behaviour of Brazed Plate Heat Exchangers. Dissertation/Thesis.
[7] Ma H., He B., Liu Y., Gao X., Lan S., Han L., Li C., Hou C., Han X. (2019). Influence of brazed joints on stress of plate-fin structures in LNG heat exchanger. Journal of Harbin Institute of Technology, 51(2), 172-178.
[8] Zibart A., Spang B., Kenig E.Y. (2023). Numerical analysis of burst pressure and maximum inflation height of aluminium made pillow plates. International Journal of Pressure Vessels and Piping, 203, 104878.
[9] Santiago R.S., Possamai T.S., Donati D.C.X., Martins G.S.M., Oba R., Tascheck B.L., Paiva K.V., Oliveira J.L.G. (2024). Structural behavior and mechanical fatigue of plate and shell heat exchangers through finite element analysis. International Journal of Pressure Vessels and Piping, 210, 105252.
[10] Martins G.S.M., Santiago R.S., Beckedorff L.E., Possamai T.S., Oba R., Oliveira J.L.G., Oliveira A.A.M., Paiva K.V. (2022). Structural analysis of gasketed plate heat exchangers. International Journal of Pressure Vessels and Piping, 197, 104634.

Experimental and simulation study of cutting forces on a porous workpiece based on Abaqus surfaces

Xinyu Liang[1], Zhenfeng Jiang[2], Fengyang Gao[3] and Yanan Pan[*,4]

[1]School of Mechanical University of Jinan, Jinan China
[2]Zibo TAA Metal Technology Co, Zibo China
[3]School of Mechanical University of Jinan, Jinan China
[4]School of Mechanical University of Jinan, Jinan China

*Corresponding author: e-mail: me_panya@ujn.edu.cn

Abstract: In the process of turning, it is not uncommon to encounter machined features on the surface of the workpiece, such as holes. When the tool makes contact with and subsequently disengages from these holes, a sudden change in the cutting force is typically observed. Finite element analysis is employed to examine the alteration in cutting force that occurs during the turning of workpieces with holes on their surface. Experimental evidence indicates that when the tool traverses the hole during cutting, it can precipitate sudden and pronounced fluctuations in the cutting force, which in turn can give rise to instability in the machining process and thereby affect the accuracy and surface quality of the workpiece. The experimental results corroborate the precision of the simulation model and underscore the substantial influence of holes on the cutting force. This study offers a foundation for investigations into tool longevity and surface machining quality when cutting porous workpieces with curved surfaces, thereby advancing the frontiers of machining technology for such workpieces and expanding their potential applications.

1. Introduction

During the turning process, the metal layer to be cut on the workpiece, the chips, and the metal layer on the machined surface all experience elastic and plastic strains. Furthermore, the flow of chips along the rake face of the tool and the sliding of the tool's flank face against the machined surface result in the generation of considerable friction. These phenomena give rise to forces that impede the cutting process. The force required by the tool to overcome this resistance is referred to as the cutting force. The magnitude and direction of the cutting force not only affect the efficiency of the machining process, but also directly determine the surface quality and accuracy of the workpiece. When machining rotationally symmetrical parts and achieving a certain surface quality, precise control of the cutting parameters is essential, as the cutting force will cause elastic-plastic deformation of the machined surface layer, which will seriously affect the accuracy of the workpiece. In practical machining, the presence of machined features, such as holes, on the surface of the workpiece results in a sudden change in the cutting force. Such abrupt alterations may result in the formation of burrs along the edges of the holes, thereby considerably impairing the overall surface quality. This study concentrates on the simulation-based investigation of the cutting process for shaped cast aluminium alloy components with holes. The results obtained from the simulation facilitate a more profound comprehension of the influence of structural discontinuities, such as holes, on the cutting force. This research provides a robust basis for further investigation into the influence of workpieces with diverse topographical characteristics on tool lifespan and surface machining quality during cutting. A more profound comprehension of these phenomena can facilitate the optimisation of the cutting process, enhance the precision and quality of complex parts, prolong the operational lifespan of cutting tools, reduce production costs, and meet the demand for high-precision, high-quality machining, thereby enhancing overall production efficiency.

2 Finite element simulation of cutting

2.1 Introduction to Abaqus

Finite element analysis is an advanced numerical analysis method that employs a mathematical model to calculate the behaviour of a continuous object by discretising it into a finite number of small units. This approach allows for the accurate prediction of the object's overall behaviour. Abaqus is a robust finite element software with a strong reputation within the industry. It is capable of analysing intricate solid mechanics and structural mechanics systems, and is particularly adept at handling highly non-linear problems. The software offers a comprehensive range of analysis capabilities, including static and dynamic analysis, to address diverse engineering requirements. In dynamic cutting simulations, Abaqus employs the principle of explicit dynamics to perform comprehensive analyses through the explicit dynamics analysis module. It is capable of efficiently handling complex contact, material nonlinearity and geometric nonlinearity, which makes it an invaluable tool for addressing intricate engineering challenges, optimising designs and enhancing product performance. This simulation utilises the ABAQUS dynamic explicit analysis module for analysis.

2.2 Basic information on workpiece material and tool

The physical properties of the cast aluminium alloy ZL205A are shown in Table 1, and the Johnson-Cook constitutive model is shown in Table 2.

Table 1 Physical properties of ZL205A.

density ρ (g/cm^3)	Conductivity Γ (%LACS)	melting point T (°C)	linear expansivity α (10-6·K)	Thermal conductivity Λ [W/((m·K))]
2.82	25	544-633	21.6	113

Table 2 Johnson-Cook constitutive model.

A (MPa)	*B* (MPa)	*n*	*C*	*m*
297.94	735.56	0.66	0.00672	1.303

2.3 3D model creation

In the context of cutting simulations, two-dimensional cutting is typically employed for the purpose of machining simulation. Nevertheless, for components with intricate geometries, two-dimensional simulation is inadequate for accurately reproducing the surface characteristics of the workpiece that the tool encounters during the cutting process. Accordingly, this paper employs a three-dimensional cutting simulation methodology. The initial step involves the creation of a workpiece and tool model with hole features through the utilisation of 3D modelling software, followed by the precise assembly of these models within the same software. Subsequently, the model is imported into the Abaqus simulation software for analysis[1]. In the simulation settings, the tool was defined as a discrete rigid body and solid extraction was performed on it, with the objective of ensuring the accuracy of the calculation. The workpiece was meshed with local refinements in critical areas to enhance the precision of the analysis. It was then designated as a mesh component with fixed constraints at the bottom end[2]. In order to simulate the actual cutting movement, a plane motion pair is established between the workpiece's centre of rotation and the plane where the tool tip and cutting path are located.

The trajectory of the tool's movement is accurately defined in the form of an amplitude curve by establishing the parametric equation of a circle. This ensures that the simulation can truly reflect the various complex morphological characteristics encountered by the tool during the cutting process. The simulation model is shown in Figure 1, which demonstrates the results of this detailed modelling and simulation process and provides a reliable basis for further analysis and optimisation.

Figure 1 Simulation model.

2.4 Extraction and analysis of cutting forces

By establishing a reference point at the tool tip position and defining the output of the force and reaction force at this point in the numerical simulation output, it is possible to obtain more accurate information on the cutting force and stress distribution. Accordingly, a stress cloud map and cutting force curve can be generated at the tool tip reference point, as illustrated in Figure 2. These visualised results encompass not only the details of the stress distribution but also the trend of the cutting force throughout the cutting process. This facilitates a more profound comprehension of the stress state and wear mechanism of the tool.

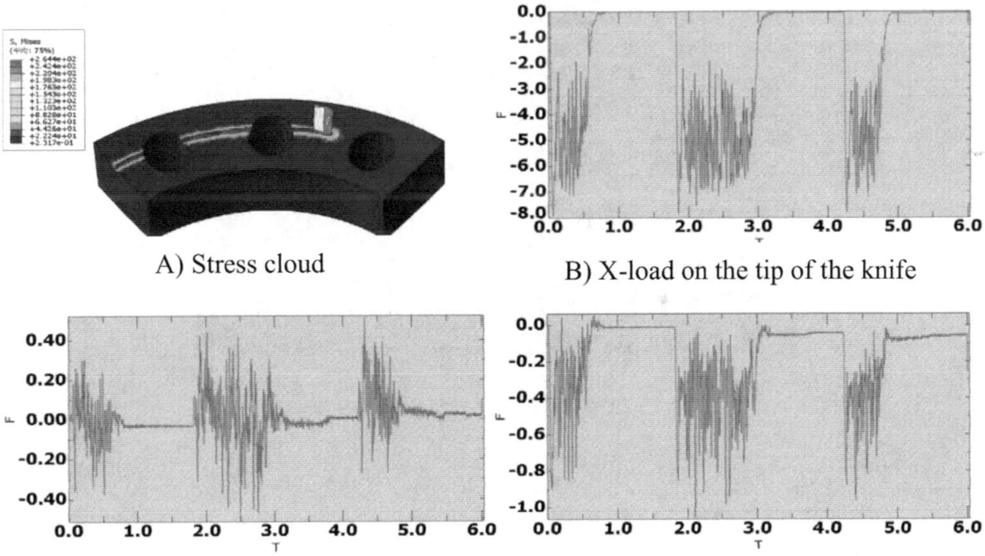

A) Stress cloud B) X-load on the tip of the knife

C) Y-load on the tip of the knife D) Z-load on the tip of the knife

Figure 2 Simulation results.

As illustrated in Figure 2 B), the cutting force undergoes a sudden and significant change from zero to a larger value at the moment the tool makes contact with the workpiece. During the cutting process, the cutting force exhibits fluctuations within a certain range, with a considerable degree of variability. As illustrated in the process animation of the simulation results, the chips do not form a continuous chip. Additionally, the friction between the chip and the tool varies with the formation and peeling of the chip. This prevents the formation of a stable cutting zone, resulting in significant fluctuations in the cutting force during stable cutting. It is evident that the turning process is subject to transient impact forces. It can be observed that the cutting tool is subjected to impact loads from cutting in and cutting out on three occasions during one revolution. The maximum sudden change in the x-direction load occurs when the tool tip cuts into the hole.

3 Cutting force test

3.1 Experimental plan

A dynamometer was employed to quantify the cutting force throughout the cutting process of the inner surface of the sample. The data obtained from the dynamometer not only reflects the magnitude of the cutting force, but also captures its dynamic changes during the cutting process, thus providing a comprehensive insight into the cutting phenomenon. The data obtained from these measurements are of great significance for the analysis of the mechanical behaviour during the cutting process, optimisation of cutting parameters and improvement of machining quality.

In this experiment, a CNC lathe was employed, and the cutting process utilised a PCD diamond blade with the designation DCMT11T308 and a tip radius of 0.8 mm. The cutting force was quantified utilising a general-purpose cutting force measurement system (GDFMS). The workpiece material was ZL205A, and the experimental apparatus is illustrated in Figure 3.

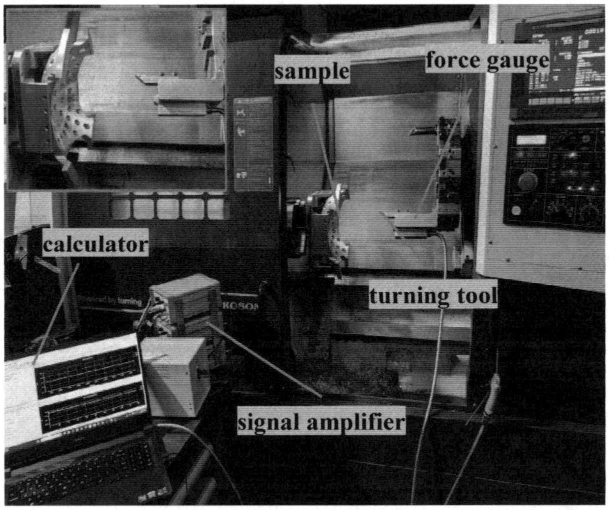

Figure 3 Cutting test equipment.

3.2 Experimental results

As the tool traversed the aperture during the cutting process, the experiment revealed that the cutting force exhibited pronounced fluctuations, akin to the simulation outcomes. These fluctuations render the machining process highly unstable, resulting in significant toolholder vibration. This unstable vibration has a detrimental impact on both the accuracy of the machining process and the quality of the resulting workpiece surface. The experimental results corroborate the precision of the simulation model, demonstrating that the presence of holes during machining exerts a considerable influence on the cutting force and system stability. As illustrated in Figure 4 of the experimental results, the cutting process can be divided into three distinct stages: stable cutting, dry cutting, and plunge cutting. During the stable cutting stage, the cutting force exhibits slight fluctuations within a defined range. The source of the cutting force during this stage is analogous to that observed in continuous cutting. During the instantaneous cutting-in stage, the cutting force increases instantaneously, resulting in an impact force. This impact force is generated by the collision between the workpiece and the cutting tool. During intermittent turning, the tool life and workpiece machining quality are significantly influenced by the presence of this periodic impact force. During the cutting-out stage, the cutting force is relatively low. During the dry cutting stage, the theoretical cutting force should be zero; however, in practice, due to machine vibration and zero drift in the cutting force measurement system, a value comparable to that observed during the cutting-out stage may be recorded[3].

Consequently, the existence of holes can precipitate abrupt alterations in the cutting force. Such abrupt fluctuations in force can precipitate pronounced vibration between the tool and the workpiece, which in turn accelerates tool wear and the formation of surface imperfections on the workpiece. Consequently, when formulating the subsequent machining plan, it is imperative to optimize the cutting parameters to mitigate the detrimental impact of sudden changes in cutting force and vibration on the overall machining outcomes. This finding offers invaluable guidance for enhancing the machining stability and surface quality of intermittent surface parts. The numerical results of the cutting force are shown in Table 3.

Table 3 Results of cutting forces.

F \ Dir	X	Y	Z
Max	7.4	0.6	0.0
Avg	1.66	-2.45	-2.3
Volatility range	7.4 ~ -1.6	0.6 ~ -6.0	0.0 ~ -5.2

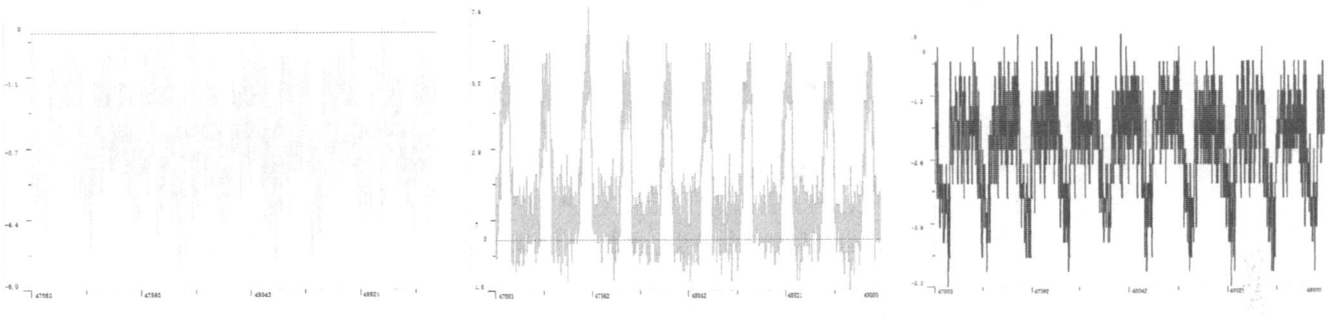

Figure 4 Experimental results.

4 Analysis of cutting force and its edge surface quality when cutting holes with tools

4.1 Cutting force analysis in the negative shear area

As illustrated in Figure 2 D), the simulation analysis reveals a sudden reversal in the cutting force when the tool reaches the edge of the hole. This phenomenon occurs because as the tool approaches the edge of the hole, the material thickness between the cutting edge and the edge of the hole diminishes, thereby reducing its supporting rigidity. When the cutting force surpasses the limit that the end material can withstand, the material at the edge of the hole deforms and rotates, resulting in negative shearing and a counter-directional cutting force generated on the cutting edge of the tool[4].

4.2 Surface quality analysis of hole edges

The occurrence of negative shear is accompanied by the formation of burrs, which reduce the surface quality of the hole edge. Each time the tool cuts the hole area, it will leave metal residues or defects on the cutting edge of the hole edge, i.e. cutting burrs[5]. There are two main types of burrs in the direction of the main motion exit, usually referred to as positive burrs and negative burrs. In actual cutting, burrs

Journal of Physics: Conference Series **2992** (2025) 012039 doi:10.1088/1742-6596/2992/1/012039

usually appear in the form of coexistence of positive and negative burrs, as shown in Figure 5 A). Under stable cutting conditions, the deformed area is usually divided into the first, second and third deformed areas[6]. However, as the tool moves towards the edge of the hole, the cut enters an unstable state and a new deformed zone, the negative shear zone, is formed below the cutting surface. The distribution of the deformed areas is shown in Figure 5 B).

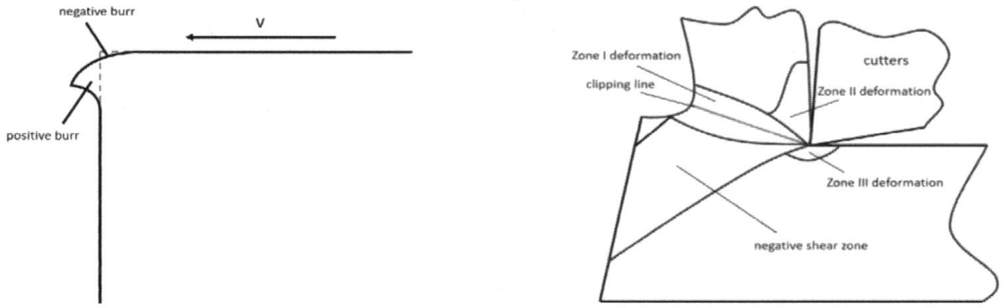

A) Schematic diagram of burr pattern B) Theoretical Modelling of Negative Shear Zones

Figure 5 Schematic diagram of burr pattern and negative shear zone.

Microscopic observation of the edges of the holes revealed that under normal circumstances, the edges have both positive and negative burrs. The scanning results of the electron microscope are shown in Figure 6.

Figure 6 Image of the edge of the hole.

5 Conclusions

Based on Abaqus software, a three-bit cutting simulation model of cast aluminium alloy was established, the cutting force was extracted, the surface morphology was observed, actual cutting force experiments were carried out, and the three-directional cutting force was measured by piezoelectric ceramic force gauge. The following conclusions were drawn.

(1) The cutting force changes from zero to a larger value at the moment of tool contact with the workpiece, the cutting force fluctuates within a certain range during the cutting process, and there is a reverse mutation of the cutting force due to the negative shear effect when the tool moves to the edge of the hole.

(2) When the tool cuts into the cavity, the edge of the hole will produce burrs because of the negative shear effect, the shape of the burr is related to the cutting parameters and material parameters, and burrs lead to the poor surface quality of the edge of the hole.

(3) The peculiarity of intermittent cutting is that there is a cutting impact force, which is larger than the steady state force, and this paper makes experimental research on its size and change rule.

(4) Intermittent cutting due to the presence of impact force makes the tool life decrease.

Journal of Physics: Conference Series **2992** (2025) 012039 doi:10.1088/1742-6596/2992/1/012039

References

[1] İynen O, Ekşi A K, Akyıldız H K, et al. Real 3D turning simulation of materials with cylindrical shapes using ABAQUS/Explicit [J]. Journal of the Brazilian Society of Mechanical Sciences and Engineering, 2021, 43(8):1-18.
[2] Ceretti E, Attanasio A, Giardini C, et al. Investigation and FEM-based simulation of tool wear in turning operations with uncoated carbide tools [J]. Wear, 2010, 269(5): 344-350.
[3] Shuang Y, Li J, Zhu J, et al. Investigation of machining Ti-6Al-4V with graphene oxide nanofluids: Tool wear, cutting forces and cutting vibration [J]. Journal of Manufacturing Processes, 2019, 49:35-49.
[4] Poór D I, Geier N, Pereszlai C, et al. A critical review of the drilling of CFRP composites: Burr formation, characterisation and challenges [J]. Composites Part B: Engineering, 2021, 223: 109155.
[5] Pang X, Zeng Y, Zhang J, et al. Analytical model and experimental verification of Poisson burr formation in ductile metal machining [J]. Journal of Materials Processing Technology, 2021, 290: 116966.
[6] Han J, Hao X, Li L, et al. Investigation on surface quality and burr generation of high aspect ratio (HAR) micro-milled grooves [J]. Journal of Manufacturing Processes, 2020, 52: 35-43.

Research on explosion dispersion based on near-field and far-field coupling calculation method

Yanyu Jiang [1, 2], Zheng Mei [2] and Zhenggang Xiao [1]*

[1] School of Chemistry and Chemical Engineering, Nanjing University of Science and Technology, Nanjing 210094, Jiangsu, China
[2] Hunan Vanguard Group Co., Ltd, Institute of Guided Weapons and Equipment, Changsha 410137, Hunan, China

*Corresponding author's e-mail: Z G Xiao: xiaozhg@njust.edu.cn

Abstract: In order to study the influence of different structures on the liquid explosion-throwing process and cloud mass morphology, this paper proposes a liquid explosion-throwing process and morphology prediction method based on near and far field coupling calculation. In this method, the near-field finite element throwing jet data of the explosive device is used as input, the trajectory of the droplet group is reconstructed based on the aerodynamic evolution law and the mapping of the jet to the droplet group, and the concentration field formed by droplet peeling and evaporation is constructed. Finally, the simulation calculation of the whole process of liquid explosion and evaporation and the high-fidelity morphology prediction of cloud and fog mass are realized. In this paper, a set of simulation and calculation software for the whole process of liquid explosion and dispersion based on this method is developed, which can calculate and predict the far-field process and cloud morphology in real-time based on the near-field finite element calculation results. The method and software are used to calculate and predict the influence of radial reinforcing rings on the effect of explosion throwing. The simulation results show that the wider the width of the reinforcing ring is, the greater the interval of jet delamination in the near-field stage is, the jet becomes sharper, and the effect of the steel reinforcing ring is more obvious than that of the Al reinforcing ring. Compared with the throwing device without a reinforcing ring, the cloud inhomogeneity produced by the steel reinforcing ring is significantly reduced, the volume of the cloud is increased by about 1 times, the diameter of the cloud is reduced by 23%, and the thickness of the cloud is increased by 78%. With the increase of the width/thickness of the Al reinforcing ring, the cloud inhomogeneity showed a decreasing trend, the volume of the cloud increased slightly, the diameter of the cloud decreased slightly, and the thickness of the cloud increased by about 26%. The simulation results help to optimize the structural design of the blast-throwing device.

1. Introduction

Liquid explosive discharging is the process of using explosives to explode to generate a driving force that drives the liquid to spread to the surrounding area and mix with the air to form a cloud of mist. It has a wide range of application value in both agricultural production processes and industrial fields. The liquid explosive-throwing process involves complex laws and effects such as explosion mechanics, fluid dynamics, and multiphase flow. The cloud mass is a complex dynamic system composed of the explosion-throwing device and the surrounding air, so the liquid explosion throwing has always attracted the attention of the academic community [1-3].

During the period from 1970 to 2000, numerical simulations of liquid explosion projections established a general framework for near-field and far-field calculations of "divide and conquer", focusing on the overall behavior of clouds and the description and analysis of multiphase matter interactions [4]. In recent years, Xue et al. [5] and Ding and Liu [6] conducted research on the different characteristics of the liquid explosion and scattering process based on the near and far fields, but due to computer performance limitations, the former did not describe the interaction process between droplets and air, which was a qualitative study, while the latter was based on one-dimensional and two-dimensional models for the formation and movement of droplets. As a result, the models of both cannot be directly used in real explosive-throwing devices. Recently, dimensional analysis, mathematical derivation, simulation calculation, experimental fitting, and other methods have been used to explore the structure-performance relationship of real throwing devices, and a certain overall prediction ability has been formed [7-9]. Xu et al. [10] used dimensional analysis combined with experimental data fitting to obtain a cloud and fog radius prediction formula. Li and Liu [11] studied the throwing effect of square devices. The above studies reveal the key process and mechanism of explosive dispersion and obtain the influence of various structural factors on the effect of explosive dispersion. The simulation mainly focuses on the influence of the structural characteristics of the device on the properties of the near-field jet, while the experimental research focuses on the movement of liquid droplets and the overall morphology of cloud masses.

In order to solve the above problem of the connection between the near field and the far field of explosion ejection, the liquid explosion discharging process and morphology prediction method based on the coupling calculation of near and far fields open up the near-field and far-field simulation path. Based on this, the corresponding calculation software was developed, and the effectiveness of the program in predicting the effect of liquid explosion and scattering was verified. For the sake of structural stability during mounting, it is necessary to design a reinforcing ring for the device to prevent damage or deformation of the device during loading/transfer. Finally, under the influence of different reinforcing ring structures, various throwing effects and rules of throwing devices were studied.

2. Theory and computational methods

2.1. Calculation principles of near-field and far-field simulation

The liquid explosion throwing process can be divided into three parts: near-field process, transition process, and far-field process, as shown on the left in Figure 1. The near-field-far-field transition process is more complex. The liquid jet is broken into droplets under the combined action of air resistance and explosion propulsion, which involves cross-scale effects and multiphase flow interactions. Its behavior is difficult to calculate and predict, which makes the gap between the near-field process and the far-field process for a long time, which hinders the realization of the whole process simulation calculation.

In view of this, this paper proposes to establish a mapping algorithm from near-field jet to droplet group, so as to open up the near-field-far-field simulation pathway. The mapping is as follows:

a. Jet velocity = muzzle velocity of the droplet.

b. Jet direction = direction of initial motion of the droplet.

c. Jet position = initial droplet position.

d. Jet real-time flow = droplet mass × number of droplets emitted in real-time.

This mapping is based on the following basic assumptions:

a. The liquid jet is completely converted into droplets in a short period of time. Its transformation time is negligible compared with the overall throwing time.

b. In the process of breaking the liquid jet into droplets, the kinetic energy lost is negligible, and the kinetic energy and momentum are conserved.

c. The droplets move independently after formation and do not interact with other droplets.

Based on this mapping, the characteristic transmission path of the throwing device, liquid jet, droplet group, and cloud mass is opened up, so as to realize the simulation and calculation of the whole process of the action of the throwing device with different structures, as shown on the right in Figure 1.

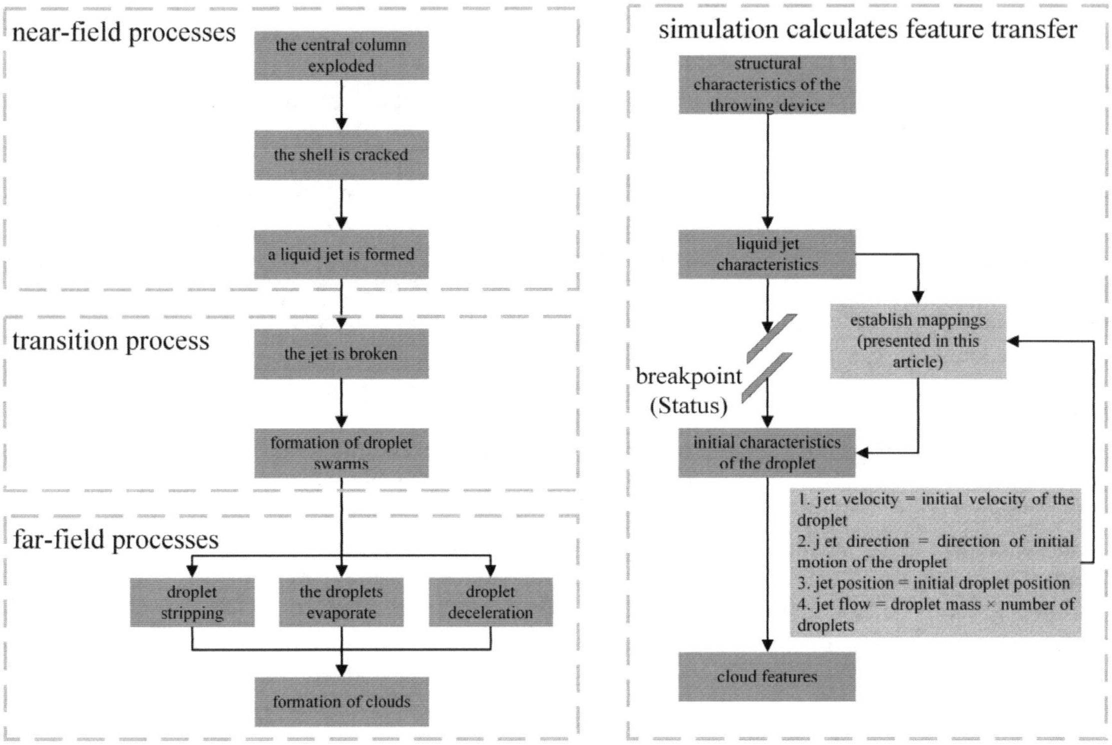

Figure 1. Schematic diagram of liquid explosion and dispersion process.

2.2. Jet-droplet group mapping algorithm

2.2.1 Get jet data. In order to obtain the jet information, commercial software such as ANSYS was used for structural modeling and near-field explosion projection simulation calculations. An array of observation points is set up near the enclosure of the device. Observations are static geometric points. Each observation point can record the density and velocity of matter flowing through it independently. Figure 2 shows the density recorded at an observation and the velocity in the X direction (the same is true for the velocity in the Y and Z directions) over time. In the density plot (Figure 2), you can see the area of a plateau-shaped curve formed by the flow of liquid (water) through the observation point, as well as the density fluctuations due to the shock wave compressed air. In the velocity graph (Figure 2), you can see the fluctuations in jet velocity due to the effects of the shock wave being confined and reflected by the shell, the shell breaking, etc. Since the observation points do not record the type of material flowing through them separately, the velocity data needs to be filtered by the density data to select the velocity data that belongs to the liquid (rather than the shell/air). After selection, the velocity data is a description of the movement of the liquid jet. According to the experience of this study, the velocity dataset in the range of the liquid density ± 20% in the standard state can be selected as the jet velocity set, because it is difficult for the shock wave to compress or stretch the liquid by more than 20%. The air density is much lower than 80% of the liquid density, and the density of the shell (generally Al or steel) is much greater than 120% of the liquid density. When the air, shell and other substances pass through the observation point, the recorded real-time density will exceed the range of the liquid density ± 20%. This method is a good way to pick out the parts of the velocity data that belong to the liquid that is being thrown, so as to obtain the jet motion data near the observation point.

2.2.2 Creation of a primordial droplet population based on the jet. Based on the time-varying data of jet velocity, it can be divided into several droplets in the form of data segmentation. The droplets are assigned the formation time and initial velocity according to the real-time velocity of the jet. As shown in Figure 3, the X-velocity of the liquid jet obtained from an observation point is shown in the curve. The area under the curve is the total length of the jet in the X direction (the same is true for the Y and Z directions). The area below the curve at different time points represents the volume of liquid flowing near the observation point perpendicular to the direction of the jet at different times. According to the assumptions in Section 2.1, the time and kinetic energy loss taken to split the liquid jet into a group of droplets are negligible. Thus, the area below the jet velocity curve can be divided into small rectangles of the same width and different heights, each of which represents a droplet. The height of the rectangle represents the velocity of the droplet, the area represents the size of the droplet, and the position represents the formation time of the droplet. Through this segmentation, the mapping relationship between the jet and droplet group is established, which lays a foundation for the simulation and calculation of far-field droplet behavior based on the near-field jet.

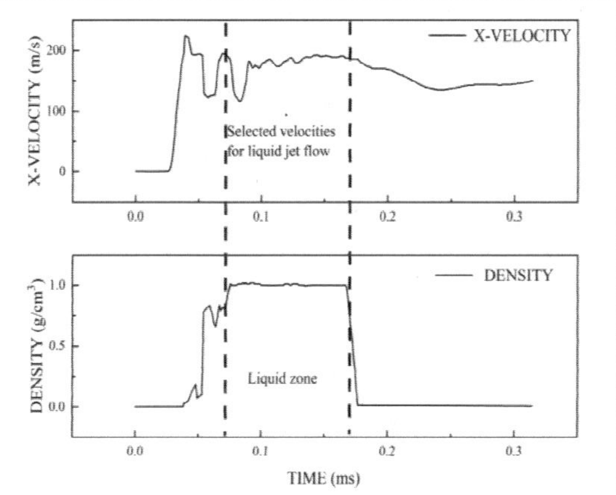

Figure 2. Schematic diagram of material density and X-velocity recorded at a single observation point.

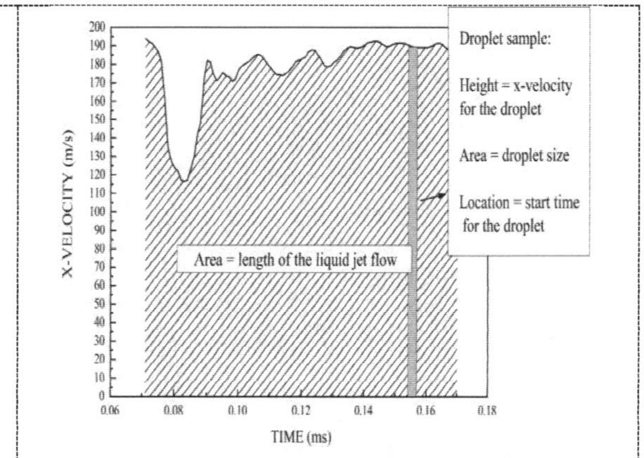

Figure 3. Schematic diagram of the process of decomposing a jet into droplets.

2.3. Full-process simulation and calculation framework

The general framework of the whole process simulation and calculation of liquid explosion and dispersion is shown in Figure 4. Firstly, the real model of the explosion and dispersion device is established. Its structural characteristics are fully considered. The data of the near-field jet are obtained through finite element simulation calculation. Secondly, by establishing the mapping of the jet to the droplet group, the initial time, position, and motion velocity of the droplet group were obtained. Based on this, the droplet group was created in the far-field large-scale simulation computing domain. Finally, through the governing equations of droplet stripping, evaporation, and aerodynamic deceleration, the real-time situation and motion trajectory of peeling and evaporation of each droplet were calculated and tracked. The temporal and spatial distribution of droplets, strippings, and gasifications in the throwing area was obtained. Finally, the dynamic morphology characteristics of cloud masses were obtained. The first step can be implemented using existing finite element calculation software, such as Autodyn, LS-dyna, etc. The second and third steps can be implemented by self-programming.

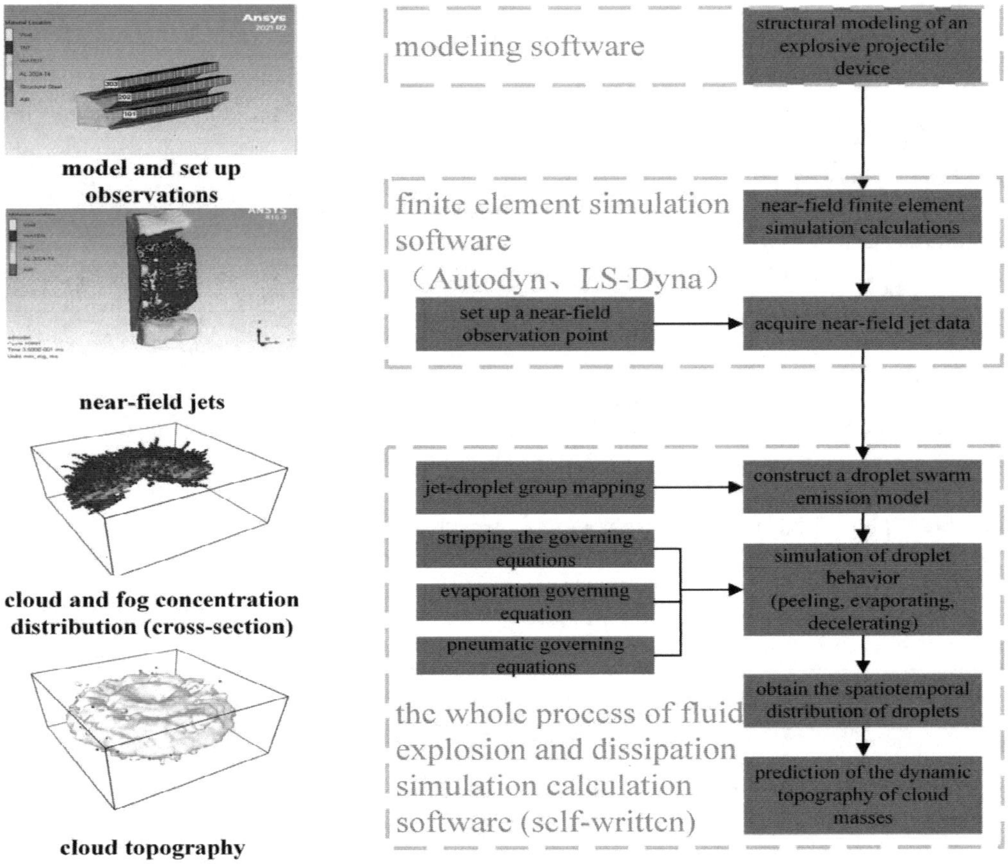

Figure 4. Simulation calculation framework for the entire process of liquid explosion dispersion.

According to the equations of peeling, evaporation, aerodynamic deceleration, and volume change of the droplets, the spatiotemporal trajectory of each droplet and its retention on the trajectory can be programmed to calculate, especially the retention of the droplet at the current position after evaporation, and the hysteresis effect of the tiny droplets formed after peeling that is easier to decelerate than the main droplet. From this, the overall spatial and temporal distribution of the projectile can be calculated [12-14]. According to the jet-droplet group mapping algorithm and the droplet control equation, the simulation calculation software "Explosion Throwing Device Simulation Calculation Expert System-V1.6 Version" was written. It contains 3 function tabs, namely "Global Settings", "File Path", and "Cloud Throw".

3. Simulation calculation examples

3.1. Structural modeling

The throwing device is a cylindrical structure. A reinforcing ring part is added on the inner side of the cylinder wall. The structure and size are shown in Figure 5.

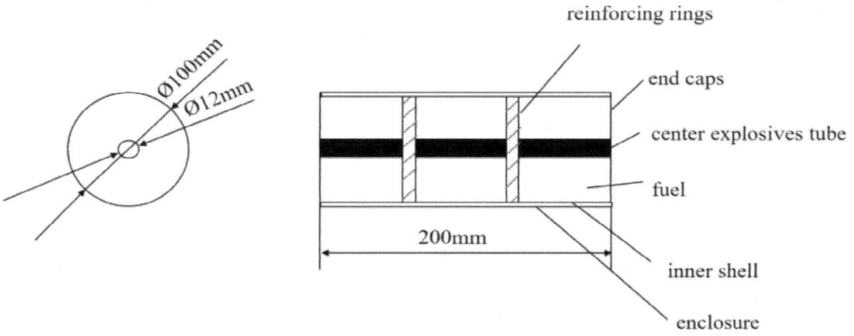

Figure 5. Explosion dispersal device structure with reinforced ring.

All device parameters are set the same, but the material and size of the reinforcing ring are different. Table 1 shows the material parameters of the relevant throwing device.

Table 1. Material parameters of the scattering device.

Parts	Material	ρ/ kg·m^{-3}	Equation of state	Material model
End caps	Al 2024 T-4	2785	Shock	Johnson-Cook
Inner shell	Al 2024 T-4	2785	Shock	Johnson-Cook
Center charge	TNT	1630	Shock	Johnson-Cook
Fuel blends	WATER	998	Shock	none
Reinforcing ring (Al)	Al 2024 T-4	2785	Shock	Johnson-Cook
Reinforcing ring (steel)	Structural steel	7850	Shock	Johnson-Cook

3.2. Near-field simulation analysis

In Figure 6, the AL-T2W10/STEEL-T2W10 model represents an Al/steel reinforcing ring with a thickness of 2 mm and a width of 10 mm. The different colors in Figure 6(a) represent different materials, with dark blue being the TNT explosion product, green being Water, light blue being Al, and red being AIR. In Figure 6(b), the red is Steel, and the pink is AIR, and others are consistent with Figure 6(a). Under the same driving force, the following phenomena are observed:

(1) There is a stronger constraint in the part with the reinforcing ring. The strength of the part is enhanced, resulting in the detonation force concentrated in the non-reinforcing shell part. The weaker strength part ruptures first, which makes the shell rupture unevenly, showing a "segmented" rupture, resulting in jet stratification and uneven ejection of liquid on the Z-axis.

(2) By comparing the width of the reinforcing ring, it is evident that a wider reinforcing ring provides greater strength. This increased width leads to a larger interval of jet delamination and a sharper jet formation. The effect is particularly pronounced with steel reinforcing rings, which offer even greater strength. As a result, cloud resistance increases, leading to more significant delamination in the far field. However, this may also result in an expanded throwing radius.

(3) By comparing the thickness of the reinforcing ring, it can be seen that the influence of the thickness of the steel reinforcing ring is similar to the width but not as obvious as the width. However, the width/thickness of the Al reinforcing ring has no obvious effect, because the entire casing of the throwing device is also Al. The binding force generated when the width/thickness of the reinforcing ring is small, is negligible.

(4) The influence of the near-field throwing process on the far-field throwing effect can only be briefly judged. The far-field throwing simulation needs to be further carried out by using the jet data to obtain the cloud mass morphology.

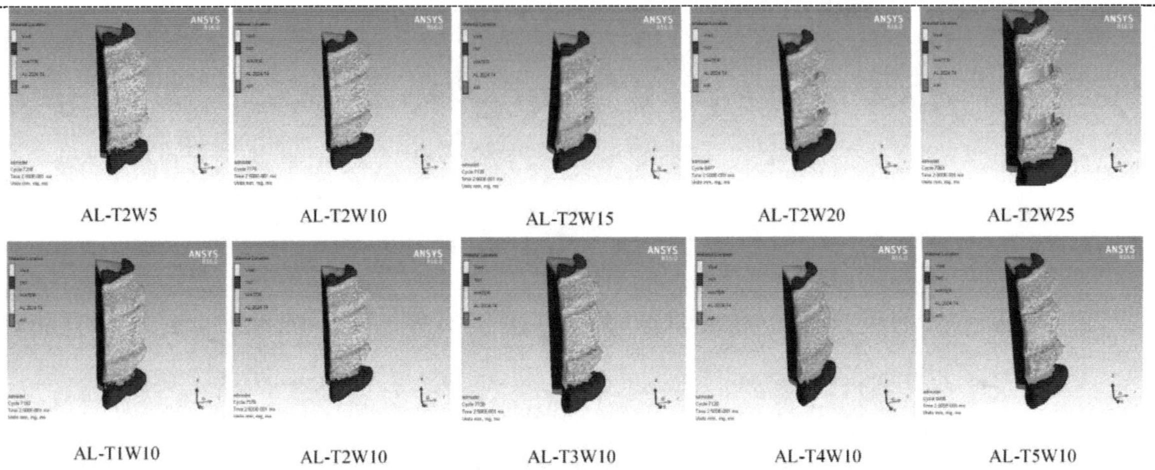

(a) Near field simulation of Al reinforced ring at 0.25ms time.

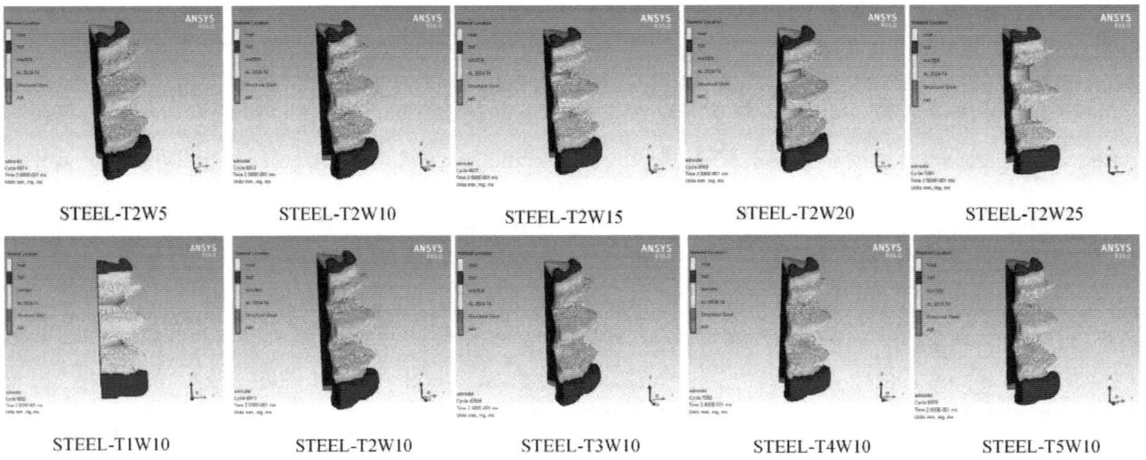

(b) Near field simulation of steel reinforced ring at 0.25ms time

Figure 6. Comparison diagram of the effect of strengthening the ring shell at the same time of rupture.

3.3. Far-field simulation analysis

Figure 7 shows a comparison of the characteristic parameters of cloud formation in different models, including cloud inhomogeneity (variance of global sampling point concentration), cloud volume, cloud diameter, and cloud thickness. The patterns obtained through the analysis are as follows:

(1) Effect of the reinforcing ring on cloud heterogeneity (variance of global sampling point concentration): The cloud inhomogeneity produced by the Al reinforcing ring is similar to that of the non-reinforcing ring at first, but with the increase of width/thickness, the cloud inhomogeneity decreases overall. The cloud inhomogeneity of the steel reinforcing ring is very low. The width and thickness of the reinforcing ring have little effect. The reason is that when the binding force of the reinforcing ring is small or unconstrained, the projectiles accumulate near the Z=0 plane. With the enhancement of the constraint, the projectiles in the aggregation area gradually disperse to the upper and lower sides, resulting in a decrease in the global concentration gradient.

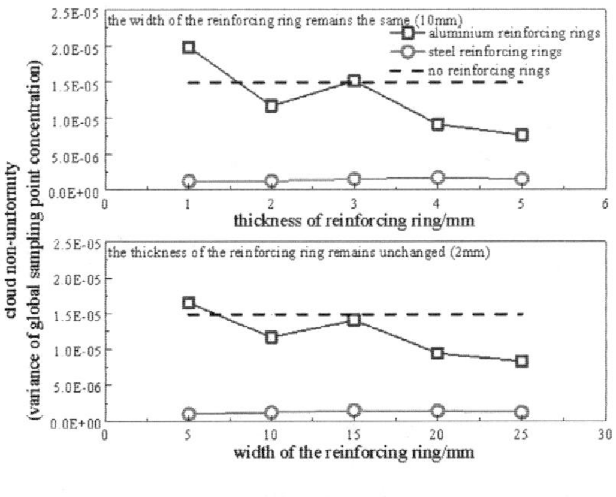

(a) Cloud non-uniformity.

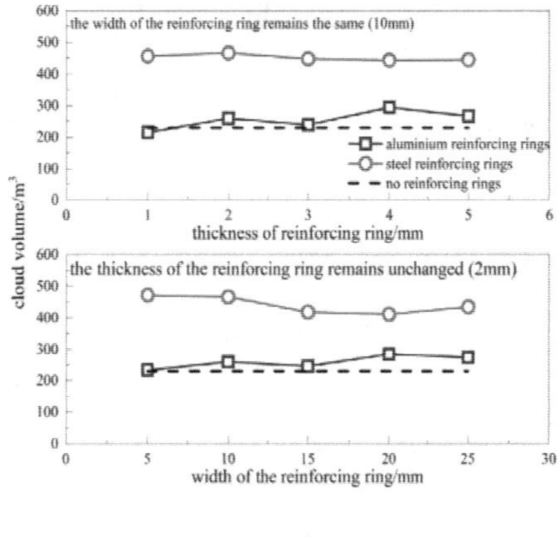

(b) Cloud volume.

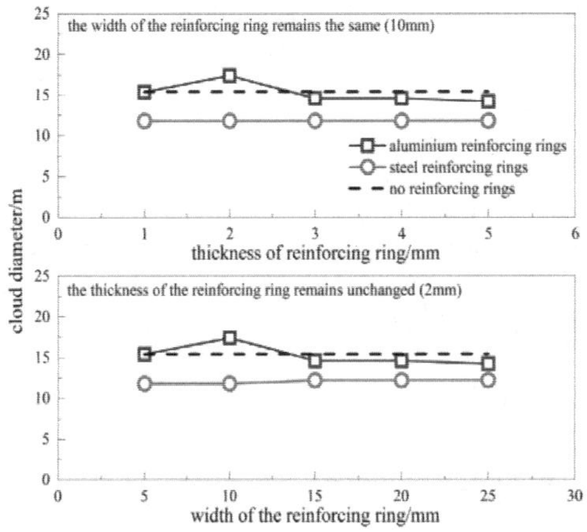

(c) Cloud diameter.

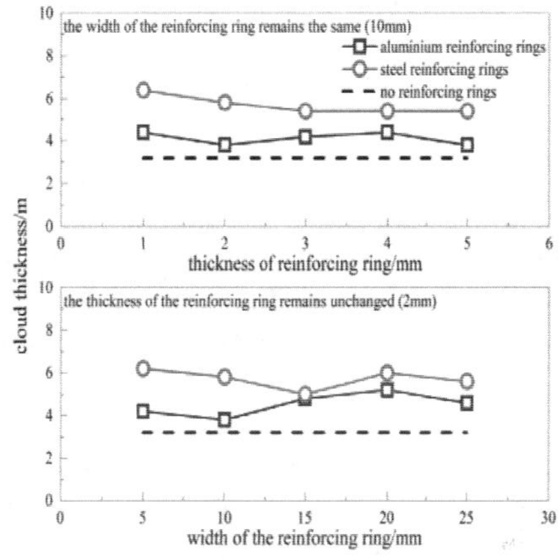

(d) Cloud thickness.

Figure 7. Comparison of characteristic parameters of cloud clusters formed by different models.

(2) Effect of the strengthening ring on the volume of clouds: The volume of the cloud produced by the Al reinforcing ring is similar to that of the non-reinforcing ring at first, but with the increase of width/thickness, the volume of the cloud is gradually slightly larger than that of the non-reinforcing ring. The cloud volume of the steel reinforcing ring is about twice that of the non-reinforcing ring, and the change in the width and thickness of the reinforcing ring has little effect on the cloud volume. The reason for this is that the Al reinforcing ring ruptures together with the shell, which has less constraint and less influence on the cloud thickness. However, the strong constraint of the steel reinforcing ring leads to a significant increase in the thickness of the cloud.

(3) Effect of the reinforcing ring on the diameter of the cloud: The diameter of the cloud produced by the Al reinforcing ring is initially comparable to that of the non-reinforcing ring, but with the increase of the width/thickness of the reinforcing ring, the diameter of the cloud is gradually slightly smaller than that of the non-reinforcing ring. The cloud diameter of the steel reinforcing ring is about 77% of that of the non-reinforcing ring. The width and thickness of the reinforcing ring have little effect on the cloud diameter. The reason is that when the reinforcing ring constraint is large. The projectile is thrown obliquely to the upper and lower sides. The horizontal component of its velocity decreases, resulting in a decrease in the diameter of the cloud.

(4) Effect of the strengthening ring on cloud thickness: The Al and steel reinforcing rings increased the cloud thickness by 26% and 78%, respectively, compared with the non-reinforcing rings. The reason for this is that the weak and strong constraints cause the clouds to disperse weakly and strongly in the axial direction, respectively.

4. Conclusion

In this paper, a method for predicting the process and morphology of liquid explosion and dispersion based on near-field and far-field coupling calculation is proposed. A set of simulation and calculation software for the whole process of liquid explosion dispersion based on this method is developed. Using this method and software, the near-field and far-field action processes of the explosive throwing device with reinforcing ring structure and the characteristic parameters of the cloud formed are obtained through simulation calculation, and the main conclusions are as follows:

(1) The results of near-field throwing show that the shell of the reinforcing ring device is unevenly fractured, showing a "segmented" rupture, resulting in jet stratification and uneven ejection of liquid on the Z-axis. As the width of the reinforcing ring becomes wider, the interval between the jet layers becomes larger, and the jet becomes sharper, especially the effect of the steel reinforcing ring. The effect of the thickness of the steel reinforcing ring is similar to the width but not as obvious as the width. The Al reinforcing ring has a weak effect on the jet.

(2) The results of far-field scattering show that the characteristic parameters such as cloud non-uniformity, cloud volume, cloud diameter, and cloud thickness produced by the Al reinforcing ring are comparable to those of the non-reinforcing ring. With the increase of the width/thickness of the Al reinforcing ring, the cloud inhomogeneity of the cloud mass decreases in general, the volume increases slightly, the cloud diameter decreases slightly, and the cloud thickness increases by about 26%. Compared with the throwing device without a reinforcing ring, the cloud inhomogeneity produced by the steel reinforcing ring is significantly reduced, the volume of the cloud is increased by about 1 times, the diameter of the cloud is reduced by 23%, and the thickness of the cloud is increased by 78%. The main principle is that as the reinforcing ring confinement changes from weak to strong (from Al to steel), the jet changes from horizontal ejection to upper and lower sides, and the projectile accumulation area changes from near the Z=0 plane to disperse in a thicker area.

Based on the above conclusions, the feasibility of the liquid explosion throwing process and morphology prediction method based on the coupling calculation of near and far fields proposed in this paper is verified, as well as the usability of the simulation calculation software for the whole process of liquid explosion and displacement. The method and software can be effectively applied to the forward design and optimization of the structure of the explosive throwing device.

Acknowledgement
National Natural Science Foundation of China "Ye Qi Sun" Joint Fund

References

[1] Li H B, He C, Du H W, et al. (2024) The influence of the cross-sectional shape of the throwing charge on the blast throwing process. Journal of Projectiles, Arrows and Guidance, 44(3): 29-34. DOI: 10.15892/j.cnki.djzdxb.2024.03.004

[2] Shi Y, Li G P, Qi J, et al. (2023) Explosive projectile study based on near-field dynamics. Mechanical Design & Manufacturing Engineering, 52(4): 87-92. DOI: 10.3969/j.issn.2095-509X.2023.04.018

[3] Wang Y X, Xie L F, Jia X L, et al. (2020) Experimental study of the explosion of FAE fuel with a charge of 300 kg thrown into fog. Explosives, 49(2): 23-28. DOI: 10.3969/j.issn.1001-8352.2020.02.004

[4] Gardner D R. (1990) Near field dispersal modeling of liquid fuel-air explosives. DOI: 10.2172/6540334

[5] Xue S S, LIU J C, Qin C S, et al. (2001) Experimental and numerical research of fuel explosion into fog. Explosion and impact, 21(4): 5. DOI: 10.3321/j.ISSN:1001-1455.2001.04.006
[6] Ding J, Liu J C. (2001) Duration of liquid fuel explosions and numerical research on the entire process of formation. Journal of Fire Fried Drug, 20(1): 20-23. DOI: 10.3969/j.issn.1007-7812.2001.01.007
[7] Guo X Y, Hui J M, Xie L F. (2005) Experimental research on the process of fuel explosion and sprinkle. Journal of High -pressure Physics, 19(2): 120-126. DOI: 10.11858/gywlxb.2005.02.004
[8] Zhang T, Wu J, Hui J M. (2010) Experimental research on the impact of FAE Cloud Boom Boarding Features and Power of Explosion. Bouncing and guidance Journal, 30(1): 137-140. DOI: 10.3969/j.issn.1673-9728.2010.01.042
[9] Yan S L, Liu F, Yue Z W, et al. (2007) The impact of the amount of dosage on the water explosion is sprinkled into the fog motion characteristics. Journal of Harbin University of Technology, 39(11): 1825-1828. DOI: 10.3321/j.ISSN:0367-6234.2007.11.036
[10] Xu Q M, Xie L F, Wang Y X, et al. (2021) Calculation of fuel cloud mist radius based on measical analysis. Journal of Weapons and Equipment Engineering, 42(5): 168-172. DOI: 10.11809/bqzbgcxb2021.05.031
[11] Li J P, Liu S Q. (2020) Optimization of the structure of square explosion and throwing device. Journal of High -pressure Physics, 34(2): 31-40. DOI: 10.11858/gywlxb.20190835
[12] Wang S Y, Wang Z Q, Ji D K. (2018) Under the dynamic conditions of the Yunban Battle Department, the research of the cloud blasting the cloud group of clouds. Bouncing and guidance Journal, (4): 1-5. DOI: 10.15892/j.cnki.djzdxb.2018.04.001
[13] Wang Y, Bai C H, Li J P. (2017) Research on numerical simulation of the hood structure on the rate of throwing of fuel installation. Journal, 38(1): 43-49. DOI: 10.3969/j.issn.1000-1093.2017.01.006
[14] Yan C X, Zhang Y, Liu C Y, et al. (2020) The movement characteristics of spray solution drip in air environment. Atomic energy science and technology, 54(1): 66-71. DOI: 10.7538/yzk.2019.youxian.0060

Numerical simulation analysis of rolling groove forming of fire cooling pipe for offshore platforms based on ANSYS/LS-DYNA

Wei Chen *[1], Guohe Jiang [1] and Zhongliang Zhou [2]

[1] School of Business Ship, Shanghai Maritime University, Shanghai 201306, China
[2] Shanghai Zhenhua Heavy Industries Co., Ltd. (Group), Shanghai 200125, China

*Corresponding author's e-mail: 18143774511@163.com

Abstract. Under the background of the country's vigorous development of offshore wind power, the offshore substations and step-up substations equipped in wind farms have gradually popularized the pipe connection method of the crimp clamp type. Correspondingly, it is necessary to roll grooves on the pipes to meet the installation requirements of the crimp clamp. The size and accuracy of the crimp need to meet the clamp's requirements to ensure the safe and reliable operation of the pipeline. Currently, the quality of the rolling groove equipment in the market varies, and even leakage and other situations occur after installation. This paper conducts a simulation analysis of the optimal rolling depth and number of turns required for seamless steel pipes through finite element software simulation, proposes methods to improve the quality of rolling grooves, and based on the actual equipment situation, aims to improve the rolling groove accuracy and quality of the existing equipment as much as possible.

1. Introduction

1.1 Background introduction

The clamp type, also known as the grooved form of pipeline clamp connection, has been widely used in onshore fire protection and water supply and drainage system pipeline systems due to its advantages such as convenient installation, wide application range, and low space requirements. Under the background of the country's vigorous development of offshore wind power, the offshore substations and step-up substations equipped in deep and far-sea wind farms have gradually popularized the grooved clamp form of pipeline connection. To match this, grooves must be rolled on the purchased seamless steel pipes to meet the installation requirements of grooved pipeline clamps. The size and accuracy of the grooves need to meet the requirements of the clamps to ensure the safety, reliability, and long-term operation of the pipeline. Currently, the quality of grooving equipment in the market varies in terms of grooving accuracy and quality, and leakage and other problems often occur after installation.

Domestic scholars' research on grooves is currently mainly focused on the grooving problem of welded pipe products. Duan Lingyu took ERW welded pipe as the research object and analyzed in detail the causes of grooving cracking from aspects such as the weldability of the material, welding process parameters, welded pipe forming, surface galvanizing, and grooving processing process, and proposed some solutions and preventive measures. On the one hand, it is necessary to start from the control of raw materials and production process management to improve the weld strength; on the

Journal of Physics: Conference Series **2992** (2025) 012041 doi:10.1088/1742-6596/2992/1/012041

other hand, efforts should be made to improve the test methods, especially to strengthen the first-piece inspection to expose weld defects promptly. In addition, the characteristics of welded steel pipes must also be understood during use, and careful protection should be taken during the grooving process to avoid damage to the weld seam. Bong Zhuangzhuang et al. studied the influence of factors such as welding residual stress, the number of grooving rings, the feed per single grooving pass, and the initial position of grooving on weld cracking during grooving of HFW welded pipe. Based on the software simulation results and experimental verification, quantitative guiding data were obtained. The number of grooving rings should be as large as possible when grooving, and a uniform downward pressure scheme should be adopted during actual grooving [1].

1.2 Research introduction of this paper

In this paper, the pressure analysis of Q235 pipes with different thicknesses is conducted through finite element simulation[2] technology to confirm the different feed rates applicable to the corresponding wall thicknesses, and the methods for improving the quality of the grooved pipe of the lifting equipment are proposed. Although more and more units tend to use finite element simulation technology[3] to predict product performance during the product development process, the accuracy of the simulation results determines its true role in the product development process. The simulation accuracy is affected by many factors, such as the degree of model simplification, mesh size, quality, and material parameters[4]. Among them, material parameters are the most important factor affecting the simulation accuracy. This paper mainly studies the rolling feed rates required for different materials, and the adapted single feed rate is determined through simulation analysis using ANSYS/LS-DYNA software.

2. Establishment of the finite element model

2.1 Schematic diagram of the principle and structure of groove pressing

The processing principle of the groove is that the motor drives the pressure wheel to rotate. By taking advantage of the extrusion and friction between the pressure wheel of the grooving machine and the welded pipe, the shape and structure of the pressure wheel press a groove at the fixed position of the welded pipe. The groove after rolling is shown in Figure 1 and Figure 2 below.

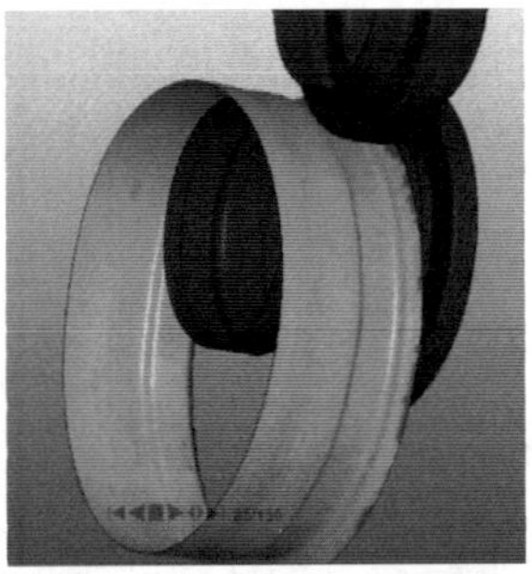

Figure 1 Simulation effect of rolling groove.

Figure 2 Actual effect of rolling groove.

2.2 Size selection
In this study, seamless steel pipes are adopted, and the material of the pipe body to be selected is Q235[5]. As shown in Table 1.

Table 1. Dimensions of steel pipes and trenches.

Indicator name	Numerical value
Outer diameter of steel pipe (mm)	70
Wall thickness	2.4
Size from pipe end to groove edge A	15.88
Trench width B	7.14
Minimum allowable wall thickness for grooves	1.24
The total length of steel pipe L/	200

2.3 3D model construction
To reduce the finite element calculation amount and ensure optimization efficiency, in this experiment, only the rolling module in the overall three-dimensional model is taken out separately for finite element simulation calculation. As shown in Figure 3 and Figure 4.

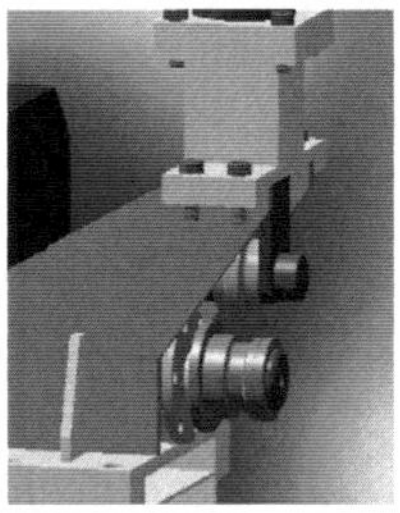

Figure 3 Rolling execution module.

Figure 4 Rolling execution status.

2.4 Setting of simulation operating parameters
According to the actual processing process, the pipe fittings and extrusion roller models were imported into the LS-DYNA module [1] ANSYS Workbench. After setting the relevant parameters, the simulation running parameters were set, including the depth of a single groove pressing and the rotational speed, both of which were set to 60 rpm.

This simulation has set up five situations with the groove depth being 2 mm, 3 mm, and 5 mm, analyzing different rolling pressures [6].

3. Simulation analysis results
Compared with the grooved welded pipes studied by other scholars, the seamless steel pipe grooved problem studied in this paper faces fewer crack problems, but the stress concentration effect is more

obvious. By simulating and analyzing the stress conditions under different groove depths, the suitable single groove depth is determined. Figure. 5, Figure. 6 are the parameter Settings, and Figure. 7, 8, and 9 are the cloud images of simulation results.

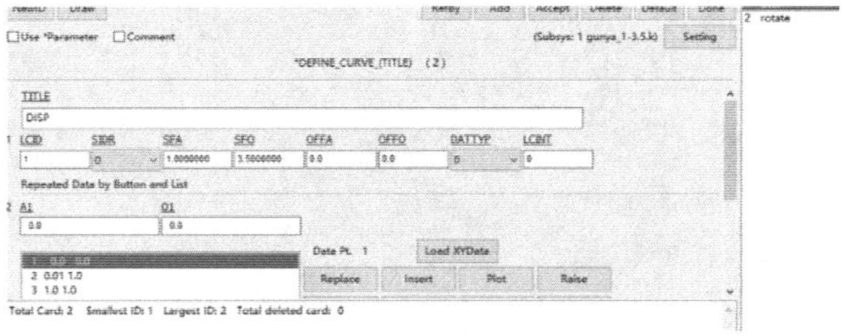

Figure 5 Groove depth control.

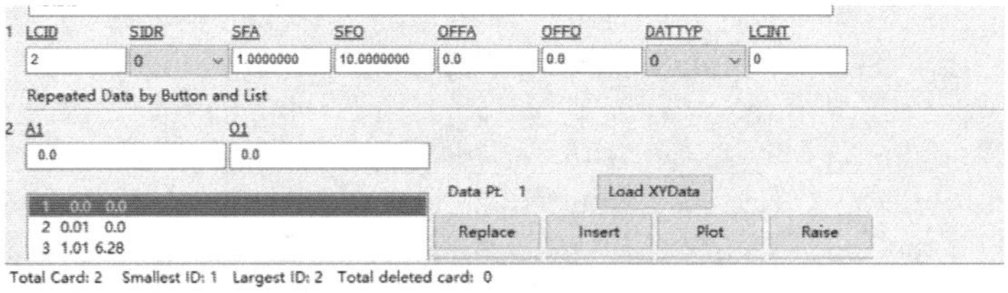

Figure 6 Pressure groove speed control.

Figure 7 Equivalent stress cloud map of groove depth 2 mm.

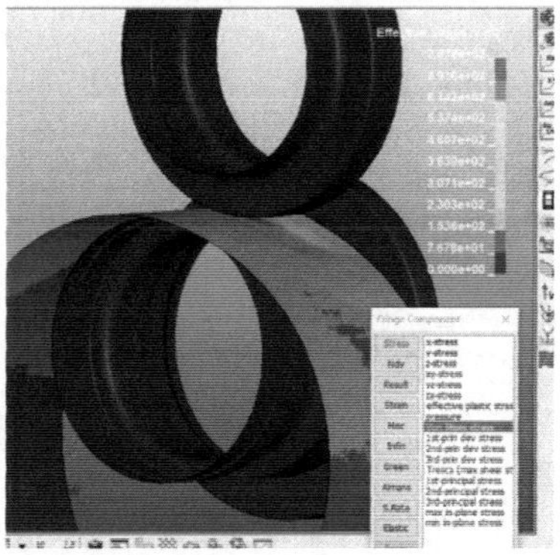

Figure 8 Equivalent stress cloud map of groove depth 3 mm.

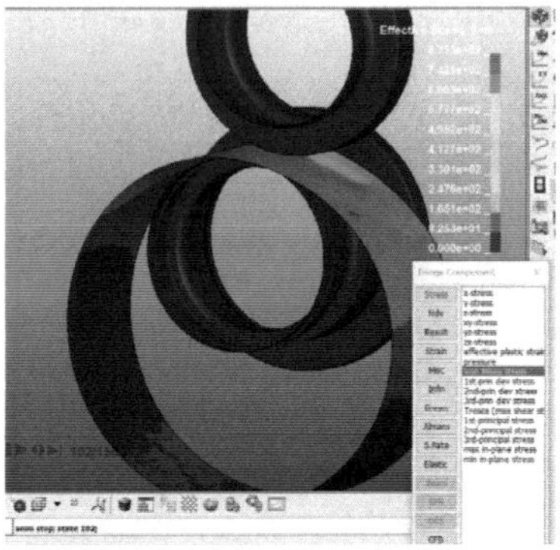

Figure 9 Equivalent stress cloud map of groove depth 4 mm.

4. Discussion

According to ANSYS/LS-DYNA simulation and rolling groove experiment, we found that for seamless steel pipe, the stress concentration in the pipeline during rolling groove processing can be effectively alleviated while reducing the depth of the single ring pressing groove. It is suggested that the depth of the groove should be at least less than the standard requirements when the groove is pressed for the first time, which can significantly reduce the stress concentration and crack problems caused by the pipeline during the rolling process. At the same time, the depth of the single-ring groove should not be too shallow, because the marginal effect will not be more improved, which will affect the processing efficiency.

5. Conclusion

To ensure that the compressive performance of the pipeline is not reduced [7] and avoid the occurrence of side wall tearing, it is recommended that the final groove depth should not exceed the original pipe wall thickness. It is hoped that the results of this simulation experiment can provide a reference for the actual rolling groove production, reduce the actual test cost of subsequent projects, and ensure the research and development of new rolling groove equipment.

References

[1] Feng Zhuangzhuang, et al. Effect of HFW welding pipe rolling process on weld cracking [J]. Welding Pipe 2022, 45 (12): 46-52. 10.19291/j.cnki.1001-3938.2022.12.008.

[2] Wang Shoucai, Sun Ang, and Liu Rugang. Numerical Simulation and Experimental Verification of the Riveting Process of a New Type Lugless Plate Nut [J]. Mechanical Engineering, 2023, 34 (259): 875-881. DOI: 10.3969/j.issn.1004-132X.2023.07.014.

[3] STOPFORD M. Maritime economics [M]. 3rd ed. London: Routledge, 2009: 372-373.

[4] LIU Xiu, JIN Xia, Lou Hangfei, et al. Study on tensile properties of 304 stainless steel foil at different strain rates [J]. Materials Science and Technology, 2019, 27 (5): 59-65. 10.11951/j.issn.1005-0299.20180137.

[5] GUO Zitao, et al. Dynamic Constitutive Relationship of Q235 Steel Based on J-C Model [J]. Explosion and Shock: 2018, 38 (4), 804-810. 10.11883/bzycj-2016-0333.

[6] IU Xiao. Surface rolling densification effect and microstructure evolution of iron-based powder metallurgy materials [J]. Powder Metallurgy Technology, 2024, 42 (05): 456-463. 10.19591/j.cnki.cn11-1974/tf.2024060021.

[7] Chen Wei. Design and performance of internal air supply pipe for dehumidification of the main cable of suspension bridge [J]. Journal of Jilin University (Engineering and Technology Edition). 2021, 51 (05): 1, 749-1, 755. 10.13229/j.cnki.jdxbgxb20200508.

Journal of Physics: Conference Series **2992** (2025) 012042 doi:10.1088/1742-6596/2992/1/012042

Dynamic modeling and analysis of tool magazine reversing mechanism

Wenbo Zheng [1], Manxin Wang [1,*], Kun Liu [1], Jianguo Zhu [2] and Kangjun Xia [2]

[1] School of Mechanical Engineering, Nanjing University of Science and Technology, Nanjing, Jiangsu, 210094, China
[2] Okada Intelligent (Jiangsu) Co., Ltd., Nanjing, Jiangsu, China

[*] Corresponding author's e-mail: mxwang@njust.edu.cn

Abstract: Tool magazine is the core functional component of CNC machine tools to store and exchange tools. Its tool reversing mechanism is responsible for converting the tool to the tool change state or indexing state, which is a key part of tool magazine tool change. In order to complete the optimization of the driving and structural parameters of the inverted tool mechanism, this paper carries out the dynamic modeling and analysis of the inverted tool mechanism. Firstly, the vector method is used to establish the mapping relationship between the position, velocity, and acceleration of each moving member of the mechanism and the input; the principle of virtual work is used to establish the mapping model between the driving force and the motion of the mechanism; finally, combined with the arithmetic example, the influence laws of different input forces and scale parameters on the end motion of the mechanism are analyzed and obtained. The parameter optimization is carried out.

1. Introduction

CNC machine tool [1] is the industrial mother machine. The tool magazine system [2] is the core functional component of tool stocking and management in CNC machine tools. This system can efficiently carry out the rapid switching of tools [4] through precise cooperation with the tool changer manipulator [3] and the spindle, so as to realize the fast and uninterrupted production process to meet the diversified machining needs. The reversing mechanism [5] is a crucial component of the tool magazine. Its primary function is to facilitate the tool change during the movements of the inverted knife and the returning knife. This ensures smooth replacement and accurate positioning of the tool. The performance of the reversing mechanism directly influences the stability of subsequent actions such as tool pulling [6] and rotating [7]. Therefore, studying the reversing tool mechanism is of great significance.

Previous research on knife reversing mechanisms has primarily focused on two areas: structural design and accuracy analysis. In their work, Bian et al. [8] proposed a method for measuring the reversing accuracy of the knife magazine using a laser displacement sensor. They calculated the two error angles associated with the knife magazine reversal using matrix transformation methods. Through testing, they found that the reversing accuracy deviation of the knife magazine under rated load falls within a range of ±100μm. The accuracy deviation range of a tool magazine under rated load is ±100μm, which provides a theoretical database for the construction of the reliability test bench of the tool magazine. In terms of structural design, an external self-locking mechanism for a tool magazine reversing cylinder is designed [9], which reduces the production cost without expanding the cylinder size and simplifies the composition structure. A tool magazine fast reversing tool return mechanism is invented [10], which realizes the fast reversing tool return through the cooperation of guide shaft and linear bearing and reduces the time of tool change in the tool magazine. A self-lubricating knife magazine reversing knife device is designed [11]. The grease can enter the opening position from the discharge end to realize the self-lubrication of the knife sleeve in the inverted knife process.

Mechanism dynamics modeling and analysis is mainly concerned with the stability of the mechanism, performance analysis in different states, etc., and has been widely used in several fields [12-13]. In robotics, Wang et al. [14] carried out dynamics modeling and analysis of a series-parallel hybrid quadruped bionic robot, established the explicit leg mechanism dynamics equations and the inverse dynamics equations [15], and analyzed the influence law of the step length on the maximum output force of the driving hydraulic cylinder through examples. In aerospace, Bai [16] carried out kinetic modeling and analysis on the pressurized delivery system of a launch vehicle, revealing the influence law of phase change on the operating characteristics of the pressurized delivery system. However, little research has been reported on the dynamics of the tool magazine reversing mechanism.

This paper takes the tool magazine reversing mechanism as the research object, establishes the mapping relationship between the position, velocity, and acceleration of each moving member of the mechanism and the input through the vector method [17], establishes the relationship between the driving force and the mechanism motion by using the principle of virtual work, and analyzes the influence law of different driving parameters and scale parameters on the end motion of the mechanism, which then provides the basis for the structural optimization of the reversing mechanism.

2. Introduction to inverted tool mechanism and establishment of the coordinate system

The three-dimensional model of the tool magazine reversing mechanism is shown in Figure 1, which can convert the up and down movement of the cylinder piston rod into the back-and-forth rotation of the toolset, thus realizing the reversing and returning action.

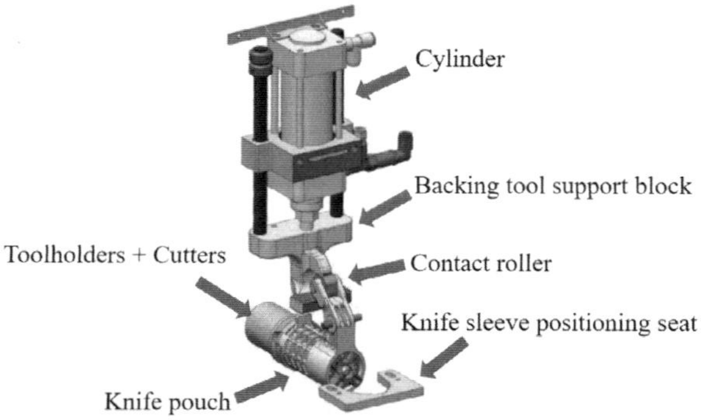

Figure 1. Three-dimensional model of inverted tool mechanism.

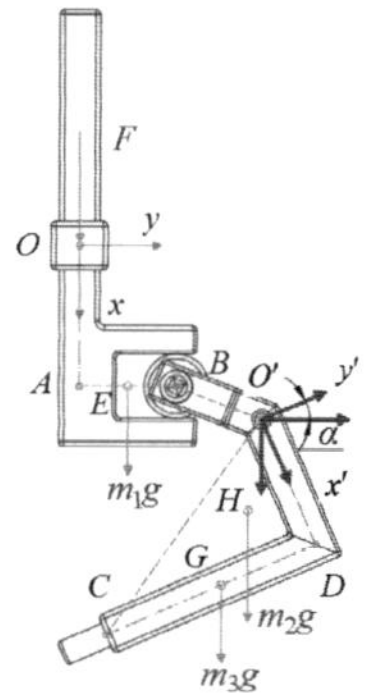

Figure 2. Motion diagram of the reverse tool mechanism.

To facilitate the description of the motion process of the reversing tool mechanism, the mechanism motion diagram shown in Figure 2 is drawn. Point O is the center of the cylinder, point O' is the tool sleeve rotation point, point A is the center of the reverse tool support block, point B is the center of the roller, and point C is the reference point at the end of the tool sleeve. We establish a fixed coordinate system $\{O-xy\}$ at point O, an x axis along the direction of the cylinder push rod, and y axis horizontally to the right. The tool sleeve implicated coordinate system $\{O'-x'y'\}$ is established on O', the x' axis is perpendicular to the tool sleeve axis, and the y' axis is parallel to the tool sleeve axis. We make the x' axis vertical line from point C and cross the x' axis at point D, and set the angle between the x' axis and x axis as α . α can represent the rotation angle of the tool sleeve. The rotation matrix of the system $\{O'-x'y'\}$ relative to the system $\{O-xy\}$ can be expressed as:

$$\boldsymbol{R} = \begin{bmatrix} \cos\alpha & -\sin\alpha \\ \sin\alpha & \cos\alpha \end{bmatrix} \tag{1}$$

3. Kinematic modeling

The purpose of the position modeling of the reversing tool mechanism [18] is to establish the mapping relationship between the displacement of the cylinder piston rod and the rotation angle of the tool sleeve. As shown in Figure 2, the metric $\boldsymbol{r} = \left(r_x \quad r_y\right)^{\mathrm{T}}$ of reference point O' in system $\{O-xy\}$ can be expressed as:

$$\boldsymbol{r} = q\boldsymbol{x} + p\boldsymbol{y} - \boldsymbol{r}_B \tag{2}$$

In Formula (2), $\boldsymbol{x}$ and $\boldsymbol{y}$ represent the unit vector on the x axis and y axis respectively. q is the expansion length of the cylinder piston rod, and $q = |OA|$. p is the distance between reference points A and B, and $p = |AB|$. $\boldsymbol{r}_B = \boldsymbol{R}\boldsymbol{r}'_B$. $\boldsymbol{r}'_B = l_B\left(\cos\theta \quad \sin\theta\right)^{\mathrm{T}}$ is the measure of point B in a coordinate system $\{O'-x'y'\}$. θ is the angle between $O'B$ and the x' axis. l_B is the length of $O'B$.

The relationship between cylinder displacement and tool sleeve rotation angle can be obtained by multiplying the two ends of Formula (2) by $\boldsymbol{x}$:

$$q = r_x + l_B\cos\theta\cos\alpha - l_B\sin\theta\sin\alpha = r_x + l_B\cos\left(\theta+\alpha\right) \tag{3}$$

If the size of q is known, the sleeve rotation angle can be obtained from Formula (3):

$$\alpha = 2\pi - \arccos\left(\frac{q-r_x}{l_B}\right) - \theta \tag{4}$$

The mapping relationship between piston rod speed and cutter sleeve rotation angular speed can be established by modeling the velocity of the reversing tool mechanism. Derivation of Formula (3) gives the mapping relationship as follows:

$$\dot{q} = v = -\omega l_B \sin\left(\theta+\alpha\right) \tag{5}$$

In Formula (5), v is the speed of the piston rod, and ω is the rotational angular speed of the cutter sleeve.

When the piston rod speed is known, the sleeve angular speed is:

$$\omega = -\frac{v}{l_B\sin(\theta+\alpha)} \tag{6}$$

The acceleration modeling of the reversing tool mechanism is to establish the mapping relationship between the motion acceleration of the piston rod and the rotation angular acceleration of the tool sleeve. Derivation of Formula (3) gives the mapping relationship as follows:

$$\ddot{q} = \dot{v} = a = -\varepsilon l_B\sin\left(\theta+\alpha\right) - \omega^2 l_B\cos\left(\theta+\alpha\right) \tag{7}$$

In the formula, a is the piston rod motion acceleration, and ε is the knife sleeve rotation angular acceleration.

When a is known, the sleeve angular acceleration is:

$$\varepsilon = -\frac{a}{l_B\sin(\theta+\alpha)} - \frac{v^2\cos(\theta+\alpha)}{l_B^2\sin^3(\theta+\alpha)} \tag{8}$$

4. Modeling and simulation of rigid body dynamics

4.1. Rigid body dynamics modeling

In order to facilitate the analysis, the reversing tool mechanism is divided into two components. Component 1 is the cylinder piston rod and inverted tool support block for linear motion. Component 2 is the roller, tool set, tool shank and tool for rotary motion. The force on the mechanism is shown in

Figure 2, with point E being the center of gravity of Member 1, point H being the common center of gravity of the roller and the toolset, and point G being the common center of gravity of the shank and the tool.

The combined force F_1 of gravity and inertia forces in Member 1 can be expressed as:

$$F_1 = m_1 g + m_1 a = m_1 g + m_1 \ddot{q} \quad (9)$$

In the formula, m_1 is the mass of Member 1. In Member 2, the combined moment T_2 of its gravitational moment and moment of inertia about the fixed point O' can be expressed as:

$$\begin{aligned} T_2 &= m_2 g l_{O'H} \cos(\beta_2 + \alpha) + m_3 g l_{O'G} \cos(\beta_3 + \alpha) - (J_2 + J_3)\varepsilon \\ &= m_2 g l_{O'H} \cos(\beta_2 + \alpha) + m_3 g l_{O'G} \cos(\beta_3 + \alpha) - (J_2 + J_3)\ddot{\alpha} \end{aligned} \quad (10)$$

In the formula, m_2 is the total mass of the roller and the knife sleeve in Member 2; m_3 is the total mass of the shank and tool in Member 2; $l_{O'H}$ is the distance between point H and point O'; $l_{O'G}$ is the distance between point G and point O'; J_2 is the rotational moment of inertia of the roller and knife sleeve to O' in Member 2; J_3 is the rotational inertia of the shank and tool to O' in Member 2; β_2 is the angle between $O'H$ and x' axis; β_3 is the angle between the $O'G$ and the x' axis.

In summary, according to the principle of virtual work [19], the rigid body dynamics equations of the reversing tool mechanism can be constructed as:

$$F\delta q + F_1 \delta q + T_2 \delta \alpha = 0 \quad (11)$$

In the formula, F is the force of the cylinder. It can be obtained by substituting Formulas (9) and (10):

$$F\delta q + (m_1 g - m_1 a)\delta q + [m_2 g l_{O'H} \cos(\beta_2 + \alpha) + m_3 g l_{O'G} \cos(\beta_3 + \alpha) - (J_2 + J_3)\varepsilon]\delta\alpha = 0 \quad (12)$$

The relationship between the imaginary displacements δq and $\delta\alpha$ can be expressed by the following equation:

$$\delta\alpha = -\frac{1}{l_B \sin(\theta + \alpha)}\delta q = \frac{1}{\sqrt{l_B^2 - (q - r_x)^2}}\delta q \quad (13)$$

This is then combined with the above to simplify the final result:

$$F = \frac{m_2 g l_{O'H} \cos(\beta_2 + \alpha) + m_3 g l_{O'G} \cos(\beta_3 + \alpha) - (J_2 + J_3)\varepsilon}{l_B \sin(\theta + \alpha)} + m_1 a - m_1 g \quad (14)$$

From this, a function between F and (α, a, ε) can be established.

4.2. Cylinder drive force simulation

The main scales and structural parameters of the reversing tool mechanism are shown in Table 1.

Table 1. Main scales and structural parameters of the chamfering mechanism.

Parameter name	Parameter size	Parameter name	Parameter size
l_B	50mm	J_2	6923.86kg·mm^2
r_x	108mm	J_3	31214.65kg·mm^2
θ	225°	$l_{O'H}$	48.11mm
m_1	2.24kg	$l_{O'G}$	125.28mm
m_2	0.192kg	β_2	326.9°
m_3	2.1kg	β_3	298.61°

The given tool sleeve trajectory is as follows:

$$\alpha = 4.448t^3 + 36.66t^2 + 0.5227t - 0.02342 \quad (15)$$

Setting the rotation angle of the knife sleeve as 0~90°, the reversing time can be obtained as 1.43s. Based on the kinetic equation given before, the driving force of the cylinder of the reversing mechanism can be obtained. Figure 3 illustrates the comparison of the driving forces obtained from calculations and simulations based on the previous kinetic theory model.

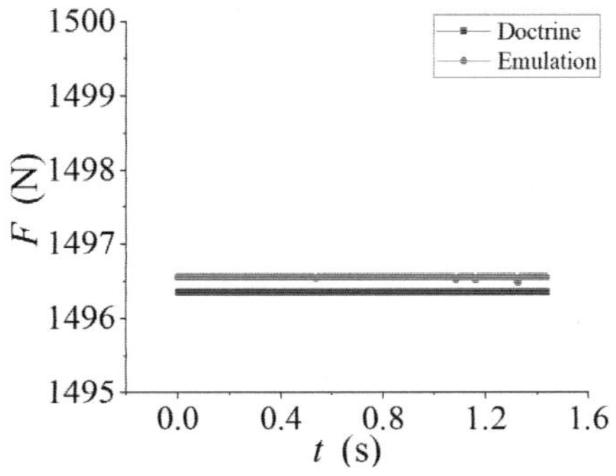

Figure 3. Drive Comparison Chart.

As can be seen from Figure 3, the theoretical model and the driving force changes obtained from the simulation are basically the same, which proves the correctness of the theory.

5. Example

5.1. Influence law of different cylinder parameters on the motion of reversing tool mechanism

When different input forces are set, the output motion of the end will also be different. For this reason, the effect of different input forces (i.e. cylinder output force) on the output motion will be explored below. The cylinder output force F can be expressed as:

$$F = \frac{\pi}{4} D^2 p \eta_c \tag{16}$$

In the formula, D is the cylinder block diameter, P is the cylinder working pressure, η_c is the cylinder working efficiency, and η_c is taken as 0.8 in this paper.

5.1.1. Effect of cylinder diameter D on the motion of the reversing tool mechanism. The cylinder diameters were 50mm, 63mm, and 80mm. The cylinder working pressure is set to 0.6MPa. The three values will be substituted into the kinetic model built above, which can be obtained under the three kinds of D reversing tool mechanism of the movement. Figure 4 shows the cylinder piston rod movement. Figure 5 shows the end of the mechanism movement.

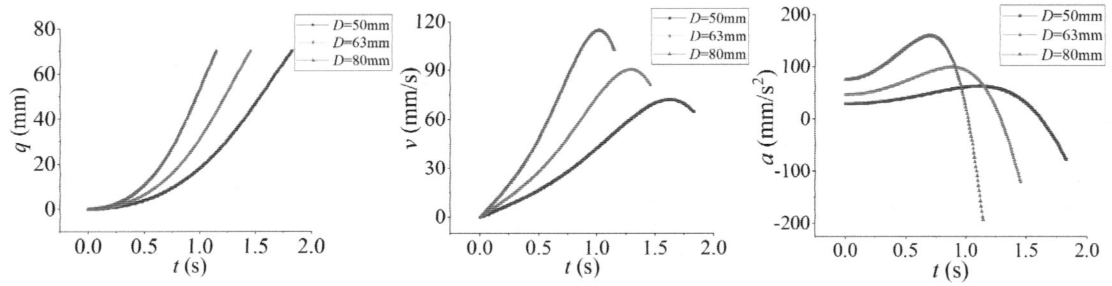

(a) Displacement (b) Velocity (c) Acceleration

Figure 4. Cylinder input curves for three-cylinder diameters.

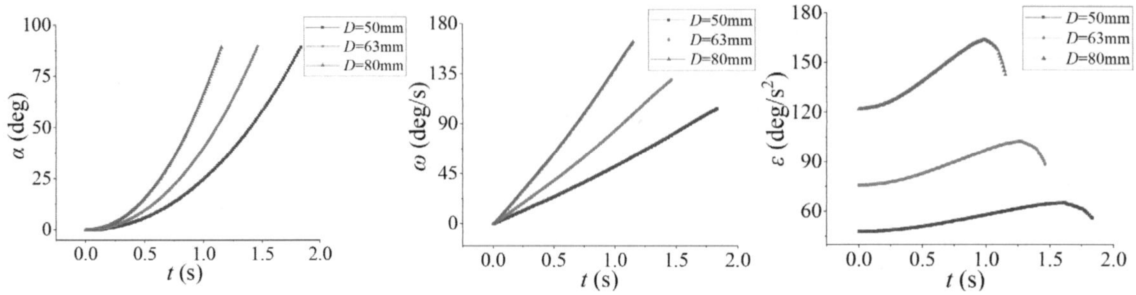

(a) Angular displacement (b) Angular velocity (c) Angular acceleration

Figure 5. Mechanism output curves for three diameters.

From Figures 4 and 5, it can be seen that the reversing time decreases with the increase of D. Moreover, the trends of the input motion and the input motion are basically the same in all three cylinder diameters. The magnitude of the change increases with the increase of the diameter.

5.1.2. Influence of cylinder operating pressure on the motion of the reversing tool mechanism. Taking the values of P as 0.4MPa, 0.6MPa, and 0.8MPa, respectively, and the cylinder diameter is set to 63mm, substituting the three values into the model constructed above, the motions of the reversing tool mechanism can be obtained under the three kinds of P. The motion of the piston rod of the cylinder is shown in Figure 6. The motion of the mechanism's end is shown in Figure 7.

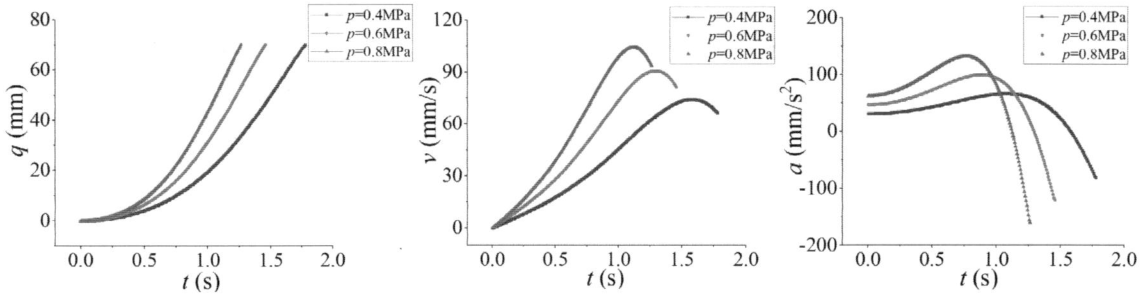

(a) Displacement (b) Velocity (c) Acceleration

Figure 6. Cylinder input motion curves at three pressures.

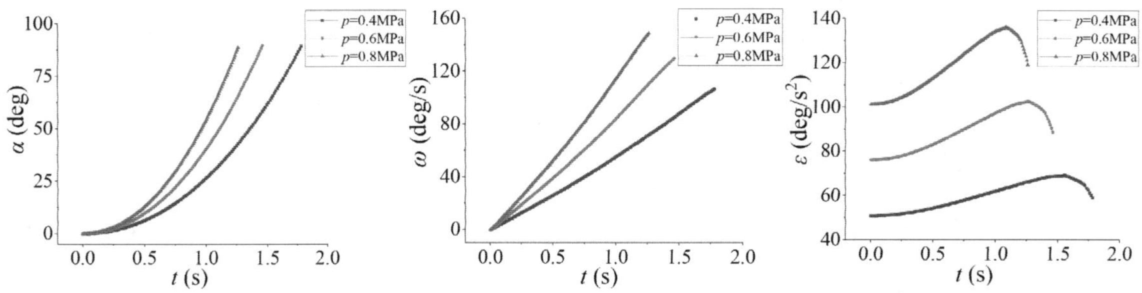

(a) Angular displacement (b) Angular velocity (c) Angular acceleration

Figure 7. Output motion curves of the mechanism at three pressures.

As can be seen from Figure 6, the reversing time also decreases with the increase of the effect of the change in cylinder working pressure on the input and output motions which is consistent with the effect produced by the change in cylinder diameter.

5.2. Influence law of different structural parameters on the motion of reversing tool mechanism

From the kinematic analysis in the previous section, it is known that the parameters affecting the end motion are variable parameters such as length l_B and distance r_x in addition to the driving force. The above parameters are modified and calculated using the kinetic model respectively to analyze their influence laws on the end motion.

5.2.1. Effect of l_B on end motion. As can be seen from the fit of the components of the reversing tool mechanism, when the rest of the structural parameters are not changed, the roller will interfere with the reversing tool support block when l_B reaches a certain value. The reversing tool support block will also interfere with the tool sleeve when l_B is lower than a certain value. We let the upper and lower limits of l_B be $l_{B\max}$ and $l_{B\min}$, respectively, with the following expressions:

$$l_{B\max} = \frac{r_y - R_1 - R_2}{\cos(\theta - \pi)} \quad , \quad l_{B\min} = \frac{r_y - l_{\max}}{\cos(\theta - \pi)} \tag{17}$$

In the formula, R_1 is the radius of the cylinder actuator, R_2 is the radius of the roller, and $l_{\max}$ is the maximum length of the inverted knife support block that can be extended along the y axis. Analysing the reversing tool mechanism shows that $r_y = 79\text{mm}$, $R_1 = 8\text{mm}$, $R_2 = 14\text{mm}$ and $l_{\max} = 58\text{mm}$. From Formulas (17), we get $l_{B\max} = 80\text{mm}$ and $l_{B\min} = 40\text{mm}$.

Combining the above conditions, the values of l_B are taken as 40mm, 50mm, 60mm, 70mm and 80mm, which are substituted into the dynamics model for solving. The curves of the input and output motions of the mechanism at five lengths are obtained, as shown in Figures 8 and 9.

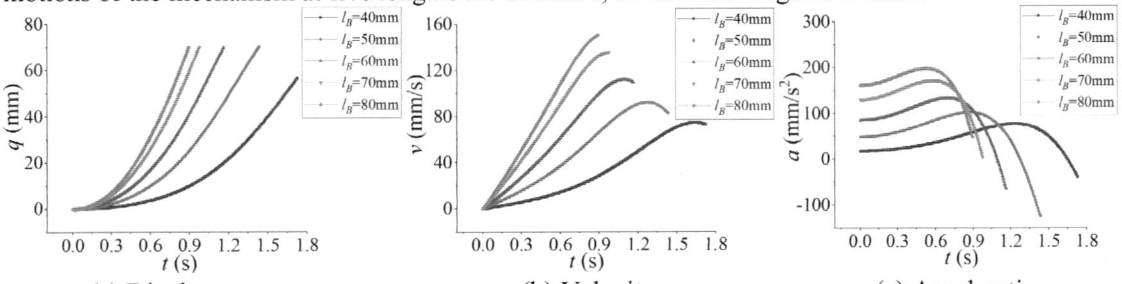

(a) Displacement (b) Velocity (c) Acceleration

Figure 8. Input motion curves for each l_B lower mechanism.

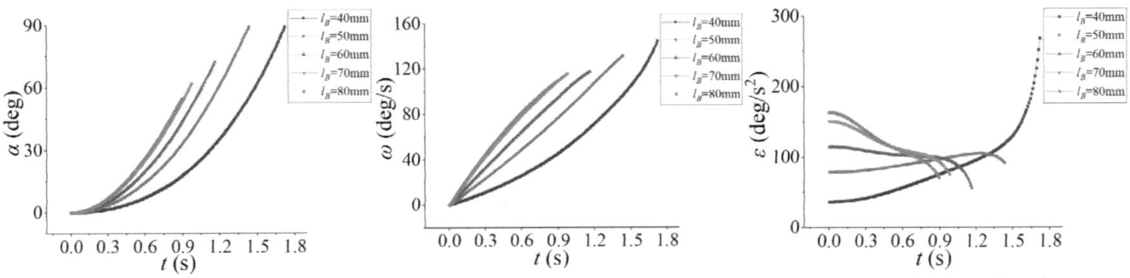

(a) Angular displacement (b) Angular velocity (c) Angular acceleration

Figure 9. Output motion curve of the mechanism under each l_B.

From Figures 8 and 9, it can be seen that the increase of l_B can reduce the chamfering time and the acceleration at the end of the movement at the moment of the end of the movement, but there will be phenomena such as chamfering is not in place. There is an upper limit of l_B. Therefore, l_B should be increased but not more than 80mm. The rest of the parameters need to be adjusted to ensure that the reversing tool movement is reliable.

Journal of Physics: Conference Series **2992** (2025) 012042 doi:10.1088/1742-6596/2992/1/012042

5.2.2. Effect of r_x on end movement. When r_x is increased or decreased, the cylinder actuator extends or retracts accordingly. The pusher must not be shortened indefinitely, or the reversing tool support block will interfere with the cylinder; the pusher itself is limited in length and must not be extended indefinitely. We let the upper and lower bounds of r_x be $r_{x\min}$ and $r_{x\max}$, respectively, with the following expressions:

$$r_{x\min} = r_x - l_0 \quad , \quad r_{x\max} = r_x - l_0 + l_1 \tag{18}$$

In the formula, l_0 is the pre-extension length of the cylinder actuator under the existing mechanism, and l_1 is the maximum pre-extension of the cylinder actuator with the existing mechanism. From the structural parameters of the reversing tool mechanism: $l_0 = 14\text{mm}$ and $l_1 = 40\text{mm}$. $r_{x\min} = 94\text{mm}$ and $r_{x\max} = 134\text{mm}$ are calculated from Formulas (18).

Again, the r_x values are set to be 94mm, 104mm, 114mm, 124mm and 134mm, which are substituted into the dynamics model and solved to obtain the input motion and output motion curves of the mechanism at each r_x, as shown in Figures 10 and 11.

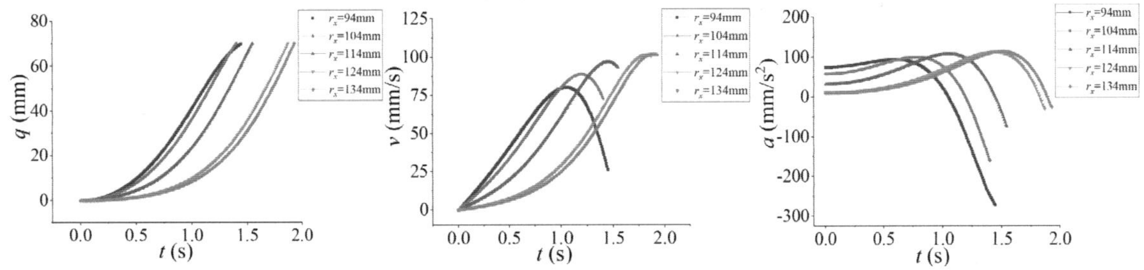

(a) Displacement (b) Velocity (c) Acceleration

Figure 10. Input motion curves for each r_x lower mechanism.

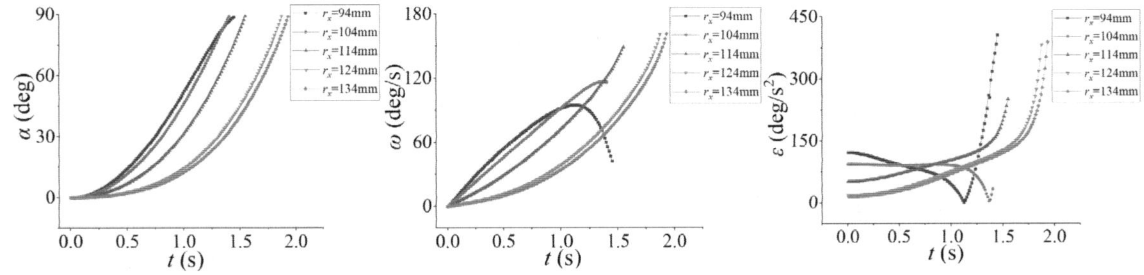

(a) Angular displacement (b) Angular velocity (c) Angular acceleration

Figure 11. Output motion curve of the mechanism under each r_x.

From Figures 10 and 11, it can be seen that when r_x decreases, the reversing time decreases and the terminal acceleration also decreases. However, when r_x decreases from 104mm, the end angular acceleration will rise instead. In summary, r_x should be as low as possible, but not lower than 104mm.

5.2.3. Parameter preference and comparison. Combined with the previous section, the input motion parameters and scale parameters are optimized. The optimization results are shown in Table 2.

Table 2. Optimized input and scale parameters.

Parameter name	Parameter size
D	63mm
p	0.6MPa
l_B	80mm
r_x	104mm

The inverted tool mechanism before and after optimization is analyzed to obtain the output motion curves, as shown in Figure 12, where the original mechanism curves are shown in black and the optimized mechanism curves are shown in red.

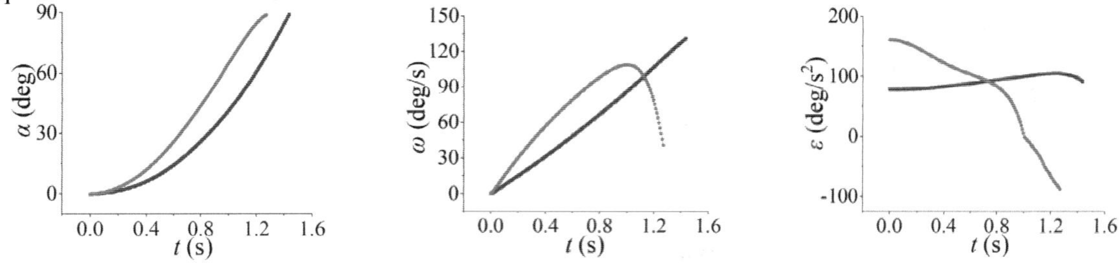

(a) Angular displacement (b) Angular velocity (c) Angular acceleration

Figure 12. Output motion curve of the mechanism under each r_x.

As can be seen from Figure 12, the optimized mechanism has a shorter reversing time, the overall value of the end angular velocity is smaller, and the size of the angular acceleration at the end of the movement is close to that of the original mechanism, which does not produce a more violent impact. It can be seen that the optimized mechanism has a better motion performance.

6. Conclusion

Based on the results and discussions presented above, the conclusions are obtained as:

(1) The kinematic analysis of the reversing tool mechanism is carried out by the vector method. The mapping relationship between the input and output of the mechanism with respect to position, velocity, and acceleration is obtained.

(2) The force of the reversing tool mechanism is analyzed, the rigid body dynamics equation of the mechanism is established, and the accuracy of the theory is verified by simulation.

(3) Through the arithmetic example, the influence laws of different input motion parameters and structural parameters on the output motion are obtained. The variation ranges of each parameter in optimizing the performance of the reversing tool mechanism are obtained.

(4) Preliminarily, the parameters of the reversing tool mechanism are preferred. The motion of the mechanism before and after optimization is compared. The results show that the mechanism performs better after optimization.

References

[1] Mazzuchi T A, Soyer R. Assessment of machine tool reliability using a proportional hazards model [J]. Naval Research Logistics, 1989, 36(6): 765–777. DOI: 10.1002/1520-6750(198912)36: 63.0.CO;2-C.

[2] LI Yinglan. Reliability improvement of tool magazines for horizontal machining centers [J]. Metalworking (Cold Processing), 2023(3): 67-70. DOI: 10.3969/j.issn.1674-1641.2023.03.016.

[3] Ryuh B S, Park S M, Pennock G R. An automatic tool changer and integrated software for a robotic die polishing station [J]. Mechanism & Machine Theory, 2006, 41(4): 415-432. DOI: 10.1016/j.mechmachtheory.2005.06.004.

[4] Xiao Z. Case analysis of tool magazine failure of CNC machine tools [J]. Metal Processing (Cold Processing), 2024(7): 80-83. DOI: 10.3969/j.issn.1674-1641.2024.07.024.
[5] Shenzhen Dejie Machinery Equipment Co. A kind of cylinder linkage reversing tool mechanism: CN202223396827.8 [P]. 2023-05-30.
[6] Zhang Jiangshuai. Typical fault prediction of automatic tool change system of machining center based on tool pulling force monitoring [D]. Jilin: Jilin University, 2022.
[7] ZHANG Jun, WU Yangfang. Improved design and analysis of rotary indexing mechanism of tool magazine [J]. Mechanical Design and Manufacturing, 2019(8): 197-200. DOI: 10.3969/j.issn.1001-3997.2019.08.051.
[8] BIAN Chengfei, HAN Jun, WANG Manxin, et al. Measurement method of tool magazine reversing accuracy based on laser displacement sensor [J]. Mechanical Design and Manufacturing Engineering, 2022, 51(5): 76-84. DOI: 10.3969/j.issn.2095-509X.2022.05.015.
[9] Kunshan North Vantage Machinery Co. External self-locking mechanism of tool magazine reversing cylinder: CN202020626563.2 [P]. 2020-12-15.
[10] Zongxin Industrial Equipment (Weifang) Co. Servo tool magazine fast backward tool return mechanism: CN202320020537.9 [P]. 2023-05-26.
[11] Okada Intelligent (Jiangsu) Co. A self-lubricating tool magazine reversing device and a circular tool magazine: CN202310379547.6 [P]. 2023-09-08.
[12] WANG Xuejun, ZHANG Fan. Modeling and analysis of climbing robot dynamics [J]. Mechanical Science and Technology, 2023, 42(1): 38-45. DOI: 10.13433/j.cnki.1003-8728.20200601.
[13] CHEN Meng, MA Yanhong, LIU Shuguo, et al. Rotor dynamics analysis of aero-engine complete machine with finite element model [J]. Journal of Beijing University of Aeronautics and Astronautics, 2007, 33(9): 1013-1016. DOI: 10.3969/j.issn.1001-5965.2007.09.003.
[14] WANG Xiaolei, JIN Zhenlin, LI Xiaodan, et al. Dynamics modeling and analysis of series-parallel hybrid quadruped bionic robot [J]. Journal of Agricultural Machinery, 2019, 50(4): 401-412. DOI: 10.6041/j.issn.1000-1298.2019.04.046.
[15] Cui Xing. Dynamic modeling simulation and analysis of launch vehicle pressurization and delivery system [D]. National University of Defense Technology, 2020.
[16] Bai, W.-J. Dynamic modeling and analysis of xk2120 CNC machine tool [D]. Liaoning: Northeastern University, 2010. DOI: 10.7666/d.J0104950.
[17] Mustafa Ilhan Gökler, Murat Bidotlgidotn Koç. Design of an automatic tool changer with disc magazine for a CNC horizontal machining center [J]. International Journal of Machine Tools & Manufacture, 1997, 37(3): 277-286.DOI: 10.1016/S0890-6955(96)00066-1.
[18] Dong Chenglin. Research on parametric modeling and integrated design method of a new five-degree-of-freedom hybrid robot [D]. Tianjin: Tianjin University, 2020.
[19] Mei Fengxiang. The universality of universal equations of dynamics—Analytical Mechanics Notes No. 31 [J]. Mechanics and Practice, 2020, 42(2): 209-213. DOI: 10.6052/1000-0879-19-333.

Journal of Physics: Conference Series **2992** (2025) 012043 doi:10.1088/1742-6596/2992/1/012043

Application of constitutive model considering stress state and spatial gradient in the simulation of cast aluminum wheel testing

Yazhou Jiang [1], Wuguo Chen [2,3], Taisong Cui [1], Zihan Jiang [2,3], Bin Zhang [1], Dai Sun [2], Feng Gao [2,3] and Tengteng Wang [2,3,*]

[1] State Key Laboratory of Intelligent Vehicle Safety Technology, Chongqing 401133, China

[2] SydoTech Co. Ltd., Chongqing 401100, China

[3] Beijing Institute of Technology Chongqing Innovation Center, Chongqing 401100, China

[*] Corresponding author: tengteng.wang@sydo.com.cn

Abstract: The differences in micro-structure at different locations of cast aluminum alloy wheels lead to variations in the spatial gradient of mechanical properties. Additionally, the yield stress and fracture behavior during the deformation of cast aluminum materials exhibit pronounced tensile-compressive asymmetry. The ignorance to consider those two factors or insufficient consideration of them results in inadequate prediction accuracy in the service testing simulation of cast aluminum wheels. This study develops a rheological behavior law considering both stress state and spatial gradient and implements a user-defined subroutine based on LSDYNA. Simulation results demonstrate that the new rheological behavior law significantly improves prediction accuracy for both material-level and component-level testing, compared to traditional material models that neglect these factors.

1. Introduction

Cast aluminum (Al) alloys are widely used for wheels, hubs, shock absorber towers, front/rear floor panels, subframes, and even integrated body structures in cars owing to their high specific strength, low cost, and strong forming capability [1]. However, the casting process characteristics of these alloys introduce defects in components, such as porosity, segregation, and micro-structure inhomogeneity, which significantly affect their mechanical performance during service [2]. The microstructure inhomogeneity then turns to the variability of mechanical property wrt. spacial gradient, which has been observed in lots of research in the case of cast Al wheels [3, 4, 5, 6, 7, 8]. Such variations pose multiple challenges for the simulation and prediction of component collision safety performance during service.

Firstly, collision simulation analysis via LSDYNA or ABAQUS requires the input of both constitutive and fracture data for materials which typically assume material homogeneity and continuity, meaning that all elements in the mesh model of a component are assigned the same material properties [9, 10, 11, 12]. This default assumption makes it difficult to replicate the spatial gradient variation in the mechanical properties of cast components in simulations. Even if such variations are considered, the complex geometries of cast components prohibit simple sampling for the characterization of constitutive models [13, 14]. Besides, cast Al alloys possess relevant high tensile-compressive asymmetry, i.e. the yield strengths are different when they undergo tension, shear, or compression. The most-used constitutive models based on the von Mises yield criterion with isotropic strain hardening, such as Mat_24 in LSDYNA, naturally have difficulty in accurately predicting material behaviors under multiple stress states with a single set of model parameters.

The consideration of spatial gradient property in collision simulation could be realized by the part-division method, but such a method depends still on a set of elements, so it is not thoroughly detailed [15]. In the view of obtaining the spatial gradient property itself, complicated tests and simulations such as RVE could be used, which is difficult in industry applications [16, 17, 18, 19].

The aim of the present work is to assign an independent constitutive model for each element to realize the consideration of spatial gradient. Stress triaxiality is introduced as an indicator for asymmetric yield. What is more, a simple experimental matrix is designed to obtain the spacial gradient properties. Such simulation work is accomplished via the implementation of a user-defined material subroutine LSDYNA.

2. Experimental and simulation

2.1. Reference matrix

The reference matrix covers quasi-static tests including uni-axial tension, shear, notched tension and cupping tests, and dynamic uni-axial tension tests, whose sample geometries can be referred to in Figure 1. The specimens are prepared at the wheel rim, with the length direction along the wheel's circumference of 2 mm thick. The quasi-static tests are conducted on Zwick 100kN ALLROUND. Dynamic tests are performed on Zwick HTM 5020 to obtain the load information. The displacement is extracted from a numerical extensometer of 10mm via digital image correlation powered by GOM3D. Loading speeds for dynamic tests are 0.0125ms^{-1}, 0.125ms^{-1}, 2.5ms^{-1}, and 12.5ms^{-1} for nominal strain rates of 1s^{-1}, 10s^{-1}, 200s^{-1}, and 1000s^{-1}, respectively.

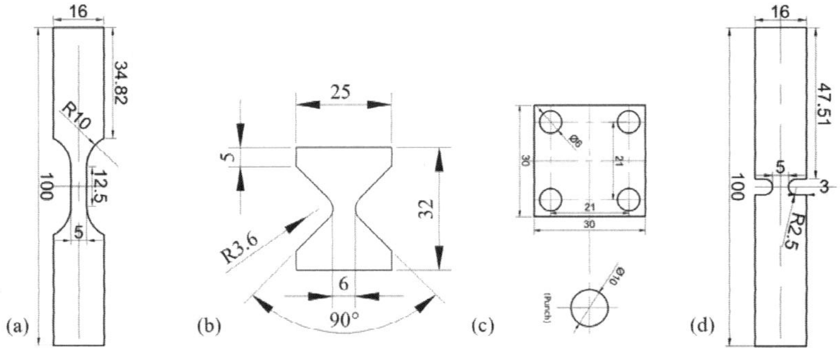

Figure 1. Sample geometry of reference matrix where (a) uni-axial tensile, (b) shear, (c) cupping, and (d) notch tensile (unit: mm).

2.2. Normalization matrix

The normalization matrix covers quasi-static uni-axial compression and Leeb hardness tests, which are used to assess the spatial gradient of the mechanical properties wrt. positions of the wheel. The quasi-static compression cylinders are prepared along the radial direction from the center of the wheel, whose diameter is normally 15mm, and height is 5.5mm and 3.5mm for samples prepared at spokes and rim, respectively. The sampling positions for the compression test and Leeb hardness test are shown in Figure 2.

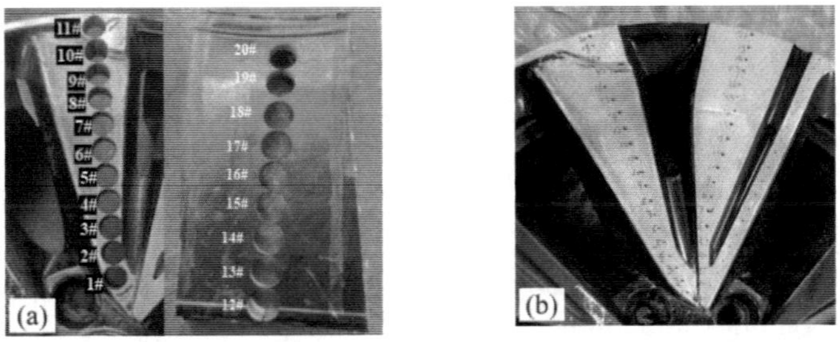

Figure 2. Sampling positions of (a) compression and (b) Leeb hardness of the normalization matrix.

2.3. Verification matrix

The verification matrix includes dynamic radial drop-weight tests with two different weights and static axial compression tests, as shown in Figure 3. A cylindrical flat indenter with a diameter of 150 mm is used in the axial static compression test. Counterweights of 0.45t and 0.6t are adopted in the radial drop-weight test for the crash of a rectangle indenter of 380mm×280 mm freely dropped from 2.1m high. The axial static compression test is carried out on Zwick 100kN ALLROUND to obtain information on displacement versus load. Time and acceleration are recorded using a patch-type force sensor in the radial drop-weight test.

Figure 3. (a) Axial static compression and (b) 450kg/600kg radial drop-weight in the verification test matrix.

2.4. Simulation routine

All the tests in the above three matrices are simulated on LSDYNA R11. The tetrahedral formulation of ELFORM=13 with a characteristic length of ~5mm is used for the explicit analysis. Boundary conditions consistent with the test are applied. Contact is realized via AUTOMATIC_SURFACE_TO_SURFACE. References, normalization, and verification matrices are simulated step by step. The simulation of the verification matrix is shown in Figure 4.

The default Mat_24 and a user-defined material subroutine Mat_41 are used as constitutive models (strictly "rehological law"). The "ADD_DAMAGE_GISSMO" is used for fracture. All the parameters for the material models are input as curves or curve tables. The main difference between Mat_24 and Mat_41 is given below.

$$\bar{\sigma}_{M24} = f_1(\bar{\varepsilon}) \cdot f_2(\dot{\varepsilon}) \tag{1}$$

In Equation (1), $\bar{\sigma}_{M24}$ represents the von Mises equivalent stress, $\bar{\varepsilon}$ is the corresponding equivalent plastic strain, $\dot{\varepsilon}$ denotes the equivalent strain rate, and f_1 and f_2 are the non-coupled scalar functions of strain hardening and strain rate effect, respectively.

$$\bar{\sigma}_{M41} = f_1(\bar{\varepsilon}) \cdot f_2(\dot{\varepsilon}) \cdot f_3(\eta) \cdot f_4(\boldsymbol{u}) = \bar{\sigma}_{M24} \cdot f_3(\eta) \cdot f_4(\boldsymbol{u}) \tag{2}$$

In Equation (2), $\bar{\sigma}_{M41}$is also von Mises equivalent stress, η is the stress triaxiality, and $f_3(\eta)$ is a non-monotonic continuous single-valued function with a range of ~[0.5, 2.0] providing asymmetric yield. $f_4(u)$ is a position dependent function with a range of~[0.5, 1.0].

The GISSMO model related to the stress triaxiality and Lode angle [20, 21, 22, 23] is used to predict fracture. The damage accumulates in a nonlinear manner. The damage factor D is expressed as:

$$D = \left(\frac{\bar{\varepsilon}}{\bar{\varepsilon}_f(\eta, \bar{\theta})}\right)^n \tag{3}$$

where n is the damage accumulation index. $\bar{\varepsilon}_f(\eta, \bar{\theta})$ represents the equivalent plastic strain to fracture as a function of stress triaxiality η and Lode parameter $\bar{\theta}$. Fracture occurs when D arrives at 1.

The characterization of the rehological model is done by minimizing the difference between experiment and corresponding simulation of the reference matrix, which is defined by a cost function (Fc) as follows:

$$Fc = \frac{1}{4}\sum_{i=1}^{4}\left(\frac{\sum_{k=1}^{k=300}\left|F_{i,k}^{\exp} - F_{i,k}^{\text{sim}}\right|}{\sum_{k=1}^{k=300}\left|F_{i,k}^{\exp}\right|}\right) \tag{4}$$

where i represents the 4 types of tests in the reference matrix. k represents the equal interval interpolation points along the x-axis, with 300 points for each test. $F^{\exp}$ represents the force value obtained from the test, and F^{sim} represents the predicted force value from the corresponding simulation. The algorithm for calibration of the reference matrix is as follows: (1) We set parameters and calculate $f_1(\bar{\varepsilon}), f_2(\dot{\varepsilon}), f_3(\eta)$ and $\bar{\varepsilon}_f = \bar{\varepsilon}_f(\eta, \bar{\theta})$; (2) We run all the corresponding simulations of the reference matrix, and output the information of force versus displacement; (3) We calculate Fc. The calculation terminates if Fc converges else we go to step (1).

It is notable that $\bar{\varepsilon}_f = \bar{\varepsilon}_f(\eta, \bar{\theta})$ is obtained through inverse distance weight (IDW) algorithm [24, 25, 26]. Specifically, within a rectangle domain defined by $\eta \in [-1, 1]$ and $\bar{\theta} \in [-1, 1]$, the anchor points $(\eta, \bar{\theta}, \bar{\varepsilon}_f)$ are iteratively tested, and interpolation is performed in the $\eta - \bar{\theta} - \bar{\varepsilon}_f$ space. Afterward, 100 equal-interval planes normal to $\bar{\theta}$ are used to slice the fracture surface to generate 100 curves, which are then written in DEFINE_TABLE.

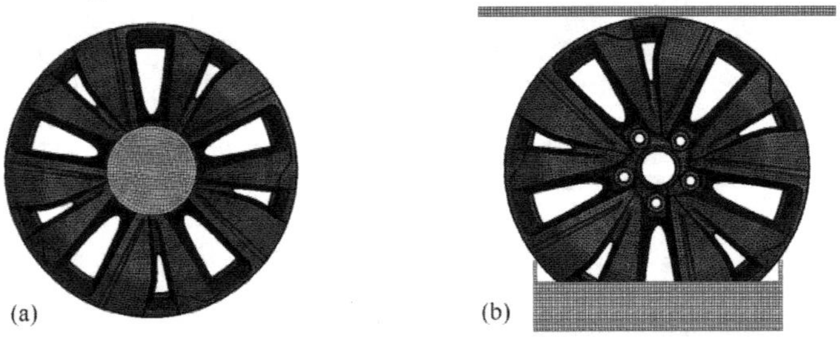

Figure 4. Numerical model of the verification matrix: (a) Axial static compression (b) Radial drop-weight.

3. Results and discussion

3.1. Experimental and numerical results of reference matrix

Force versus displacement results of the four tests are shown as dashed lines in Figure 6. Significant curve dispersion of the samples taken from the same position is observed due to the feature of cast processing. Similar results are also reported in others' studies [10, 11, 12, 13].

The results of simulation without fracture of quasi-static tests in the reference matrix are also presented in Figure 6. The curves for constitutive models are given in Figures 5 (a) and (c). The experimental result of uni-axial is found well matched with the numerical ones via either Mat_24 or Mat_41 (with only $f_3(\eta)$), because those two material models are equivalent in uni-axial case. However, Mat_24 behaves insufficiently when the precision of shear and/or cupping is considered. Such weak points are greatly overcome by introducing the modification of triaxiality, i.e. the $f_3(\eta)$, in which function value of $f_3(\eta)$ is smaller than 1 at positions of $\eta = 0$ and $\eta = 2/3$ for shear and cupping, respectively. Correspondingly, a function value 1 is selected for notch tensile because the simulation with $f_1(\bar{\varepsilon})$ alone matches well. Such modification of isotropic hardening of $f_1(\bar{\varepsilon})$ via $f_3(\eta)$ is referred as asymmetric yielding. It is proved necessary for metals such as mild steel and cast Al alloy.

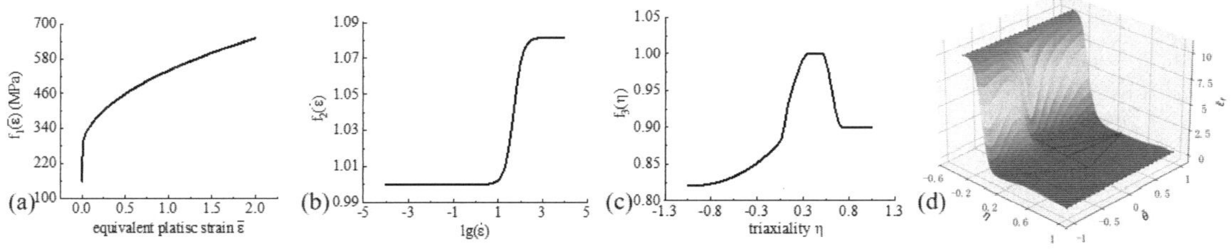

Figure 5. Input curves or curve tables for the material models used in simulation of reference matrix, where (a) strain hardening $\mathbf{f_1(\bar{\varepsilon})}$, (b) strain rate effect $\mathbf{f_2(\dot{\varepsilon})}$, (c) triaxiality multiplier $\mathbf{f_3(\eta)}$ and (d) strain to fracture surface.

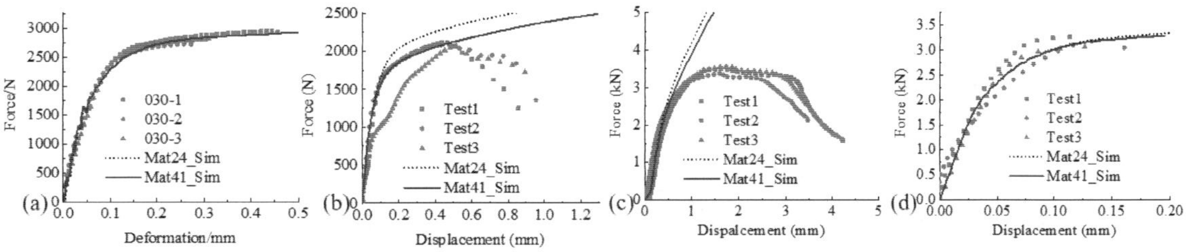

Figure 6. Comparison of experimental and simulation (without fracture) results of (a) uni-axial, (b) shear, (c) cupping and (d) notch tension of quasi-static tests of reference matrix.

The strain-stress relations of the dynamic tests are given in Figure 7(a). Four typical curves representing the four strain rate levels are selected as shown in Figure 7(b). Then the rate-effect of $f_2(\dot{\varepsilon})$ is extracted by matching the calculated strain-stress relation and the experimental ones, as given in Figure 5(b). $f_2(\dot{\varepsilon})$ is named as Dynamic Impact Factor (DIF). A reasonable matching is found by such DIF.

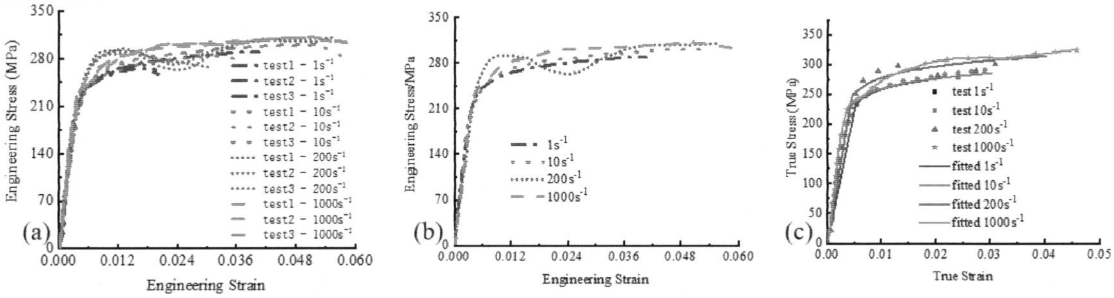

Figure 7. Steps to treat the dynamic uni-axial tensile tests (a) Extracted engineering strain v.s. engineering stress curves (b) Selected curves exhibiting the typical strain rate effect (c) A comparison of fitted strain-stress curves and the selected experimental ones.

The $\eta - \bar{\theta} - \bar{\varepsilon}_f$ fracture surface is shown in Figure 5(d), which holds the black dots as an anchor point for the IDW algorithm and additionally works for each time step of calculation with Mat_24 or Mat_41. The simulation results via Mat_41 + GISSMO provide good accordance with experiments, which is better than that via Mat_24 + GISSMO, as shown in Figure 8. Thus Equation (2) is a reasonable extension of Equation (1). What is more, one can intuitively find that IDW interpolation can generate a greater flexible surface than analytical expressions such as MMC or DIEM models.

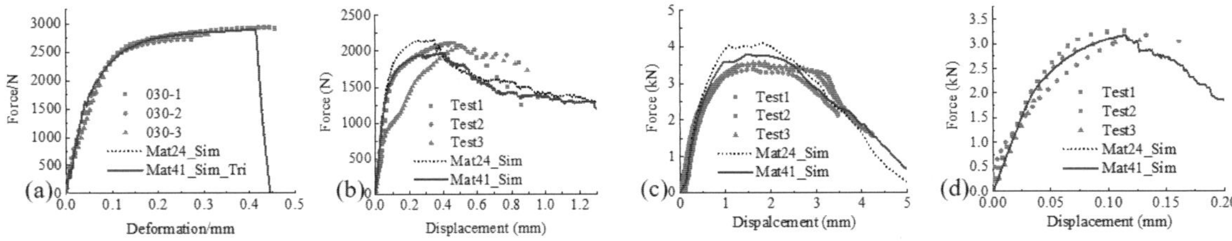

Figure 8. Comparison of experimental and simulation (with fracture) results of (a) uni-axial, (b) shear, (c) cupping and (d) notch tension of quasi-static tests of reference matrix.

3.2. Normalization matrix results

The stress-strain curves of uni-axial compression tests along the wheel are shown in Figure 9(a). The strength at the wheel rim center is lower than at the ends. The strength at the center of the spokes is also lower than at both ends, which is similar to the Leeb hardness trend, as shown in Figure 9(b). Such observations demonstrate that the mechanical properties of the wheel exhibit a clear spatial gradient. Figure 9(c) is the fitted curves of compressive tests based on stress at strain 0.1, and the normalization coefficient field in the range [0, 1] is then projected to the wheel, as shown in Figure 9(d). The projection is realized by MAT_41 during initialization, where the red arrows in Figure 9(d) represent the fitting path in Figure 9(c). At this stage, the wheel part in simulation is capable of representation of the spacial gradient material property.

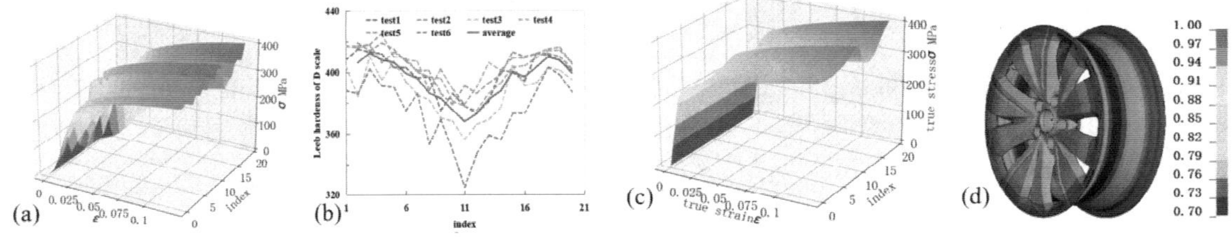

Figure 9. (a) Strain-stress relation of compressive tests (b) Leeb hardness along rim (c) Fitted curves of strain-stress curves (d) Projected normalization field $\mathbf{f_4(u)}$.

3.3. Experimental and numerical results of verification matrix

The results of the axial static compression test and corresponding simulations are shown in Figure 10. The force-displacement curve by Mat_41 (with $f_3(\eta) \cdot f_4(u)$) exhibits better prediction than Mat_24, as shown in Figure 10(a). However, the morphology does not differ much between the simulation results by Mat_24 and Mat_41.

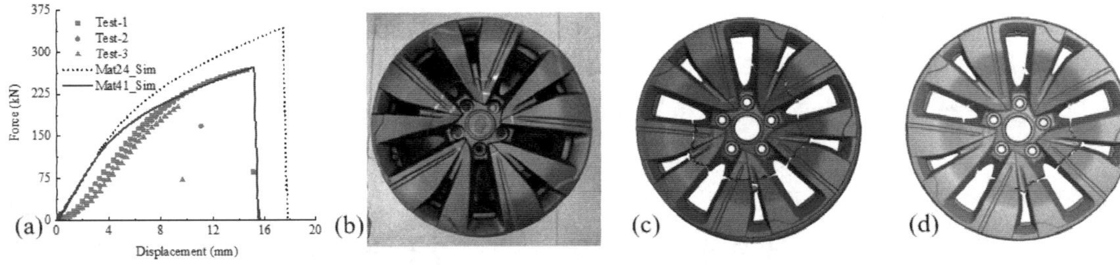

Figure 10. (a) Comparison of force-displacement curves from experiment and simulations. (b), (c) and (d) are fracture morphology of experiment, Mat_24 simulation and Mat_41 simulation of axial static compression test, respectively.

The results of the radial drop-weight test and corresponding simulations are shown in Figure 11 and Figure 12. One can easily see that the crack length and propagation trend prediction by Mat_41 (with $f_3(\eta) \cdot f_4(u)$) is closer to the experimental results than Mat_24, especially in case of 600kg test. Besides, the acceleration-time curve predictions by Mat_41 match the experimental ones better than those by Mat_24 as well. The improvements in the simulation are additionally powered by $f_4(u)$, i.e. the spatial gradient correction factor obtained via the normalization tests, which can be deemed as a precise reproduction of real material property in the virtual world of simulation.

Figure 11. Comparison of experimental and simulation results of 450kg case, where (1) is acceleration-time curves, and (b)~(d) are fracture morphology of experimental, simulation with Mat_24 and simulation with Mat_41, respectively.

Figure 12. Comparison of experimental and simulation results of 600kg case, where (1) is acceleration-time curves, and (b)~(d) are fracture morphology of experimental, simulation with Mat_24 and simulation with Mat_41, respectively.

4. Conclusion

The mechanical properties of the cast wheel of Al alloy were tested and simulated, leading to the following conclusions:

Cast Al alloy exhibits significant asymmetric yield in case of compression, tension, or shear in the view of the material itself. Such asymmetry should be considered during the numerical simulation.

The mechanical properties along the wheel part evolve due to the cast processing. Such evolution or spacial gradient property should be also reproduced in the simulation in order to get either better force or better morphology prediction.

The feature of symmetric yield and spacial gradient property is implemented in a user-defined material model in LSDYNA, which could be a certain solution for similar simulation cases.

Acknowledgments

This work is supported by open fund project IVSTSKL-202307 provided by the State Key Laboratory of Intelligent Vehicle Safety Technology.

Reference

[1] Guo Y, Zhu X, Yang Y and Xiong N 2015 Research state of lightweight materials and manufacturing processes in the automotive industry Forging & Stamping Tech 40 6

[2] Li Y, Liu J, Zhang Q and Huang W 2021 Casting defects and microstructure distribution characteristics of aluminum alloy cylinder head with complex structure Mater. Today Commun 27 102416

[3] Ransenigo C, Tocci M, Viscardi C, Serafini M and Pola A 2022 Development of a model for the prediction of mechanical properties for Al-Si-Mg castings L Metall. Ital 114 18

[4] Jian Y, Bo L, and Hai Y H 2024 Research on composition-process-property prediction of die casting Al alloys via combining feature creation and attention mechanisms Journal of Materials Research and Technology 28 335-346

[5] Ozaydin O 2019 Influence of different Si levels on mechanical properties of aluminium casting alloys Eur. J. Eng. Nat. Sci 3 165-72

[6] Xun Q W, Zhi G D, Si P L and Li M P 2024 Computation method for mechanical performances of die-cast aluminum alloys based on non-uniform material properties Materials Today Communications 40 109683

[7] Manente A and Timelli G 2011 Optimizing the heat treatment process of cast aluminium alloys Recent Trends Process. Degrad. Al. Alloys 9 197-220

[8] Shang R, Altenhof W, Li N, and Hu H 2005 Wheel impact performance with consideration of material inhomogeneity and a simplified approach for modeling Int. J. Crashworthiness 10 137-50

[9] Cerit M 2010 Numerical simulation of dynamic side impact test for an aluminium alloy wheel Sci. Res. Essays 5 2694-701

[10] Kumar R and Meher S 2013 Topology optimization of aluminium alloy wheel Int. J. Mod. Eng. Res 3 1548-53.

[11] Jian Y, Bo L, Dongwei S, Qin Y, and Tiegang H 2025 Local stress/strain field analysis of die-casting Al alloys via 3D model simulation with realistic defect distribution and RVE modelling Engineering Failure Analysis 168 109104

[12] Lee J, Kim S, Park H, Bong H, and Kim D 2018 Metal plasticity and ductile fracture modeling for cast aluminum alloy parts J. Mater. Process. Tech 255 584-95

[13] Léost Y, Sonntag A and Haase T 2017 Modeling of a cast aluminum wheel for crash application Conf. on 11th European LSDYNA (Salzburg)

[14] Jian Y, Bo L, Yunbo Z, Yiben Z, and Haiyou H 2024 Data extension-based analysis and application selection of process-composition-properties of die casting aluminum alloy Engineering Applications of Artificial Intelligence 133

[15] Ballo F, Previati G, Mastinu G and Comolli F 2020 Impact tests of wheels of road vehicles: a comprehensive method for numerical simulation Int. J. Impact Eng 146 103719

[16] Cheng Y T and Cheng C M 1999 Scaling relationships in conical indentation of elastic-perfectly plastic solids Int. J. Solids Struct 36 1231-43

[17] Jie H, Guo J D, Shi D L, Jian Z and Ji W 2024 Uneven distribution of cooling rate, microstructure and mechanical properties for A356-T6 wheels fabricated by low pressure die casting Journal of Manufacturing Processes 127 196-210

[18] Byun T S, Hong J H, Haggag F M, Farrell K and Lee E H 1997 Measurement of through-the-thickness variations of mechanical properties in SA508 Gr.3 pressure vessel steels using ball indentation test technique Int. J. Press. Vessels Pip 74 231-38

[19] Jin Z and Zhao J 2018 Yield strength and strain hardening exponent of metal materials evaluated by micro-load continuous ball indentation method Mater. Mech. Eng 42 72-7

[20] Effelsberg J, Haufe A, Feucht M, Neukamm F and Du B. Du 2012 On parameter identification for the GISSMO damage model Int. Conf. on 12th LSDYNA Users (Dearborn)

[21] Neukamm F, Feucht M and Haufe A 2008 Consistent damage modeling in the process chain of forming to crashworthiness simulations Conf. on 7th German LSDYNA Forum (Bamberg)

[22] Neukamm F, Feucht M, and Haufe A 2009 Considering damage history in crashworthiness simulations Conf. on 7th European LSDYNA (Salzburg)
[23] Basaran M, Wolkerlin S D, Feucht M, Neukamm F and Weichert D 2010 An extension of the gissmo damage model based on lode angle dependence Conf. on 9th LSDYNA Forum (Bamberg)
[24] Daly C 2006 Guidelines for assessing the suitability of spatial climate data sets Int. J. Climatol 26 707-21
[25] Luo W, Wen L Y, Yao H L, Jia D, and Heng H 2022 Research on microstructure and mechanical properties of die casting Mg-4Al-1Si-3RE (Ce, La) and AS41 alloys Materialstoday 33 104625
[26] Zhang J, Guo L, and Zhang X 2012 Effects of interpolation parameters in inverse distance weighted method on DEM accuracy J. Geomatics Sci. Tech 29 51-6

Numerical study on flow and temperature rise characteristics of low-pressure cylinder of steam turbine under variable working conditions

Hongwu Wang*, Huan Liu, Tingshan Ma, Tian Xie, Ting Wang, Rongzu Yang, Yaowen Wang, Pengcheng Zhai, Longwen Yu and Xinjing Jing

Xi'an Thermal Power Research Institute Co., LTD, Xi'an, China

*Corresponding author's e-mail: wanghongwu@tpri.com.cn

Abstract—The internal flow dynamics and temperature rise within the low-pressure cylinder of the steam turbine's final stage at reduced flow rates present substantial complexities. These intricacies pose significant challenges for effective thermal power peak regulation and the transformation to zero output. This study focuses on the low-pressure cylinder of a power plant, establishing a computational model for the single-channel flow field in a five-stage cascade. Through numerical analysis, we investigate the operational efficiency, steam flow dynamics, and temperature elevation features under varying working conditions. When the flow diminishes to 3.84% of the design condition, the low-pressure turbine is incapable of generating power. Under extremely low flow rates, the cylinder transitions to a choking state, resulting in the formation of various flow structures in the final cascade. In low-flow conditions, the heating effect attributed to choking at the tip of the final cascade's static clearance is pronounced. As the flow rate decreases to 2.23% of the specified design parameters, the mean surface temperatures of the final static and rotor blades surge by 62.65% and 72.20%, respectively. Under choking conditions, both the operational performance and internal flow structure experience significant alterations. The elevated temperatures of the final cascade adversely affect the blades' working environment, necessitating careful consideration during low-flow operation.

1. Introduction

Since the commencement of the 13th Five-Year Plan, the installed capacity of grid-connected renewable energy has seen rapid expansion [1]. However, this surge has concurrently exacerbated the challenge of peak regulation within the power grid system [2]. Specifically, during the winter heating season in China's "three Northern regions," numerous cogeneration units encounter significant difficulties in peak regulation due to heating demands, resulting in substantial light-related issues [3]. Wind and water abandonment has led to a pressing need to investigate the potential of deep peak regulation as a means to address the challenges associated with the peak regulation of power grids. This approach aims to enhance the operational flexibility of thermal power units [4]. To meet this requirement, numerous power plants have implemented zero-output modifications to the low-pressure cylinder of steam turbines. Such modifications ensure the safe and stable functioning of the low-pressure cylinder under conditions of reduced flow rates, thereby improving the depth of peak regulation capacity of thermal power units [5]. Specifically, at low flow rates, a significant occurrence of three-dimensional vortex formations may manifest on the final stage blades, alongside pronounced backflow phenomena at the exit of the final stage cascade channel. Moreover, the final stage blades may experience conditions leading to a "blast" state, where the surface temperature of the blades escalates considerably due to the effects of blast heating. This escalation poses a detrimental risk to the structural integrity of the long blades, thus jeopardizing the safe operation of the turbine [6]. Statistics indicate that blade-related accidents account for approximately 40% of the total safety incidents reported in steam turbine units [7]. Consequently, it is imperative to investigate the operational state of steam turbines under variable working conditions, as it holds substantial engineering significance.

Research conducted by both domestic and international scholars has highlighted that the phenomenon of steam flow separation at the final rotor blade intensifies as the flow of the working medium in the low-pressure cylinder diminishes. As this occurs, the affected area progressively extends towards the tip of the blade and the upstream cascade [8]. In addition, Gong et al. [9] categorized the vortex structures in the last stage of the low-pressure cylinder experienced during off-design turbine conditions into three types. Wang and Lyu [10] provided a comprehensive overview of the flow characteristics and structural dynamics of the low-pressure cylinder in a steam turbine operating under conditions of reduced volumetric flow. The study delves into analyses concerning the turbulence scales and frequency scales associated with this setup. Complementarily, the research team conducted a numerical investigation into the instability features of the flow within the final seven stages [11].

In circumstances where the outputs are zero, the steam flow experience is significantly below the design specifications. This discrepancy results in diminished steam velocity in certain regions, leading to blast losses induced by blade agitation, which in turn reduces the effective enthalpy drop across all stages and prompts the occurrence of blast heating [12]. Ren et al. [13] indicated that when the flow rate approaches a critical threshold, a pronounced eddy current can form at the apex of the gap between the rotor and stator blades in the final stage, subsequently transitioning the system into a blast state. Regarding the phenomenon of temperature rise observed in the low-pressure cylinder during the steam turbine's blast operation, Xie et al. [14] identified that blast heating is most pronounced in the region between the blades of the final stage.

2. Calculation method

2.1 Computational models

In this paper, we conduct a numerical analysis of a five-stage cascade within the low pressure cylinder of the CJK350/269-24.2/0.4/566/566 model. This unit features a supercritical, single intermediate reheat, single-shaft, double-cylinder, and double-exhaust steam indirect air-cooled condensate heating system, manufactured by the Harbin Turbine Plant. A three-dimensional model is developed to represent the five-stage cascade. The meridional plane alongside the three-dimensional model depicting the area in Figure 1. The color blue denotes the static blades, while the orange color indicates the moving blades. The geometric parameters of the steam turbine cascade are in Table 1.

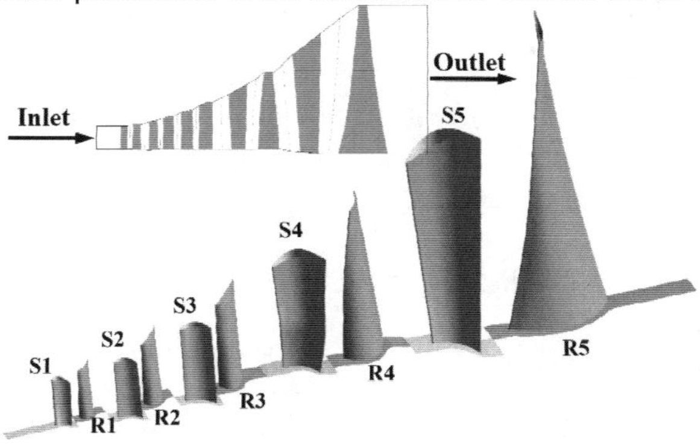

a) Three-dimensional model of blade

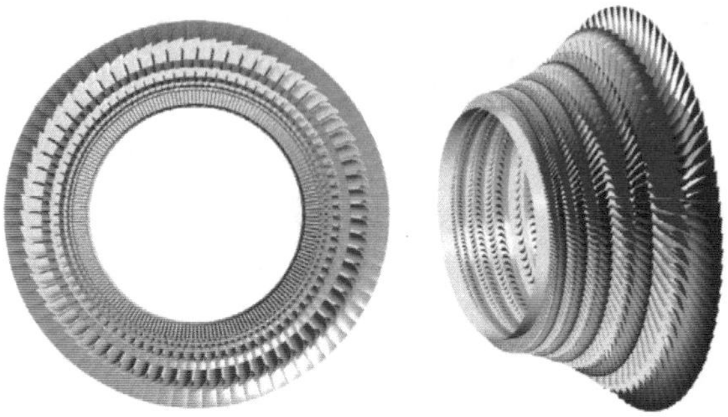

b) Three-dimensional model of the whole cycle blade

Figure 1. Three-dimensional model of the whole cycle blade.

Table 1. Geometric parameters of steam turbine cascade.

Progression		Number	Height/mm	Diameter/mm
Stage 1	S1	192	104.81	1860.12
	R1	180	113.92	1865.46
Stage 2	S2	142	135.72	1879.90
	R2	120	151.92	1901.61
Stage 3	S3	122	181.08	1925.78
	R3	100	214.97	1963.50
Stage 4	S4	84	297.23	2042.50
	R4	98	350.32	2080.20
Stage 5	S5	68	548.02	2263.40
	R5	80	678.34	2311.56

2.2 Numerical method

The hexahedral mesh computational model is generated using NUMECA Autogrid5. To accommodate the geometric shape of the blades, the H-O-H topology strategy is employed, which incorporates O-type grid generation within the near-wall regions. For mainstream area, H-type is utilized, and the computational grid is refined in the near-wall regions. Figure 2 provides a schematic representation of the computational grid.

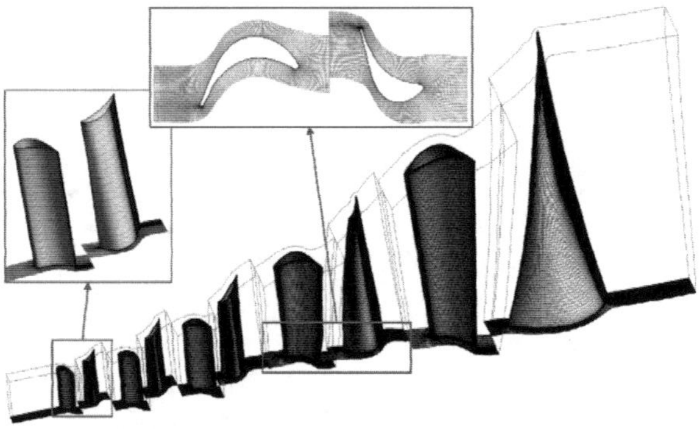

Figure 2. Schematic diagram of the computational grid.

The hexahedral mesh generation process for the calculation model is performed using NUMECA Autogrid5. This process leverages the H-O-H topology strategy to implement O-type grid generation in the blade's near-wall areas. In contrast, a grid as H-type is applied to the stream flow area, where the computational grid is further densified near the walls. The schematic diagram of the computational grid is presented in Figure 2.

Using the overall weekly flow as key indicators, we calculated the Grid Convergence Index (GCI) for three sets of computational grids, each with a different number of nodes, under design operational conditions. The findings are summarized in Table 2. When the grid consists of a total of 11,791,06 nodes, both indices yield GCI values below 0.5%, indicating that the grid size is sufficiently refined to yield grid-independent solutions. Consequently, this grid has been selected for subsequent calculations.

Table 2. GCI values of different grid numbers.

Number of nodes/10^6	Flow/kg·s^{-1}	GCI/%	Temperature/K	GCI/%
7.75	100.256	1.131	319.985	0.673
11.79	101.685	0.440	322.464	0.277
17.94	102.245	-	323.486	-

For the numerical simulation, we utilized Ansys CFX. To accurately model the phase transition, we opted for Steam 1vl, a real steam model that adheres to the IAPWS IF97 industrial standard. The Reynolds time-averaged Navier-Stokes (N-S) model is commonly employed for turbulence simulations. In this study, we applied the SST k-ω model within the RANS framework to effectively predict flow separation.

2.3 *Boundary conditions*

In the context of numerical simulation, the total inlet pressure and total temperature within the calculation domain are specified, alongside an exhaust back pressure of 12 kPa. The boundaries flanking both the static blade and dynamic blade cascade channels are implemented with rotating periodic boundary conditions, while the remaining walls are treated as adiabatic non-slip boundaries. To manage the rotor interface, the freeze-rotor method is employed, operating at a rotor speed of 3000 r/min. Table 3 presents the boundary conditions corresponding to the eight operational scenarios evaluated in this study, with Case 1 representing the design operating condition.

Table 3. Boundary conditions of each working condition.

Working Conditions	Pressure/kPa	Temperature/°C
Case one	489	276.6
Case two	457	277.4
Case three	354	283.3
Case four	252	259.0
Case five	209	245.9
Case six	24	258.3
Case seven	21	258.6
Case eight	15	258.2

3. Results and discussion

3.1 *Work performance*

The in-stage power N_u is:

$$N_u = Gh_u \tag{1}$$

where G represents flow. h_u denotes the enthalpy drop. Stage static efficiency η_s denotes:

$$\eta_s = \frac{H_{01} - H_{03}}{H_{01} - H_{3s}} \tag{2}$$

where H_{01} denotes the total enthalpy at the entrance. H_{03} denotes the total enthalpy at the exit. H_{3s} denotes the isentropic exit static enthalpy. Additionally, the total stage efficiency η_t denotes:

$$\eta_t = \frac{H_{01} - H_3}{H_{01} - H_{3s}} \tag{3}$$

where H_3 denotes the exit static enthalpy.

Figure 3 illustrates the numerical performance indicators across various situations. As depicted in Figure 3(a), a decrease in total inlet pressure results in a continuous decline in flow rate. Specifically, in the lowest operational condition, approximately 2.23% of that observed in the design operating condition, Case 1. Simultaneously, the total power generation declines. Significantly, when the flow rate reduces to 3.84% of Case 7, the cylinder displays a negative power output. At this critical juncture, cylinder transitions from a power-generating machine to a power-consuming device that requires external energy to sustain its operational speed. Regarding the observed trends, the variations in output power and flow rate exhibit a similar pattern. Conversely, two kinds of efficiency initially remain stable before experiencing a sharp decline corresponding with the reduction in steam flow rate. This behavior suggests that in cases where the total inlet pressure remains relatively high (Cases 1 to 5), the typical flow dynamics is sustained, thereby allowing efficiency to remain above 90%.

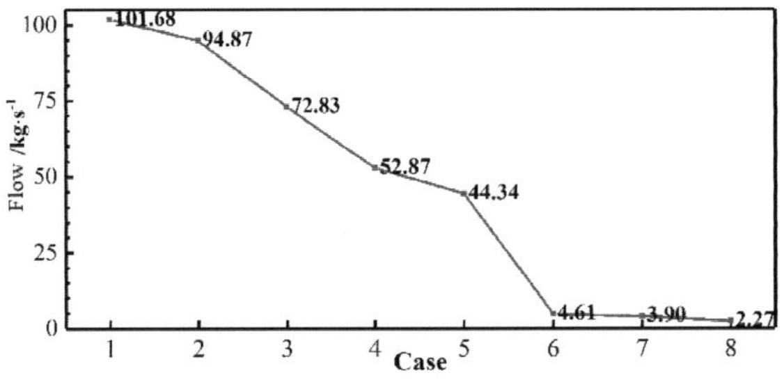

a) Flow

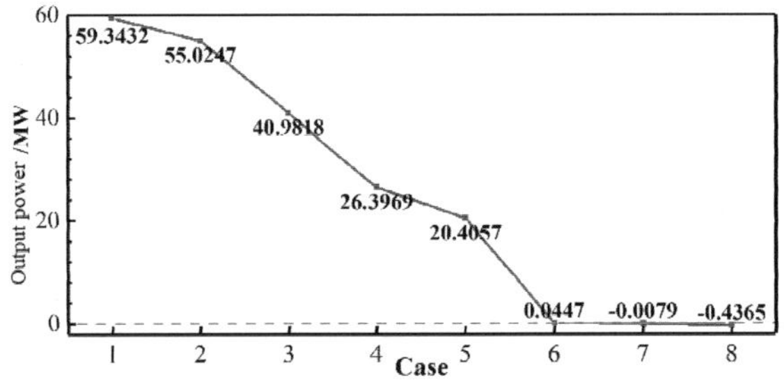

b) Output power

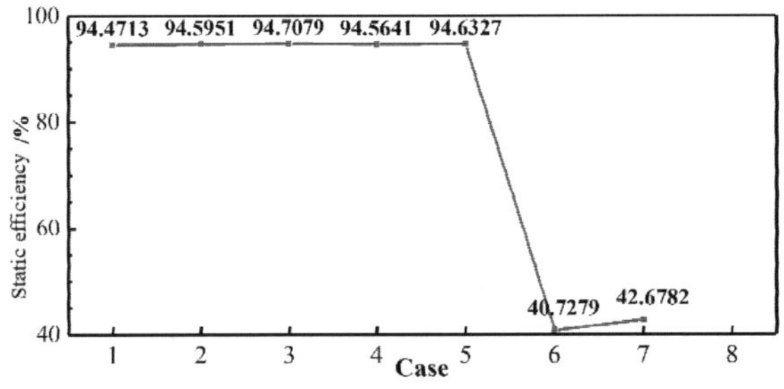

c) Static efficiency

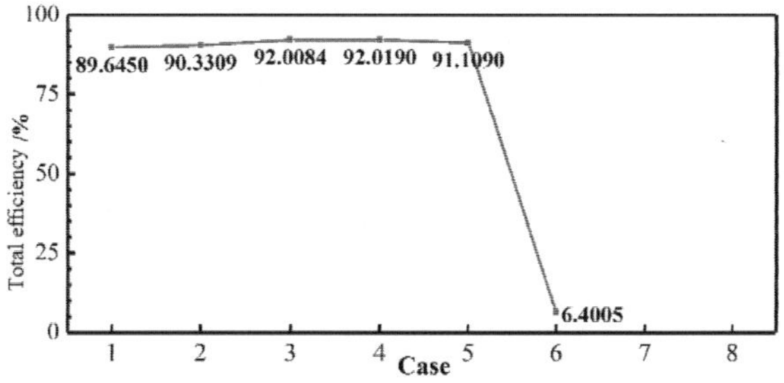

d) Total efficiency

Figure 3. Performance under different working conditions.

3.2 Flow analysis

Figure 4 illustrates the distribution of velocity vectors across various operating situations. In Cases 1 to 3, where the flow rate is elevated, a region characterized by higher exhaust velocities manifests downstream of the rotor tip. With a gradual decrease in total inlet pressure, this region shrinks progressively, ultimately vanishing completely by Case 4, as depicted in Figure 4(d). In Case 5, when the steam flow reaches merely 43.6% of the design condition, flow separation is observed, as illustrated by flow structure (1) in Figure 4(e). The extent of the separated flow gradually extends toward the tip, ultimately culminating in the formation of a hub separation area, identified in Figure 4(f), at the lower end wall. The presence of this separation zone causes nearly all outflow during blast conditions to be expelled from the tip of the final rotor blade stage.

The typical flow path of steam under these blast conditions is indicated by a black arrow in Figure 4(f). Initially, the steam moves towards the blade tip during the final static stage of the blade, accumulates at the base of the rotor blade within the reverse flow gap, and eventually exits from the tip. During this last stage, the steam flow experiences significant deflection, and the working media moving in divergent directions collide. This interaction leads to energy losses and a heating effect due to the blast, which will be analyzed in greater detail in the following section. In this final phase, the steam flow undergoes considerable deviation, causing the working medium moving in divergent paths to collide. This interaction leads to energy losses and a heating effect due to the blast, which will be analyzed in greater detail in the following section.

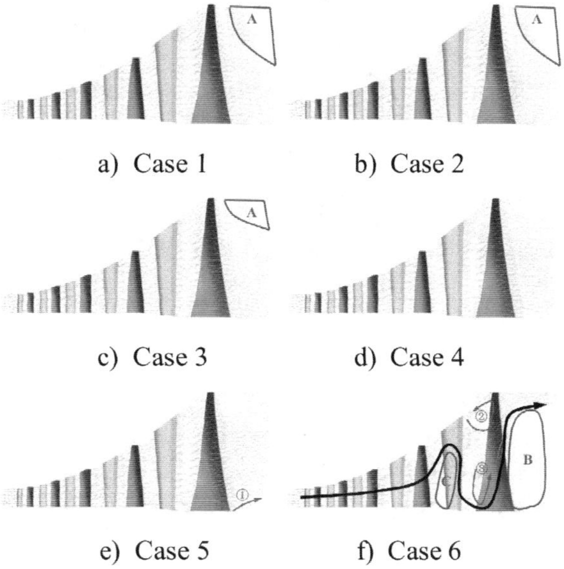

a) Case 1 b) Case 2

c) Case 3 d) Case 4

e) Case 5 f) Case 6

g) Case 7 h) Case 8

Figure 4. Velocity vector diagram of low-pressure cylinder through flow area.

3.3 Temperature rise characteristics

Figures 5 and 6 illustrate the temperature of the blade surface and the final rotor blade surface (specifically the suction surface) of the turbine's low-pressure cylinder. As depicted in Figures 5(a) through e, prior to the flow decreasing to 43.61% of Case 5, the temperature profile within the low-pressure cylinder exhibits a consistent downward trend, transitioning from the supply air temperature to approximately 300 K upon discharge. Concurrently, the area characterized by elevated temperature at the edge of the final stage rotor blade outlet progressively diminishes with a reduction in the inlet flow rate. However, upon further decline in flow rate, the final stage experiences a state of blast, resulting in significant alterations to the temperature distribution: in Cases 6 and 7, where the flow rate remains relatively high, there is no discernible decrease in temperature across the entire basin. In contrast, Case 8, which represents the lowest flow rate, reveals a marked increase in temperature at the corresponding position on the final cascade. This "blast" condition distinctly manifests heating effects at both the dynamic and static tip locations of the final stage, as illustrated in Figures 6(f) to (h), where the red dotted line indicates the specific positions. Notably, as the flow rate of the working medium decreases, the extent of the high-temperature zone's influence expands, accompanied by a larger temperature increase.

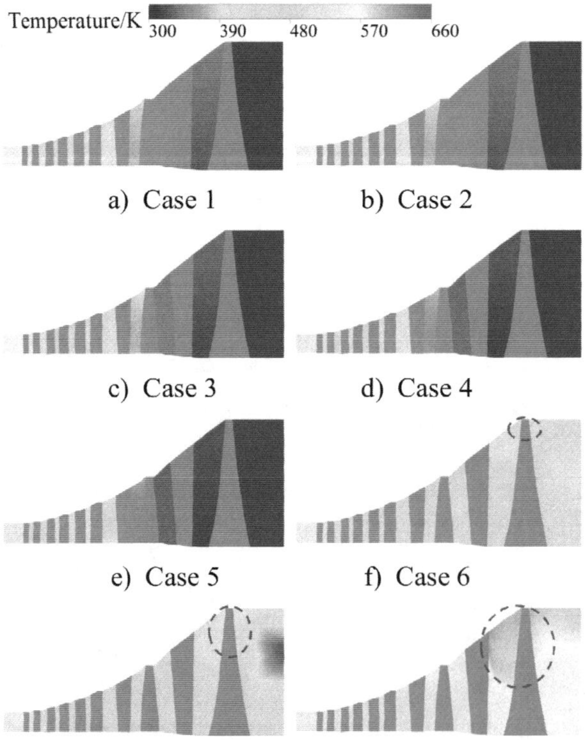

a) Case 1 b) Case 2

c) Case 3 d) Case 4

e) Case 5 f) Case 6

g) Case 7 h) Case 8

Figure 5. Meridional temperature distribution of low-pressure cylinder.

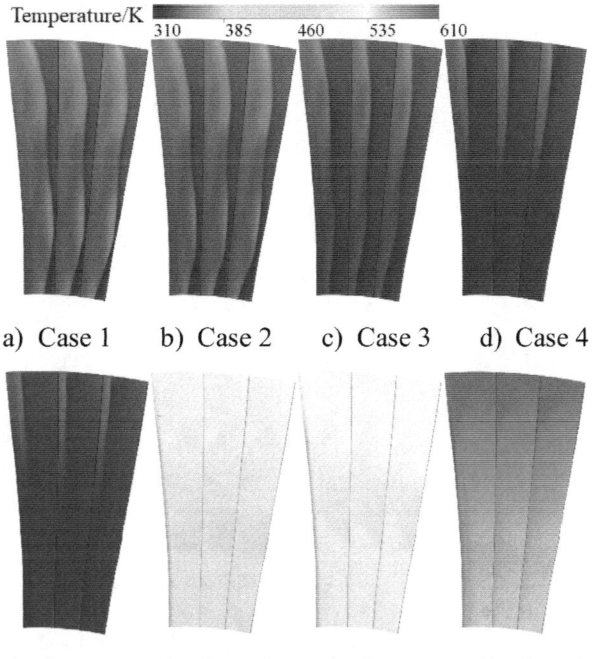

a) Case 1 b) Case 2 c) Case 3 d) Case 4

e) Case 5 f) Case 6 g) Case 7 h) Case 8

Figure 6. Surface temperature distribution of the final rotor blade.

Figure 7 illustrates the variations in average and maximum surface temperatures of both the final stage static and moving blades across various operating conditions. As demonstrated in Figure 7, when the turbine cascade functions normally, an inverse relationship is observed between the blade surface temperature and the flow rate; specifically, as the flow rate decreases, the surface temperature of the blades also diminishes. This finding is in alignment with the trends depicted in Figures 6(a)-(e). Furthermore, a comparative analysis of the temperature fluctuations reveals that the surface temperature of the moving blades exhibits greater sensitivity to changes in flow rate. Upon transitioning into the blowing state, a significant spike in the surface temperature of the moving blades is observed. Although the average surface temperature of the moving blades remains marginally higher than that of the stationary blades in Case 6-8, the highest temperature recorded occurs at the surface of the stationary blades. As indicated in Figure 5(h), the thermal effect of the blowing condition is most pronounced during the final cascade transition to static clearance. This observation suggests that the temperature distribution across the static blades is notably more uneven, resulting in heightened thermal stress during the blowing state. Notably, when compared to Case 1 the temperatures of the final stage blades experience significant increases of 62.65% and 72.20%, respectively, under a flow condition of 2.23% (Case 8). Additionally, the peak temperatures rise by 72.32% and 57.92%, respectively.

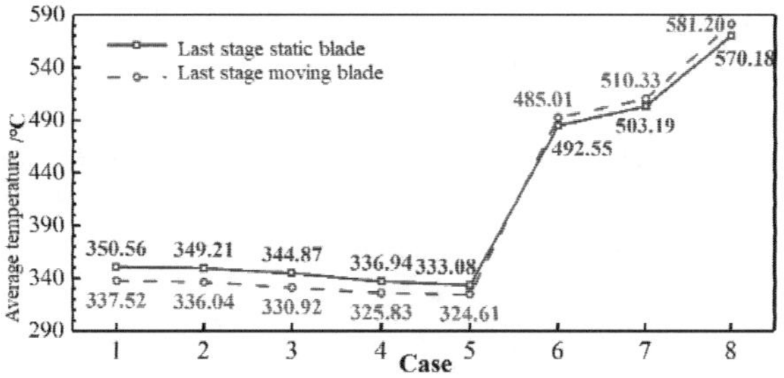

a) Average temperature

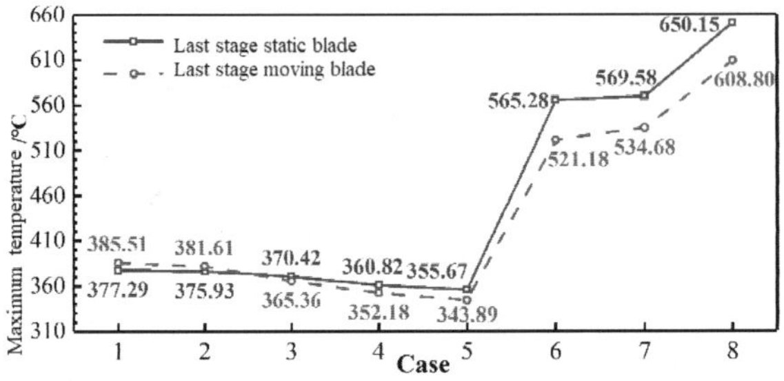

b) Maximum temperature

Figure 7. Average and maximum temperature of final blade surface under different working conditions.

4. Conclusions

The flow rate and electrical power generated by the turbine's low-pressure cylinder show a steady reduction as the starting pressure drops. The cylinder is unable to produce power when its flow rate drops to 3.84% of the design situation because the growing work done by the steam working fluid is not enough to keep the rotor blades spinning. Initially, the efficiency remains relatively stable; however, it experiences a significant drop upon reaching a critical flow state.

A decrease in flow rate induces flow separation at the bottom wall of the final blade. In this final phase, the steam flow undergoes considerable deviation, causing the working medium moving in divergent paths to collide.

In low flow scenarios, the cascade passage associated with the final stage blade exhibits a blast heating effect. The top portion of the last cascade within the equilibrium reduction exhibits the most noticeable temperature increase. When the flow rate diminishes to 2.23% of the design situation, the temperatures of the final-stage static and rotor blades rise by 62.65% and 72.20%.

References

[1] SHU Yinbiao, ZHAO Yong, ZHAO Liang, et al. Study on low carbon energy transition path toward carbon peak and carbon neutrality [J]. Proceedings of the CSEE 2023, 43(5):1663-1672.

[2] WANG Yumei, ZHOU Yongxin, ZHANG Jiqin. Low-carbon scheduling of virtual power plants for integrated energy systems considering the demand side [J]. Electrotechnology, 2023(18):73-76.

[3] ZHAO Jinfeng, FENG Zhongbao, CAO Lihua, et al. Study on Flow Separation Characteristics of the Last Stage in Steam Turbine Under Small Flow Rate Conditions [J]. Turbine Technology, 2019, 61(3): 192-194.

[4] SHI Rong, WANG Xueyan, LU Xin. et al. Study on coordinated control of source-grid-load-storage for clean energy internet based on improved clustering algorithm [J]. Grid and Clean Energy, 2023, 39(7):134-139.

[5] Diurno T, Andreini A, Facchini B, et al. Numerical Investigation of Potential Flow Induced Vibrations of Steam Turbine Last Stage Rotor at Low Load Operation - Part 2: Rotating Instabilities Detection [C]. ASME Turbo Expo, Rotterdam, 2022: GT2022-81921.

[6] Oliver Pütz. Prediction of Rotating Instabilities in Low Pressure Steam Turbines Operating at Low Load [C]. ASME Turbo Expo, Rotterdam, 2022: GT2022-80368.

[7] IANG Haiyang, DU Ershun, JIN Chen, et al. Optimal Planning of Multi-time Scale Energy Storage Capacity of Cross-national Interconnected Power System With High Proportion of Clean Energy [J]. Proceedings of the CSEE, 2021, 41(6): 2101-2114.

[8] LI Shuming, LIU Qingsong, ZHU Xiaodong, et al. Flexibility transformation analysis of 350 MW supercritical cogeneration unit [J]. Power Generation Technology, 2018, 39(5):449-454.

[9] GONG Sheng, SHI Qiguang, MAO Yuchen, et al. Present Situation and Development of Flexible Technology of Thermal Power Units in China [J]. Applied Energy Technology, 2017(5): 1-6.
[10] WANG Shudong, LYU Weizhi. Quantitative valuation of coal-fired units under deep peak regulation conditions [J]. Journal of Chinese Society of Power Engineering, 2020, 40(9):701-706.
[11] ZHANG Xuhui, ZHAO Zhonghua, CUI Fuxing, et al. Experimental research on deep peak regulation characteristics of 1 030 MW ultra-super critical unit [J]. Shandong Electric Power, 2021, 48(2):58-62.
[12] JU Wenping, LYU Kai, MA Tingshan, et al. Comparison of thermo-electric decoupling techniques for heating units [J]. Thermal Power Generation, 2018, 47(9):115-121.
[13] REN Lin, XU Wen, PENG Wenxiang. Heating retrofitting for 1 000 MW ultra super-critical unit [J]. Energy and Energy Conservation, 2021(12):94-99.
[14] XIE Tian, YANG Rongzu, CHENG Dongtao, et al. Numerical simulation of internal flow in connecting pipe of IP and LP cylinder under zero-output heating conditions of LP cylinder [J]. Thermal Power Generation, 2021, 50(11):99-106.

Journal of Physics: Conference Series **2992** (2025) 012045 doi:10.1088/1742-6596/2992/1/012045

Numerical simulation and experimental study on heat treatment of automotive riveted nuts

Le Yu[1,*] and Mingliang Hu[1]

[1]Shanghai Vocational College of Science and Technology, Shanghai, China

[*]Corresponding author's e-mail: 3010575510@qq.com,

Abstract: In actual production and subsequent use feedback, the mechanical properties of automotive rivet nuts, such as strength, hardness, guaranteed load, and head binding force, are important parameters to reflect the product performance. In addition to the influence of raw materials on these mechanical properties, the heat treatment process is also the most important control means. In this paper, the heat treatment process of automotive riveted nuts is simulated by using the numerical simulation software Simulact Forming, the heat treatment process parameters are optimized by orthogonal experimental design method, and finally, a better combination of heat treatment process parameters is obtained. The test results show that the optimized combination of heat treatment process parameters can significantly reduce the residual stress of automotive riveted nuts and improve product quality.

1. Introduction

Automotive rivet nuts as fasteners play a vital role in automotive manufacturing [1]. Its mechanical properties, such as strength, hardness, guaranteed load, and head binding force, directly affect the safety and reliability of the vehicle. Heat treatment is one of the important means to improve the mechanical properties of automotive rivet nuts [2]. However, the traditional heat treatment process often depends on experience and testing, which has the problems of unstable process parameters and inconsistent product quality. Therefore, using numerical simulation technology to simulate the heat treatment process and optimize the heat treatment process parameters has important practical significance and application value.

2. Heat treatment process and numerical simulation method of automobile rivet nut

2.1 Heat treatment process of automobile rivet nut

The heat treatment process of automotive rivet nuts mainly includes heating, insulation, cooling and other steps. In the heating stage, the workpiece is heated to a certain temperature in the heat treatment equipment to change its internal structure and performance. In the insulation stage, the workpiece is kept at the set temperature for a certain time to ensure that its internal organization is fully transformed. In the cooling stage, the workpiece is rapidly cooled to obtain the required microstructure and properties. In actual production, the heat treatment of automotive rivet nuts is to stack the semi-finished workpieces formed by the previous step of cold heading into the heat treatment equipment in large quantities.

Journal of Physics: Conference Series **2992** (2025) 012045 doi:10.1088/1742-6596/2992/1/012045

2.2 Numerical simulation method
Numerical simulation is a method to simulate and analyze the heat treatment process through computer simulation technology. In this paper, the heat treatment process of automotive riveted nuts is simulated by using the numerical simulation software Simulact Forming. The software has powerful pre-processing, solving and post-processing functions, and can simulate the changes in the temperature field, microstructure field and stress field during heat treatment. Through numerical simulation, the temperature distribution, microstructure transformation and stress change of the workpiece during heat treatment can be predicted, which provides a theoretical basis for optimizing the heat treatment process parameters [3].

3. Numerical simulation of heat treatment of automobile rivet nut
The heat treatment process of automotive rivet nuts is after the cold heading process. If you want to control the reference value of the numerical simulation of the heat treatment process, you should pay attention to the fine simulation of the cold heading process before, and the cold heading simulation results that fit the actual situation can lay the foundation for the simulation setting and operation of the next heat treatment process.

3.1 Insert cold heading model
Since the workpiece is an axisymmetric model, in order to reduce the calculation time, the workpiece is imported into the 1/16 3D model. The STP format file saved by CATIA in the form of CAD geometry is imported successively in Simulink Forming, and it is assigned to each component, as shown in Figure 1.

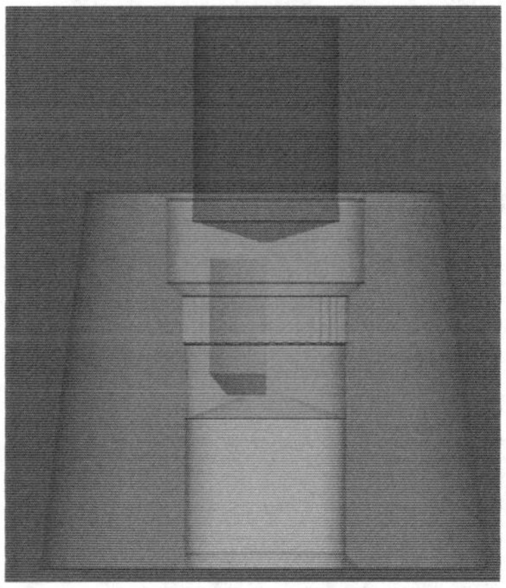

Figure 1. Cold heading forming assembly model after import.

3.2 Generation and import of workpiece materials
The automobile rivet nut material selected in this paper is ML08Al carbon steel. Before creating the material file of ML08Al carbon steel, the chemical composition of the raw material needs to be detected by electron probe microscopy, and then the complete results are recorded. Then, in the general steel module of the JMatPro software, the Simpact Prompt submodule is selected, as shown in Figure 2. The chemical composition, grain size grade and austenitizing temperature of ML08Al carbon steel are input, and through calculation, the JMT format file that can be imported into Simpact Forming software is obtained.

Journal of Physics: Conference Series **2992** (2025) 012045 doi:10.1088/1742-6596/2992/1/012045

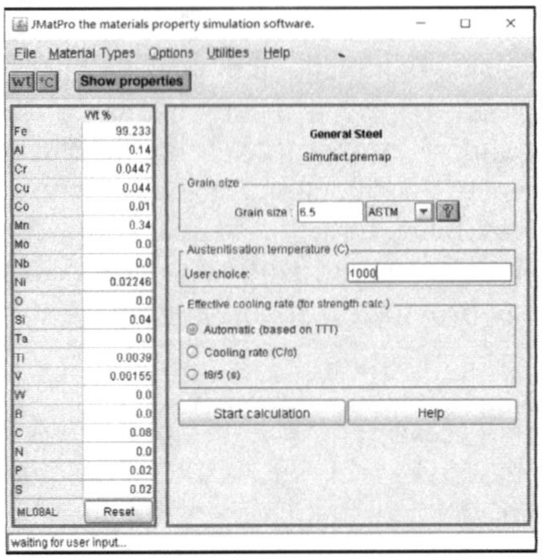

Figure 2. Distribution of workpiece temperature field.

3.3 Grid division and pretreatment

Mesh generation is one of the key steps in the numerical simulation of automotive rivet nuts [4]. In this paper, the hexahedral mesh generation option is adopted, and the simulation task is not interrupted when the finite element mesh is damaged and distorted through the automatic re-division function. At the same time, the upper and lower punch and die are set as rigid bodies, and only the workpiece needs to be meshed. The pretreatment stage also includes setting boundary conditions, loading thermal physical parameters and mechanical properties parameters. After completing the above model establishment and parameter setting, you can submit the running numerical simulation task. Thus, the model shown in Figure 3 is established.

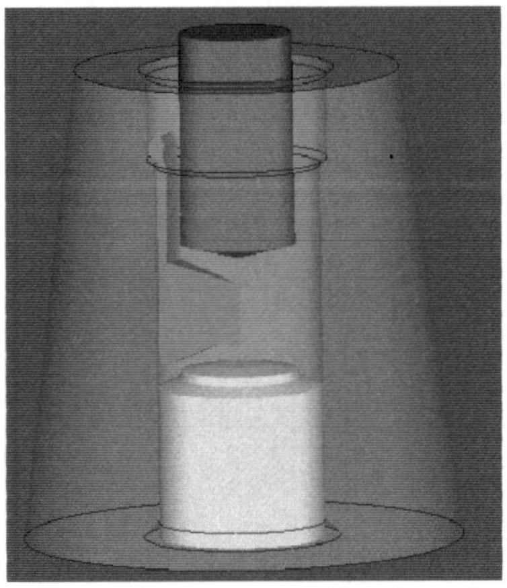

Figure 3. Cold heading model of the fourth process.

3.4 Finite element model establishment and parameter setting of heat treatment

3.4.1 Initial model setting of workpiece heat treatment. After creating a new heat treatment module task, you need to select "process and result step" in "input from result" to generate the corresponding geometry file for the model after the previous cold heading, and you need to select all options in "result conversion". Then, a geometric model with cold upsetting results will be generated.

3.4.2 Addition and setting of heat treatment process parameters. In "heattreat", according to the original process parameters of the actual heat treatment process of the riveted nut of this model of the automobile, the temperature, time and other parameters in the heating, insulation and cooling stages are set in turn. The heating temperature is 920°C, the heating time is 15 min, the insulation temperature is 920°C, the insulation time is 10 min, the cooling method is air cooling to room temperature 25°C, and the cooling time is 60 min.

3.4.3 Meshing setting of heat treatment workpiece. Although the mesh has been generated in the cold heading simulation before, in order to avoid the error stop in the middle of the operation, the heat treatment module needs to be regenerated. The mesh generator is set as a hexahedron, the cell size is 0.17 mm, the number of cells is 32949, the internal roughness level is level 2, and the refinement box needs to be loaded for local refinement.

3.4.4 Setting of simulation parameters. The ML08Al material set before is directly used as the heat treatment material. After the thermal parameters of the workpiece are defined, the initial content of ferrite and pearlite of the material is manually set. Finally, the parameters of the heat treatment control window are set to check the model and submit it for operation.

3.5 Numerical simulation results and analysis

Through numerical simulation, the temperature field, microstructure field and stress field of automobile riveted nut during heat treatment were obtained.

The temperature distribution at each stage of heat treatment is shown in Figure 4. Due to the shape and structure characteristics of the upper part of the workpiece, the temperature rises and falls faster. The temperatures at all places in the insulation stage can reach the set 920°C, and at the end of cooling, the temperatures at all places can reach the set 25°C.

Figure 5 shows the change of phase structure in the heating stage. Pearlite and ferrite will gradually decrease when reaching the critical temperature, while austenite will gradually increase; In the cooling stage, after reaching the critical temperature, austenite will gradually transform into ferrite and pearlite, that is to say, austenite is an intermediate product. The microstructure field simulation results show that the microstructure of the workpiece has changed significantly during the heat treatment process [5].

Figure 6 shows the equivalent stress changes of the workpiece before and after heat treatment. It can be seen that the equivalent stress at most positions of the whole has been reduced, the stress distribution of the workpiece after heat treatment is more uniform, and the stress concentration has been significantly improved [6].

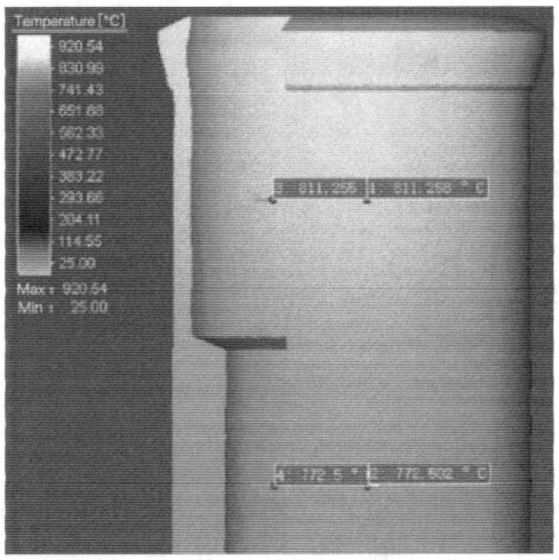

(a) Heating stage

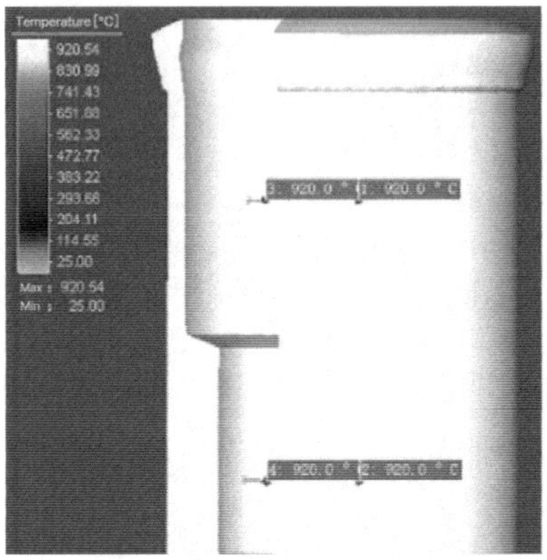

(b) Insulation stage

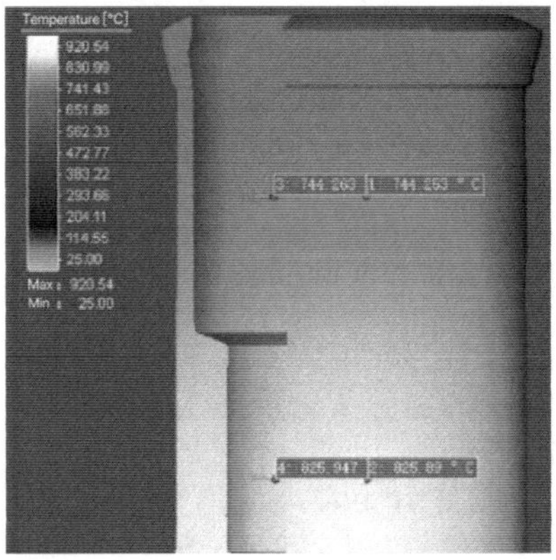

(c) Cooling phase

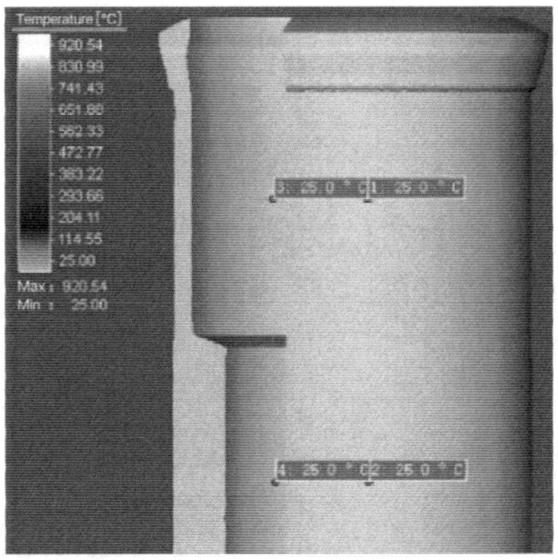

(d) Cooling completed

Figure 4. Distribution of workpiece temperature field.

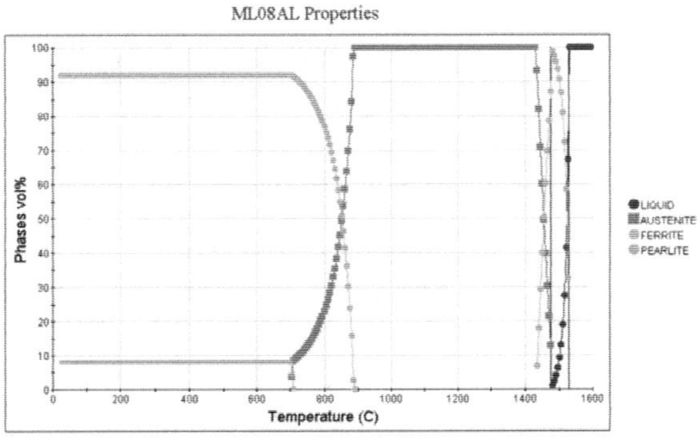

Figure 5. Phase structure change diagram.

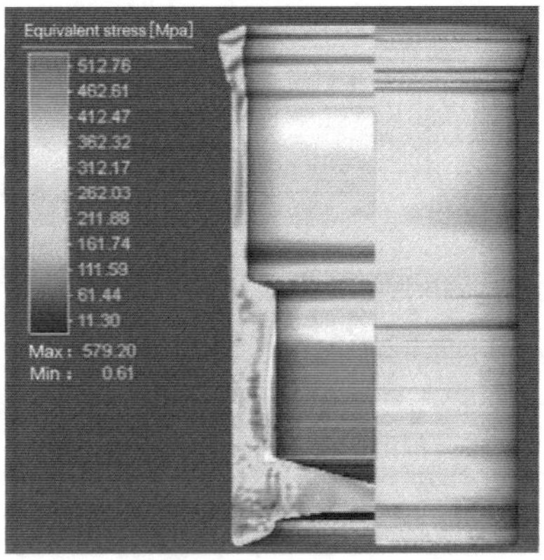

(a) Before heat treatment

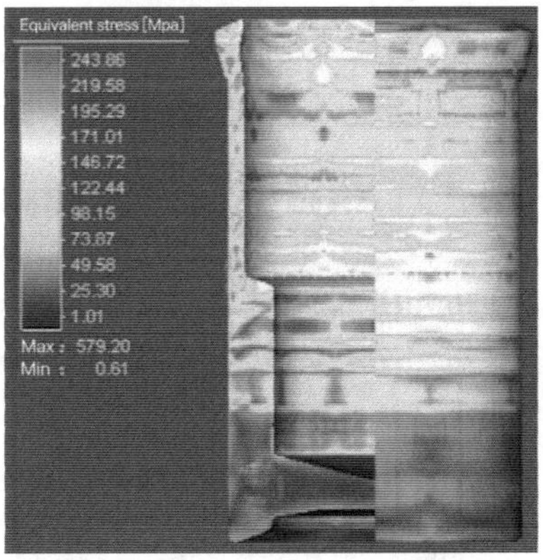

(b) After heat treatment

Figure 6. Equivalent stress change.

4. Optimization of heat treatment process parameters for 4 automobile rivet nuts

4.1 Setting of orthogonal test factors and levels

According to the numerical simulation results and analysis of the single factor variable method of heat treatment process parameters carried out previously, we will determine the three factors of this orthogonal test and their corresponding reasonable factor levels, as shown in Table 1 [7].

Table 1. Factors and levels of orthogonal test.

Level \ Factor	Test Factors		
	Annealing temperature A(°C)	***Holding time B(s)***	***Cooling method C***
1	890	190	Air cooling
2	860	160	Cooling to room temperature after cooling to room temperature
3	830	130	Cooling to room temperature with furnace

4.2 Orthogonal experimental design

In order to optimize the heat treatment process parameters of automotive riveted nuts, the orthogonal experimental design method is used in this paper [8]. Annealing temperature, holding time and cooling mode are selected, and each factor is set with three levels, as shown in Table 2. Through the L9 (34) orthogonal test table, nine groups of tests were carried out to explore the influence degree of each factor and the optimization scheme. The axial stress at a characteristic point is selected as the test index to evaluate the effect of heat treatment process parameters on residual stress.

Table 2. Orthogonal test table of three factors and three levels.

Test number \ Factor	A	B	C
	Annealing temperature	***Holding time***	***Cooling method***
1	1	1	1
2	1	2	2
3	1	3	3
4	2	1	2
5	2	2	3
6	2	3	1
7	3	1	3
8	3	2	1
9	3	3	2

4.3 Orthogonal test results and analysis

According to the results of the orthogonal test, the range and variance of each factor were calculated, and the influence degree of each factor on residual stress was obtained. By observing the range and ranking results of various factors, it can be seen that the influence of cooling mode on residual stress is the largest, followed by annealing temperature and holding time. The trend chart of residual stress results corresponding to each factor level in the orthogonal test was drawn. According to the results of the orthogonal test, the optimal combination of process parameters was determined as A1B1C3, that is, the annealing temperature was 890°C, the holding time was 190 s, and the cooling mode was cooling to room temperature with the furnace.

4.4 Verification of optimal process parameter combination

In order to verify the effectiveness of the optimal combination of process parameters, the numerical simulation was carried out by using the Simulink Forming software, and the results were compared with the original heat treatment scheme. Figure 7 shows the residual stress distribution under the optimal combination of process parameters. The corresponding test index residual stress is 46.4595 Mpa, which is less than any one of the nine orthogonal tests.

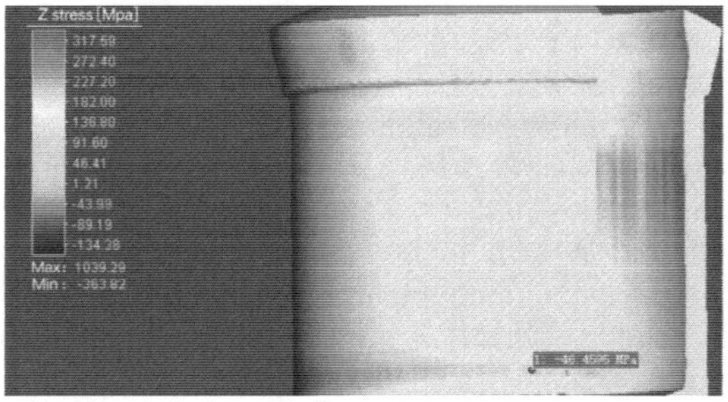

Figure 7. Results of optimization scheme.

In addition, in order to explore the changes in the best process scheme compared with the original heat treatment process scheme, the coordinates of the same feature point can be inserted by using the point tracking function. The residual stress results corresponding to the original heat treatment process scheme can be obtained, as shown in Figure 8, with a value of 95.8099 Mpa [9].

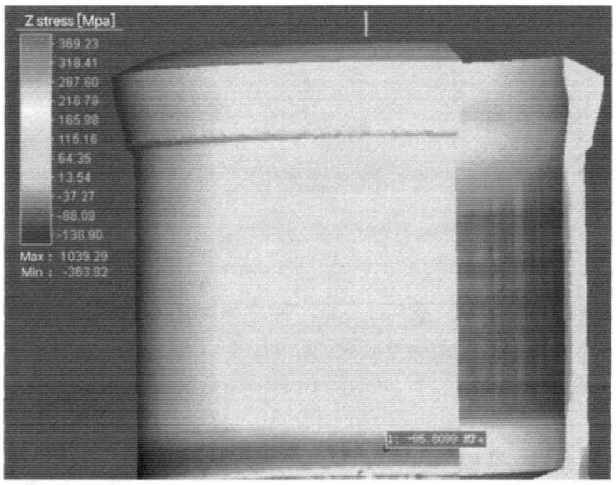

Figure 8. Results of original heat treatment scheme.

It can be seen that the residual stress after optimization is significantly reduced. The residual stress before heat treatment was 268.962 Mpa. Compared with the original heat treatment scheme, the residual stress after optimization is reduced by 51.5%, and the elimination range of the overall residual stress is increased from 64.4% to 82.7%. The rationality of the conclusion of the orthogonal test is verified, and it shows that the optimization method of heat treatment process parameters of automobile rivet nuts designed by the orthogonal test has a certain reliability.

5. Experimental verification of new heat treatment process for automotive rivet nuts

5.1 Test equipment and methods

In order to further verify the accuracy of the numerical simulation results, the heat treatment test of automotive riveted nuts was carried out in this paper. The test equipment includes a heat treatment furnace, X-ray diffractometer, metallographic cutting machine, grinding machine, metallographic inlay machine, grinding and polishing machine, constant temperature drying oven and metallographic microscope. The test method includes heat treatment test, residual stress detection and metallographic structure observation.

5.2 Residual stress test results and analysis

The residual stress of automobile riveted nuts after heat treatment was detected by X-ray diffractometer [10]. Figure 9 shows the test results of residual stress values at different positions after the new heat treatment process. It can be seen that the residual stress values at different positions after the new heat treatment process are lower than those before the heat treatment and after the original heat treatment process, and the stress concentration area decreases significantly. This confirms that the residual stress optimization effect of the new heat treatment process is better. At the same time, since the error between the simulation value and the test value does not affect the exploration of the law between the heat treatment process parameters and the results, it can be considered that the optimization results of the process parameters obtained by the numerical simulation of heat treatment with Simulink Forming software are reliable.

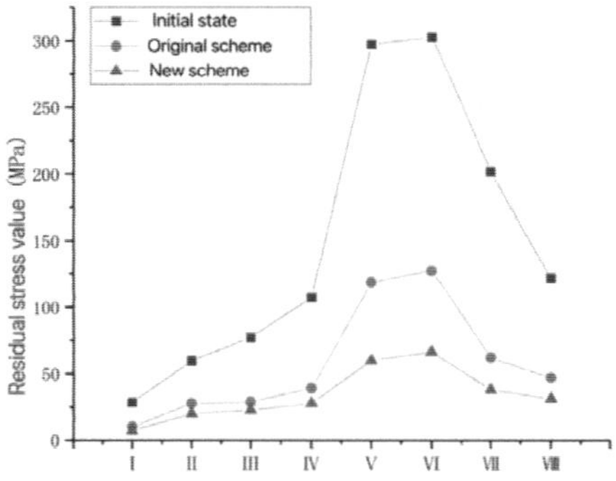

Figure 9. Test results at different section positions.

5.3 Metallographic structure observation results and analysis

The metallographic structure of automobile riveted nuts after heat treatment was observed by a metallographic microscope. Figure 10 shows the metallographic structure comparison before the annealing heat treatment, after the original heat treatment scheme and after the new heat treatment scheme. It can be seen that the microstructure of the new heat treatment scheme is more uniform and fine, the grain size is reduced, and the microstructure and properties are improved. This further verifies that the new heat treatment process is conducive to further refining the grain size and improving the plasticity and toughness of the product.

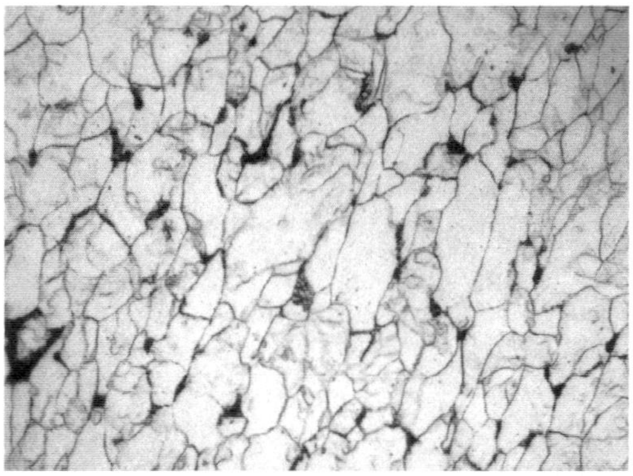

(a) Before annealing heat treatment

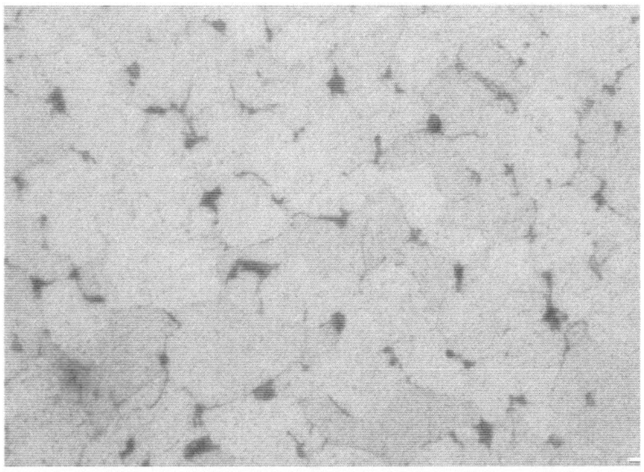

(b) After the original heat treatment scheme

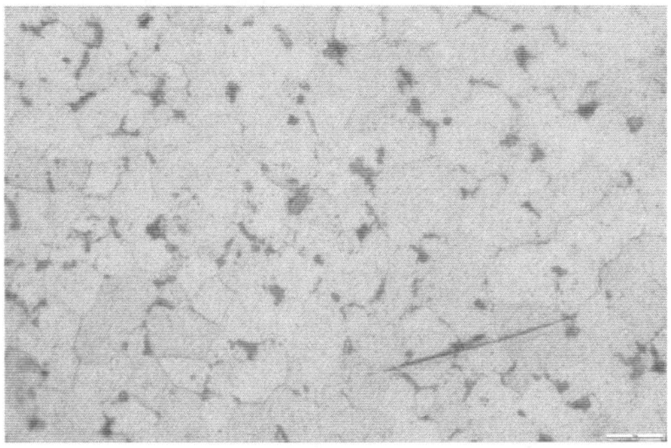

(c) After the new heat treatment scheme

Figure 10. Metallographic structure comparison diagram.

6. Conclusion

(1) The heat treatment process of automotive riveted nuts was simulated by using the numerical simulation software Simulink Forming, and the temperature field, microstructure field and stress field were obtained. Through numerical simulation, the temperature distribution, microstructure transformation and stress change of the workpiece during heat treatment can be predicted, which provides a theoretical basis for optimizing the heat treatment process parameters.

(2) The orthogonal experimental design method was used to optimize the heat treatment process parameters of automotive riveted nuts. Through range analysis and variance analysis, the optimal combination of process parameters was determined as an annealing temperature of 890°C, holding time of 190 s, and cooling mode of cooling to room temperature. After optimization, the residual stress is significantly reduced and the product quality is improved.

(3) The accuracy of numerical simulation results was verified by the heat treatment test. The test results show that the residual stress value after the new heat treatment process is lower than that before the heat treatment and after the original heat treatment process, and the stress concentration area decreases significantly. At the same time, the microstructure observation results also show that the microstructure after the new heat treatment scheme is more uniform and fine, and the microstructure and properties are improved.

Acknowledgment

This research comes from the 2024 school-level teaching and scientific research youth project of Shanghai Vocational College of Science and Technology, "Numerical Simulation and Experimental Research on Heat Treatment of Automotive Riveted Nuts", the project number is 2420216.

References

[1] Lu Xiuyu, Xia Yanhua, Zhou Yong, et al. (2015) Current situation and development of automotive fasteners in China [J]. Heat treatment technology and equipment, (6): 76-80.
[2] Zhao, H, Liu, Y. (1998) Metalology and Principles of Heat Treatment. Harbin Institute of Technology Press, Harbin.
[3] Liu, J, Wang, C, Sun, D. (2021) Simufact Metal Forming Process Simulation Standard Tutorial. Posts and Telecommunications Press, Beijing.
[4] Liu Zheng, Zhong Shaohua, GUI hang, et al. (2020) Numerical simulation and optimization of cold forging of automotive rivet nuts [J]. Forging technology, 45(10): 187-194.
[5] Smith, A., Johnson, B., & Roberts, C. (2022) Modelling the Austenite-to-Ferrite Transformation in Steel during Continuous Cooling using a Cellular Automaton Approach [J]. Journal of Materials Science, 57(1): 601-614.
[6] Khan, M. A., et al. (2023) Simulation of Stress and Microstructure Evolution during Heat Treatment of a Nickel-Based Superalloy [J]. Journal of Materials Engineering and Performance, 32(1): 72-84.
[7] S. Karthik, P. Ramanujan, et al. (2021) Optimization of Heat Treatment Process Parameters on AISI 4140 Steel using Taguchi Method and Response Surface Methodology [J]. Materials Today: Proceedings. pp. 235-242.
[8] Li Jian, Feng Zaixin, Zuo Sen, et al. (2021) The heat treatment process of Cu Ni Sn alloy was optimized by orthogonal test [J] Metal heat treatment, 46(3): 61-66.
[9] Chen Lei, Zhang Chun, Wang Leping, et al. (2017) Numerical simulation and sensitivity analysis of hot stamping die cooling system based on simufact [J]. Journal of Changjiang University, 14 (09): 28-32.
[10] Yu Kang, Sun Yafei, Chen Xiaojiang. (2015) X-ray diffraction residual stress test method and application [J]. Rocket propulsion, 41(02): 102-107.

Journal of Physics: Conference Series **2992** (2025) 012046 doi:10.1088/1742-6596/2992/1/012046

Research on the calculation method of contact stress between variable-axis diameter closed-type linear bearing and guideway

Weiming Li, Yuan Xue *, Yuhong Liu and Yukuan Xie

Harbin Institute of Technology, School of Mechanical and Electrical Engineering, Harbin, China

*Corresponding author's e-mail: xueyuan106@hit.edu.cn

Abstract: For the closed-type linear bearing and guideway at the same time by the bending moment and radial force under the action of the contact stress problem, this paper analyses the variable-axis diameter closed-type linear bearings and guideway contact region of the geometric law, making the contact stress and friction is not coupled, the guideway axis for the rigid body, the bearing for the flexible body of the assumptions, based on Winkler elastic basis model, deduced a variable-axis diameter of the closed-type linear bearings and the guideway of the contact stress calculation method. The analytical solution and finite element solution of contact stress are obtained through specific examples. The two are compared. The results show that the analytical solution and the finite element solution are basically in agreement. Therefore, the proposed contact stress calculation method is correct and reasonable, which provides a theoretical basis for the strength analysis and optimal design of the variable-axis diameter of the closed-type linear bearings.

1. Introduction

Linear bearings are widely used in all kinds of mechanical systems because of their high precision, high smoothness, small coefficient of friction low cost, etc [1-3]. They are mainly used in mobile and expandable mechanisms, for example, the mobile joints of the radar array expanding mechanism use linear bearings [4]; the sliding table device of the magnetic levitation vehicle also uses linear bearings [5]. The contact stress between a linear bearing and its guideway is usually nonlinear, and the solution calculation is more complicated [6]. Many scholars have done some research on such problems. Liu et al. [7] applied the Winkler elastic basis model to analyze the hole-shaft contact problem with a gap. They compared the results of the Hertz contact theory with the results of the finite element method. The comparison results show that the Winkler elastic basis model is the closest to the finite element method results, and the error of Hertz contact theory increases rapidly with the increase of contact angle; Antoni [8] applied the Winkler elastic basis model to analyse the mechanical model of the coordinated contact between the trunnion and the pin, and the stress distribution curves obtained from the Winkler elastic basis model matched well with the curves obtained from the finite element method. Ji et al. [9] took the ball linear bearing subjected to transverse load as the research object, combined with the Hertz contact theory and the deformation coordination relationship, and established the mechanical model of linear bearing subjected to transverse load. Zou et al. [10] developed a nonlinear stiffness model for linear

Journal of Physics: Conference Series **2992** (2025) 012046 doi:10.1088/1742-6596/2992/1/012046

bearings using Hertz contact theory, but only the simple loading case of pendant loading was considered. Gao et al. [11] established a theoretical model of contact stresses in the shaft-hole connection of hydraulic bracket based on Hertz contact theory, and the results of the contact stress distribution have strong nonlinearity, but also only consider the case of the pin is subjected to radial load.

Based on the above analysis, for this type of problem, although different methods are used to establish the mechanical model, only the radial force is considered to calculate and analyze its stress. In fact, such bearings are subject to bending moment in addition to radial load. Therefore, this paper proposes a method for calculating the contact stresses between a variable-axis diameter closed-type linear bearing and the guideway under the action of bending moment and radial load at the same time.

2. Variable-axis diameter closed-type linear bearing and guideway contact area geometry law analysis

The linear bearing mounted load will occur after the contact deformation, as shown in Figure 1. If the guide rail is regarded as a rigid body, the linear bearing is regarded as an elastomer. The contact process can be assumed to be divided into two steps:

(1) The linear bearing rotates around the point of rotation O at an angle of rotation of θ.

(2) The linear bearing is translating along the vertical axis, and the translating displacement is ε.

The angle of rotation θ is generated by the bending moment M acting on the center of rotation O. The translational displacement ε is generated by the load G along the component force F perpendicular to the axis. That is, the final state of deformation of the linear bearing in contact with the guideway can be described by only two independent parameters θ and ε.

In addition, the presence or absence of clearance between the bearing and the guideway results in two different contact geometries [12]. When there is a clearance $\Delta R = 0$ between the linear bearing and the guideway, the contact area at both ends of the linear bearing is continuous and the circumferential contact range is half a circle. When there is a clearance $\Delta R > 0$ between the linear bearing and the guideway, the contact area at both ends of the linear bearing is discontinuous and the circumferential contact range varies with height. This paper only discusses the situation when $\Delta R = 0$.

On the basis of the above analyses, the following assumptions are made for the contact between the linear bearing and the guideway in order to facilitate the study:

(1) The guideway is a rigid body and the linear bearing is an elastomer. Both are isotropic materials.

(2) The part of the linear bearing in coordinated contact with the guideway is treated as a Winkler elastic basis model.

(3) Friction and contact stresses are not coupled.

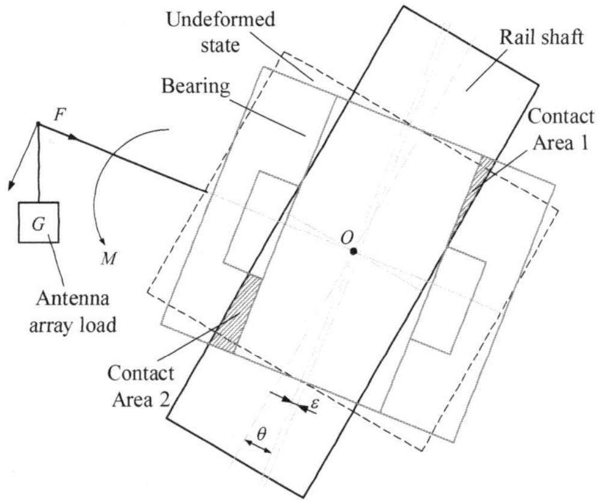

Figure 1. Closed-type bearing and guideway contact state.

3. Derivation of an analytical solution for the contact stress distribution law between a variable-axis diameter closed-type linear bearing and guideway

When the guide rail is placed at an angle β, the middle section of the closed bearing is loaded with a heavy load, and the symbols of the physical quantities are shown in Figure 2. Load gravity G is equivalent to the geometric center of the closed bearing. O can be decomposed into bending moment M , axial force F_t along the axis direction and perpendicular to the axial direction of the radial force F, which in the bending moment M and radial force F under the action of the closed bearing and the guideway axis diameter coordinated contact. It occurs at the upper and lower ends of the bearing to produce two independent contact areas, the friction on the contact area and the balance of axial force F_t. The physical quantity of the upper contact area Ω_1 is labeled 1. The physical quantity of the lower contact area Ω_2 is labelled 2.

Contact region Ω_1 in the height of h at the cross-section of the contact state shown in Figure 3, according to the figure shown in the established cross-section height positive direction $h_{(h_1)}$ and $h_{(h_2)}$, taking into account the $\tan\theta \approx \theta$, can be obtained by closed bearing two contact regions Ω_1 and Ω_2 at the cross-section of the maximum deformation.

$$\delta_{h1\max} = (H_1 - h)\tan\theta \approx (H_1 - h)\theta \tag{1}$$

$$\delta_{h2\max} = (H_2 - h)\tan\theta \approx (H_2 - h)\theta \tag{2}$$

where H_1 and H_2 are the contact start heights of the two contact areas Ω_1 and Ω_2 respectively. It is relative to the origin of the respective contact area coordinate system O_{h_1} and O_{h_2}, considering the rotation angle θ as an infinitesimal quantity, which is expressed as:

$$H_1 = \frac{1}{2}R_2\tan\theta + \frac{\varepsilon}{\tan\theta} \approx \frac{\varepsilon}{\theta} \tag{3}$$

$$H_2 = \frac{1}{2}R_2\tan\theta - \frac{\varepsilon}{\tan\theta} \approx -\frac{\varepsilon}{\theta} \tag{4}$$

where R_2 is the inner diameter of the bearing, H is the bearing length, h_1 is a part of the height of the second inner diameter R_3 of the bearing, and h_2 is another part of the height. The upper contact region Ω_1 is distributed in the range along the bearing height direction $-\frac{H}{2} \le h \le -h_1$. The lower contact region Ω_2 is distributed in the range along the bearing height direction $-\frac{H}{2} \le h \le -h_2$.

According to the Winkler elastic basis model [13], the maximum values of contact stresses at the height h of the contact region on both sides of the intermediate section can be obtained as:

$$\sigma_{n1\max} = E\frac{\delta_{h1\max}}{R_1 - R_2} = \frac{E}{R_1 - R_2}(H_1 - h)\theta \tag{5}$$

$$\sigma_{n2\max} = E\frac{\delta_{h2\max}}{R_1 - R_2} = \frac{E}{R_1 - R_2}(H_2 - h)\theta \tag{6}$$

where R_1 is the outer diameter of the bearing and E is the modulus of elasticity.

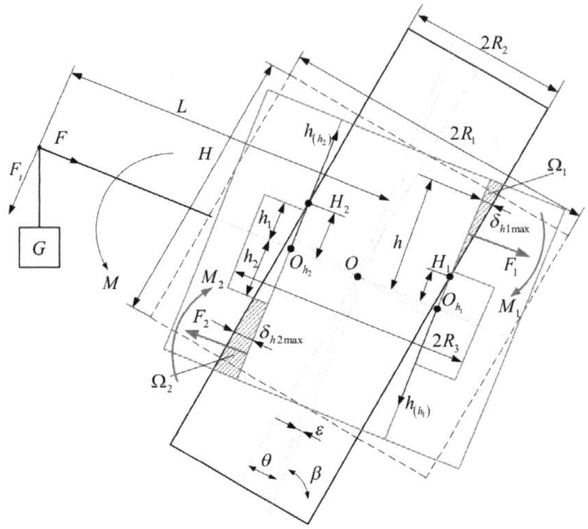

Figure 2. Closed-type bearing force state and symbols.

In ΔOO_1K, the cosine theorem gives the contact half angle φ as:

$$\cos\varphi = \frac{(H_1 - h)\tan\theta}{2R_2} \tag{7}$$

The bearing rotation angle θ is very small and can be regarded as an infinitesimal quantity. $\cos\varphi \approx 0$, which gives $\varphi = 90^{\circ}$. The contact region is approximated as half a circular arc region.

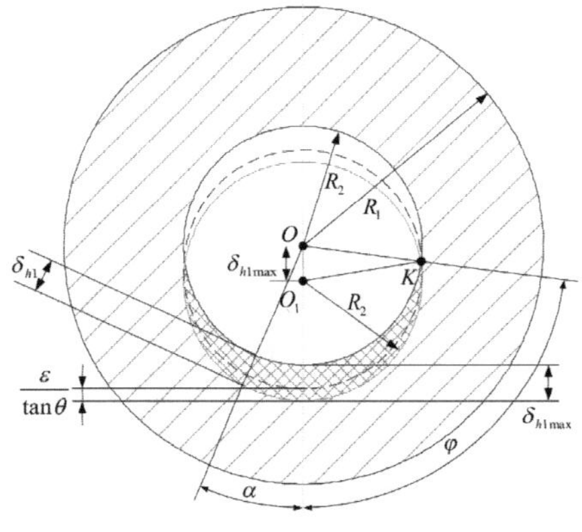

Figure 3. Contact state of the cross section at height h in the contact region Ω_1.

The combined forces on the cross-section at a height of h in the contact region on both sides are:

$$P_{h1} = \int_{-\varphi}^{\varphi} \sigma_{n1\max} R_2 \cos\alpha d\alpha = \frac{E\pi R_2}{2(R_1 - R_2)}\theta(H_1 - h) \tag{8}$$

$$P_{h2} = \int_{-\varphi}^{\varphi} \sigma_{n2\max} R_2 \cos\alpha d\alpha = \frac{E\pi R_2}{2(R_1 - R_2)}\theta(H_2 - h) \tag{9}$$

Let $B=\frac{E\pi R_2}{2(R_1-R_2)}$, we integrate the above equation along the linear bearing height direction, respectively, to obtain the magnitude of the combined force in the two contact regions as:

$$F_1=\int_{-\frac{H}{2}}^{-h_1} P_{h1}dh=\frac{B\theta(H-2h_1)(H+4H_1+2h_1)}{8} \tag{10}$$

$$F_2=\int_{-\frac{H}{2}}^{-h_2} P_{h2}dh=\frac{B\theta(H-2h_2)(H+4H_2+2h_2)}{8} \tag{11}$$

Then the magnitude of the combined contact moment on the two contact areas towards the centre of rotation O can be obtained as:

$$\begin{aligned}M_1&=\int_{-\frac{H}{2}}^{-h_1}\left|P_{h1}h\right|dh=-\int_{-\frac{H}{2}}^{-h_1}B\theta(H_1-h)hdh\\&=\frac{B\theta(H-2h_1)(H^2+2Hh_1+3H_1H+4h_1^2+6H_1h_1)}{24}\end{aligned} \tag{12}$$

$$\begin{aligned}M_2&=\int_{-\frac{H}{2}}^{-h_2}\left|P_{h2}h\right|dh=-\int_{-\frac{H}{2}}^{-h_2}B\theta(H_2-h)hdh\\&=\frac{B\theta(H-2h_2)(H^2+2Hh_2+3H_2H+4h_2^2+6H_2h_2)}{24}\end{aligned} \tag{13}$$

Then we select the bearing as the object of study column equilibrium equations, bearings subjected to various forces and moments in the direction shown in Figure 1, to obtain the equilibrium equations are as follows:

$$\begin{cases}F_1-F_2+F=0\\-M_1-M_2+M=0\end{cases} \tag{14}$$

Substituting each of the above force and moment expressions into the equilibrium equations gives the expressions for H_1 and H_2:

$$H_1=\frac{-2B\theta H^3+3B\theta Hh_1^2-3B\theta Hh_2^2-6FH+8B\theta h_1^3+6B\theta h_1^2h_2+2B\theta h_2^3-12Fh_2+24M}{6B\theta(H-2h_1)(H+h_1+h_2)} \tag{15}$$

$$H_2=\frac{-2B\theta H^3-3B\theta Hh_1^2+3B\theta Hh_2^2+6FH+2B\theta h_1^3+6B\theta h_1h_2^2+12Fh_1+8B\theta h_2^3+24M}{6B\theta(H-2h_2)(H+h_1+h_2)} \tag{16}$$

Based on the relationship satisfied by H_1 and H_2: $H_1+H_2=0$, $H_2-H_1=2\frac{\varepsilon}{\theta}$, the angle of rotation θ and the translational displacement ε of the bearing can be found as follows.

$$\theta=\frac{F(6h_2^2-6h_1^2)}{\lambda_1}+\frac{M(12H-12h_1-12h_2)}{\lambda_1} \tag{17}$$

$$\varepsilon=\frac{-\lambda_2\theta+F(6H^2-12h_1^2-12h_2^2)+M(24h_2-24h_1)}{\lambda_3} \tag{18}$$

where

$$\begin{cases}\lambda_1=BH^4-BH^3h_1-BH^3h_2-4BHh_1^3-4BHh_2^3+Bh_1^4+4Bh_1^3h_2+6Bh_1^2h_2^2+4Bh_1h_2^3+Bh_2^4\\\lambda_2=-2BH^3h_1+2BH^3h_2+3BH^2h_1^2-3BH^2h_2^2+2Bh_1^4-8Bh_1^3h_2+8Bh_1h_2^3-2Bh_2^4\\\lambda_3=3B(H-2h_1)(H-2h_2)(H+h_1+h_2)\end{cases} \tag{19}$$

IOP Publishing
doi:10.1088/1742-6596/2992/1/012046

For ease of representation, we let:

$$\begin{cases} k_{\theta F} = \dfrac{\left(6h_2^{\,2} - 6h_1^{\,2}\right)}{\lambda_1} \\ k_{\theta M} = \dfrac{\left(12H - 12h_1 - 12h_2\right)}{\lambda_1} \\ k_{\varepsilon F} = \dfrac{\left(6H^2 - 12h_1^{\,2} - 12h_2^{\,2}\right)}{\lambda_3} \\ k_{\varepsilon M} = \dfrac{\left(24h_2 - 24h_1\right)}{\lambda_3} \\ k_{\varepsilon\theta} = -\dfrac{\lambda_2}{\lambda_3} \end{cases} \tag{20}$$

Based on the above equation, the closed-type bearing cylindrical vice flexibility matrix can be obtained as expressed below:

$$\begin{bmatrix} \theta \\ \varepsilon \end{bmatrix} = \begin{bmatrix} k_{\theta F} & k_{\theta M} \\ k_{\varepsilon F} + k_{\varepsilon\theta} k_{\theta F} & k_{\varepsilon M} + k_{\varepsilon\theta} k_{\theta M} \end{bmatrix} \begin{bmatrix} F \\ M \end{bmatrix} \tag{21}$$

The contact stresses at the contact angle α in the cross-section at height h in the contact region on both sides are:

$$\sigma_{n1} = \sigma_{n1\max} \cos\alpha = \frac{E}{R_1 - R_2}(H_1 - h)\theta \cos\alpha \tag{22}$$

$$\sigma_{n2} = \sigma_{n2\max} \cos\alpha = \frac{E}{R_1 - R_2}(H_2 - h)\theta \cos\alpha \tag{23}$$

4. Comparison of Analytical and Finite Element Solutions For Variable-Axis Diameter Closed-Type Linear Bearings In Contact With Guideway

The correctness and reasonableness of the analytical solution of the contact stress in this paper are verified by an example. The known data of calculation example 1 are shown in Table 1. L is the distance between the load centre of mass and the axis. The meanings of the rest of the symbols have been described in the previous section.

Table 1. Known data for calculation example 1.

R_1/mm	R_2/mm	H/mm	h_1/mm	h_2/mm	E/Mpa	G/N	L/mm	β/deg
40	20	70	15	20	2e5	2000	200	70

4.1 Analytical solution for contact stress

Substituting the data into Equations (10)-(23) gives the angle of rotation $\theta = 2.802420 \times 10^{-3}\,\text{deg}$, the translational displacement $\varepsilon = 3.689783 \times 10^{-4}\,\text{mm}$, the combined forces in the contact region on both sides as $F_1 = 6523.8\text{N}$, $F_2 = 7207.9\text{N}$, and the combined moments as $M_1 = 173339.4\text{Nmm}$ and $M_2 = 202537.7\text{Nmm}$.

The expressions for the contact stresses at the contact angle α at the height h section of the contact area on both sides in their respective coordinate systems are given by:

$$\sigma_{n1} = -0.489115(h + 3.771901)\cos\alpha \tag{24}$$

where $-35\text{mm} \le h \le -15\text{mm}$ and $-90^\circ \le \alpha \le 90^\circ$.

$$\sigma_{n2} = -0.489115(h - 3.771901)\cos\alpha \tag{25}$$

where $-35\text{mm} \le h \le -20\text{mm}$ and $-90^\circ \le \alpha \le 90^\circ$.

4.2 Finite element solution of contact stress

Ansys workbench finite element analysis software is selected to solve the contact problem. Based on the previous assumption, friction and contact stress are not coupled, so the contact type of the bearing and the guideway shaft is set as frictionless. The guideway shaft is set as a rigid body and the bearing is set as a flexible body. The finite element model and solution results are shown in Figure 4.

4.3 Comparison of analytical and finite element solutions

By manipulating the internal tools of Ansys workbench, the data related to the finite element solution can be extracted and processed computationally. The rotation angle, translational displacement, contact combined force, and combined moment can be obtained and compared with the results of the algorithm in this paper, as shown in Table 2. The contact stress values at sections $h = -35\text{mm}$, $h = -27.5\text{mm}$ and $h = -21.5\text{mm}$ in the contact region on both sides of the finite element model were extracted respectively and compared with the analytical solution of the contact stresses. The comparison curves are shown in Figure 5.

As can be seen from the comparison of the results of the two solution methods, the results of the rotation angle, translational displacement, contact combined force and combined moment obtained by the two methods are basically the same. The resultant curves of the contact stress on different cross-sections in the contact region are also in good agreement. Especially the results of the two algorithms are basically the same at $h = -27.5\text{mm}$. From the trend of change, the results of the analytical solution are larger than those of the finite element when $h < -27.5\text{mm}$; and the results of the finite element are larger than those of the analytical solution when $h > -27.5\text{mm}$. Overall, the correctness and reasonableness of the contact stress calculation method proposed in this paper are verified by comparing the results with those of finite element analysis.

5. Conclusions

This paper calculates and analyses the contact stress between the variable-axis diameter closed-type linear bearing and guideway under the joint action of bending moment and radial force. It draws the following conclusions:

(1) The geometric law of the contact area between the variable axis diameter closed linear bearing and the guideway is analysed. It is concluded that the final contact deformation state of the linear bearing and the guideway under the action of bending moment and radial force at the same time can be described by two independent parameters θ and ε only.

(2) Based on Winkler elastic basis model, the contact mechanics model of variable-axis diameter closed-type linear bearing and guideway is established. The contact stress calculation formula is deduced.

(3) Using specific examples, the comparison of analytical solution and finite element solution shows that the maximum relative error of deformation, combined force, and combined moment is less than 5%. The contact stress curves at $h = -35\text{mm}$, $h = -27.5\text{mm}$ and $h = -21.5\text{mm}$ sections are basically coincident, which indicates that the proposed calculation method is correct and effective. It provides a theoretical basis for the calculation and analysis of the contact stress of the cylindrical vice of the same kind with variable axial diameters, as well as for the optimization of the calculation and analysis.

(a) Finite element models

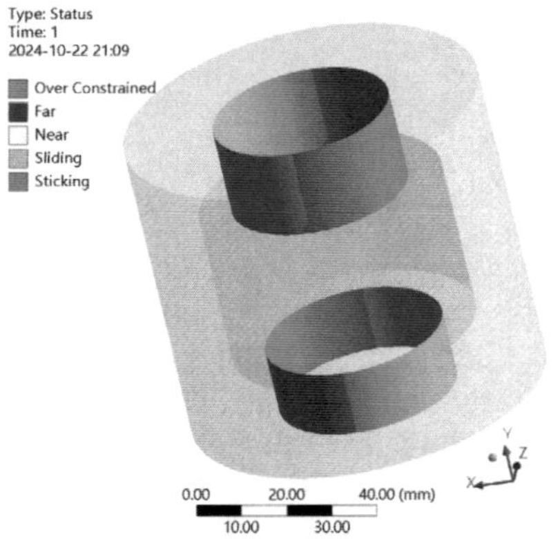

(b) Cloud diagram of contact state

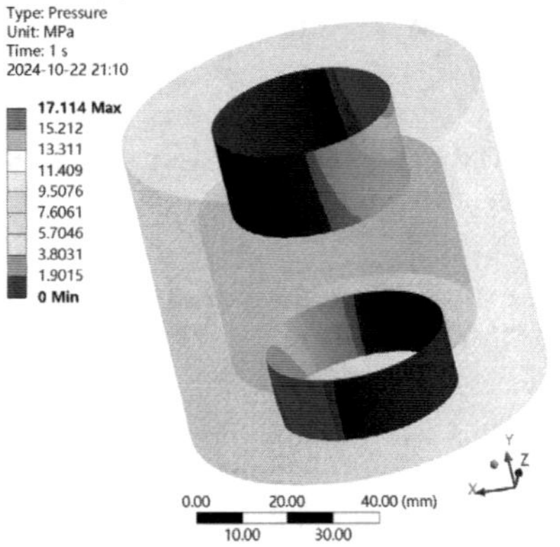

(c) Cloud diagram of contact stress

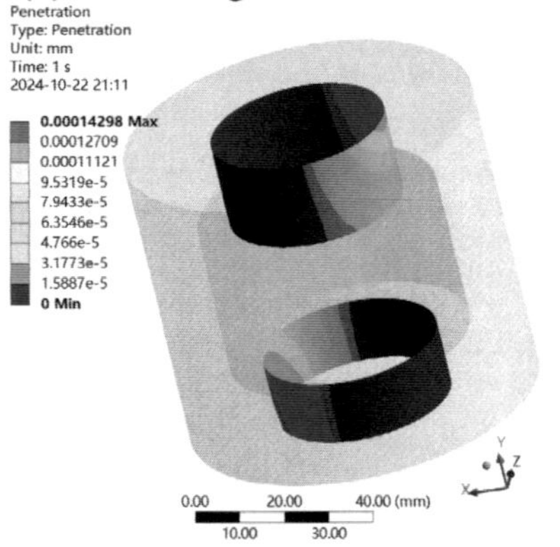

(d) Cloud diagram of penetration

Figure 4. Finite element solution results of contact stress.

Table 2. Comparison between finite element solution and analytical solution.

	θ / deg	ε / mm	F_1 / N	F_2 / N	M_1 / Nmm	M_2 / Nmm
Finite element solution	2.762802E-3	1.768993E-4	6584.2	7268.2	173580.0	202300.0
Analytical solution	2.802420E-3	1.844892E-4	6523.8	7207.9	173339.4	202537.7
Relative error	1.414%	4.114%	0.926%	0.837%	0.139%	0.117%

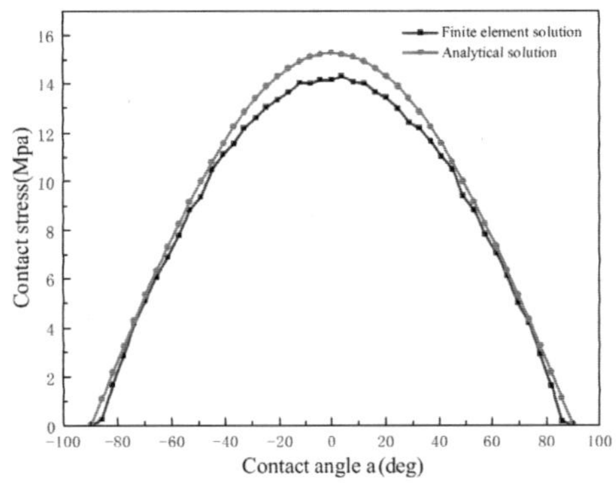

(a) Contact region Ω_1 at height $h=-35\text{mm}$

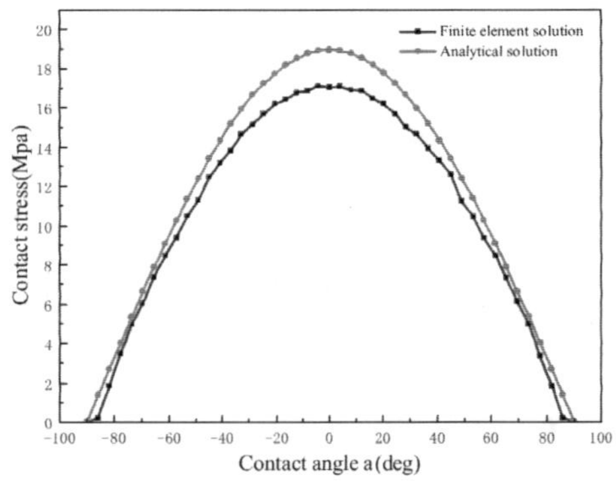

(b) Contact region Ω_2 at height $h=-35\text{mm}$

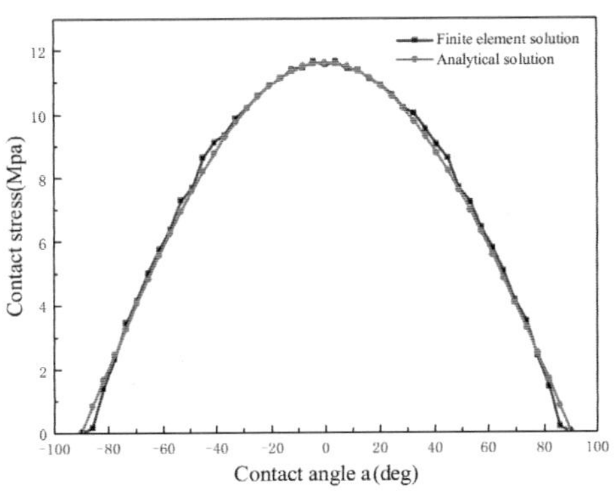

(c) Contact region Ω_1 at height $h=-27.5\text{mm}$

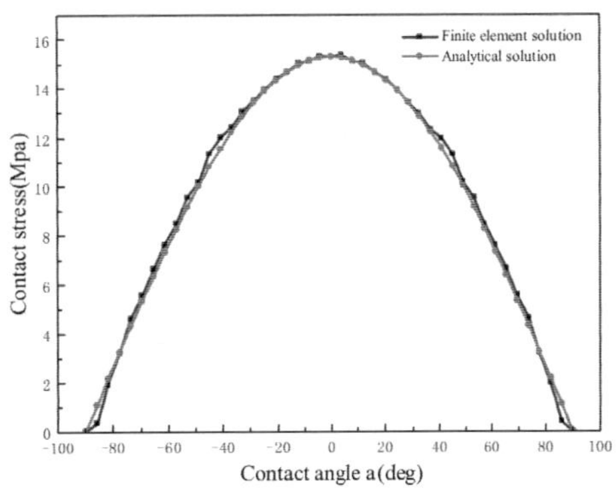

(d) Contact region Ω_2 at height $h=-27.5\text{mm}$

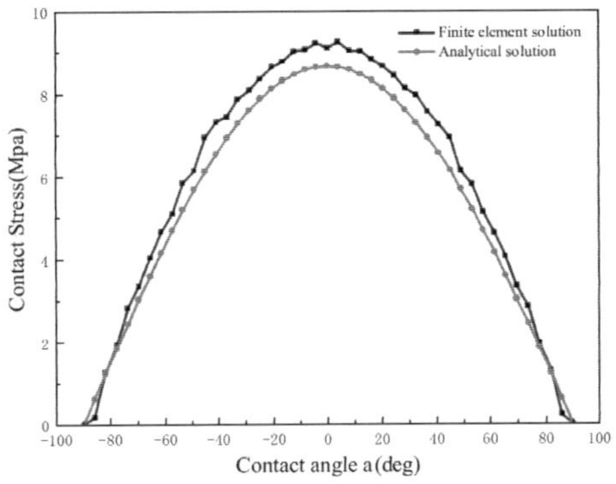

(e) Contact region Ω_1 at height $h=-21.5\text{mm}$

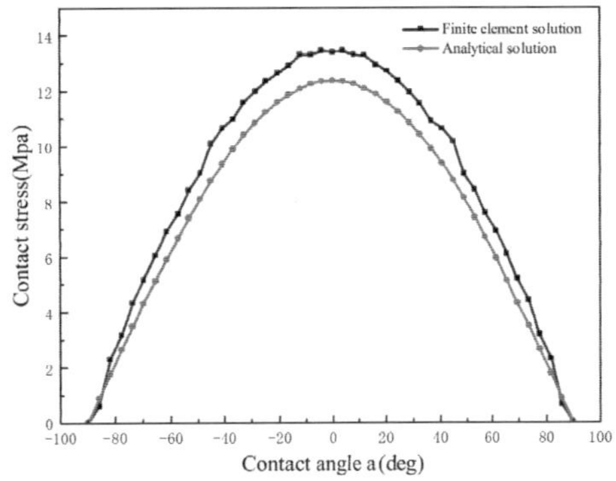

(f) Contact region Ω_2 at height $h=-21.5\text{mm}$

Figure 5. Comparison of contact stresses for the two algorithms.

References

[1] Cao H, Niu L, Xi S and Chen X. Mechanical model development of rolling bearing-rotor systems: A review [J]. Mechanical Systems and Signal Processing, 2018, 102: 37-58.

[2] Wang W, Shen G, Zhang Y, et al. Dynamic reliability analysis of mechanical system with wear and vibration failure modes [J]. Mechanism and Machine Theory, 2021, 163: 104385.

[3] Viana C A A, Alves D S, Machado T H. Linear and nonlinear performance analysis of hydrodynamic journal bearings with different geometries [J]. Applied Sciences, 2022, 12(7): 3215.

[4] Meng Q, Liu X J, Xie F. Structure design and kinematic analysis of a class of ring truss deployable mechanisms for satellite antennas based on novel basic units [J]. Mechanism and Machine Theory, 2022, 174: 104881.

[5] Hu Palu, Zhao Chunfa, Cai Wenfeng, Gong Junhu and Feng Yang Numerical Simulation and Comparative Analysis on Dynamic Performances of Two Types of Medium-low Speed Maglev Vehicles [J]. Journal of Mechanical Engineering, 59(10): 311-322.

[6] Liu Z, Xu M, Zhang H, et al. Modeling and analyzing of nonlinear dynamics for linear guide slide platform considering assembly error [J]. Nonlinear Dynamics, 2022, 108(3): 2193-2221.

[7] Liu C S, Zhang K, Yang R. The FEM analysis and approximate model for cylindrical joints with clearances [J]. Mechanism and machine theory, 2007, 42(2): 183-197.

[8] Antoni N. A study of contact non-linearities in pin-loaded lugs: Separation, clearance and frictional slipping effects [J]. International journal of non-linear mechanics, 2014, 58: 258-282.

[9] JI Yue, AN Qi. Mechanical performance analysis of linear bearings considering rolling element size error and radial clearance [J]. Journal of East China University of Science and Technology (Natural Science Edition), 2022, 48(01): 120-129.
[10] Zou H T, Wang B L. Investigation of the contact stiffness variation of linear rolling guides due to the effects of friction and wear during operation [J]. Tribology international, 2015, 92: 472-484.
[11] Gao Hongwei, Lv Kaibo, Li Jian, et al. Contact stress and failure analysis of hydraulic bracket pin connection [J]. Machine Tools and Hydraulics, 2023, 51(01): 152-157.
[12] Fang B, Wan S, Zhang J and Hong J. Research on the influence of clearance variation on the stiffness fluctuation of ball bearing under different operating conditions [J]. Journal of Mechanical Design, 2021, 143(2): 023403.
[13] Pantelidis L. The equivalent modulus of elasticity of layered soil mediums for designing shallow foundations with the Winkler spring hypothesis: A critical review [J]. Engineering Structures, 2019, 201: 109452.

Journal of Physics: Conference Series **2992** (2025) 012047 doi:10.1088/1742-6596/2992/1/012047

Gap flow field and experiment in electrical discharge machining using water-in-oil dielectric

Renpeng Bian, Andong Sun, Yonghong Liu* and Bo Qiao

College of Mechanical and Electronic Engineering, China University of Petroleum (East China), Qingdao, China

* Corresponding author: liuyhupc@163.com

Abstract—Electrical discharge machining (EDM) is a non-contact specialized machining technology that uses the thermal energy generated by pulse discharge between an electrode and a workpiece to erode material. As an important part of EDM, the flow characteristics of the dielectric in the machining gap directly affect debris removal. Since the dielectric affects the debris removal in the EDM gap and affects the machining efficiency and electrode loss, based on the simulation software Fluent, the influence of flushing velocity on the gap flow field is investigated through simulation analysis and designed for experimental verification of EDM. Simulation and experimental results show that increasing the flushing velocity of the dielectric is conducive to the removal of debris in the gap and the improvement of the discharge state, thus improving the machining efficiency. However, too high a flushing velocity will lead to an unstable discharge state and reduce processing efficiency, and the recommended flushing velocity is 10 m/s.

1. Introduction

Electrical Discharge Machining (EDM) is a technology that relies on the electrical-thermal phenomenon generated by pulse discharge between an electrode and a workpiece to erode material [1]. Since there is no direct interaction force between the electrode and the workpiece, EDM is classified as a non-contact machining technology. The difficulty of machining with EDM depends on the thermal-electrical characteristics of the workpiece, as the heat generated during the pulse discharge process is the source of the machining energy. Therefore, EDM is suitable for machining any conductive material, regardless of the material's hardness [2,3]. EDM technology is widely used for machining conductive materials that are difficult to cut, such as stainless steel, hardened steel, titanium alloys, and cemented carbides [4].

EDM is a spark discharge process carried out in dielectric, and if the debris removal between the electrode and the workpiece is inefficient, leading to debris accumulation, it can disrupt discharge stability and reduce the energy utilization of the pulse power supply, thereby affecting material removal rates and electrode wear [5]. Enhancing the flowability of the dielectric is an effective method to address the issue of debris accumulation. However, it is difficult to observe how different stroke rates of the dielectric affect the flow field distribution in the discharge gap during EDM. With the increasing maturity of finite element simulation technology, using finite element software to simulate and analyze related EDM flow field issues has become a research hotspot.

Tricarico et al. [6] established a geometric simulation model for EDM, considering the relationship between material removal rate and the machining gap, and simulated the electrode wear shape after machining. Kunieda et al. [7] developed a new geometric simulation model for EDM, correlating workpiece removal with gap width, and simulated the shape after EDM. Cetin et al. [8] studied the effects of the lifting height during electrode lifting on the velocity field and eroded particle movement in the gap flow field through numerical simulations. Zhang et al. [9] developed a gap flow field model in EDM, analyzed the impact of the electrode on the flow velocity in micro-hole gaps, and simulated the movement of fluid and debris in the micro-hole gap flow field. These simulations for the EDM gap flow field are of significant reference, but the influence of the flushing velocity of the dielectric on the EDM gap flow field with experimental verification has not been investigated.

Based on the simulation software FLUENT, this paper establishes the EDM gap flow field model, explores the influence of the flushing velocity on the EDM gap flow field characteristics through the combination of simulation analysis and experiment, and designs the experiment to verify it. The simulation and experimental results show that increasing the flushing velocity of the dielectric is conducive to enhancing the flow velocity of the gap flow field and effectively removing the etching debris in the gap, thus improving the machining efficiency of EDM.

2. Material and methods

2.1 Assumptions and simplifications

Before establishing the gap flow field model, the following assumptions and simplifications were made [10]: (1) The water-in-oil dielectric is assumed to be incompressible, and the characteristic parameters of the flow field do not change with time. (2) Only the liquid phase and solid phase are considered in the flow field. (3) It is assumed that the heat generated by the discharge machining does not affect the dielectric. (4) The impact of electrode wear on the geometry and dimensions of the gap flow field is neglected.

2.2 Model establishment and parameter settings

The machining model for internal flushing in EDM is shown in Figure 1. The dielectric is a water-in-oil dielectric, and the electrode material is copper with an outer diameter of 16 mm, and an inner diameter of 6 mm. During EDM, the dielectric enters the electrode tube from the top, flows through the bottom and side gaps, and finally exits through the side gaps of the electrode. This internal flushing method effectively removes debris, thereby enhancing machining efficiency.

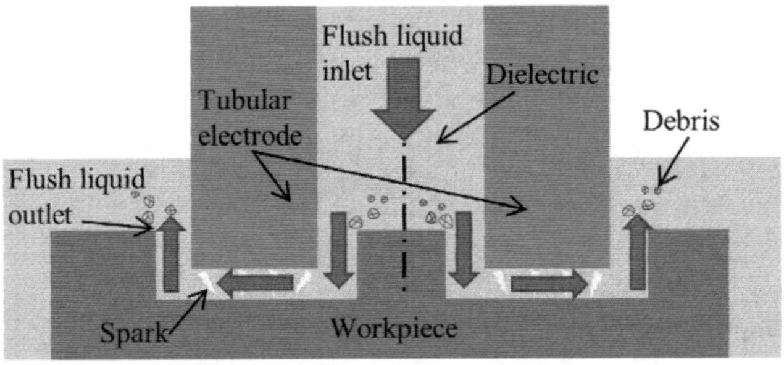

Figure 1. Diagram of internal flushing liquid process.

The flow field model of EDM is shown in Figure 2(a). Based on this flow field model, ICEM CFD is used to perform mesh generation. In the Part settings, the maximum mesh size for the inlet and outlet is set to 0.5, ensuring that the mesh in these regions is sufficiently fine to accurately simulate fluid flow. The maximum mesh size for the wall surfaces is set to 1. Although this makes the mesh relatively coarse, it is sufficient to support the calculation requirements for wall surface flow. The generated mesh is shown in Figure 2(b).

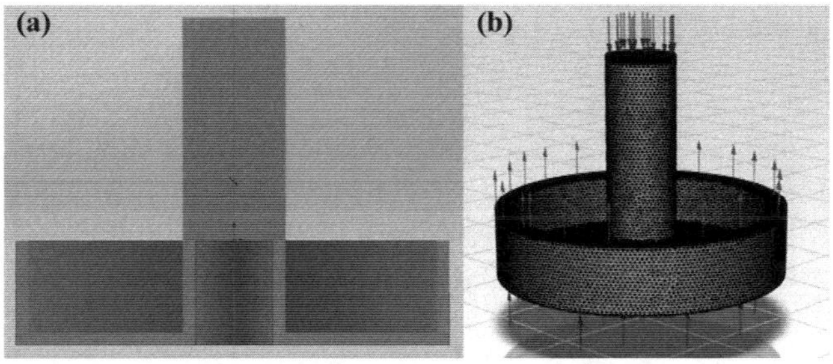

Figure 2. Cross-sectional diagram of the internal flushing liquid model and mesh division diagram.

To evaluate the quality of the generated mesh, the CFD file was imported into Fluent for mesh quality assessment. After inspection, it was confirmed that both the minimum volume and minimum face area are greater than zero, meeting the required criteria.

Water-in-oil dielectric is selected in EDM, with a density of 900 kg/m³ and a viscosity of 5.5 mPa·s. The machining depth is set to 1 mm. The fluid properties are set as water-liquid, and the workpiece material is S316L stainless steel, with a density of 7980 kg/m³. It is assumed that the removed particles are spherical inert particles with a diameter of 100 μm.

2.3 Experiment

The EDM experiment used the Sanguang D7140P machine. The specific machining parameters for this experiment are shown in Table 1, and the average of three repetitions of each experiment.

Table 1. Specific experimental parameters.

Categories	Parameter	
	Title	Norm
Workpiece	Stainless steels	316L
Electrode	Copper	Outer diameter 16 mm, inner diameter 6 mm
Dielectric	Water-in-oil	Viscosity 5.5 mPa·s
Electrical parameter 1	Peak current	3 A
Electrical parameter 2	Pulse width	30 μs
Electrical parameter 3	Pulse interval	20 μs
Non-electrical parameter 1	Flushing velocity	5 m/s; 7 m/s; 10 m/s; 12 m/s
Non-electrical parameter 2	Machining depth	1 mm

In this paper, experiments are carried out to investigate the effect of different flushing velocities on the machining efficiency and electrode loss of EDM. The machining efficiency is expressed as material removal rate (MRR) and the electrode loss is characterized by relative electrode wear rate (REWR).

MRR is calculated as shown in Equation (1).

$$MRR = \frac{m_{w2}-m_{w1}}{\rho_w \cdot t} \tag{1}$$

where ρ_w is the density of the workpiece, g·cm^{-3}. m_{w2} and m_{w1} are the mass of the workpiece before and after machining, g. t is the total machining time, s.

The REWR is calculated as shown in Equation (2).

$$REWR = \frac{(m_{e2}-m_{e1})/\rho_e}{(m_{w2}-m_{w1})/\rho_m} \times 100\% \tag{2}$$

where ρ_e is the density of the electrode, g·cm^{-3}, and m_{e2} and m_{e1} are the mass of the electrode before and after machining, g, respectively.

3. Results and discussion

3.1 Velocity distribution in the gap flow field

Through simulation, a vector diagram of the dielectric flow in the gap flow field was obtained, as shown in Figure 3. The direction indicated by the arrows in Figure 3 is the flow direction of the dielectric, which enters from the middle gap flow field, then flows through the bottom gap flow field, and finally flows out from the two side gap flow fields.

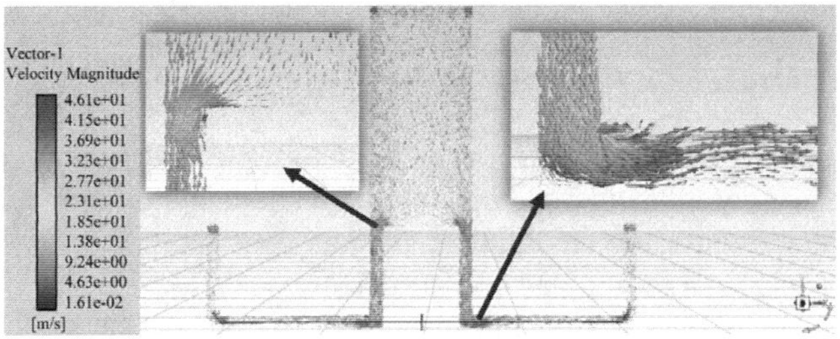

Figure 3. Diagram of internal flushing liquid process.

The velocity distribution contour in the gap flow field is obtained by simulation and is shown in Figure 4. From Figure 4, it can be observed that when the dielectric enters the gap between the electrode and the workpiece, the cross-sectional area of the flow channel decreases, resulting in an increase in flow velocity. At the bottom corner, the velocity reaches as high as 35.7 m/s, which helps in flushing the debris out of the gap. In the lateral gap region, due to the influence of wall friction, the fluid velocity gradually decreases.

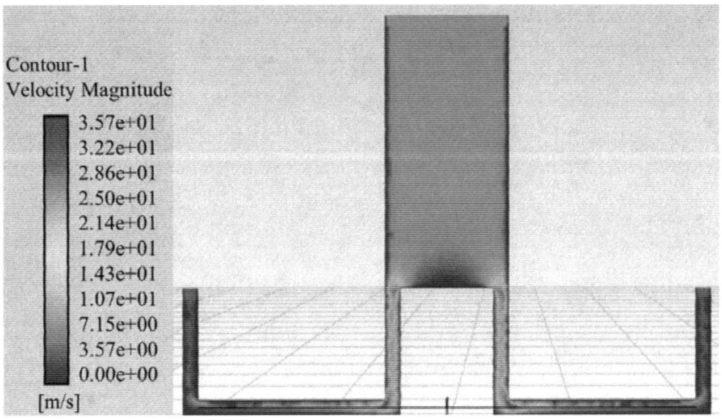

Figure 4. Diagram of internal flushing liquid process.

3.2 Impact of flushing velocity on flow field

In this paper, in analyzing the effect of the flushing velocity of the dielectric on the EDM gap flow field, inlet velocities were set to 5.02 m/s, 7.12 m/s, 9.82 m/s, and 11.82 m/s, respectively. Figure 5 illustrates the velocity distribution at the bottom surface for different flushing velocities. From Figure 5, it can be observed that at the same flushing velocity, the flow velocity in the bottom gap flow field decreases as it moves toward the bottom. Additionally, as the flushing velocity increases, the flow velocity in the bottom gap flow field also increases. Consequently, it can be concluded that increasing the flushing velocity enhances the flow velocity in the gap flow field, which aids in removing debris particles, improves the discharge state in the gap, and consequently increases machining efficiency.

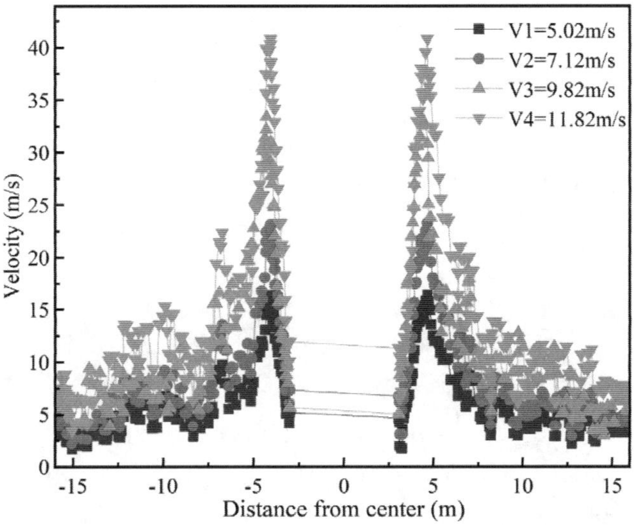

Figure 5. Velocity distribution diagram at the bottom surface for different flushing velocities.

3.3 Experimental investigation of the effect of flushing velocity in EDM

Based on the simulation conclusions, increasing the flushing velocity of the dielectric in the gap flow field can significantly enhance the removal effect of the etched debris, which can achieve the improvement of the machining efficiency of EDM by improving the discharge state between the workpiece and the electrode. To verify the reliability of these simulation conclusions, EDM experiments were conducted using a water-in-oil dielectric, and the specific experimental parameters are shown in Table 1.

The material removal rate (MRR) and relative electrode wear rate (REWR) for different flushing velocities are shown in Figure 6. From Figure 6, it can be seen that as the flushing velocity increases, the MRR first increases and then decreases, while the REWR continues to increase. This phenomenon is consistent with simulation results. It is well established that increasing the flushing velocities of the dielectric facilitates debris removal and thus increases the material removal rate. However, excessively high flushing velocities can disrupt the stability of the discharge state, adversely affecting material removal efficiency. Increased electrode wear is attributed to both the enhanced discharge that raises material removal rates and the removal of carbides adhering to the electrode surface, which undermines its protective effect.

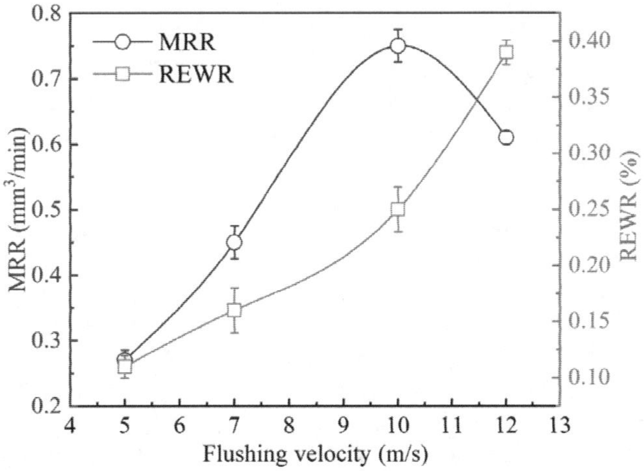

Figure 6. MRR and REWR at different flushing velocities.

Figure 7 shows the morphology of the recast layer on the workpiece for different flushing velocities during EDM. As illustrated in Figure 7(a), when the flushing velocity is 5 m/s, the recast layer is the thickest and has an uneven outer profile, indicating the poorest discharge state. In Figures 7(b-d), as the flushing velocity increases to 7-12 m/s, the thickness of the recast layer decreases, and the outer profile becomes smoother, reflecting an improvement in the discharge state. At a flushing velocity of 10 m/s, the recast layer has the flattest outer profile, indicating a stable discharge state. At this velocity, the debris removed from the workpiece surface is adequately flushed away, resulting in a surface with minimal adhesion. Therefore, selecting an appropriate flushing velocity is beneficial for achieving a thinner and more uniform recast layer.

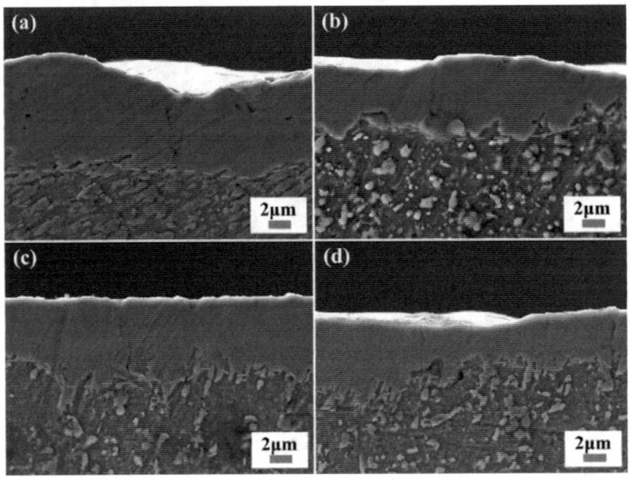

Figure 7. Recast layer on EDM workpieces at different flushing velocities: (a) 5 m/s; (b) 7 m/s; (c) 10 m/s; (d) 12 m/s.

4. Conclusions

This study investigates the characteristics of EDM flow fields and machining performance at different flushing velocities. The conclusions are as follows:

(1) A gap flow field model for internal flushing in EDM was developed and analyzed using Fluent. A comparison of simulation results at different flushing velocities shows that increasing the flushing velocity enhances the flow velocity in the gap, thereby improving the MRR.

(2) Experiments reveal the impact of flushing velocity on MRR and REWR, with results consistent with simulations. In EDM, increasing the flushing velocity initially increases the MRR, but it then decreases, while the REWR increases.

(3) Higher flushing velocities facilitate debris removal from the gap and improve the discharge state, leading to increased MRR. However, excessively high flushing velocities can destabilize the discharge state and reduce machining efficiency. A flushing velocity of 10 m/s is recommended.

Acknowledgment

This work was supported by the Equipment Pre-research Collaborative Foundation for Innovation Team from Ministry of Education of the People's Republic of China (Grant No. 8091B042209).

References

[1] Xu C, Xu F, Sun F, Zhang X, Jin J, Luan B, Oka K. Research on an intelligent control method of a magnetic actuator for micro electrical discharge machining. Actuators. 2022, 11(12), 11. https://doi.org/10.3390/act11120371.

[2] Li Z, Tang J, Li Y, Bai J. Investigation on surface integrity in novel micro-EDM with two-dimensional ultrasonic circular vibration (UCV) electrode. J. Manuf. Process. 2022, 76:828-840. https://doi.org/10.1016/j.jmapro.2022.03.004.

Journal of Physics: Conference Series **2992** (2025) 012047 doi:10.1088/1742-6596/2992/1/012047

[3] Bian R, Sun A, Liu Y, Wu X, Experimental investigations and process optimization of SKD61 for sinking EDM, Mater. Manuf. Process. 2024, 2368541. https://doi.org/10.1080/10426914.2024.2368541.
[4] Li Z, Tang J, Li Y, Bai J. Investigation on surface integrity in novel micro-EDM with two-dimensional ultrasonic circular vibration (UCV) electrode, J. Manuf. Process. 2022, 76:828–840. https://doi.org/10.1016/j.jmapro.2022.03.004.
[5] Ahmed S, Speidel A, Murray J.W, Ahmed N, Cuttell M, Clare A.T, Electrolytic-dielectrics: a route to zero recast electrical discharge machining, Int. J. Mach. Tools Manuf. 2022, 181: 103941, https://doi.org/10.1016/j.ijmachtools.2022.103941.
[6] Tricarico C, Delpretti R, Dauw D.F. Geometrical Simulation of the EDM Die-Sinking Process[J]. CIRP Ann. Manuf. Technol. 1988, 37(1):191-196. https://doi. org/10.1016/s0007-8506(07)61616-3.
[7] Kunieda M, Kiyohara M. Simulation of die-sinking EDM by discharge location searching algorithm[J]. Int. J. Electr. Mach. 1998, 3:79-85. https://doi. org/10.2526/ijem.3.79.
[8] Cetin S, Okada A, Uno Y. Effect of debris distribution on wall concavity in deep-hole EDM[J]. JSME Int. J. 2004, 47:553-555. https://doi. org/10.1299/jsmec.47.553.
[9] Zhang P, Zhen Yi, Yu M, Tao D, Yu D, Zhang Q, Li H. Investigating mechanisms of debris removal in ultrasonic vibration-assisted EDM drilling. Int. J. Mech. Sci. 2024, 279. https://doi. org/10.1016/j.ijmecsci.2024.109486.
[10] Philip T. E, Mukund R. P, Maria A. B, Bedri B. Theoretical models of the electrical discharge machining process. III. The variable mass, cylindrical plasma model. J. Appl. Phys. 1993. 73(11): 7900-7909. https://doi.org/10.1063/1.353942.

Journal of Physics: Conference Series **2992** (2025) 012048 doi:10.1088/1742-6596/2992/1/012048

Study on the drag and heat transfer characteristics of near-wall oblate spheroids

Ting Xiong[1, 2], Yongkang YANG [1], Chenchen Chen [1], Jingdun Pang [3], Hongsheng Zhang [3], and Yong Chen [1, 2, *]

[1] School of Transportation and Logistics Engineering, Wuhan University of Technology, Wuhan, Hubei, 430070, China
[2] State Key Laboratory of Maritime Technology and Safety, Wuhan, Hubei, 430070, China
[3] National Engineering Research Center for Dredging Technology and Equipment, Shanghai, 200000, China

[*] Corresponding author's e-mail: yongchen@whut.edu.cn

Abstract: In this study, computational fluid dynamics was used to calculate the drag coefficient and Nusselt number for oblate ellipsoids placed near the wall. The simulation results indicate that the drag coefficient and Nusselt number increase with increasing particle-wall distance. Compared to the cases in unbounded uniform flow, the maximal relative difference in drag coefficient and Nusselt number are 75.61% and 36.64%, respectively. Thus, the wall has a significant effect on the momentum and heat transfer between particles and fluids.

1. Introduction
In recent years, oblate spheroidal particles have been increasingly used in industrial applications due to their unique shape and excellent hydrodynamic performance [1], [2]. Although previous studies have investigated the flow and heat transfer behavior of oblate spheroids through experiments and numerical simulations, they have mainly focused on particles in uniform flow [3]. Geonzon et al. [6] used microfluidic technology to study the drag on spherical particles in near-wall shear flow, providing drag information that can be used to optimize microfluidic designs. Rousta et al. [7] investigated three mechanisms of the heat transfer between solid particles and walls when the distance between them is less than half the particle radius and summarized the optimal heat transfer modes. Considering that particles are often in a near-wall state in practical applications [8], [9], this study seeks to address this gap by analyzing the forces and heat transfer characteristics of oblate spheroids near walls under different Reynolds numbers, shapes, and orientations through detailed computational fluid dynamics (CFD) simulations [10].

2. Research method

2.1. Computational setup
The study used commercial software ANSYS Fluent to conduct numerical simulations for incompressible flow, examining the forces and heat transfer characteristics of oblate spheroids near walls. The particles were kept stationary during the simulation. Four Reynolds numbers (50, 100, 200, and 500) were modeled. The Reynolds number was based on the volume equivalent diameter of the

oblate. A parameter (H) representing the distance between the particle and the wall was introduced to describe how the wall affects the forces and heat transfer behavior.

As shown in Figure 1, the computational domain was a cuboid, and the particle was located near the bottom of the domain. The orientation angle (θ), which was defined as the angle between the particle's rotational axis and the flow direction, varied from 0° to 180°. Uniform fluid velocity and temperature were applied at the inlet, and at the outlet Neumann boundaries in the flow direction were assigned both for velocity and temperature. The boundary conditions for the bottom wall were non-slip and adiabatic. Symmetric boundary conditions were applied to other side boundaries. Dirichlet boundary conditions for velocity and temperature at the particle surface were used.

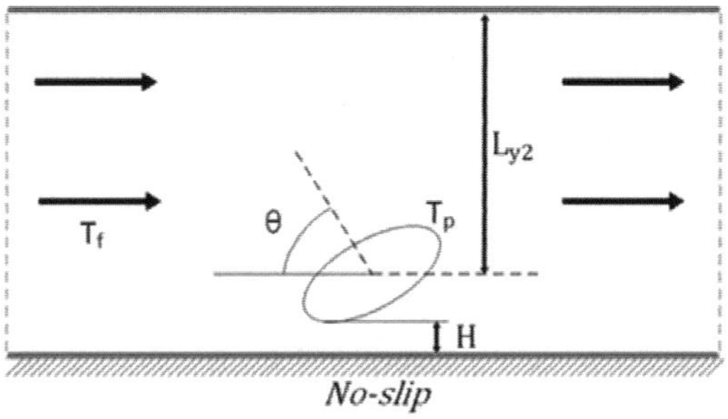

Figure 1. Schematic description of computational setup.

2.2. Grid independence verification

A structured mesh was used to discretize the computational domain, with mesh refinement applied near the particle to accurately resolve the hydrodynamic and thermal boundary layers. To ensure the accuracy of the results, the effect of domain and grid sizes was assessed. As shown in Figure 2, when L_{y2} = $15d_v$, the variation for both c_d and Nu were both within 0.32%, indicating that the effect of domain size on the simulation results could be neglected. Therefore, the final domain size was set to $30d_v \times 15d_v \times 15d_v$.

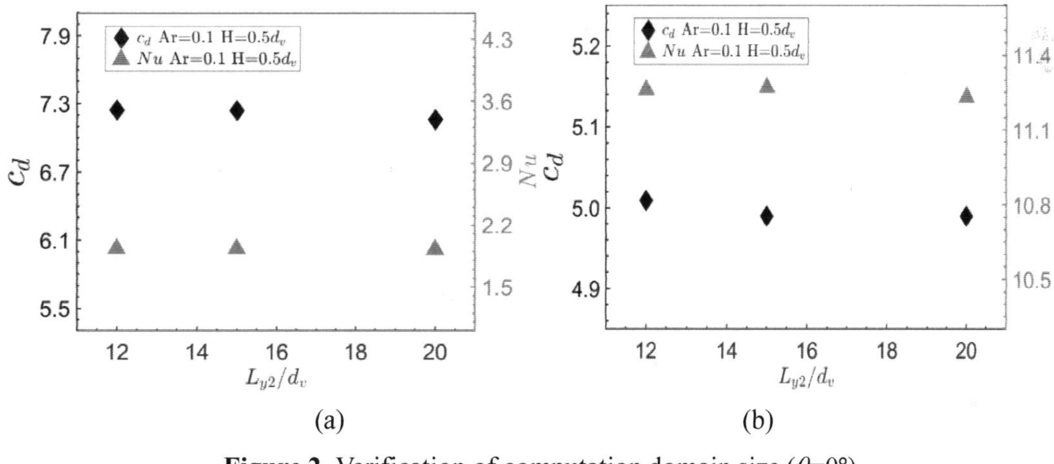

Figure 2. Verification of computation domain size (θ=0°).

After determining the size of the computational domain, grid independence verification was further studied, as shown in Figure 3. By comparing the variations in the drag coefficient and Nusselt number under different grid sizes, it was found that increasing the number of grids under the current refinement conditions had a minimal effect on the results. The drag coefficient variation was only 0.3%. The Nusselt number variation was around 1.7% across different grid sizes, indicating that the grid size was sufficient to resolve the boundary layer. Based on these results, a grid scheme with 3.66 million cells was selected.

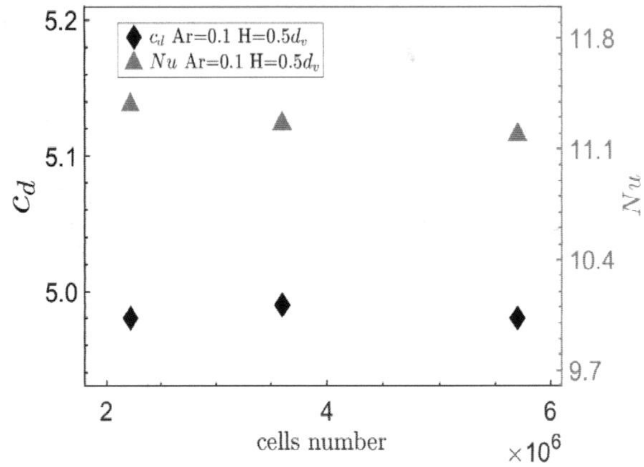

Figure 3. Verification of grid independence (*Re*=500).

2.3. Validation
To validate the accuracy of the model established in this study, spherical particles were selected for simulation to obtain the drag coefficient and Nusselt number under conditions of Re = 10, 50, 100, and 200. The results were compared with the formula-based results from Richter et al. [11], as shown in Figure 4. A good comparison is observed. Previous studies [12] have demonstrated that the simulation results for prolate spheroids using this model have negligible error, indicating that the model established in this paper can be used for subsequent simulations of oblate spheroids.

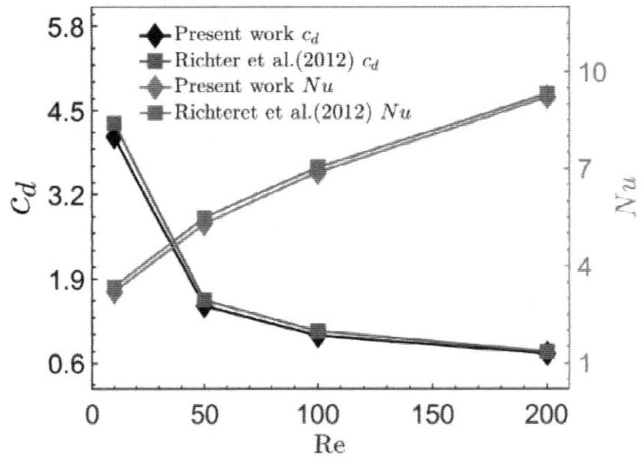

Figure 4. Comparison of *cd* and Nu for a sphere between current simulations and the one proposed by Richter et al. [11].

3. Results and discussions

3.1. Effect of wall on the drag coefficient
Figure 5(a) shows the drag coefficient of an oblate with Ar = 0.5 at different particle-wall distances when θ=90°. The numerical results indicate that the presence of the wall significantly affects the drag force acting on the oblate spheroids. As expected, the drag coefficient decreases with increasing Reynolds number. Compared to the case of $H = 0.25d_v$, the drag coefficient is 75.61% larger than that of $H = d_v$. When the distance (H) is small ($H = 0.25d_v$), the side facing the wall is within the boundary layer. The fluid velocity in this layer is slower, which reduces viscous friction on the wall-facing side of the particle, leading to lower drag. As the distance (H) increases (e.g., $H = d_v$), the particle is out of the boundary layer induced by the wall. At this stage, the influence of the wall on the surrounding fluid diminishes significantly. As a result, the drag force increases. In contrast, the drag force for

oblate spheroids of incident angle θ=0° does not increase with increasing particle-wall distance, as shown in Figure 5(b). This may be due to the fact that the jet caused by the gap induces flow separation behind the particle, which enhances the drag acting on the particle.

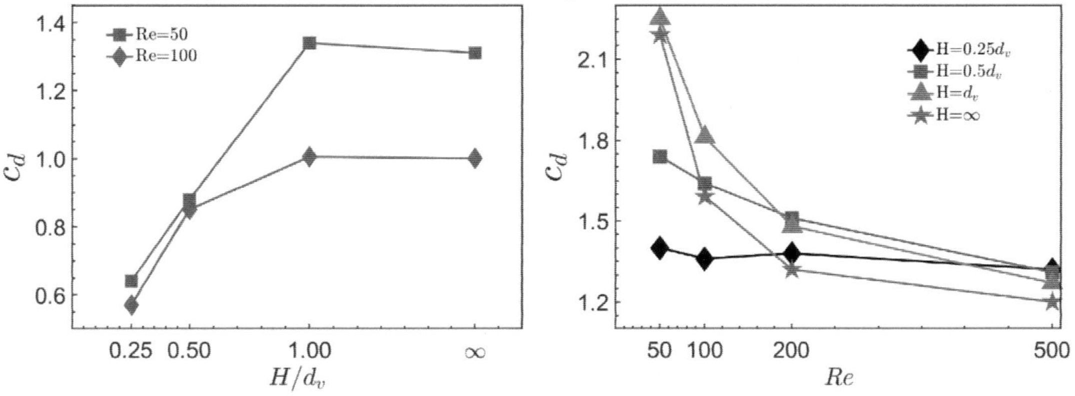

(a) Ar=0.5, θ=90° (b) Ar=0.5, θ=0°

Figure 5. Drag coefficient as a function of particle-wall distance.

3.2. *Effect of wall on the nusselt number*

The Nusselt number (Nu) is a dimensionless quantity that quantifies the intensity of convective heat transfer. A higher Nusselt number indicates greater convective heat transfer intensity, signifying an enhanced heat transfer capability. Figure 6 shows the relationship between particle-wall distance and Nu. When $H = 0.25d_v$, the presence of the wall severely hinders the particle's heat transfer, resulting in the minimum Nu. As the particle-wall distance increases, the interference of the wall on the interphase heat transfer decreases, leading to an increase in Nu.

The Nusselt number at a very large particle-wall distance ($H = \infty$) is only 0.86% greater than that at $H = d_v$, suggesting that when $H = d_v$, the wall has little effect on the heat transfer. At this distance, the particle's heat transfer capability is essentially consistent with that under uniform flow conditions.

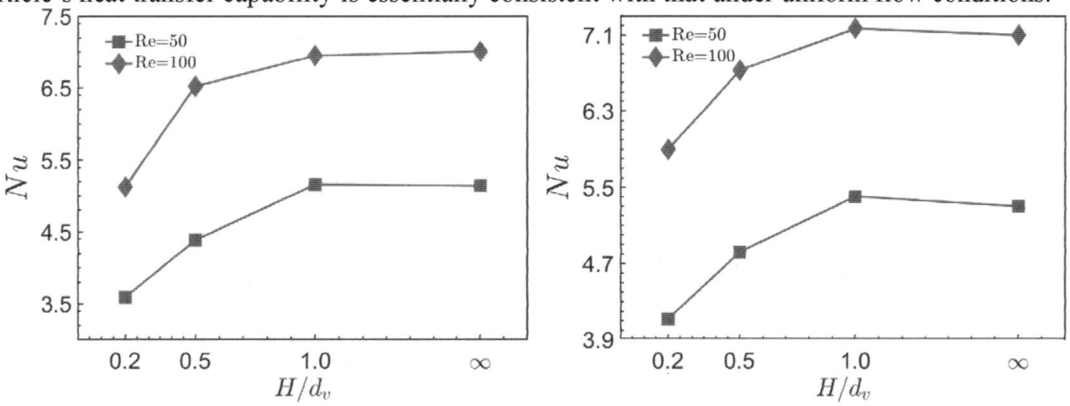

(a) Ar=0.5 (b) Ar=0.8

Figure 6. Nu of an oblate ellipsoid at different particle-wall distances (θ=90°).

As shown in Figure 7, the orientation angle (θ) of an oblate spheroid significantly impacts its heat transfer. The orientation determines the shape and size of the gap between the particle and wall, influencing fluid flow and heat transfer in this space. When the incident angle is larger (e.g. θ=30°), the gap is short and widens gradually, which promotes fluid movement and enhances heat transfer. In contrast, a larger angle creates a narrower, elongated gap that restricts fluid flow, causing more fluid to remain near the wall and increasing thermal resistance. This resistance is greatest when the particle's major axis is parallel to the wall (θ=90°), at which the gap is at its narrowest and longest, retaining the most fluid. In contrast, the particle incident angle has little effect on the Nusselt number for the oblates in uniform flow.

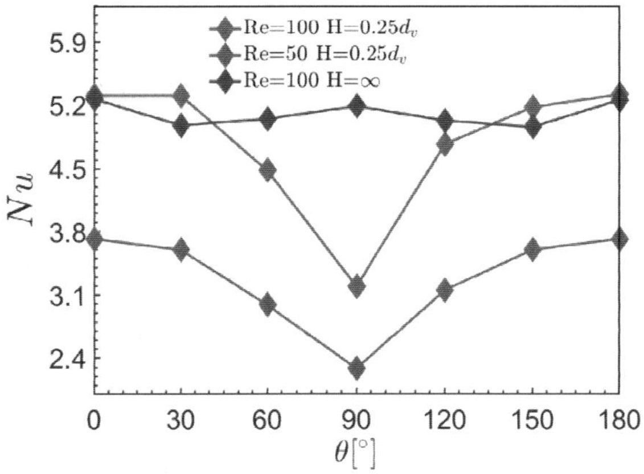

Figure 7 Effect of particle orientation on the Nusselt number of oblate ellipsoid (Ar=0.1).

4. Conclusion

This study provides an in-depth exploration of the drag and heat transfer characteristics of oblate spheroids near walls. The following main conclusions can be drawn:

(1) As the distance between the particle and the wall decreases, the wall effect gradually intensifies, leading to a reduction in the drag, with a relative difference of up to 75.61%.

(2) As the particle-wall distance increases, the interference of the wall with the particle's heat transfer process gradually diminishes, resulting in an increase in the Nusselt number (Nu) up to 36.64%.

(3) For $\theta \leq 90^{\circ}$, the Nusselt number decreases with increasing θ, whereas it increases with increasing θ, implying that the integrated effect of particle orientation will have a very important impact on the interphase heat transfer.

References

[1] Dong K. Study on nano-catalytic hydrogenation of cinnamaldehyde enhanced by rotating packed bed [D]. 2022.

[2] Gan Lin, Xu Mao-sheng, ZHU Bingchen. Heat transfer parameters of annular cylindrical particle packed bed [J]. Acta Chemologica Sinica, 2000, (6): 778-783.

[3] Ke C, Shu S, Zhang H, et al. On the drag coefficient and averaged Nusselt number of an ellipsoidal particle in a fluid [J]. Powder Technology, 2018, 325: 134–144.

[4] Chen Y, Jiang P, Xiong T, et al. Drag and heat transfer coefficients for axisymmetric nonspherical particles: A LBM study [J]. Chemical Engineering Journal, 2021, 424: 130391.

[5] Ouchene R. Numerical simulation and modeling of the hydrodynamic forces and torque acting on individual oblate spheroids [J]. Physics of Fluids, 2020, 32(7): 073303.

[6] Geonzon LC, Kobayashi M, Adachi Y. Effect of shear flow on the hydrodynamic drag force of a spherical particle near a wall evaluated using optical tweezers and microfluidics [J]. Soft Matter, 2021, 17(34): 7914–7920.

[7] Rousta F, Lessani B. Near-Wall heat transfer of solid particles in particle-laden turbulent flows [J]. International Communications in Heat and Mass Transfer, 2020, 112: 104475.

[8] D. Toghraie, M. Afrand, A. D. Zadeh, et al. Numerical investigation on the flow and heat transfer of a multi-lobe particle and equivalent spherical particles in a packed bed with considering the wall effects [J]. International Journal of Mechanical Sciences, 2018, 138-139: 350-367.

[9] M. Zare, S. H. Hashemabadi. Particle-fluid heat transfer close to the bed wall: CFD simulation and experimental study of particle shape influence on the formation of hot zones [J]. International Journal of Thermal Sciences, 2020, 150:106223.
[10] Guo Xue-Yan. Application of CFD method to heat transfer in fixed bed reactor [J]. Acta Chemologica Sinica, 2008, 59(8): 1914-1922.
[11] Richter A, Nikrityuk PA. Drag forces and heat transfer coefficients for spherical, cuboidal and ellipsoidal particles in cross flow at sub-critical Reynolds numbers [J]. International Journal of Heat and Mass Transfer, 2012, 55(4): 1343–1354.
[12] Xiong T, Chen C, Chen Y. Heat transfer characteristics of packed bed comprising oblate ellipsoids [J]. CHEMICAL ENGINEERING (CHINA), 2023, 51(10): 48–52.

Journal of Physics: Conference Series **2992** (2025) 012049 doi:10.1088/1742-6596/2992/1/012049

Thermal dynamic simulation study of the aircraft ECS laboratory air source

Zheng Dai [1,2*] **and Yi Cui** [1,2]

[1] Shanghai Aircraft Design and Research Institute, Commercial Aircraft Corporation of China Ltd., No. 5188 Jinke Road, Shanghai, China
[2] State Key Laboratory of Airliner Integration Technology and Flight Simulation

*Corresponding author's e-mail: daitseng@126.com

Abstract. The Environmental Control System's working conditions are complex, and the regulation and control are challenging. Moreover, the bleed air system has characteristics such as large thermal inertia, high nonlinearity, and strong coupling between air supply systems, making it difficult to design and verify its dynamic characteristics. This study aims to evaluate the structure of the air source facility in the aircraft ECS laboratory to meet the dynamic performance requirements of engine bleed air. Through the coordinated operation of various components, the proposed air source configuration can adjust the output air temperature T1 and pressure P1 within a certain range. A dynamic Matlab-based model was established, and typical component parameters were assumed for preliminary studies. The results show that the proposed air source facility configuration performs well in controlling temperature T1 and pressure P1. The P1 control performance is close to ideal, and the T1 control is also close to ideal when the heater heats the air stream temperature T5 to 1, 500oF. Additionally, the control interactions between T1 and P1 are small, indicating good controllability.

1. Introduction

The environmental control system is an important airborne system that affects aircraft energy consumption and creates a comfortable cabin environment. The simplified commercial aircraft ECS block diagram is as shown in Figure 1. It is responsible for temperature and pressure regulation, and the dynamic characteristics are key factors that cannot be ignored in the function and performance of the environmental control system [1]. In particular, the bleed air system directly draws high-temperature and high-pressure gas from the engine, with the highest temperature at the bleed port approaching 1, 200°F and the highest pressure at about 550 psia. At the same time, the temperature and pressure of the bleed air will change rapidly with the variation of engine thrust and downstream air supply configuration, posing a challenging regulation and control "mission" for the bleed air system. Additionally, the bleed air system itself has inherent characteristics such as large thermal inertia, a high degree of nonlinearity, and strong coupling effects between air supply systems. It is a key technology that needs to be addressed in the field of environmental control.

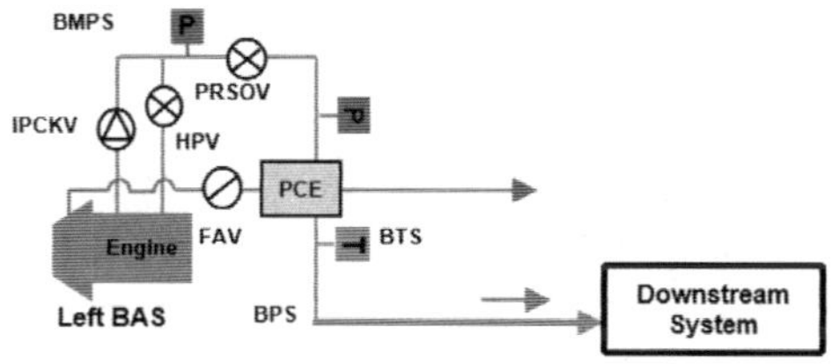

Figure 1. Simplified commercial aircraft ECS block diagram.

The objective of the ECS Laboratory (ECSLab) Air Source is to provide airflow to simulate engine bleed air and ram air sources for testing the ECS system. The ECSLab supply pressure and temperature are therefore necessarily fast responding. This simulation study is to evaluate possible ECSLab facility structures that could meet the demanding dynamic behavior of engine bleed air.

2. Air source performance requirements

For ECS applications, the engine bleeds air pressure, and the temperature from the engine high-pressure port typically ranges from 40 to 500 psia and 450 to 1, 200°F. During takeoff, the pressure and temperature could change from 55 to 550 psia and 500°F to 1, 200°F rapidly. During the cruise at 40, 000 ft, the pressure and temperature are around 160 psia and 1, 050°F. At the same altitude for descend, they are about 50 psia and 680°F.

For the low-pressure port, engine bleed air pressure and temperature from the engine high-pressure port typically range from 10 to 150 psia and 200 to 700°F. During takeoff, the pressure and temperature could change from 10 to 150 psia and 110°F to 670°F rapidly. During the cruise at 40, 000 ft, the pressure and temperature are around 35 psia and 540°F. At the same altitude for descend, they are about 10 psia and 270°F.

However, during rapid engine pressure transient like take-off operation, both pressure and temperature can change rapidly and form one engine operation state to another operation state in about 10 to 20 seconds. Hence, it would be difficult to emulate this rapid pressure and transient performance in a laboratory. Reasonable and achievable performance requirements for the air source to emulate the engine performance are in order, as shown in Table 1.

Table 1. Reasonable and achievable performance requirements.

Parameter	Control accuracy	Overshoot	Settling time
Pressure	± 2 to 5 psi	Less than 5 to 10 psi	Less than 30 sec
Temperature	± 5 to 10°F	Less than 10 to 30°F	45 to 75 sec

To meet these requirements, the following suggests an Air Source Facility structure intending to meet these requirements and to be examined by using dynamic analysis.

3. Air source configuration and operation

The proposed configuration shown in Figure 2 intends to minimize total energy consumption and achieve good controllability. The ECSLab Air Source outputs are the air temperature T_1 and air pressure P_1. T_1 can vary from 200 and P_1 can vary from 25 psia to 600 psia. The system operating airflow rate is about 260 ppm maximum. To recover the heat energy leak due to V_3 valve leakage or during system warm-up as required, the proposed Regen Heat Exchanger intends to recover a part of these energy losses. However, the dynamic model developed can also simulate the cases without the Regen Heat Exchanger actions for comparison purposes to determine any advantage that could be gained by using the heat exchanger. Without the Regen Heat Exchanger, during the warm-up period, the warming air may need to be discharged directly into the atmosphere via V_3.

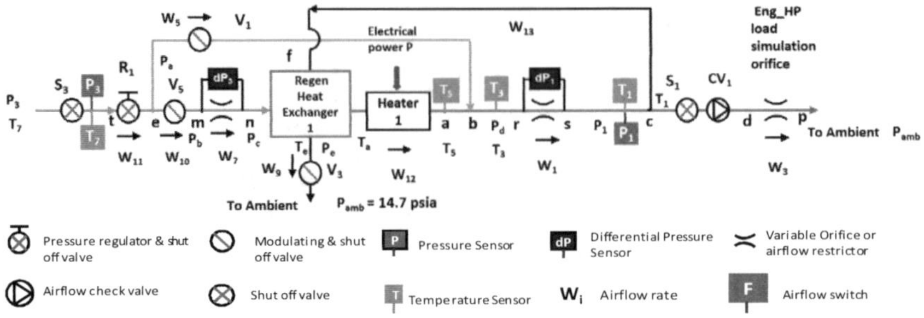

Figure 2. The proposed air source configuration.

The air supply pressure P_3 and supply temperature are assumed to be 650 psia and 100°F. These values can be changed to study the effects of supply variations on system performance. P_1 is regulated by R_1. T_1 is regulated by modulating V_1 and V_5. When operating, V_1 and V_5 are moving in opposite directions, to achieve rapid T_1 and T_3 responses (i.e., V_1 opens (closes), then V_5 closes (open)). Since R_1 essentially controls the total airflow W_{11} which is the sum of W_{10} (i.e., through V_5) and W_5 (i.e., through V_1), the W_1 to W_5 airflow ratio can be changed to obtain T_1 without affecting much on the total flow W_{11}. Hence, the system P_1 and T_1 open loop dynamics are essentially decoupled and are necessary to achieve good dynamic controllability properties at various P_1 and T_1 operating points.

The Heater is an electrical heater and the Heater discharge temperature T_5 is controlled by using the bang-bang control method. T_5 can be controlled to various values. For this study, T_5 is either controlled to 1, 500°F or 1, 100°F. 1, 500°F is preferred to overcome the expected thermal inertia of the ECSLab ducting. V_3 and V_5 are controlled to ensure there is always a minimum amount of airflow through the Heater to protect the Heater.

4. Simulation model and component data

Typical aircraft ECS-type components are assumed and scaled with suitable thermal mass for the preliminary studies. In this case, all valves are butterfly-type with specific leakage areas, the heat exchanger is cross-flow type, and the ducting is made of stainless steel. The heater is represented by a simple nonlinear model based on an energy balance approach with a time constant of about 3 min at maximum flow rates.

4.1 Valve data

The valve actuation dynamics will be represented by using a first-order transfer function with a time constant. The valve component data used is shown in Table 2. For this study, the minimum valve leakage is assumed to be 0.1 sq in or about 2.6% of the maximum valve area.

Table 2. The valve component data used.

Valve	Dia (in)	The time constant (Sec)
V_1	3.9	5
V_3	3.9	5
V_5	3.9	5
R_1	3.9	3
S_1	3.9	3
S_3	3.9	3
CV_1	3.9	1

4.2 Air duct data

The air duct component data used is shown in Table 3.

Table 3. The air duct component data used.

Air duct	Dia (in)	Length (ft)	Thickness (in)	Volume (ft^3)
e-b	4.9	29.5	0.39	3.90
e-m	4.9	3.3	0.39	0.43
m-n	4.9	6.6	0.39	0.87
a-b	4.9	16.4	0.39	2.17
b-c	5.9	32.8	0.39	6.24
c-f	4.9	32.8	0.39	4.33
c-d	5.9	6.6	0.39	1.25
t-e	5.9	6.6	0.39	1.25

4.3 Regen heat exchanger data

The Regen Heat Exchanger is made of stainless steel with a mass of 1, 200 lbm. The regen heat exchanger data used is shown in Table 4.

Table 4. The regen heat exchanger data used.

Heat exchanger vol.	Vol. (ft^3)	Heat exchanger dP	Pressure drop (psi)
Hot side vol.	5	dP_h	$0.008W_h^{1.5987}$
Cold side vol.	5	dP_c	$0.008W_c^{1.5987}$

4.4 Dynamic model

A Matlab-based ECSLab Air Source dynamic model is developed and the modeling approach refers to the commonly used formulas [2-5]. This model is developed by using identical variable names as shown in the Air Source Model configuration above to facilitate the tracing of system performance and variables at various points within the system structure. The model can be set to simulate the Air Source structure without the Regen Heat Exchanger. This feature allows the comparison of energy consumption and dynamic performance for the cases with and without the Regen Heat Exchanger.

As shown in Figure 3, T_1 performance for T_5=1, 300°F is as shown below. The temperature spikes are required to heat the duct to achieve fast responses. However, the system is stable. T_1 dynamic performances are close to ideal and can emulate the engine performance quite closely. Hence for start-up operation, some form of preheating the system is necessary. This is to allow the system to warm up to about 1, 000°F with system shut-off valve S_1 closed during warm-up and allow the warm-up air flow through the system and either discharge to the atmosphere or return to the Regen Heat Exchanger via V_3. Once T_1 reaches about 1, 000°F, S_1 opens completely to simulate the engine take-off transient.

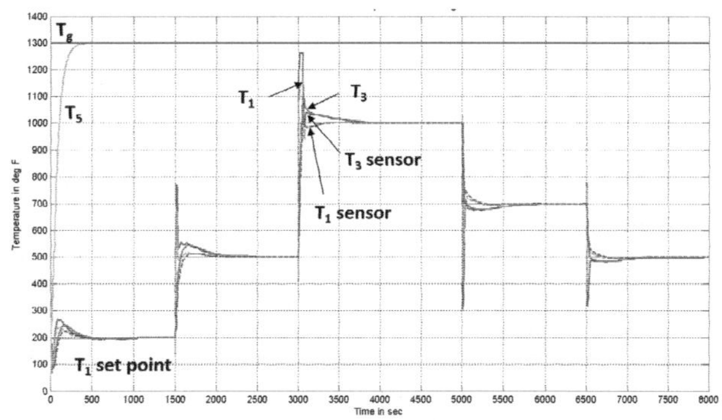

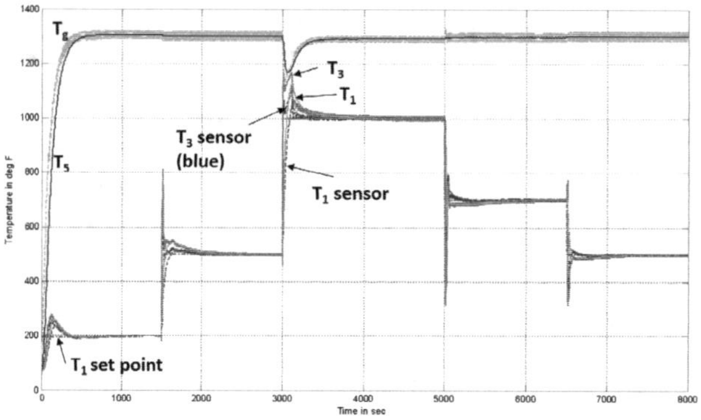

Figure 3. T_1 performance for T_5=1, 300°F with and without RHX configuration.

5. Conclusion

For the proposed Air Source Facility configuration, the major temperature T_1 dynamic performance limitation is caused by the duct b-c which is assumed to be 10 m long. However, it does not have too much effect on the P_1 pressure dynamics. The proposed Air Source Facility exhibits very good controllability properties and P_1 and T_1 controls do not have many adverse control interactions. This is illustrated by the simple control structure that can be used to control the system.

This simple dynamic model may be used to perform trade studies to optimize component sizing for performance and cost comparison purposes. An alternate structure that uses trim air temperature control may also be studied to determine possible advantages and disadvantages when compared to this proposed configuration.

References

[1] SAE ARP 217D: Testing of Airplane Installed Environmental Control Systems. Warrendale, PA: Society of Automotive Engineers, (1999).

[2] Zheng Dai and Yi Cui. Study on the Fast Air Heating Method for the Testbed of the Environmental Control System of the Aircraft. In: Journal of Physics: Conf. Series., Vol. 1, 060, pp. 012072. IOP Publishing Ltd. (2018).

[3] Zheng Dai and Yi Cui. Dynamic simulation studies of control design for engine bleed air test facility. In: MATEC Web Conf., Vol. 198, pp. 04003. EDP Sciences (2018).

[4] Zhao ZHAO, Yuhang LI, Sujun DONG, Yi CUI, and Zheng DAI. An analytic model for transient heat conduction in bi-layered structures with flexible serpentine heaters. In: Appl. Math. Mech. -Engl. Ed., 42 (9), 1, 279-1, 296 (2021).

[5] Yonggui Zheng, Meng Liu, Hao Wu, Jun Wang, Peng Xu, and Side Jin. An Available-to-Implement Thermal Facility for Dynamic Bleed Air Test of Aircraft Environmental Control System. Aerospace (2022).

AUTHOR INDEX

Bai, Haoyu 290
Bian, Renpeng 391
Cao, Huiquan 171
Chao, Yang 11
Chen, Bo 105
Chen, Chenchen 398
Chen, Kaihui 200
Chen, Qinzhu 98
Chen, Wei 336
Chen, Wuguo 351
Chen, Yajuan 11
Chen, Yong 398
Chen, Yu'Nan 184
Chi, Dao 312
Chu, Hongyuan 243
Cui, Shuai 50
Cui, Taisong 351
Cui, Yi 404
Dai, Zheng 404
Dong, Quancheng 50
Dong, Yiran 155
Fan, Tao 114
Feng, Qiaobo 191
Gao, Feng 351
Gao, Fengyang 319
Gao, Haiqiang 50
Gao, Yuehao 98
Guan, Shengzhi 122
Guo, Hao'Nan 184
Guo, Tao 184
Guo, Yuqin 132
Guo, Zhenyu 148
Han, Yancong 298
Hao, Yuhua 72
Hong, Rongjing 283
Hu, Guofang 36
Hu, Mingliang 370
Huang, Fu 237
Huang, Jing 283
Huang, Weiqing 270
Huang, Zan 177
Jia, Su 228
Jiang, Guohe 336
Jiang, Yafeng 43
Jiang, Yanyu 326
Jiang, Yazhou 351
Jiang, Zhenfeng 319
Jiang, Zhonglin 89
Jiang, Zihan 351
Jing, Xinjing 360
Kou, Xiping 228
Kuang, Jiayu 270
Li, Bo 141
Li, Dianxin 26
Li, Fuzhu 132
Li, Huaikun 50
Li, Hui 26
Li, Jingying 250
Li, Juan 250
Li, Qingshan 171
Li, Weiming 381
Li, Xianchao 237
Li, Xiangyu 184
Li, Yan 228
Li, Yong 89
Lian, Shuaimei 43
Liang, Chuanze 200, 219
Liang, Xinyu 319
Liu, Dan 26
Liu, Hua 243
Liu, Huan 360
Liu, Jianyong 26
Liu, Junwei 114
Liu, Kun 341
Liu, Yonghong 298, 391
Liu, Yu 250
Liu, Yuhong 381
Liu, Yuming 298
Liu, Zuhuang 122
Ma, Jinghua 164
Ma, Tingshan 360
Ma, Xiuhua 177
Mai, Huaming 141
Mei, Zheng 326
Meng, Decheng 105
Ming, Yang 1
Ouyang, Yan 228
Pan, Wenjie 171
Pan, Yanan 319
Pang, Jingdun 398
Peng, Haiting 191
Piao, Chengdao 155
Qiao, Bo 391
Qiu, Yafeng 200, 219
Ren, Jie 89
Ren, Peirong 270

Shang, Fei ... 305
Shen, Bo ... 148
Shen, Qiang ... 290
Shi, Yunbo ... 184
Sun, Andong ... 391
Sun, Dai ... 351
Sun, Xuan ... 50, 105
Tang, Jiaheng ... 270
Wang, Dechao ... 155
Wang, Dong ... 208
Wang, Hongwu ... 360
Wang, Hua ... 283
Wang, Jingshu ... 164
Wang, Manxin ... 341
Wang, Ning ... 36
Wang, Qingfeng ... 164
Wang, Tengteng ... 351
Wang, Ting ... 360
Wang, Tong ... 237
Wang, Wei ... 312
Wang, Xiaoping ... 11
Wang, Xuesheng ... 98
Wang, Yansheng ... 78
Wang, Yaowen ... 360
Wang, Ying ... 250
Wang, Yong ... 114
Wang, Yu ... 184
Wang, Yuefang ... 78
Wang, Zhanlei ... 191
Wang, Zhen ... 122
Wei, En ... 57
Wei, Jinggong ... 57
Wu, Hao ... 171
Wu, Huisong ... 228
Wu, Jiahua ... 208
Wu, Mingge ... 89
Wu, Senyao ... 43
Xia, Kangjun ... 341
Xia, Pengyu ... 270
Xia, Xinhua ... 171
Xiao, Hua ... 177
Xiao, Zhenggang ... 326
Xie, Tian ... 360
Xie, Xumeng ... 171
Xie, Yukuan ... 381
Xiong, Ting ... 398
Xu, Feiyun ... 1
Xu, Jiazhong ... 148
Xu, Minggang ... 122
Xu, Zhenhua ... 283
Xue, Yu ... 290
Xue, Yuan ... 381
Yang, Dongjie ... 132
Yang, Rongzu ... 360
Yang, Wentao ... 114
Yang, Yongkang ... 398
Yang, Zhe ... 305
Yao, Zhenqiang ... 260
Ye, Guoqing ... 260
Yin, Hang ... 132
Yin, Ling ... 141
Yin, Zuwang ... 208
Yu, Jianrong ... 26
Yu, Le ... 370
Yu, Longwen ... 360
Zhai, Pengcheng ... 360
Zhang, Bin ... 351
Zhang, Dongfang ... 57
Zhang, Guoqing ... 177
Zhang, Hongsheng ... 398
Zhang, Peng ... 184
Zhang, Weiwei ... 141
Zhang, Wenming ... 26
Zhang, Xiujuan ... 72
Zhang, Yanzhen ... 36
Zhang, Yue ... 50
Zhao, He ... 19
Zhao, Rui ... 184
Zhao, Wenhao ... 298
Zhao, Xuanyi ... 270
Zhao, Yadong ... 312
Zheng, Chao ... 298
Zheng, Shengkui ... 155
Zheng, Wenbo ... 341
Zhou, Weihan ... 164
Zhou, Xiaoguang ... 208
Zhou, Xin'an ... 148
Zhou, Zhongliang ... 336
Zhu, Jianguo ... 341
Zhu, Weijie ... 200, 219
Zou, Shenye ... 19